Information Systems Development

Information Systems Development

Advances in Methodologies, Components, and Management

Edited by

Marite Kirikova
Janis Grundspenkis
Riga Technical University
Riga, Latvia

Wita Wojtkowski
W. Gregory Wojtkowski
Boise State University
Boise, Idaho

Stanisław Wrycza
University of Gdansk
Gdansk, Poland

and

Jože Zupančič
University of Maribor
Kranj, Slovenia

Springer Science+Business Media, LLC

11th International Conference on Information Systems Development: Methods and Tools, Theory and Practice, Riga, Latvia, September 12–14, 2002

ISBN 978-0-306-47698-3   ISBN 978-1-4615-0167-1 (eBook)
DOI 10.1007/978-1-4615-0167-1
©2002 Springer Science+Business Media New York
Originally published by Kluwer Academic / Plenum Publishers, New York in 2002

http://www.wkap.nl/

10   9   8   7   6   5   4   3   2   1

A C.I.P. record for this book is available from the Library of Congress

PREFACE

This book is the result of the 11th International Conference on Information Systems Development - Methods and Tools, Theory and Practice, held in Riga, Latvia, September 12-14, 2002. The purpose of this conference was to address issues facing academia and industry when specifying, developing, managing, reengineering and improving information systems. Recently many new concepts and approaches have emerged in the Information Systems Development (ISD) field. Various theories, methodologies, methods and tools available to system developers also created new problems, such as choosing the most effective approach for a specific task, or solving problems of advanced technology integration into information systems.

This conference provides a meeting place for ISD researchers and practitioners from Eastern and Western Europe as well as from other parts of the world. Main objectives of this conference are to share scientific knowledge and interests and to establish strong professional ties among the participants.

The 11th International Conference on Information Systems Development (ISD'02) continues the tradition started with the first Polish-Scandinavian Seminar on Current Trends in Information Systems Development Methodologies, held in Gdansk, Poland in 1988. Through the years this Seminar has evolved into the International Conference on Information Systems Development. ISD'02 is the first ISD conference held in Eastern Europe, namely, in Latvia, one of the three Baltic countries.

ISD'02 comprised not only the scientific program represented in these proceedings, but also tutorials on "A Pattern-Based Approach to Building Organisational Memories" and "Techniques for Information Searching on the Internet" that were intended for both, the research and business communities. During ISD'02 we also held International Symposium on Research Methods and PhD Consortium.

The selection of papers was carried out by the International Program Committee. All papers were reviewed in advance by three people. Papers were evaluated according to their originality, relevance, and presentation quality. All papers were evaluated only on their own merits, independent of other submissions.

We would like to thank the authors of papers submitted to ISD'02 for their efforts. We would like to express our thanks to members of the International Program Committee and external reviewers for their essential work and for many useful comments that greatly helped authors to improve the quality and relevance of their papers. We also wish to acknowledge the support of the members of the Organising Committee. We thank the administration of Riga Technical University for their support.

Janis Grundspenkis
Marite Kirikova
Wita Wojtkowski
W. Gragory Woitkowski
Stanislaw Wrycza
Joze Zupancic

PROGRAM COMMITTEE

Organizing committee – co-chairmen

Janis Grundspenkis	Riga Technical University	(Latvia)
Marite Kirikova	Riga Technical University	(Latvia)
Wita Wojtkowski	Boise State University	(USA)
W. Gregory Wojtkowski	Boise State University	(USA)
Stanislaw Wrycza	University of Gdansk	(Poland)
Joze Zupancic	University of Maribor	(Slovenia)

International program committee

Witold Abramowicz	Economic University Poznan	(Poland)
Gary Allen	University of Huddersfield	(UK)
Janis Barzdins	Latvian University	(Latvia)
Juris Borzovs	Riga Information Technology Institute	(Latvia)
Chris Freyberg	Massey University	(New Zealand)
Hamid Fujita	Iwate Prefectural Univetsity	(Japan)
Edwin Gray	Glasgow Caledonian University	(UK)
Hele-Mai Haav	Institute of Cybernetics	(Estonia)
G Harindranath	Royal Holloway University of London	(UK)
Igor Hawryszkiewycz	University of Technology, Sydney	(Australia)
Alfred Helmerich	Research Institute for Applied Technology	(Germany)
Lech J. Janczewski	The University of Auckland	(New Zealand)
Roland Kaschek	Massey University	(New Zealand)
Marian Kuras	Cracow Academy of Economics	(Poland)
Rein Kuusik	Tallinn Technical University	(Estonia)
Robert Leskovar	University of Maribor	(Slovenia)
Henry Linger	Monash University	(Australia)
Leszek Maciaszek	Macquarie University	(Australia)
Heinrich Mayr	University of Klagenfurt	(Germany)
Sal March	University of Minnesota	(USA)
Elisabeth Metais	CNAM/CEDRIC	(France)
Murli Nagasundaram	Boise State University	(USA)
Anders G. Nilsson	Karlstad University	(Sweden)

CONTENTS

REFLECTIONS ON INFORMATION SYSTEMS DEVELOPMENT 1988-2002

David Avison and Guy Fitzgerald[*]

1. INTRODUCTION

The publication of the third edition of Information Systems Development: Methodologies, Techniques and Tools in September 2002 gives us the opportunity to look back on the previous two editions published in 1988 and 1995, as well as this new edition, and reflect on the progress of information systems development over the past 15 or so years. On reflection, the publication of the three editions seems to coincide with three eras of information systems development methodologies. We refer to these as *early methodology era, methodology era* and *era of methodology reassessment.* In this paper we develop these themes.

Avison and Fitzgerald (1988) was published when information systems was a fledgling discipline. We saw information systems development (and maintenance of systems already developed) as the main thrust of our own work as systems analysts in organizations in the period previous to us joining academia, and we also saw it as a core area for any program in information systems taught in higher education. The success of the ISD conference over the last ten years is enough evidence to illustrate well the importance of this aspect of information systems in any curriculum. Interest in this area has increased over the period.

When we were systems analysts in industry from the mid 1970s, we either followed no methodology at all (and depended on a mixture of experience, luck and advice, in order to survive) or followed a simple life cycle approach. Therefore we can also see an earlier period, the *pre methodology era* preceding the three eras discussed in this paper.

At Thames Polytechnic, we both taught the approach of the UK National Computing Centre, which was a life cycle approach. The NCC teaching package came with documentation, methods and training included. Indeed, we gave an eight-week training course for prospective systems analysts. In general, our clients were experienced computer programmers wanting to 'move up' to systems analysis. This was part of the early methodology era, one of somewhat unsophisticated and technical life cycle approaches.

---

[*] David Avison, ESSEC Business School, Paris, France and Guy Fitzgerald, Brunel University, Uxbridge, England

Information Systems Development: Advances in Methodologies, Components, and Management
Edited by Kirikova *et al.*, Kluwer Academic/Plenum Publishers, 2002

However, we were somewhat unconvinced by the single view of information systems development exemplified by these traditional life cycle approaches. It was our research of what other academics taught and researched and also our research into what practitioners do, that led us to teach a broader spectrum of methodologies in our courses and then to write the first edition of our book. We felt that teaching one narrow approach was misleading and a poor education on information systems development for our students.

In this paper we look at how information systems development has changed over the three periods that span the three editions of our book: the early methodology era to 1988, the methodology era to 1995, and the era of methodology reassessment to 2002. We also reflect on this progress and on possible developments in the future.

2. EARLY METHODOLOGY ERA TO 1988

In our book we decided to split the subject matter into themes, techniques, tools and methodologies. In 1988 the main methodologies were either simple life cycle approaches, typified by that of the NCC, or categorized as one of two themes: data oriented and process oriented. However we did suggest alternative themes – nine in total. Most of the eight techniques discussed related to the two most common methodology themes (data or process modeling). We included seven software tools and, apart from expert systems, they would be found in a CASE tool of this period, or at least an advanced one. Half of the ten methodologies described were conventional and prescriptive, either data or process oriented methodologies or blended, with both data and process elements.

Before this time, developers were technically trained but rarely fully understood the business and organizational context. This resulted in systems not being delivered to budget, nor to time and nor to need. These early methodologies aimed to address the need for control and training. They were prescriptive and methodological and their use was seen as a way to improve the track record of IS development.

But we also included ISAC, ETHICS, SSM and Multiview in this edition. These were all much less conventional. They stressed people, organizational and contingency views of information systems development. These were included as a result of our research but they were, in truth, rarely used in practice (except by their authors). These latter approaches may not be considered radical now, but at the time this provided a much broader view of the subject than that provided in most courses and in most texts.

At this time most courses in ISD were also still somewhat one-dimensional, for example, courses in SSADM in the UK and STRADIS or Information Engineering in the United States were very common. We felt this might give good training, but rather a poor education. We believed that information systems people should be aware of ethical, organizational and social issues. Information systems development is just as much about these than about data and processes and technology in general. In any case, the view that developers apply a methodology rigorously was false in our experience. Many authors have also showed this over the years at this conference. Multiview (Avison and Wood-Harper, 1990), defined in the mid 1980s, exemplified an alternative view.

Although we were aware of studies suggesting there were many more methodologies, for example, Longworth's (1985) study which identified 300 brand-name methodologies, our study suggested that most were comfortably within one of the eight types exemplified by the eight methodologies described, indeed, most within three or four themes. Further, many were not well developed, being defined, for example, only in an academic working paper.

Although we suggested 24 'good' reasons for adopting a methodology, all reasons leading to a better end product (that is a better information system), we also saw many limitations of methodologies and reasons for the poor adoption rate of methodologies.

We also suggested about 40 features that we might use to compare methodologies, though our own comparison was based on philosophy, model, techniques and tools, scope, outputs, practice and product. In our view, however, philosophy was a key issue. Did organizations want to adopt a methodology that emphasized control, emphasized technology, emphasized user's ability to take part in decision-making, stress the importance of data, stress the role of understanding the organizational context or stress flexibility (or some combination of these)?

3. METHODOLOGY ERA TO 1995

Information systems had certainly grown and matured as an academic discipline by the time the second edition of the book was published. Information systems development remained one of the core issues in the discipline; indeed, many considered it the core of the subject. We continued the same structure of the book, but there were 12 themes (as against 9 in the first edition); 11 techniques (8); 6 tools (7) and 15 methodologies (8).

One theme was dropped between the two editions, that of research. This was because this issue permeated through the book and was especially prominent in the final chapter on issues and frameworks. We added business process reengineering, object orientation and expert systems (previously the latter were included as tools). These were all hot issues at the time, and like most of these hot issues that seem to occur annually in our discipline, they have their period of great enthusiasm, then one of disappointment and finally have some longer-term impact somewhere between euphoria and depression. Although rich pictures, conceptual models and root definitions were discussed in the first edition in the context of soft systems methodology and Multiview, these were added to the techniques section of the second edition as by that time these techniques had caused some more general interest and been included in other approaches. In retrospect this illustrates the view in this period that ISD was not simply a technical, data or process issue – it had organizational, people, and other dimensions.

By the time of the second edition, CASE tools had been well accepted in the community, although there was much dispute about whether they made the positive impact that the suppliers suggested. Again, the euphoria of '1000% increased productivity through the use of CASE' had been replaced by phrases such as their use leads to 'some possible productivity gains' and 'increased adherence to common standards', but we felt that software support tools of various kinds did have the potential to support information systems development.

With the methodology era came the many methodologies evidenced in the book, some coming from academic circles, but most from practice. Many were originally based on very distinct themes – people, process, data and the rest, with appropriate techniques – but the processes of filling the gaps and expanding the scope meant that some aspects of the philosophy on which methodologies were based was frequently lost as they became just another methodology amongst many similar ones having all the well-known techniques embodied in them.

Additional methodologies in the second edition included Yourdon systems method, Merise, object-oriented analysis, process innovation, rapid applications development, KADS and Euromethod. Yourdon had become the methodology of the structured or process school, though Gane and Sarson's STRADIS was still taught. (We have always had difficulty deleting sections from the book, as some lecturers said they found even defunct methodologies useful as a teaching aid.) Merise was the equivalent methodology for francophone users as SSADM was to the British civil service and also used in larger, perhaps more bureaucratic, organizations. The discussion of object-oriented analysis reflected the interest in object modeling and, similarly, process innovation to interest in business process reengineering. Rapid applications development was also a hot topic, as many organizations wanted immediate solutions to their information systems needs. KADS and Euromethod reflected the impact of the European community on information systems. The KADS project related to developing expert systems applications and was a project funded by the EC Esprit initiative. Euromethod is a framework that combines the 'best of' European approaches, such as SSADM and Merise. It was originally designed as a European standard for information systems development, but this has proved too ambitious or inappropriate (depending on the viewpoint taken).

We argued that by 1995, the 'methodology jungle' had worsened in the sense that there were so many developments and different directions in which methodologies were going. We tried to make sense of the confusion, but we reported Jayaratna's (1994) study that suggested there were over 1000 brand-name methodologies.

To help claw through the jungle, we proposed methodology choice based on which of five classes of situation was apparent in the organization under scrutiny: well-structured with clear requirements; well-structured with unclear requirements; unstructured with unclear requirements; high user interaction systems; and very unclear situations.

We also produced an alternative way to choose between methods based on a framework of epistemology (positivism to interpretism) on one axis and ontology (realism to nominalism) on the other axis. However many choose between alternative methods, techniques and tools within one contingency approach, such as Multiview, rather than choose between alternative methodologies.

There were many pressures that led to the adoption of more formalized methodologies, for example, the requirements of certain large organizations or standards bodies. Fitzgerald (1994), for example, suggests that the ISO (International Standards Organization) and the SEI (Software Engineering Institute) were influential in this respect, as was the perceived wisdom in some quarters that 'better methods will solve the problems of IS development'.

Yet in this methodology era, there are also many reasons why organizations did not adopt any sort of methodology at all:

1   *Productivity:* The first general criticism of methodologies is that they fail to deliver the suggested productivity benefits. It is said that they do not reduce the time taken to develop a project; rather their use increases systems development lead-times when compared with not using a methodology.

2 *Complexity:* Methodologies have been criticized for being over complex. They are designed to be applied to the largest and most comprehensive development project and therefore specify in great detail every possible task that might conceivably be thought to be relevant, all of which is expected to be followed for every development project.

3 *'Gilding the lily':* Methodologies develop any requirements to the ultimate degree, often over and above what is legitimately needed. Every requirement is treated as being of equal weight and importance, which results in relatively unimportant aspects being developed to the same degree as those that are essential.

4 *Skills:* Methodologies require significant skills in their use and processes. These skills are often difficult for methodology users and end users to learn and acquire.

5 *Tools:* The tools that methodologies advocate are difficult to use and do not generate enough benefits. They increase the focus on the production of documentation rather than leading to better analysis and design.

6 *Not contingent:* Methodologies are not contingent upon the type of project or its size. Therefore the standard becomes the application of the whole methodology, irrespective of its relevance.

7 *One-dimensional approach:* Methodologies usually adopt only one approach to the development of projects and whilst this may be a strength, it does not always address the underlying issues or problems.

8 *Inflexible:* Methodologies may be inflexible and may not allow changes to requirements during development. This is problematic as requirements, particularly business requirements; frequently change during the long development process.

9 *Invalid or impractical assumptions:* Most methodologies make a number of simplifying yet invalid assumptions, such as a stable external and competitive environment. Many methodologies that address the alignment of business and information systems strategy assume the existence of a coherent and well-documented business strategy as a starting point for the methodology. This may not exist in practice.

10 *Goal displacement:* It has frequently been found that the existence of a methodology standard in an organization leads to its unthinking implementation and to a focus on following the procedures of the methodology to the exclusion of the real needs of the project being developed. In other words, the methodology obscures the important issues. De Grace and Stahl (1993) have termed this 'goal displacement' and talk about the severe problem of 'slavish adherence to the methodology'. Wastell (1996) talks about the 'fetish of technique', which inhibits creative thinking. He takes this further and suggests that the application of a methodology in this way is the functioning of methodology as a social defense, which he describes 'as a highly sophisticated social device for containing the acute and potentially overwhelming pressures of systems development'. He is suggesting that systems development is such a difficult and stressful process, that developers often take refuge in the intense application of the methodology in all its detail as a way of dealing with these difficulties. Developers can be seen to be working hard and diligently, but this is in reality goal displacement activity because they are avoiding the real problems of effectively developing the required system.

11 *Problems of building understanding into methods:* Introna and Whitley (1997) argue that some methodologies assume that understanding can be built into the method process. They call this 'method-ism' and believe it is misplaced. Method-ism assumes that the developers need to understand little or nothing about the problem situation and

that the method will somehow 'bring to light' all the characteristics that need to be discovered. Thus all that needs to be understood is the method itself. This, it is argued, is far too constraining and prevents real understanding of the problem situation emerging and being acted upon. It also inhibits the contingent use of methodologies. Introna and Whitley are not against methods as such, just this underlying assumption and its implications.

12 *Insufficient focus on social and contextual issues:* The growth of scientifically based highly functional methodologies has led some commentators to suggest that we are now suffering from an overemphasis on the narrow, technical development issues and that not enough emphasis is given to the social and organizational aspects of systems development. Hirschheim, et al. (1996), for example, argue that changes associated with systems development are emergent, historically contingent, socially situated, and politically loaded and that, as a result, sophisticated social theories are required to understand and make sense of IS development. They observe that these are sadly lacking in most methodologies.

13 *Difficulties in adopting a methodology:* Some organizations have found it hard to adopt methodologies in practice. They have found resistance from developers who are experienced and familiar with more informal approaches to systems development and see the introduction of a methodology as restricting their freedom and a slight on their skills.

14 *No improvements:* Finally in this list, and perhaps the acid test, is the conclusion of some that the use of methodologies have not resulted in better systems, for whatever reasons. This is obviously difficult to prove, but nevertheless the perception of some is that 'we have tried it and it didn't help and it may have actively hindered'.

We thus find that for some, the great hopes in the 1980s and 1990s, that methodologies would solve most of the problems of information systems development have not come to pass.

Strictly speaking, however, a distinction should be made in the above criticisms of methodologies between an inadequate methodology itself and the poor application and use of a methodology. Sometimes a methodology vendor will argue that the methodology is not being correctly or sympathetically implemented by an organization. Whilst this may be true to some extent, it is not an argument that seems to hold much sway with methodology users. They argue that the important point is that they have experienced disappointments in their use of methodologies.

4. ERA OF METHODOLOGY REASSESSMENT TO 2002

The new third edition of the book presents an even greater development from the second edition than that itself was on the first edition. We have continued the same structure, but have seven parts reflecting the seven chapters of the previous editions. However, in order to make the book readable, these seven parts have been split into 26 chapters. There are now 28 themes (as against 12 themes in the second edition and 9 in the first edition); 29 techniques (11, 8); and 25 methodologies (15, 8). We have considered tools in a different way. In the third edition we look at some specific brand-name software tools in one chapter, and also toolsets, specifically IEF, Select and Oracle, in another.

In this third edition we have divided the 28 themes into six categories: organizational, modeling, engineering and construction, people, external development and software. New themes include stages of growth, flexibility, legacy systems, evolutionary development, method engineering, web development, end-user development, knowledge management, customer orientation, application packages, enterprise resource planning, outsourcing, software engineering, and component development/open source.

The new techniques (at least for this text) include cognitive mapping, UML, case-based reasoning, risk analysis, lateral thinking, critical success factors, scenario planning, future analysis, SWOT and stakeholder analysis. The specific tools described include MS Project, Ventura, Dreamweaver, Visio and Access. Each of these exemplify a theme in the text, that is, project management, group decision support systems, web site development, drawing tools and database management systems, respectively in these cases.

Similarly, methodologies have been split into different categories: process, blended, object-oriented, rapid, people, organizational and frameworks. New methodologies include Welti's ERP development, RUP, DSDM, extreme programming, WISDM (for web applications), CommonKADS, SODA, CMM, PRINCE, and Renaissance (for legacy systems).

But this present era is characterized by a serious reappraisal of the concepts and practicalities of the methodology era. Although they have achieved, at least to some extent, some objectives, perhaps project control, user involvement and some discipline in the process, they are certainly not seen now as potential panaceas for correcting all problems of information systems development!

As we showed at the end of the previous section, productivity has not necessarily been improved, they can be far too complex, require significant skills, require expensive tools, be inflexible, inhibit creative thinking, suggest more than they can deliver, give insufficient focus on social and contextual issues and the rest. This has led many organizations turning away from methodologies.

A survey conducted in the UK by Fitzgerald et al. (1999) found that 57% of the sample were claiming to be using a methodology for systems development, but of these, only 11% were using a commercial development methodology unmodified, whereas 30% were using a commercial methodology adapted for in-house use, and 59% a methodology which they claimed to be unique to their organization, i.e. one that was internally developed and not based solely on a commercial methodology.

Thus the picture seems to emerge that the majority of organizations were using some kind of methodology, but that most of these were developed or adapted to fit the needs of the developers and the organization. Thus, although there is no large-scale use of commercial methodologies, we argue that the influence of commercial methodologies is considerably larger than their use.

Nevertheless, many organizations have turned away from formal methodologies. Many are turning to ad-hoc approaches, contingency approaches, component development, packages and outsourcing:

1   *Ad-hoc development:* This might be described as a return to the approach of the pre-methodology days in which no formalized methodology is followed. The approach that is adopted is whatever the developers understand and feel will work. It is driven by, and relies heavily on, the skills and experiences of the developers (or perhaps just trial-and-error and guesswork). This is perhaps the most extreme reaction to the backlash

against methodologies and in general terms it runs the risk of repeating the problems encountered prior to the advent of methodologies (missed cutover dates, poor control, poor communications and poor training). One area where, in the authors' experience, methodologies are not being used is in the development of web-based applications. No methodology has become a standard for web development. Another group of organizations are pinning their faith on the evolution of toolsets to increasingly guide and automate the development process.

2   *Further developments in the methodology arena:* For others there is the continuing search for the methodology holy grail. Methodologies will continue to be developed and existing ones evolve. For example, object-oriented techniques and methodologies have been gaining ground over process and entity modeling approaches for some time, although whether this is a fundamental advance is debatable. It may be that component-based development, which envisages development from the combination and re-combination of existing components, will make a long-term impact. But this may simply be a current fashion to be overtaken by the next panacea at some point in the future. It may be that the RAD approaches will prevail, or perhaps the need for flexibility will favor prototyping approaches. The current emphasis on knowledge, rather than information, may make approaches like CommonKADS popular. With the importance of web applications, focusing on customers as stakeholder, might make Customer Relationship Management (CRM) the future 'silver bullet'. But these are conjectures made at the time of writing and it is difficult to predict the future. What we do know, based on past experience, is that proposed new solutions will come and go, some will be easily forgotten whilst others will probably stand the test of time and make a genuine contribution. However, we believe it unlikely that any single approach will ever provide the solution to all the problems of information systems development.

3   *Contingency:* Most methodologies are designed for situations that follow a stated or unstated 'ideal type'. The methodology provides a step-by-step prescription for addressing this ideal type. However, situations are all different and there is no such thing as an 'ideal type' in reality. We therefore see a contingency approach to information systems development where a structure is presented but tools and techniques are expected to be used or not (or used and adapted), depending on the situation as being a third movement of this present era. Situations might differ depending on, for example, the type of project and its objectives, the organization and its environment, the users and developers and their respective skills. The type of project might also differ in its purpose, complexity, structuredness, and degree of importance, the projected life of the project, or its potential impact. The organization might be large or small, mature or immature in its use of IT. Different environments might exhibit different rates of change, the number of users affected by the system, their skills, and those of the analysts. All these characteristics could affect the choice of development approach that is required. A contingent methodology allows for different approaches depending on situations. This is a reaction to the 'one methodology for all developments' approach that some companies adopted, and is a recognition that different characteristics require different approaches. There are, however, potential problems of the contingent approach as well. First, some of the benefits of standardization might be lost. Second, there is a wide range of different skills that are required to handle many approaches. Third, the selection of approach requires experience and skills to make the best judgments. Finally, it has been suggested that

certain combinations of approaches are untenable because each has different philosophies that are contradictory. Multiview aims to provide a framework that helps people make such contingent decisions and WISDM is an adaptation of Multiview applied to web development.

4   *External development:* We also see a movement towards external development in a variety of ways. In particular we discuss the use of packages and outsourcing. Some organizations are attempting to satisfy their systems needs by buying packages from the marketplace. Clearly the purchasing of packages has been commonplace for some time, but the present era is characterized by some organizations deciding not to embark on any more in-house system development activities but to buy-in all their requirements in the form of package systems. This is regarded by many as a quicker and cost-effective way of implementing systems for organizations that have fairly standard requirements. Only systems that are strategic or for which a suitable package is not available would be considered for development in-house. The package market is becoming increasingly sophisticated and more and more highly tailorable packages are becoming available. Integrated packages, which address a wide range of standard business functions, purchasable in modular form, known as Enterprise Resource Packages (ERPs) have emerged in the last few years and have become particularly popular with large corporations. The key for these organizations is ensuring that the correct trade-off is made between a standard package, which might mean changing some elements of the way the business currently operates, and a package that can be modified to reflect the way they wish to operate. There are dangers of becoming locked-in to a particular supplier and of not being in control of the features that are incorporated in the package, but many companies have taken this risk. For others, the continuing problems of systems development and the perceived failure of methodologies to deliver, has resulted in them outsourcing systems development to a third party. The client organization is no longer so concerned with how a system is developed, and what development approach or methodology is used, but with the end results and the effectiveness of the system that is delivered. This is different to buying-in packages or solutions, because normally the management and responsibility for the provision and development of appropriate systems is given to a vendor. The client company has to develop skills in selecting the correct vendor, specifying requirements in detail and writing and negotiating contracts rather than thinking about system development methodologies.

The above features of the present era of methodology reappraisal as we see it are not mutually exclusive and some organizations are moving to a variety of these approaches. Some aspects are being absorbed or incorporated into some existing methodologies, i.e. the 'filling the gaps' and 'blending' process is still continuing. This present era is not one where all methodologies have been abandoned. It is an era where there is diversity and perhaps a more realistic view of the limitations of methodologies. For some organizations, however, it is about the abandonment of methodologies altogether. For others, it is seeking improved methodologies, but moving away from the highly bureaucratic types of the methodology era. For still others it is about moving out of in-house systems development altogether. But it should also not be forgotten that even in the era of methodology reappraisal, some organizations are still using methodologies effectively and successfully.

5. REFLECTIONS ON THE METHODOLOGY SCENE

It will be evident that this new edition is both much broader in scope and much more comprehensive than previous editions. But this reflects the methodology scene itself. There are many more choices available. This obviously includes choices in terms of techniques, tools and methodologies, but also choices with regard to application types (for example, transaction processing, decision support, enterprise resource planning, knowledge based and web-based); whether to develop in-house or partly or wholly develop applications externally (via software packages, ERP or outsourcing); what to do about legacy systems (maintain, merge with an ERP system, or replace with new systems); choices about who develops the applications (expert groups, users, mixed groups, etc.); who is involved (experts, users, customers and other stakeholders); whether development is evolutionary or revolutionary (BPR); whether the organization should aim to follow some sort of stages of growth or capability maturity model ... the list of choices could go on and on. Indeed, reading over the proceedings of this conference over the years makes us realize that our list of themes, techniques, tools and methodologies only scratches the surface of what we could have included.

Our identification and characterization of these methodology eras has been done to provide a more categorized view of the history and evolution of methodologies and to make such a history more understandable. However, some could criticize it because they do not recognize the concept of the methodology era itself. They argue there was never a period when methodologies proliferated, particularly in terms of their use. We disagree, but as with any historical categorization it is open to debate and interpretation. Our hope is that we have engendered, and contributed to, such a debate.

In view of the fact that we classify the present period as one of methodology circumspection (rather than a methodology era), it might be surprising that information systems development is still central to the discipline of information systems. There are a number of reasons for this. The first reason is that even if methodologies are not used as they are 'meant to be used' they influence practice. They might be adapted or other techniques and tools used. But they nevertheless make a useful contribution to practice. The second reason is that they are important to training and education in information systems. They teach good practice and form a good basis for discussions on information systems development. Thirdly, and conversely, it may be more of a 'methodology period' than supposed. It is true that information systems development methodologies are not adopted by all organizations, but nor were they in any period since 1988 when the first edition of the book was published. We might even claim that methodologies are in fact used now more than ever (albeit from a low base). Organizations are much more likely to find an appropriate approach for their information systems development work, even though there is rarely one clear strategy for developing information systems.

REFERENCES

Avison, D.E. and Fitzgerald, G, (1995, 1998, 2002) *Information Systems Development: Methodologies, Techniques and Tools,* McGraw-Hill, Maidenhead.

Avison, D. E. and Wood-Harper, A. T. (1990) *Multiview: An Exploration in Information Systems Development.* McGraw-Hill, Maidenhead.

Fitzgerald, B. (1994) The systems development dilemma: whether to adopt formalized systems development methodologies or not? In: Baets, W. R. J. (ed.) *Proceedings of the Second European Conference on Information Systems,* Nijenrode University Press, Breukelen, the Netherlands.

Fitzgerald, G., Philippides, A. & Probert, P, Information Systems Development, Maintenance and Enhancement: Findings from a UK Study, *International Journal of Information Management,* 40 (2), 319-329, 1999.

Hirschheim, R., Klein, H. K. and Lyytinen, K. (1996) Exploring the intellectual structures of information system development: A social action theoretic analysis, *Accounting, Management and Information Technologies,* 6, 1/2

Introna. L. and Whitley, E. (1997) Against method-ism: Exploring the limits of method, *Information Technology & People,* 10, 1, 31-45.

Jayaratna, N. (1994) *Understanding and Evaluating Methodologies: NIMSAD a Systemic Framework,* McGraw-Hill, Maidenhead.

Longworth, G. (1985) *Designing Systems for Change.* NCC, Manchester.

Wastell, D. (1996) The Fetish of Technique: Methodology as a Social Defense, *Information Systems Journal,* 6, 1, 25-30.

ISD AS FOLDING TOGETHER HUMANS & IT

Towards a revised theory of Information Technology development & deployment in complex social contexts

Dr. Larry Stapleton*

1. INTRODUCTION

This paper identifies a gap in ISD research regarding the philosophy of information technology as it relates to social impact in complex organisational contexts. It recognises that this will lead to problems of organisational stability, and that too often technology and knowledge transfer is accompanied by a one-sided approach resulting in a loss of local context. It posits a revised philosophical position based upon the work of current thinkers in the philosophy of technology/human relations and applies this position to ISD. This revised perspective challenges researchers to review their working assumptions about research in general and technology development and deployment in particular.

2. BACKGROUND

It has become apparent that traditional thinking regarding the creation and deployment of advanced information technologies requires some revision (Stapleton et. al. (2001b)). One major area of opportunity for progress is in the theory of ISD as a means by which new organisational realities can be created. Instead of regarding ISD as the creation of new information technology artefacts, it is becoming evident that, in many cases, ISD has more to do with social reconfiguration and transfer (and challenging) of knowledge and assumptions in order to create a new social space (Moreton & Chester (1999), Stapleton (2001)).

Some of the major issues raised by scientists concerned with the transfer of technology and techniques across cultures include:

1. Cultural imperialism (Banerjee (2001)): ISD can be regarded as reflecting a particular view of the world which may (or may not) be culturally located outside of the IT deployment context

---

* Information Systems & Organisational Learning Research Group, Waterford Institute of Technology, Main Campus, Cork Road, Waterford, Republic of Ireland. Email: lstapleton@wit.ie

Information Systems Development: Advances in Methodologies, Components, and Management
Edited by Kirikova *et al.*, Kluwer Academic/Plenum Publishers, 2002

2. Economic colonisation and the derailing of democracy (Chomsky (1993)): again IT is not a passive artefact but involves a cultural transfer of knowledge and ideas through global corporate business.

The specific local context in which techniques and technology are deployed is ignored, leading to major problems on the ground (a good example is Cronk (2000)).

Philosophically, engineering and technology deployment literature is strongly influenced by twentieth century positivists such as A.J. Ayer (e.g. Ayer (1936)). Functional Rationalism is a term coined in the literature to describe positivist influences in Engineering theory and practise (Bickerton & Siddiqi (1993)). Most information system development approaches are based upon functionally rationalist premises. These premises have dominated advanced technology research and practice, and has created serious problems for the study of social impact, a fact which is well documented elsewhere (Galliers (1992), Myers (1995), Stapleton (2001)). Whilst positivist science has delivered many wonderful discoveries, and has placed a human on the moon, on earth problems of social impact remain acute and poorly understood in spite of a great deal of research on socio-technical design and related areas.

Given the difficulties and criticisms associated with the functionally rational approach in inter-cultural exchange (such as technology transfer) researchers urgently need a new set of assumptions in order to guide work in this area. A new theory of technology transfer and deployment is needed which identifies and informs issues which remain poorly understood. Such a theory needs to be incorporated into research in this space. In our search for revised philosophical foundations it is important to note that alternative philosophical positions have been employed in other disciplines to address problems with positivist science in social domains.

However, these revised positions have been criticised for their own, inappropriate, assumptions when it comes to the deployment of advanced technology in culturally diverse spaces. They have also been criticised for weaknesses in the accompanying research approaches, which attempt to understand the particular cultural and social settings under scrutiny. For example, Naturalism has informed ethnographic approaches and ethno-methodology in information systems development and deployment (Suchman (1987), Bentley et. al. (1992), Simonsen (1995)). This approach has been deprecated by leading social thinkers for ignoring the intervention of researchers in the culture under scrutiny (Hammersley (1990)).

Interpretivism has also been mooted as a possible way forward. This focuses upon the idea that reality is socially constructed inter-subjectively i.e. on the basis of the sharing of subjective realities amongst participants in a social group. This has lead to ISD research trajectories based upon phenomenology and hermeneutics, which focus upon dialog and the inter-subjective construction of 'narratives' (Boland (1985), Myers (1995)). Social Constructivists also argue that reality is socially constructed and again emphasise the important role of narrative.

In IS research a body of literature has built up around soft-systems and the socio-technical design of computer artefacts which has been highly influenced by interpretivism. These have been characterised by Winograd (1995) as Heideggerian, although this view can be contested. Certainly, a primary philosophical underpinning is provided by the stream of thought which developed following Wittgenstein's later work on language games and Husserl's work on the development of a position now referred to as Phenomenology. This may or may not be entirely in tune with Heidegger's ideas and certainly we see a reduced emphasis upon embodiment in recent IS literature and a strong influence of the 'rampant textuality' criticised by Ihde[†] which shall be further discussed later in this paper.

In the 1980's this work culminated in publication by, for example, the Scandinavian researchers involved in the DEMOS and UTOPIAN projects, characterised by published work such as Ehn (1988) and Dahlbom & Mathiassen (1993). Here researchers combined a political position with radical new ideas concerning participative design in ISD. Researchers attempted to establish language games which provided a space for inter-disciplinary and multi-function systems design and examined ideas which later became embodied in approaches such as prototyping and user participatory design. Whilst this work did focus upon discourse and the creation of participative, intersubjective spaces, researchers like Ehn also tried to explore the spaces in which people lived. As Ehn pointed out 'this took us away from the academic mainstream, the reason being that this is not where our research subjects live' (Ehn (1988) p. 21). However, the constant across soft-systems thinking and other similar approaches as seen in the Scandinavian's work is the influence of phenomenology, a highly interpretivist view, criticised by some philosophers of technology as having an emphasis upon discourse and language, but leaving humans disembodied: in essence losing the humans in the text (Ihde (1998)). This has resulted in criticisms of Soft Systems and related approaches by Ciborra (1997), Stapleton (2001) and others.

Dahlbom & Mathiassen (1993) state that the issues surrounding the 'fundamental questions' of ISD require a discussion of 'the things we work with' and they see development as 'the activity in which systems are being produced' whilst quality is 'the raison d'etre of our profession and practice'. For these researchers these are 'the ingredients we see in a philosophy of systems development'.

A reading of Dahlbom & Mathiesen (1993), Checkland & Scholes (1990) and other related literature reveals the development project to be fixed upon the creation of a *technical* artefact at a *certain point* in time. It is possible to see, in this emphasis, a latent functional rationalism, with the recognised faults of positivism counterbalanced by an emphasis upon interpretivist approaches heavily influenced, in particular, by phenomenology (Stapleton (2001), Ciborra (1997), Flynn (1992)).

These postures have been criticised on the basis that organisational behaviour involves more than interpretation. It involves creation as well as discovery and authoring as well as interpreting. Interpretivism has been described in organisational literature as being too passive (Weick (1995)). Some philosophers of technology and culture have argued that interpretivism and social constructivism over-emphasise the world as narrative, something referred to as a contemporary 'rampant textuality' prevalent in scientific research of social settings (Ihde (1993) p. 91). The world is not merely a text to be interpreted. It is a space within which we find and invent ourselves, discover possibilities and engage in experience. A focus on action and creation has been lacking in interpretivist and social constructivist theory. Philosophers of technology have recently argued for a re-emphasis upon the concept of 'embodiment': humans (and indeed technology) seen as solid, rather than only locations of narrative (Ihde (1998)).

It is self-evident that one general criticism of all of the above philosophies is that they do not attempt to bring the worlds of technology and humans into a coherent analytical model for use by researchers and practitioners. This is a deep problem as it goes to the heart of the ISD discipline itself, and therefore requires a serious re-evaluation of the base assumptions of ISD. As Ciborra (1997) shows, these approaches remain, in essence, functionality driven.

[†] See for example the emphasis upon Deconstruction and Discourse as per Derrida in Rose & Truex (2000). Interestingly, this is one of the few papers which argues for the important contribution to IS theory of Latour's Agent Network Theory.

It is readily apparent that gaps exist in the theory of technology & social impact, particularly in the context of inter-cultural exchange. This will necessarily have a major impact upon ideas and concepts concerning social stability as it relates to IT development and deployment methodologies. ISD concerns itself with both development and deployment practises. Indeeed, from the earliest days IEEE Software Development standards see the 'installation phase' as 'the period of time in the software life cycle during which a software product is integrated into its operational environment and tested.... so that it performs as required' (IEEE (1983) p. 21).

However, the ISD literature has generally paid far less attention to deployment aspects of IT, than to the development aspect. Consequently, post-implementation (deployment) activities have received little attention, often to the detriment of ISD effectiveness (Stapleton (2000), Willcocks, Feeny, & Islei (1997)). Whilst the development phase leads to the structuring of a new technical artefact, the deployment phase is the critical phase in terms of social impact. Empirical studies show that the ISD deployment approach is critical for the overall effectiveness of ISD, including return on investment (Stapleton (2001)). This is particular true for large-scale deployments such as Enterprise Resource Planning systems and other inter-organisational solutions.

2.1. Revisiting ISD

The question is, are there alternative approaches which may draw us down different roads – roads that are neither positivist nor interpretivist? Are Dahlbom & Mathiassen's 'ingredients' the only way of looking at ISD? These thinkers, as important as they are, do not address important issues raised by Ricouer, Baudrillard, Latour, Ihde and others. If we are to continue the kind of radicalism central to the excellent work of Ehn, Checkland, Mumford and their contemporaries, it is important that ISD continually revisits and tests core assumptions and attempts to integrate contemporary movements in philosophy into ISD theory. This paper attempts to do just that by revisiting Latour's ideas in which humans and technology fold into eachother, creating new systems and addressing these systems as primarily social systems.

The remainder of this paper sets out an alternative to the current avenues of research under consideration and provides a basis for revising the theory of social impact in complex social settings. It achieves this by suggesting commonality between Latour's Agent Network Theory and Sensemaking Theory as expounded by Weick and others in the organisational literature. It is argued that this avenue paves a way between the philosophical positions of Interpretivism, Social Constructivism, Naturalism and Positivism, and provides a basis for progress in ISD theory.

3. THE RELATIONSHIP BETWEEN HUMANS AND NON-HUMANS

In order to understand and study the intercultural social impact of technology from this new viewpoint we must revisit the essential relationships between humans and artefacts. The work of philosopher Bruno Latour deals with the relationship between humans and non-humans and therefore provides a useful basis for such a revision. Whilst Latour's work has received some attention in the organisational studies and social studies literature, it has rarely been applied in the ISD discipline.

In Latour's analysis of these relationships he introduces the idea of '*interference in the program of action*' where *program of action* refers to the active use of a technological artefact (Latour (1999)). This is best illustrated by an example: the legalisation of guns in the USA. The National Rifles association (NRA) in the USA argue that guns should remain legal because, essentially, it is not the gun which commits horrific acts of violence, but the person in control of the gun. The gun itself is a neutral object. Alternatively, the anti-gun lobby argue that the person is somehow transformed by the gun, and will act in a more criminal way if in possession of the gun. Latour argues that, from a philosophical standpoint, these positions are 'sociological' and 'materialist' respectively. In the first position, that of the NRA, this sociological position argues that the agent (gun) is a neutral carrier of the will of the actor that adds nothing to the action. It is essentially a passive conductor through which good and evil of society can flow in equal measure. It is *society* or the human which determines what will happen, not the gun. In the second, materialist, view a person is somehow transformed by the gun and is potentially far more dangerous when in possession of this weapon. It is the *material* artefact (the gun) that determines what will happen, not the human. Simplifying, in the sociological view the gun is <u>nothing</u>, in the materialist view it is <u>everything</u>. We can translate Latour's concepts directly into current discussion of advanced technologies as follows: In most of the engineering and technology research and in the general discourse of the relationship between 'humans' and 'technology', each are treated as separate entities. Either the focus is upon the 'technical' on the one hand as the important issue or the 'social/human' on the other as the important issue. Consequently, research focuses upon addressing technical issues (including techniques, methodology, etc.) on the one hand, or social issues on the other. Consequently, these approaches rarely address deployment issues associated with the implementation and post-implementation phases. 'Soft' methods and sociotechnical approaches, it has been shown that the emphasis is primarily upon the collision of two separate systems (which remain separate), rather than the folding of one into the other as is suggested for some time by researchers of ISD but which has rarely been addressed (Boland (1985), Boland & Day (1989), Hirschheim & Newman (1991), Stapleton (2001)).

It is evident that we can identify a direct correlation between the sociological and material dichotomy expounded above, and the current state of research into the social impact of technology. However, Latour shows us that these two separate entities (human and non-human) <u>interfere</u> with one another to create a hybrid. This implies a new way of thinking about social impact in general, and ISD in particular.

Latour argues that neither perspective (sociological nor material) is correct. In order to show this he asks the question 'who is the actor'? Is the actor the gun or the person holding the gun? Latour argues that it is neither and both, it is <u>someone else</u>. This someone-else he calls the citizen-gun/gun-citizen. In this argument he makes a crucial point: If we try to comprehend techniques and technology while assuming that the human psychological capacity remains fixed, we will not understand the social impact of new technology and associated processes. Also, the technique or technology is transformed by the person i.e. the gun is different with you holding it. The gun has entered 'into a relationship' with the person holding it. It is no longer the gun-in-the-drawer, in-the-armoury or in-the-holster - it is the gun-in-the-hand. Latour argues that the twin mistake of materialists and sociologists in trying to understand the relationship between humans and non-humans is their focus upon essences (artefact or human). In Latour, both are transformed into something new, as illustrated in figure 1 helping the software engineer and the information technologist understand one way in which social impact is created. The technology is no longer an essential thing, nor is the human. It is *both together*. Human and artefact are folded into each other. They are transformed into something new, a composite of social and artefact as is argued by philosophers who criticise the over emphasis of current social research upon discourse and narrative (e.g. Ihde (1998)).

We must shift our attention away from 'technology' or 'society' or 'human context' to this new combination of social and technological. Latour calls this combination the 'hybrid actor'. Once we do this, we can see that goals (or functions) change from those of the individual components (human and non-human) to the goals/functions of the hybrid actor.

This is a very important philosophical step in our base assumptions. Applying this to the work of engineering and science in the field of IT development, we now find that we must focus upon a whole new array of actors and actions – the hybrid actors and their functions. This opens a new research trajectory for the social impact of ISD artefacts. We notice that we are now dealing with, not the goals of humans or technologies, but the new, distributed, mediated and nested set of practices whose sum may be possible 'to add up' but only if we respect the importance of mediation (*interference*) in the relationship.

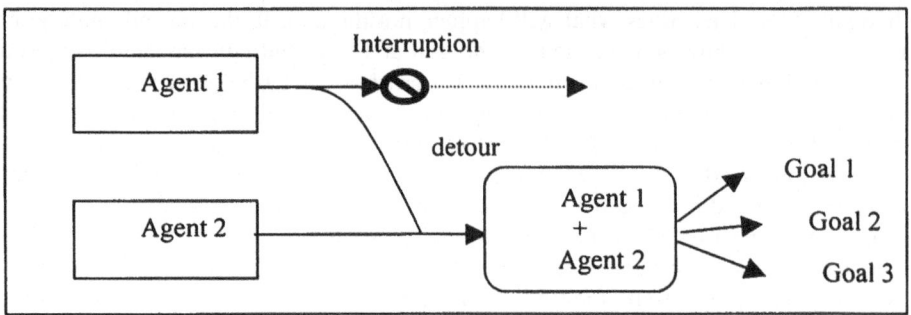

Figure 1. Interference & Goal/Function Transition (from Latour (1999) p. 170)

As this process of *interference* and *folding* develops we note how the original (perhaps explicit) goals can be lost in a maze of new goals as the entire system becomes more and more complex. For example, an early human discovers the stick, and we have a stick-human hybrid. Perhaps the human initially uses this stick to plough the ground. However, the human becomes frustrated with the stick and sharpens it thus creating a whole new set of goals and functions, such as the stick as a defensive or offensive weapon. This whole new set of goals or functions could not have been foreseen at the outset when the stick was originally discovered and deployed. It illustrates how technology deployment in human contexts must recognise that, as humans enter into and develop new relationships with the technology, goals and functions shift. This rationale directly implies that researchers of social impact in ISD must now introduce learning and adaptation theory into their armoury. Simultaneously, they must emphasise design and re-design principles for the technical component. We have not been 'made by our tools' as indicated by Marx and Hegel (homo faber fabricatus). Rather the 'association of actants' is the important thing for the researcher of social impact associated with IT deployment (Latour (1999).

Researchers must understand how
- New goals and functions appear
- New goals and functions can be understood and directed appropriately

This re-focuses our attention as ISD researchers upon processes by which organisations/societies can understand resident human/artefact hybrids within their social group. It is apparent that this requires the application of a social theory which includes organisational learning and decision making. This theory must also account for decision-making processes which are reflective, inter-subjective and iterative. Any revised theory of technology deployment must emphasise the human element of the new human-machine system and cater for humans as they attempt to make sense of the new world into which they are thrust: an inter-subjective, shifting space in which they are intricately bound with a new information technology artefact, and which often makes little sense to them (Stapleton & Byrne (2001)). Software (re-)design and deployment principles must be enhanced, or augmented, so that they can be folded into the overall management of the hybrid system. The question is, can we develop a basic theoretical model upon which these can be brought together and managed coherently? One promising social learning framework we can build upon is sensemaking theory.

3.1. Sensemaking: An Intersubjective, local Process

Sensemaking literally means the making of sense. People 'structure the unknown' (Waterman (1990) p. 41) and researchers interested in sensemaking concern themselves with how and why people create these constructions and what the affects of these structures are. This theory is a promising departure for ISD because it enables researchers to treat humans as active bodies shaping and re-shaping their world, and making sense of that same world inter-subjectively. This goes to the heart of the ISD process as those pioneers of participative systems development and design, Ehn, Mathieson, Dahlbom, Checkland, Mumford and so many others, envisioned ISD. Simultaneously, it recognises that humans act and enact, and provides a trajectory which addresses some of the criticisms of the overly discourse-based view of ISD which has emerged around participative approaches.

It is stressed in sensemaking literature that professional problem solvers such as systems engineers and managers cannot derive adequate solutions to complex, socially located, problems through to observation and analysis alone, as is typified in the dominat approaches to ISD (FitzGerald (2000)). Solutions can only be found (and re-found) by open and active experimentation. As people's interaction and learning proceeds the very basis for an analytic solution changes. Analysis and interaction are thus seen as two modes of organisational problem solving which supplement each other (Boland (1985)). In equivocal situations, such as those which prevail in IS deployment scenarios, this problem-solving mode is more potent than comprehensive data analysis (Weick (1995)).

In sensemaking a stimulus (new technology, work practices etc.) raises a series of questions, which must be explicated and understood. These questions result in actions which change the environment, resulting in new stimuli and so the cycle begins again. People involved in sensemaking activities must interact with others in order to make-sense of organisational realities. Furthermore, there is evidence that indicates that these groups of sensemakers need sensemaking support personnel to facilitate this process (Weick (1982), Stapleton (1999), Stapleton (2001)). This cooperative sensemaking indicates the *inter-*

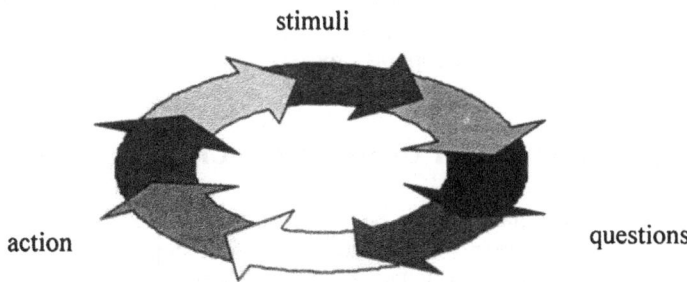

Figure 2. Sensemaking Cycles (from Stapleton (2001) p. 82)

subjective nature of technology deployment activities (Boland & Day (1989)). For example, if technology driven change occurs, many people and groups must work together in order to come to some sense of what the change means and what the appropriate responses are. This only happens as people engage with(in) the new system. In this way new goals and functions are created or discovered. Inter-subjectivity implies a high level of trust between participants in the process. Indeed, research based upon these types of activities emphasise the building of deep friendships and common understanding (e.g. Klein & Hirschheim (1991)). The convergence upon solutions implies a cyclic process during which questions we are trying to answer are progressively reviewed and understood. Sensemaking theorists argue that when the question is adequately understood then the required solutions should be obvious (Weick (1995)). This cyclic process of sensemaking is illustrated in figure 2.

In the context of technology deployment, sensemaking theory shows that the management of the introduction of technology into a social setting, and thereby the creation of a hybrid, must equally engage all major stakeholders in cooperative sensemaking. This viewpoint has important consequences for ISD. ISD methodologies generally ignore the cultural differences that exist in differing organisational settings. However, these differences are widely recognised as part of the critical backdrop that is the organisational field in which the technology will be deployed. Indeed, some philosophers of culture argue that technology is not a non-neutral artefact from a cultural perspective. Ihde (1999) shows how technology deployment involves the creation and deployment of, what he terms, 'techno-cultural' artefacts. These writers argue that technology cannot simply be transferred from one culture to another as if it were a passive, neutral object. Some argue that this cultural affect is utilised to the advantage of colonial aspirations (Banerjee (2001) and there are strong political and philosophical underpinnings for these arguments (Chomsky (1993), Baudrillard (1999)). ISD methods were created in a western intellectual space which may (or may not) be appropriate in post-socialist countries, or in so called developing nations (Stapleton et. al. (2001)). Methodology must take these local contextual issues into account. This can only be achieved by the establishment of processes which draw upon local circumstances for their energy and dynamic. Which ever approach we take to the creation of new ISD research trajectories, Bannerjee (2001), Chomsky (1993), Ihde (1999) and others show that there is a moral and professional responsibility upon ISD researchers and practitioners to recognise these techno-cultural effects. The establishment of egalitarian partnerships with associated, explicit, sensemaking processes is critical to the successful deployment of technology across inter-cultural domains. In this context, it is evident that sensemaking provides a theoretical basis for an ISD theory which enables researchers to weave local, human issues into the deployment of IT artefacts, whilst allowing us to maintain Latour's idea of folding of humans and technology into eachother. We can thus address the local, cultural contexts in which people live out their daily lives. The human does not disappear in a mist of discourse and narrative, but is centred in, and central to, the ISD support process.

4. TOWARDS AN ISD THEORY OF HUMAN-TECHNICAL HYBRIDS

The theory of sensemaking can be co-opted into Latour's vision of the human-machine hybrid. This requires a series of steps. The first step is to revise the sensemaking cycle depicted in figure 2 to a spiral. This emphasises how humans who are trying to make intersubjective sense of the new world introduced by the technology, discover new realities in their work lives as a direct result of their being part of the hybrid system. New goals and functions will emerge and must be made sense of. This is diagrammatically depicted in Figure 3. Similarly there is a spiral of redesign for the technological element of the hybrid system. As the social world changes in response to the initial impact of the human-machine hybrid, new goals and functions emerge for the technological component of the hybrid system. This requires a continuous review of how the technology operates, how it can be used in new ways, or how it must be redesigned in order for the hybrid system to remain effective. Thus the spiral in figure three is entirely appropriate for the activities

Figure 3. Sensemaking Spiral

associated with the deployment of IT, especially as regards the sensemaking support processes which are necessary for successful post-implementation (Halpin & Stapleton (2002)). Thus the weaknesses of Agent Network Theory as it has been presented in ISD (see Rose & Truex (2000) can be addressed.

In both cases the spiral represents a moving outwards to new functions and goals, and de-emphasises the more simplistic cyclic motion of the sensemaking cycle in figure 2. The centre of the spiral marks the origin of the system, the point at which the human & machine interfere with eachother. This dramatically alters Latour's view of a straight-line movement towards new goals and functions, a view which is not easily supported within theories of decision making and organisational learning (e.g. O'Keeffe (2001)).

The model remains incomplete. In our revised theory of social impact in inter-cultural contexts, a third element is needed for successful technology and knowledge transfer. Here this is termed sensemaking support and elsewhere as the 'explication process' (the term is used here in its philsopohical sense (Stapleton (2001), Blacburn (1994)). This is a spiral of continual interaction and re-interaction with both the re-engineering/re-design process and the human sensemaking process. Explication is deployed to help make sense of changes concerning the technological subcomponent, and the human process subcomponent of the hybrid. Bringing the entire model together gives figure 4.

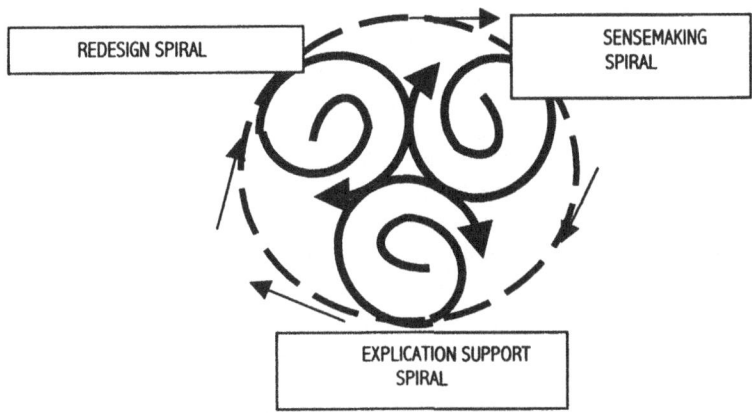

Figure 4. Revised Model of Technology and Knowledge Transfer

It is evident from the model in figure 4 that the design and deployment of a knowledge and technology transfer approach must address the entire system in a unified way. Furthermore, it must recognise that the entire hybridised system is an *open* system, i.e. there is a sharing and transference of energy and resources between the hybrid system and its environment. This is a stark omission in Latour's model, but critical if we are to begin to understand inter-cultural exchange in which very complex environments are created that impact upon the human-machine hybrid.

This model addresses the inherent ambiguities and complexities within Latour's hybrid systems by way of sensemaking support, which in turn feeds into and out of an engineering re-design process. This support feeds into, and out of, technical and non-technical elements of a hybrid system, whilst still treating it as a coherent whole.

5. CONCLUSION

The model of social impact illustrated in figure 4 can be used to drive forward theory and practice. Researchers can adopt this basic framework to identify the most effective ISD approaches. Several promising approaches have begun to appear in the literature. Firstly, at a very general level, the e-Mode2 approach (Stapleton et. al. (2001)) provides an excellent technological and organisational infrastructure within which knowledge can be produced, and in which the model presented here can be incorporated and supported. At a more operational level, the COPIS approach (Jancev & Cernetic (2000)) recognises the importance of peer relations and trust, team building and support processes in knowledge and technology transfers between EU and Post-Socialist societies. ISD researchers need to push this work forward in order to ensure that we address hybrid systems holistically rather than focussing upon the individual components. This paper also shows that it is apparent that researchers information technology development and deployment be:
1.   Made aware of the particular assumptions underpinning their work
2.   Encouraged to challenge working assumptions and identify new perspectives.
3.   Build new theories and practices upon these revised sets of assumptions
This requires a fresh impetus within ISD which actively studies philosophical positions and inquires into those positions which are useful to researchers and practitioners of ISD. This has been strenuously argued elsewhere and the call is renewed here. This is especially true in the modern organisational setting of highly complex organisational structures where managers often exist 'at the edge of chaos' (MacIntosh & MacLean (1999)). These settings are often created by the information technologies deployed. Research should pay particular attention to inter-organisational systems such as Enterprise Resource Planning and other large-scale inter-organisational solutions (Stapleton (2001), Davenport (1998)). These solutions are often accompanied by severe organisational trauma. This trauma has been directly linked to ISD practise and typically is associated with poor sensemaking support processes (Stapleton & Byrne (2001)).

E-Mode2, COPIS and other approaches mark the beginnings of a new trajectory in the study of the social impact of technology in an inter-cultural context. However, these theoretical developments largely exist outside the ISD discipline. This paper provides an important impetus for the crucial debate concerning the cultural and social impact of ISD. It recognises that new 'things' are created by ISD and attempts to understand these entities i.e. the human-machine hybrids, in a fresh way. It also provides a basis for driving this research trajectory forward. It is vital that researchers devote their efforts to moving this work forwards and uncover new pathways for research. If not, then ISD is doomed to continue creating systems that inflict themselves upon organisations, rather than enhance their effectiveness.

Latour's approach and sensemaking theory have not been brought together theoretically within the ISD literature. It is evident, however, that these two provide ISD researchers with new ways of thinking about what ISD addresses in the 21st century. ISD becomes the creation of social, hybridised systems, moving us away from the creeping functional rationalities which remain so central to the ISD domain.

6. ACKNOWLEDGEMENTS

The author gratefully acknowledges the comments and advice of the reviewers.

7. REFERENCES

Ayer, A.J. (1936). Language, Truth & Logic: The Classic Text Which founded Logical Positivism and Modern British Philosophy, Penguin Books (Reprint 1991).

Banerjee, R. (2001). Biodiversity, Biotechnology & Intellectual Property Rights: Unpacking the Violence of 'Sustainable Development', *19th Standing Conference of Organisational Symbolism (SCOS XIX)*, Dublin, (forthcoming).

Baudrillard, J. (1999). The Consumer Society: Myths and Structures, Sage: London.

Bentley, R., Hughes, J.A., Randall, D., Rodden, T., Sawyer, P., Shapiro, D., Sommerville, I. (1992). 'Ethnographically-Informed Systems Design for Air Traffic Control' in *Proceedings of Computer Supported Co-operative Work 1992*, ACM, pp. 123-146.

Bickerton, M.J. & Siddiqi, J. (1993). 'The Classification of Requirements Engineering Methods', *Proceedings of the International Symposium of Requirements Engineering*, IEEE Comp. Society Press, pp. 182-186.

Blackburn, S. (1994). Oxford Dictionary of Philosophy, OUP.

Boland, R. (1985). 'Phenomenology: A Preferred Approach to Research on Information Systems', in Mumford, E., Hirschheim, R.A., Fitzgerald, G. & Wood-Harper, A.T. (eds), *Research Methods in Information Systems*, Elsevier: Holland.

Boland, R. & Day, W. (1989). 'The Experience of Systems Design: A Hermeneutic of Organisational Action', *Scandinavian Journal of Management*, 5, 2, pp. 87-104.

Checkland, P. & Scholes, J. (1990). *Soft Systems Methodology in Action*, Wiley: NY.

Chomsky, N. (1993) 'Year 501: The Conquest Continues', A.K. Press.

Ciborra, C. (1997). 'Crisis and Foundation: An Inquiry into the Nature & Limits of Models and Methods in the IS Discipline', *Proceedings Of 5th European Conference on Information Systems*, 3, Cork Publishing: Ireland, pp.549-1560.

Cronk, L. (2000). 'Reciprocity & the Power of Giving', In *Conformity & Conflict* ed. Spradley, F. and McCurdy, D., Allyn & Bacon: MA, pp. 157-163.

Dahlbom, B. & Mathiassen, L. (1993). The Philosophy & Practice of Systems Design.

Davenport, D. (1998). 'Putting the Enterprise Back into the Enterprise systems', Harvard Business Review, August, 76, 4 pp. 121-131.

Ehn, P.(1988). Work Oriented Design of Computer Artefacts, Arbetslivscentrum.: Stockholm

Fitzgerald, B. (2000). 'System Development Methodologies: a problem of tenses', Information Technology and People, Vol.13 pp. 174-185

Flynn, P. (1992). *Information Systems Requirements*, McGraw Hill.

Galliers, R. (1992). 'Choosing Information Systems Research Approaches', in Galliers, R. (ed.), *Information Systems Research*, Blackwell, Oxford, pp.144-62.

Hammersley, M. (1990). 'What's Wrong With Ethnography? The Myth of Theoretical Description', *Sociology*, 24, 4, Nov, 1990.

Halpin. L. & Stapleton, L. (2002). 'Towards a Revised Framework of Management Theory & Practise in Uncertain, Chaotic Environments: Managing Post-Implementation in Large-Scale IS Projects', *in Proceedings of the 2002 IAMA Conference*, forthcoming.

Hirschheim, R.A. & Newman, M. (1991). 'Symbolism and Information Systems Development: Myth, Metaphor and Magic', *Information Systems Research*, 2/1.

IEEE (1983). IEEE Standard Glossary of Software Engineering Terms, ANSI/IEEE Standard 729, Institute of Electrical and Electronic Engineers, New York.

Ihde, D. (1993). *Post-Phenomenology: Essays in the Post-Modern Context*, Northwestern University Press: Ill.

Ihde, D. (1998) *Expanding Hermeneutics* Northwestern University Press: Ill.

Jancev, M. & J. Cernetic (2001). A Socially Appropriate Approach for Managing Technological Change, *8th IFAC Conference on Social Stability* (forthcoming).

Latour, B.(1999). Pandora's Hope: Essays on the Reality of Science Studies, Harvard.

Lulofs, R. & Cahn, D. (2000). *Conflict: From Theory to Action*, Allyn & Bacon: MA.

MacIntosh, R. & MacLean, D.(1999). 'Conditioned Emergence: A Dissipative Structured Approach to Transformation', *Strategic Mgt. Journal*, pp. 297-316.

Moreton, R. & Chester, M. (1999). 'Reconciling the Human, Organisational and Technical Factors of IS Development', in *Evolution and Challenge in Systems Development*, Zupancic, J., Wojtkowski, W., Wojtkowski, W.G., Wrycza, S., (eds.), Kluwer Academic/Plenum Publishers: New York, pp. 389-404

Myers, M. (1995). 'Dialectical Hermeneutics: A Theoretical Framework for the Implementation of Information Systems', *Inf.Sys.Journal*, 1, pp. 51-70.

O'Keeffe, T. (2001). 'Learning to Learn Within Dynamic Multinational Environments', *8th IFAC Conference on SWIIS* (forthcoming).

Simonsen, J. (1995). Designing Systems In An Organisational Context: An Explorative Study of Theoretical, Methodological & Organisational Issues from Action Research in Three Design Projects, Ph.D. Thesis, Dept. of Comp. Science, Roskilde University: Datalogiske Skrifter, Denmark.

Stapleton, L. (1999). 'Information Systems Development as Interlocking Spirals of Sensemaking', Zupancic, J., Wojtkowski, W., Wojtkowski, Wrycza S., (eds.), *Evolution and Challenges in Systems Development*, Plenum: NY, pp. 389-404.

Stapleton, L. (2001). Information Systems Development: An Empirical Study of Irish Manufacturing Firms, Ph.D. Thesis, Dept. of B.I.S., University College, Cork.

Stapleton, L., Cernetic, J., MacLean, D. & MacIntosh, R. (2001). 'Economic Recovery through E-Mode Knowledge Production', *8th IFAC Conference on Social Stability* (forthcoming).

Stapleton, L. & Byrne, S. (2001). 'The Illusion of Knowledge The Relationship Between Large Scale IS Integration, Head Office Decisions and Organisational Trauma', *Proceedings of the 19th SCOS Conference* Dublin (forthcoming).

Suchman, L. (1987). *Plans and Situated Actions*, Cambridge: MA.

Waterman, R. (1990). *Adhocracy: The Power to Change*, Whittle Direct Books: TE.

Weick, K. (1982). 'Management of organisational change amongst loosely-coupled systems', P.Goodman & Assocs. (eds), *Change In Organisations*, Jossey Bass:CA.

Weick, K. (1995). *Sensemaking in Organisations*, Sage Publications: CA.

Willcocks, L., Feeny, D. & Islei, G. (1997). *Managing IT As A Strategic Resource*.

Winograd, T. (1995). 'Heidegger & The Design of Computer Systems', in Feenberg, A. & Hannay, A. (eds.), *Technology & The Politics of Knowledge*, Ind. Unv. Press

DEVELOPMENT OF INFORMATION SOCIETY: PROBLEMS AND SOLUTIONS

Valentinas Kiauleikis, Audronė Janavičiūtė, Mindaugas Kiauleikis, Nerijus Morkevičius[*]

1. INTRODUCTION

The coming of the 21st century saw an increasing number of Lithuanians making use of computers.[1] However, this situation should be seen from two different perspectives: state institutions and mass media pays a lot of attention to the development of information, or knowledge society, but people involved in the practical installation of information technologies (IT) are not very optimistic. State institutions are concerned with keeping up with the progress in this field in European Union, and especially neighboring Baltic countries; the consumer, on the other hand, sees IT in the light of the possibilities of his/her enterprise that usually are quite modest.

The development of informational society is being discussed in the light of various aspects: scientific, technological, political, economic, cultural and other.[2,3] Out of a great number, three trends have been chosen for discussion here: 1) computer literacy, 2) computerized fulfillment of civil duties, and 3) professional computer competence. This article offers technologies to deal with these issues that currently are of importance in Lithuania.

The level of *computer literacy* is more and more often estimated in the light of knowledge of the ECDL program; however, a great number of computer users successfully do without skills in this program, and they by no means can be considered illiterate. The program eliminating computer illiteracy is implemented in Lithuania with moderate success, with major attention focused on students and schoolchildren. Every year the state

[*] Kaunas University of Technology, Department of Computers, Business Computing Laboratory, Studentu str. 50-215, LT-3031 Kaunas, Lithuania.

Information Systems Development: Advances in Methodologies, Components, and Management
Edited by Kirikova *et al.*, Kluwer Academic/Plenum Publishers, 2002

25

and various funds give considerable sums of money to schools, which results in a great number of young "literate" people joining the active members of society, and replacing retiring elderly people. Thanks to the latter, the number of "literate" old people increases. This tendency is illustrated in figure 1. Lithuanians living in the country form a considerable part of population, which is notable for their slow progress in the field of information technologies, yet computer is no longer ignored in the countryside. Computers are used at village schools, libraries, culture centers, and by progressive farmers who are starting using them in their professional activities. Computer skills of unemployed people, who make up 12% of Lithuanian population, are developed at different courses financed by the state as well as private and foreign funds.

Computerized fulfillment of civil duties is one part of the task to provide all citizens with the possibility to make use of informational technologies in everyday activities.[2,3] This part is a very significant direction in the development of information society, because such functions as tax payment, various references or certificates, discussions on the state decisions, or presentation of opinions, though playing a vital role in the civic society, also tend to consume a lot of time and money. Modern technologies allow the society life to reach a new level of intensity, yet this process is slower than that of eliminating computer illiteracy. It is even possible to observe that computerized fulfillment of civil duties in Lithuania is still in the very initial stage of development. The fundamental work carried out in this field consists of computer literacy and development of technical possibilities. However, it would be too optimistic to expect dramatic changes: a great number of the country's population will not start using computers in the nearest future, while civic functions have to be performed by everyone, including pensioners. Development of technical possibilities (creation of "e-government"[4]) remains only a part of strategic plans of the Government, thus greater changes here are hardly possible in this decade.

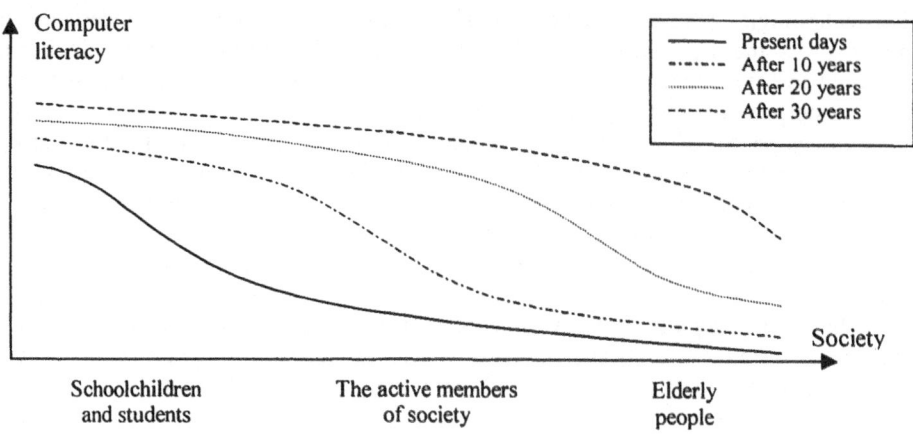

Figure 1. The hypothetical rise in computer literacy level of Lithuanian society

Professional computer competence. The issue of professional competence has been widely discussed recently; however, at the moment it is also necessary to pay the same amount of attention to the issue of professional *computer* competence. Even the best specialists – managers, engineers, economists, or accountants – they can only adapt themselves in a modern computerized organization if 1) they are computer literate; 2) they are able to use computer hardware and software related to their field of work; the latter usually already do not fall under the category of computer literacy.[5] Thus, professional computer competence could be defined as professional competence plus computer literacy plus realization of professional competence by means of computer. Professional computer competence is vital part of qualification for active members of society; updating of this competence must be a concern of an organization employing a specialist. On the whole, active members of society, creating surplus value and providing for the remaining part of society (including subsidized part of people in the countryside) must have a professional competence, regardless the field of activity.

Analytical works[3] point out the development of informational technologies during recent decades, thus giving evident illustration of the development of professional competence in the medium of developing informational technologies from calculation based scientific applications to group and communication centered computing. In Lithuania professional computer competence has been developing since approximately 1990, when first personal computers appeared on the market. A great number of then still successful enterprises were replacing big-sized ES class and mini CM class computers (ES and CM were IBM compatible computers in East European countries) with personal computers and their networks, and were employing programmers to develop new software, as well as teaching their employees to work with computers. This process could be considered an initial stage in the development of computer competence. The previous decade prepared numerous specialists using computers to perform their professional duties; this process has never stopped since. This is determined by several factors. First, computer illiteracy has not been eliminated yet. Second, young people joining the active stratum of society have a sufficient knowledge of computers, but lack proper professional competence, which is gained later, while working. Third, the notion "professional computer competence" is changing with the development of information technologies: computer skills have to be regularly updated. The basic professional computer competence can now be gained at universities and colleges; besides, first signs of it can be seen among educated Lithuanian farmers.

Here are some thoughts about the development tendencies of professional computer competence in Lithuania. The first enterprises to have introduced personal computers and their networks around the year 1990 brought up a few specialists in this field. During the following decade, a great number of enterprises switched to the computerized accounting system, which led to the increase in the number of employees with sufficient computer competence.[1] However, it is difficult to precisely indicate the level of professional computer competence; a special study is required for that purpose. Figure 2 displays two hypothetical curves of professional computer competence. The inner curve reflects the level of professional computer competence in various society strata during the first years of computerization. This curve indicates to a small number of people in the active substratum with sufficient professional computer competence. The outer curve reflects the level that should be reached by professional computer competence under the conditions of developed informational society. As the figure shows, this curve covers a lot wider stratum of society; this is due to the increasing amount of computers in organizations, as well as software meeting the needs of specific activity fields, and better computer training of specialists. However, there are limits. Studying youth can only achieve professional computer competence after gaining professional competence; also, the professional computer competence of elderly people leaving the active substratum of society relatively suffers as the latter do not actively participate in raising professional or computer competence.

It is possible to sum up the three directions in the development of information society in the following way:

1. Eliminating of computer illiteracy is just the first and necessary step towards information society. Computerized fulfillment of civil duties, and professional computer competence demands deeper knowledge of modern information technologies as well as skills of using information sources.

2. Computerization of civil functions (that is, putting authority into practice) and professional computer competence demand information literacy, that is, ability to independently define one's own informational needs, effectively use informational computer sources, create and manage one's own informational sources, use information as help in achieving aims, and understand economic, legal, social, and ethic issues of obtaining and using information.

3. Increasing resources of information necessary and widely available for the public on the Internet should become a precedent for greater public interest in the role of computer in everyday life, and also an instrument for the advancement of the **computer literacy** level of society.

4. The development of society's information literacy demands technologies and other means, the creation and development of which require computer instrumental means, specialists in information technologies, as well as favorable attitude of the state and public alike.

This article aims to survey information technologies and means used in the development of information society, and presents some of the solutions developed and being developed in Lithuania through joint effort from universities, business and the state. One of such means, created in 2001, is the Lithuanian Central Internet Gates, discussed in the next chapter of the article.

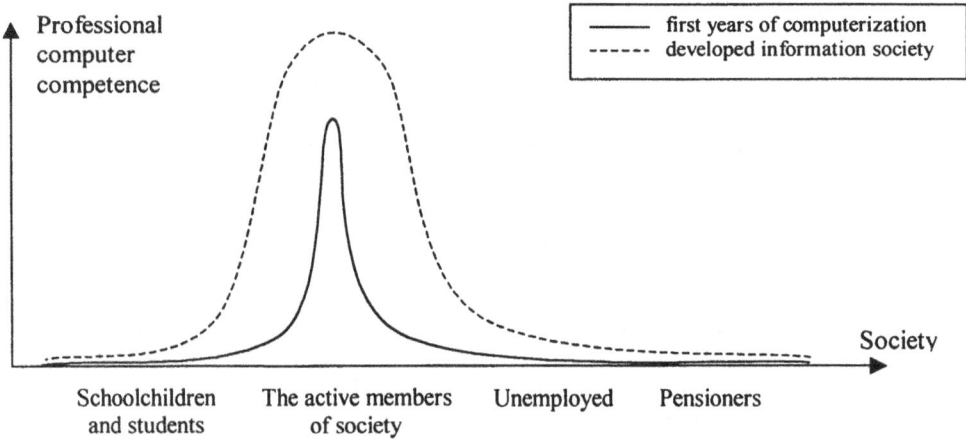

Figure 2. The progress of professional computer competence

2. LITHUANIAN CENTRAL INTERNET GATES

The following circumstances led to the development of the Lithuanian Central Internet Gates (LCIG):

1. The first information sites created by both the state institutions and private companies, aiming to specific audiences (Maps www.maps.lt, Travel www.travel.lt, e-biz www.ebiz.lt, Arts www.arts.lt and other) as well as commercial ones (Delfi www.delfi.lt, Lithuanian telecom www.takas.lt, Omnitel www.omni.lt and others) started appearing on the Internet. However, there was no site presenting the history, geography, culture, and administrative subdivision of Lithuania, especially in foreign languages.
2. In recent years, the development of information society in Lithuania has become a much-promoted idea. The Seimas, Government, and Presidency of the Republic of Lithuania established subdivisions to deal with this issue; various conceptions, strategies and plans have been developed. However, these activities have so far limited themselves to mere discussions of what is to be done.

In 2001 an initiative to deal with this situation came from business. By that moment, joint-stock company "Kraštotvarka" had already developed the methods and technologies of collecting and editing data about Lithuania, and had published cognitive books in main European languages. In 2001 this company initiated the development of the LCIG project with the aim to present cognitive information about Lithuania to users of the Internet. Kaunas University of Technology undertook dealing with technological issues. Having taken into consideration the state-level significance of the project as well as its potential for development in the future, IBM products were chosen to form the technological basis for the project. The following directives for the technological realization of the project were approved:

1. The main criterion for choosing alternatives is technology, which has to meet the highest modern requirements, and offer prospects that are possible to forecast.
2. The project must be orientated to hardware-software means and solutions of a single company.
3. Reliable operation of the system as well as high level of services must be ensured.
4. The chosen technology should not limit possibilities for the development of the system.

The architecture of the LCIG website (www.lietuva.lt) and technologies in use are shown in figure 3 (the website is currently administered by IBM Lietuva, which also rents the server IBM NetFinity5600 as well as software packages DB2 and WebSphere). A software application IMIWeb, created to administer the website and upload data to DB2, is connected to the database via ODBC.

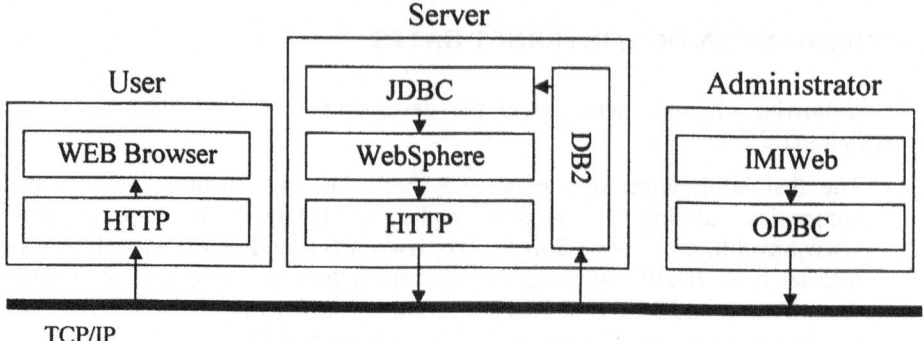

Figure 3. The architecture of the Lithuanian Central Internet Gates (www.lietuva.lt)

This is just the first part of the realization of the LCIG project, intended to present cognitive information about Lithuania. Though the data put in it is not fully structured, and its analysis is limited, it will, however, form a joint database subject to analysis together with structured data on business, services and government (Fig. 4).

Cognitive information about Lithuania is presented in six European languages: Lithuanian, English, Russian, Polish, German, and French. In the light of the development of information society, the following positive aspects of the LCIG could be pointed out:

1. The Lithuanian public has been provided the access to the systemized basic information about the country. The need for such a website became obvious when the data transfer to the database and the website operation started. A few thousand of the site visitors per day as well as letters asking for help in finding data prove this point. References from the LCIG to other sites and vice versa can be considered the starting point of the Lithuanian information system, which will allow every citizen to get the necessary information without particular efforts or delay, thus fulfilling one of the major tasks of the development of information society.

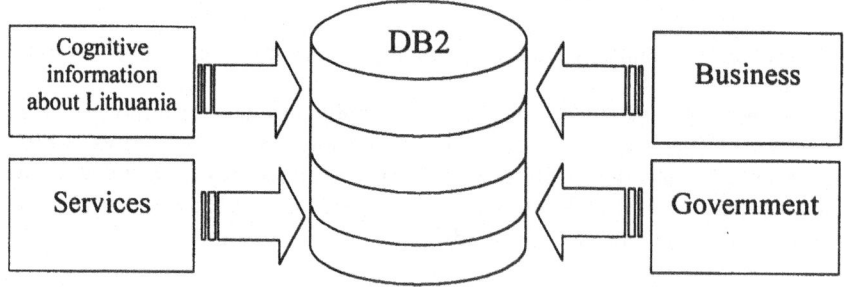

Figure 4. Functions of the Lithuanian Central Internet Gates

2. Studying youth of Lithuania is provided excellent study material on the country's geography, history, culture, and the state system. This material being presented in main European languages, the LCIG also serves as an extra means for language studies. At the same time, the computer at school becomes not just an instrument for lessons of informatics, but also a repository of information sources useful for studying of many other school subjects.

3. Systemized information about Lithuania is presented to visitors from other countries having interests in Lithuania. Even though currently the LCIG presents only cognitive information about the country, the number of the website visitors from different countries is increasing all the time.

3. THE LCIG PORTAL

Another function of the LCIG is providing services. On the whole, a lot of firms provide services via the Internet: banks accept payment transfers, inform about their activities, and allow handling of accounts; numerous companies provide information, some of them accept orders, and even payments for services and goods; state institutions also start providing their services there. The development of services tends to expand rapidly, but the lack of order in it could soon lead to chaotic situation, confusing both service providers and consumers. The LCIG, as a state website, aims to integrate the information on services. It also aims to organize the services provided by the Government, Seimas and Presidency. Such services are related to interior (for example, carrying out functions of the Ministry of Interior to interior subjects) and exterior needs (visas, customs, etc.). The LCIG architecture for the realization of the services system is shown in figure 5. The services system demands interactive communication between interior and exterior consumer and providers of services; to meet this requirement, the LCIG architecture has been expanded to the portal architecture through the IBM instrumental means. Here are some basic features of this portal.

1. The portal enables working with structured and non-structured data. Structured data are taken from regularly updated databases owned by ministries, Seimas, or other institutions. These are various information registers, structured for work with analytic programs. Service providers would perform them functions via the program application of providing services. Operation of non-structured data remains the same (Fig. 3).

2. IBM DB2 OLAP should be used to analyze data of a provided service and to realize the logic of a service in the portal architecture. A service application initiates the portal address to online transactional databases (OTDB) via the DB2 OLAP server.[6] Thus the Application server uses three sources of information to work up an application: an applicant communicating *online*, the portal database, and exterior OTDB. The role of service provider is limited by the functions of control, database maintenance and updating, and other service functions.

3. When data of services already provided via specially designed applications is also used for other services (for example, state registers), it is almost impossible to escape the problem of updating data. This problem could be dealt with by using the database integration technology and other means discussed below.

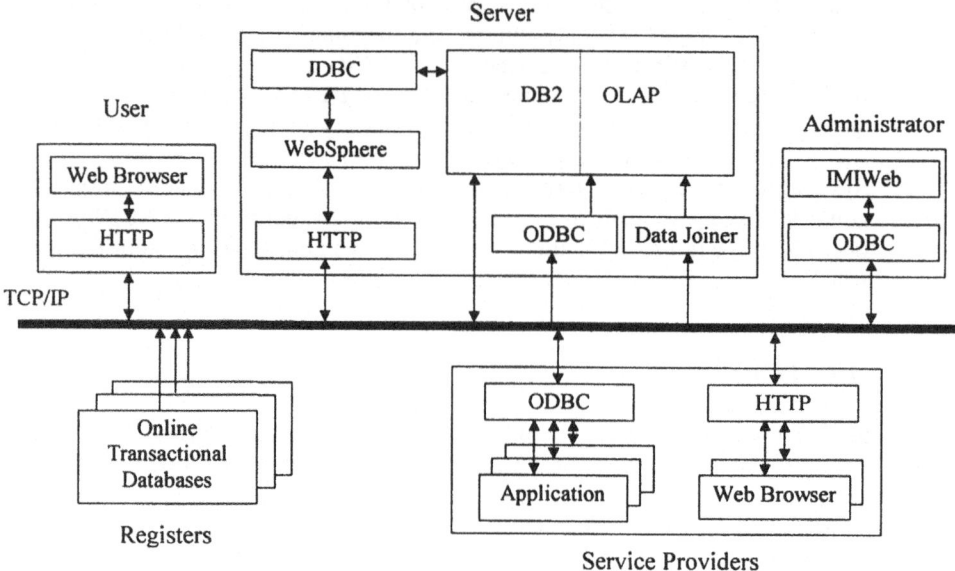

Figure 5. The LCIG Portal

In the light of the development of information society, the services provision system should activate **computerized fulfillment of civil duties**. The initiative of the portal realization is supported by the state. The Ministry of Interior undertook the supervision of the state program of the LCIG development of information society. This is one of the stages in the creation of e-government, and it will demand technological, educational and organizational solutions. One of such solutions is a project of the small-scale business enterprise portal discussed in the following section.

4. THE PORTAL OF SMALL SCALE BUSINESS ENTERPRISES

The developed technological basis could be used to create different solutions. One of major tasks in the development of information society is assistance to small-scale business. The Lithuanian Small Business Confederation unites over 2 000 small-scale business enterprises. Some of them operate some kind of software, have sufficiently developed databases, and advertise their production in websites.[7] Most of such enterprises, however, are not able to deal with such issues; thus the confederation is looking for an integrated solution enabling to provide appropriate services to small business. The suggested solution in the LCIG base is presented in figure 6.

Business Applications

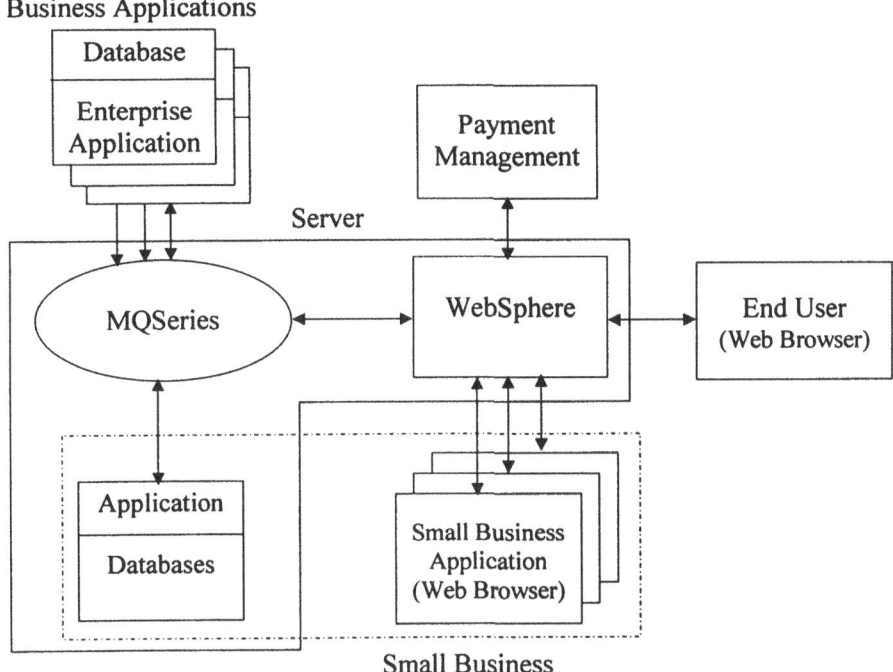

Figure 6. Portal extension for small-scale business enterprise

The portal has to solve the following three problems:

1. Support of the Internet database for small-scale business enterprises. The opinion poll of businessmen revealed, that they are most interested in the presentation of their products and services in a multi-language portal. Especially interested are those running rural tourism, as well as producers of art works (amber, ceramics, smith's, wood carving, etc.), and representatives of food, wood, flax and other industries. For the initial stage of this project, DB2 resources of the LCIG portal have been used, containing cognitive information about Lithuania, which of course includes data about localities of enterprises as well as their landscape, transport highways, and so on. In the future, if financially reasonable, Visual Warehouse package could be used to expand the solution.

2. Integration of current applications. Even those businessmen already using the Internet to present themselves are interested in the centralized multi-language access to data. The LCIG is becoming increasingly attractive in its possibilities for advertising and using the system of centralized services. However, enterprise applications have been created and introduced during the period of last 8-9 years, which naturally resulted in the formation of several generations and types of technologies that cannot interact. The project offers an IBM integration packet, MQ Series to integrate those technologies into a centralized system.[8]

3. Electronic commerce. The project realizes this function via specialized possibilities of WebSphere (Payment management). A centralized electronic store

presents customers with catalogues of products from Lithuanian small-scale businesses, as well as data on business people and their enterprises, and provides possibility to pay for chosen goods.

It is expected that the centralized portal will be implemented with the support from the state, and investments into it will be repaid via developed small business.

In the light of the development of information society, efforts of business people to present their companies, regularly update that information, analyze the market situation, and compare their possibilities with those of competitors, can be considered as development of professional computer competence and information literacy.

5. CONCLUSIONS

Elements of informational society development discussed in this article: computer literacy, computerized fulfilment of civic functions, and professional computer competence are put into practice by increasing information resources on the Internet, and developing informational technologies available to the public. In Lithuania, the Central Internet Gates have been created and are being developed to serve this purpose. They present cognitive information about Lithuania to the local and world public in six European languages. This source of information has already attracted the public attention; it is expected to become one of incentives for the advancement of computer literacy level in Lithuania. The LCIG technologies allow developing the gates into a portal with wide possibilities, including introduction of e-government functions, centralized multilingual presentation of business, and e-commerce. This allows fulfilling the tasks of informational society development: development of professional computer competence and computerized fulfilment of civic functions.

Business enterprises, universities and state institutions gradually join the development process of the LCIG.

REFERENCES

1. V. Savukynas, Lithuania in the Information Society, *Baltic IT&T Review, No 4 (23)*, 2001, pp. 53-56.
2. C. Stephanidis, G. Salvendy, Toward an Information Society for All: HCI challenges and R&D recommendations, *International Journal of Human-Computer Interaction, Vol. 11(1)*, 1999, pp. 1-28.
3. C. Stephanidis, G. Salvendy, Toward an Information Society for All: An International R&D Agenda, *International Journal of Human-Computer Interaction, Vol. 10(2)*, 1998, pp. 107-134.
4. F. Galindo, Electronic Government: From the Theory to the Action, *Proc. of The Second International Scientific Practical Conference Information Society '2000*, Vilnius, 2000, pp. 111-115.
5. E. Telesius, Skills of Information Technologies – to Everyone ECDL Program in Lithuania – Strategic Initiative of Computer Society, *The International Conference "Information Society"*, Vilnius, 1999, pp. 178-180.
6. C. Finkelstein and P. H. Aiken, *Building Corporate Portals with XML*, McGraw Hill, New York, 1999, p. 529.
7. V. Kiauleikis, A. Janavičiūtė, Information Technology of Small and Medium Business, *The 5th International Conference "Baltic Dynamics '2000" and the 6th International ICECE/SPICE Forum*, Kaunas, Lithuania, 2000, pp. 77-81.
8. B. Tseng, Managing MQSeries Systems, *EAI Journal, October 2000 Vol. 2, No. 10*, pp. 67-70.

GOAL ORIENTED REQUIREMENTS ENGINEERING

Colette Rolland[*]

1. INTRODUCTION

Motivation for goal-driven requirements engineering : In (Lamsweerde, 2000), Axel van Lamsweerde defines Requirements Engineering (RE) as "concerned with the identification of goals to be achieved by the envisioned system, the operationalisation of such goals into services and constraints, and the assignment of responsibilities of resulting requirements to agents as humans, devices, and software". In this view, goals drive the requirements engineering process which focuses on goal centric activities such as goal elicitation, goal modelling, goal operationalisation and goal mapping onto software objects, events and operations.

Many authors will certainly agree to this position or to a similar one because goal driven approaches are seen today as a means to overcome the major drawback of traditional Requirements Engineering (RE) approaches that is, to lead to systems technically good but unable to respond to the needs of their users in an appropriate manner. Indeed, several field studies show that requirements misunderstanding is a major cause of system failure. For example, in the survey over 800 projects undertaken by 350 US companies which revealed that one third of the projects were never completed and one half succeeded only partially, poor requirements was identified as the major source of problems (Standish, 1995). Similarly, a recent survey over 3800 organisations in 17 European countries demonstrate that most of the perceived problems are related to requirements specification (>50%), and requirements management (50%) (ESI, 1996).

If we want better quality systems to be produced i.e. systems that meet the requirements of their users, requirements engineering needs to explore the objectives of different stakeholders and the activities carried out by them to meet these objectives in order to derive *purposeful system requirements*. Goal driven approaches aim at meeting this objective.

As shown in Figure 1, these approaches are motivated by establishing an *intentional relationship* between the *usage world* and the *system world* (Jarke and Pohl, 1993). The *usage world* describes the tasks, procedures, interactions etc.

---

[*]
Colette Rolland , Université de Paris 1, Panthéon Sorbonne 75013 Paris Cedex 13,
rolland@univ-paris1.fr

Information Systems Development: Advances in Methodologies, Components, and Management
Edited by Kirikova *et al.*, Kluwer Academic/Plenum Publishers, 2002

35

performed by agents and how systems are used to do work. It can be looked upon as containing the objectives that are to be met in the organisation and which are achieved by the activities carried out by agents. Therefore, it describes the activity of agents and how this activity leads to useful work.

The *subject world*, contains knowledge of the real world domain about which the proposed system has to provide information. It contains real world objects which are to be represented in the system.

Requirements arise from both of these worlds. However, the subject world imposes domain- requirements which are facts of nature and reflect domain laws whereas the usage world generates user-defined requirements which arise from people in the organisation and reflect their goals, intentions and wishes.

The *system world* is the world of system specifications in which the requirements arising from the other two worlds must be addressed. The system world holds the modelled entities, processes, and events of the subject and usage worlds as well as the mapping from these conceptual specifications to the design and implementation levels of the software system.

These three worlds are interrelated as shown in Figure 1. User-defined requirements are captured by the *intentional relationship*. Domain-imposed requirements are captured by the *representation relationship*.

Understanding the *intentional relationship* is essential to comprehend the reason why a system should be constructed. The usage world provides the rationale for building a system. The purpose of developing a system is to be found outside the system itself, in the *enterprise,* or in other words, in the context in which the system will function. The relationship between the usage and system world addresses the issue of the system purpose and relates the system to the goals and objectives of the organisation. This relationship explains *why* the system is developed. Modelling this establishes the conceptual link between the envisaged system and its changing environment. *Goal-driven approaches* have been developed to address the semiotic, social link between the usage and the system world with the hope to construct systems that meet the needs of their organisation stakeholders.

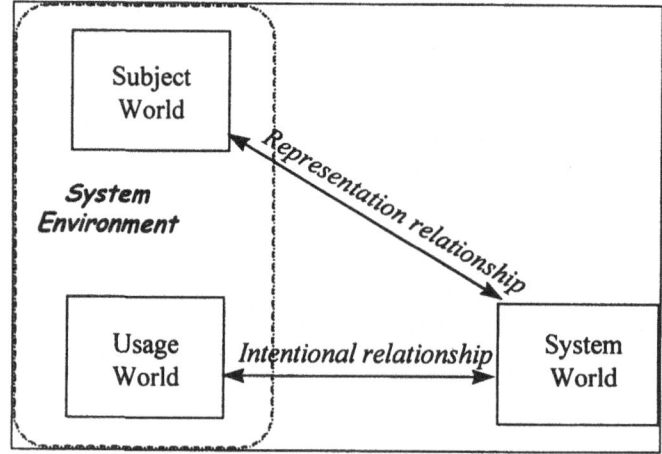

Figure 1. The relationships between the usage, subject and system worlds.

Roles of goal in requirements engineering : Goal modelling proved to be an effective way to *elicit requirements* (Potts, 1994; Rolland et al, 1998; Dardenne et al., 1993; Anton, 1994; Dubois et al., 1998; Kaindl, 2000; Lamsweerde, 2000). The argument of goal driven requirements elicitation being that the rationale for developing a system is to be found outside the system itself, in the enterprise (Loucopoulos, 1994) in which the system shall function.

Requirements engineering assumes that the To-Be developed system might function and interact with its environment in many alternative ways. Alternative goal refinement proved helpful in the systematic *exploration of system choices* (Rolland et al, 1999; Lamsweerde, 2000; Yu, 1994).

Requirements completeness is a major RE issue. Yue (Yue, 1987) was probably the first to argue that goals provide a criterion for requirements completeness : the requirements specification is complete if the requirements are sufficient to achieve the goal they refine.

Goals provide a means to ensure *requirements pre-traceability* (Gotel et al., 1994; Pohl, 1996; Ramesh, 1995]. They establish a conceptual link between the system and its environment, thus facilitating the propagation of organisational changes into the system functionality. This link provides the rationale for requirements (Bubenko et al., 1994; Sommerville and Sawyer, 1997; Ross, 1977; Mostov, 1985; Yu, 1993) and facilitates the explanation and justification of requirements to the stakeholders.

Stakeholders provide useful and realistic viewpoints about the To-Be developed system but requirements engineers know that these viewpoints might be conflicting (Nuseibeh, 1994). Goals have been recognised to help in the *detection of conflicts* and their resolution (Lamsweerde, 2000; Robinson, 1989).

Difficulties with goal driven approaches : However, several authors (Lamsweerde et al., 1995; Anton, 1998; Rolland et al, 1998; Haumer et al, 1998) also acknowledge the fact that dealing with goal is not an easy task. We have applied the goal driven approach as embodied in the EKD method (Bubenko et al., 1994; Kardasis, 1998; Loucopoulos, 1997; Rolland et al., 1997b) to several domains, air traffic control, electricity supply, human resource management, tool set development. Our experience is that it is difficult for domain experts to deal with the fuzzy concept of a goal. Yet, domain experts need to discover the goals of real systems.

It is often assumed that systems are constructed with some goals in mind (Davis, 1993). However, practical experiences (Anton, 1996; ELEKTRA, 1997) show that goals are not given and therefore the question as to where they originate from (Anton, 1996) acquires importance.

In addition, enterprise goals which initiate the goal discovery process do not reflect the actual situation but an idealised environmental one. Therefore, proceeding from this may lead to ineffective requirements (Potts, 1997). Thus, goal discovery is rarely an easy task.

Additionally, it has been shown (Anton, 1996) that the application of goal reduction methods (Dardenne et al., 1993) to discover the components goals of a goal, is not as straight-forward as literature suggests. Our own experience in the F3 (Bubenko et al., 1994) and ELEKTRA (Rolland et al., 1997a) projects is also similar. It is thus evident that help has to be provided so that goal modelling can be meaningfully performed.

Paper outline : The objective of this paper is (a) to highlight some of the issues of goal driven approaches in requirements engineering, (b) to provide an overview of the state-of-the art on these issues and (c) to illustrate how L'Ecritoire approach deals with them.

In section 2 we briefly introduce the L'Ecritoire, a goal driven approach developed in our group (Rolland et al, 1998; Tawbi, 2001; Ben Achour, 1999; Rolland et al, 1997b; Rolland et al, 1999) to support requirements elicitation, specification and documentation. The presentation of this approach in section 3 will be used as the means to raise issues in goal driven requirements engineering, and to provide a state-of-the art on these issues. Section 4 concludes and considers some additional issues.

2. AN OVERVIEW OF L'ECRITOIRE

L'Ecritoire is a tool for requirements elicitation, structuring, and documentation. Figure 2 shows that the approach underlying L'Ecritoire uses *goal-scenario coupling* to discover requirements from a computer-supported analysis of textual scenarios. L'Ecritoire produces a requirements document which relates system requirements (the functional & physical levels in Figure 2) to organisational goals (behavioural level in Figure 2).

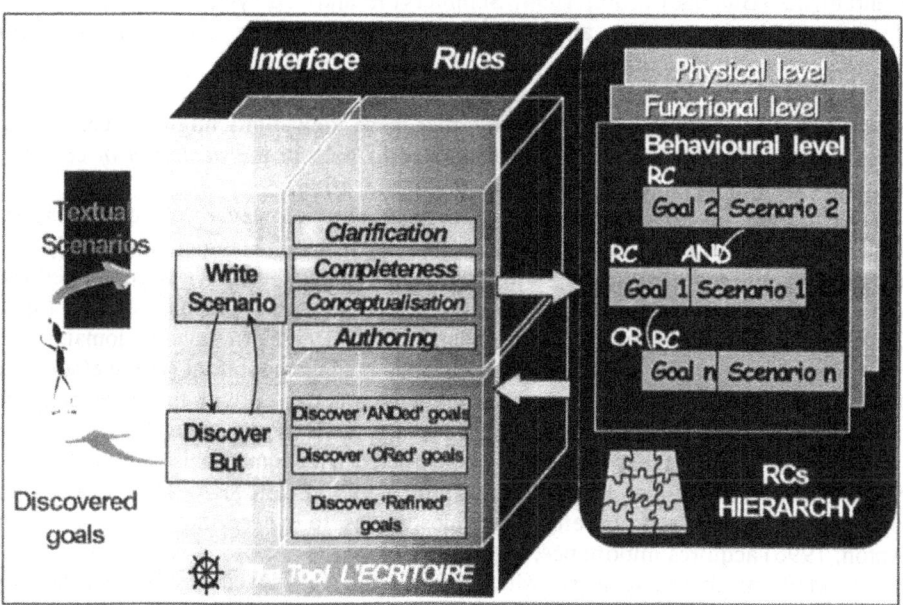

Figure 2. The L'Ecritoire architecture & functionality

Central to the approach is the notion of a *requirement chunk* (RC) which is a pair <goal, scenario>. A goal is 'something that some stakeholder hopes to achieve'(Plihon, 1998) whereas a scenario is a possible behaviour limited to a set of purposeful interactions taking place among agents'(CREWS, 1998). Since a goal is intentional and a scenario operational in nature, a requirement chunk is a possible way of achieving the goal.

L'Ecritoire aims at eliciting the collection of requirements chunks through a *bi-directional coupling* of goals and scenarios allowing movement from goals to scenarios and vice-versa. As each goal is discovered, a scenario is authored for it. In this sense the goal-scenario coupling is exploited in the forward direction from goals to scenarios. Once a scenario has been authored, it is analysed to yield goals. This leads to goal discovery by moving along the goal-scenario relationship in the reverse direction. By exploiting the goal scenario relationship in the reverse direction, i.e. from scenario to goals, the approach proactively guides the requirements elicitation process. The sequence of steps of the process is as follows :

Initial Goal Identification
repeat
Goal Formulation
Scenario Authoring
Goal Elicitation Through Scenario Analysis
until all goals have been elicited.

Each of the steps of the cycle is supported by mechanisms to *guide* the execution of the step. The guidance mechanism for goal formulation is based on a linguistic analysis of goal statements. It helps in reformulating a narrative goal statement according to a goal template. The mechanism for scenario authoring combines style/content guidelines and linguistic devices. The former advise authors on how to write scenarios whereas the latter provide automatic help to check, correct, conceptualise, and complete a scenario. Finally, three different goal discovery strategies for goal elicitation are used.

The next section introduce the approach in more detail with the aim to raise general issues in goal driven approaches to requirements engineering and to present the related state-of-the art. General issues are introduced with the ❖ symbol whereas the L'Ecritoire concepts are presented under the ● symbol.

3. DISCUSSING ISSUES IN GOAL DRIVEN APPROACHES THROUGH L'ECRITOIRE PRESENTATION

At the core of the L'Ecritoire approach is the notion of a *Requirement Chunk*. We define a *Requirement Chunk* (RC) as a pair <G, Sc> where G is a goal and Sc is a scenario. Since a goal is intentional and a scenario is operational in nature, a requirement chunk is a possible way in which the goal can be achieved. Let us introduce the notions of goal, scenario and requirement chunk and discuss issues related to them.

3.1. The Notion Of A Goal

- A *goal* is defined (Plihon, 1998) in L'Ecritoire as 'something that some stakeholder hopes to achieve in the future'.
- In (Lamsweerde, 2001), a goal is an objective the system under consideration should achieve. Goals thus, refer to intended properties referred to in (Jackson, 1995; Lamsweerde, 2001) as *optative* properties by opposition to *indicative* ones.

3.2. Goal Formulation

- In L'Ecritoire, a goal is expressed as a clause with a main verb and several parameters, where each parameter plays a different role with respect to the verb. For example in the goal statement :

$$\textit{'Withdraw }_{verb}\textit{ (cash)}_{target}\textit{ (from ATM)}_{means}\textit{'},$$

'*Withdraw*' is the main verb, '*cash*' is the parameter target of the goal, and '*from ATM*' is a parameter describing the means by which the goal is achieved. We adopted the linguistic approach of Fillmore's Case grammar (Fillmore, 1968), and its extensions (Dik, 1989; Schank, 1973) to define goal parameters (Prat, 1997). Each type of parameter corresponds to a case and plays a different role with respect to the verb, e.g. target entities affected by the goal, means and manner to achieve the goal, beneficiary agent of the goal achievement, destination of a communication goal, source entities needed for goal achievement etc.

❖ Goal: statements are often texts in natural language (Anton, 1996; Cockburn, 1996) and may be supplemented as suggested by (Zave, 1997) with an informal specification to make precise what the goal name designates.

The motivation for semi-formal or formal goal expressions is to be the support of some form of automatic analysis. We will see later in the paper how the L'Ecritoire goal template helps reasoning about goals. Typical semi-formal formulations use some goal taxonomy and associate the goal name to a predefined type (Anton, 1998; ELEKTRA, 1997; Dardenne et al., 1993).This helps clarifying the meaning of the goal. For instance, in (Mylopoulos, 1992) a non functional goal is specified by the specific sub-type it is instance of. Similarly, in Elektra (Elektra, 1997), goals for change are pre-fixed by one of the seven types of change: *Maintain, Cease, Improve, Add, Introduce, Extend, Adopt* and *replace*. Graphical notations (Chung et al., 2000; Mylopoulos, 1992; Lamsweerde, 2001) can be used in addition to a textual formulation.

Formal specifications of goals like in Kaos (Dardenne et al, 1993) require a higher effort but yield more powerful reasoning.

3.3. Coupling Goal And Scenario

• In L'Ecritoire, a *goal* is coupled with a *scenario*. In this direction, from goal to scenario, the relationship aims to concretise a goal through a scenario. A *scenario* is 'a possible behaviour limited to a set of purposeful interactions taking place among several agents' (CREWS, 1998). Thus, the scenario represents a possible behaviour of the system to achieve the goal. In L'Ecritoire, a scenario is defined as composed of one or more *actions* which describe a unique path leading from an *initial* to a *final state* of agents. Figure 3 is an example of scenario associated to the goal '*Withdraw cash from the ATM'*.

The *initial state* defines the preconditions for the scenario to be triggered. For example, the scenario '*Withdraw cash from the ATM*' cannot be performed if the initial state '*The bank customer has a card*' and '*The ATM is ready*' is not true. The *final state* is the state reached at the end of the scenario. The scenario '*Withdraw cash from the ATM*' leads to the compound state '*The user has cash*', and '*The ATM is ready*'.

Actions in a scenario are of two types, atomic actions and flows of actions. *Atomic actions* are interactions '*from*' an agent '*to*' another, which affects some '*parameter objects*'. The clause '*The bank customer inserts a card in the ATM*' is an example of an atomic action involving two different agents '*The bank customer*' and '*the ATM*' and having the '*card*' as parameter.

> The user inserts a card in the ATM.
> The ATM checks the card validity.
> If the card is valid a prompt for code is given by the ATM to the user, the
> user inputs the code in the ATM.
> The ATM checks the code validity.
> If the code is valid, the ATM displays a prompt for amount to the user.
> The user enters an amount in the ATM.
> The ATM checks the amount validity.
> If the amount is valid, the ATM ejects the card to the user and then the ATM
> proposes a receipt to the user.
> The user enters the user's choice in the ATM.
> If a receipt was asked the receipt is printed by the ATM to the user but
> before the ATM delivers the cash to the user.

Figure 3. Scenario associated to the goal '*Withdraw cash from the ATM*'.

Flows of actions are composed of several actions and can be of different types, *sequence*, *concurrent*, *iterative* and *conditional*. The sentence *'The bank customer gets a card from the bank, then the bank customer withdraws cash from the ATM'* is an example of a sequence comprising two atomic actions. The flow of actions *'While the ATM keeps the card, the ATM displays an "invalid card" message to the bank customer'* is concurrent; there is no predefined order between the two concurrent actions.

❖ Many authors suggest to combine goals and scenarios (Potts, 1995; Cockburn, 1995; Leite et al, 1997; Kaindl, 2000; Sutcliffe, 1998; Haumer et al., 1998; Anton, 1998; Lamsweerde et Willemet, 1998). (Potts, 1995) for example, says that it is « unwise to apply goal based requirements methods in isolation » and suggests to complement them with scenarios. This combination has been used mainly, to make goals concrete, i.e. to operationalise goals. This is because scenarios can be interpreted as containing information on how goals can be achieved. In (Dano et al., 1997; Jacobson, 1995; Leite, 1997; Phol and Haumer, 1997), a goal is considered as a contextual property of a use case (Jacobson, 1995) i.e. a property that relates the scenario to its organisational context. Therefore, goals play a documenting role only. (Cockburn, 1995) goes beyond this view and suggests to use goals to structure use cases by connecting every action in a scenario to a goal assigned to an actor. In this sense a scenario is discovered each time a goal is. Clearly, all these views suggest a unidirectional relationship between goals and scenarios similarly to what we introduced in L'Ecritoire so far. We will see later on, how L'Ecritoire exploits the goal/scenario coupling in the reverse direction.

3.4. Relationships Among Goals

- In L'Ecritoire, requirement chunks can be assembled together through *composition, alternative and refinement* relationships. The first two lead to AND and OR structure of RCs whereas the last leads to the organisation of the collection of RCs as a hierarchy of chunks of different granularity.

  AND relationships among RCs link complementary chunks in the sense that every one requires the others to define a completely functioning system. RCs linked through *OR relationships* represent alternative ways of fulfilling the same goal. RCs linked through a *refinement relationship* are at different levels of abstraction.

  The goal *'Fill in the ATM with cash'* is an example of *ANDed* goal *to 'Withdraw cash from the ATM'* whereas *'Withdraw cash from the ATM with two invalid code capture '* is *ORed* to it. Finally *'Check the card validity'* is linked to the *goal 'Withdraw cash from the ATM'* by a *refinement* relationship.

- Many different types of *relationships* among goals have been introduced in the literature. They can be classified in two categories to relate goals: (1) to each other and (2) with other elements of requirements models. We consider them in turn.

  AND/OR relationships (Bubenko et al, 1994; Dardenne et al, 1993; Rolland et al, 1998; Loucopoulos et al, 1997; Mylopoulos 1999) inspired from AND/OR graphs in Artificial Intelligence are used to capture goal decomposition into more operational goals and alternative goals, respectively. In the former, all the decomposed goals must be satisfied for the parent goal to be achieved whereas in the latter, if one of the alternative goals is achieved, then the parent goal is satisfied.

  In (Mylopoulos, 1992; Chung et al., 2000), the inter-goal relationship is extended to support the capture of negative/positive influence between goals. A sub-goal is said to *contribute* partially to its parent goal. This leads to the notion of goal *satisfycing* instead of goal *satisfaction*. The 'motivates' and 'hinders' relationships among goals in (Bubenko et al, 1994) are similar in the sense that they capture positive/negative influence among goals.

  Conflict relationships are introduced (Bubenko et al, 1994; Dardenne et al 1993; Nuseibeh, 1994; Easterbrook, 1994) to capture the fact that one goal might prevent the other to be satisfied.

  In addition to inter-goal relationships, goals are also related to other elements of requirements models. As a logical termination of the AND/OR decomposition, goals link to operations which ensure them (Anton, 1994; Anton and Potts, 1998; Kaindl, 2000; Lamsweerde et Willemet, 1998). Relationships between goals and system objects have been studied in (Lee, 1997) and are inherently part of the KAOS model (Lamsweerde et al., 1991; Dardenne et al., 1993)).

  Relationships with agents have been emphasized in (Yu 1993; Yu 1997) where a goal is the object of the dependency between two agents. Such type of link is introduced in other models as well (Dardenne et al, 1993; Lamweerde et al., 1991; Letier, 2001) to capture who is responsible of a goal. As discussed earlier, goals have been often coupled to scenarios (Potts, 1995; Cockburn, 1995; Leite, 1997; Kaindl, 2000; Sutcliffe, 1998; Haumer et al., 1998; Anton, 1998; . et al., 1998). In (Bubenko et al, 1994) goals are related to a number of concepts such as *problem*, *opportunity* and *thread* with the aim to understand better the context of a goal. Finally the interesting idea of *obstacle* introduced by (Potts, 1995) leads to obstructions and resolution relationships among goals and obstacles (Lamweerde, 2000a; Sutcliffe, 1998).

3.5. Levels Of Abstraction In Goal Modelling

- The L'Ecritoire approach identifies three levels of requirements abstraction, namely the *contextual, functional* and *physical* levels.
  The aim of the *contextual level* is to identify the services that a system should provide to fulfil a business goal. *'Improve services to our bank customers by providing cash from ATM'* is an example of contextual goal for satisfying the business goal *'Improve services to our bank customers'*. The scenario attached to this goal identifies the services of the To-Be system.
  At the *functional level* the focus is on the interactions between the system and its users to achieve the services assigned to the system at the contextual level. Thus, the contextual level is the bridge between business goals and system functional requirements. *'Withdraw cash from the ATM'* is an example of functional goal and the scenario of Figure 3 attached to it describes a flow of interactions to achieve this goal.
  The *physical level* focuses on what the system needs to perform the interactions selected at the system interaction level. The *'what'* is expressed in terms of system internal actions that involve system objects but may require external objects such as other systems. This level defines the software requirements to meet the system functional requirements.
- ❖ As in L'Ecritoire goals many approaches suggest to formulate goals at different *levels of abstraction*. By essence goal centric approaches aim to help in the move from strategic concerns and high level goals to technical concerns and low abstraction level goals *'Improve services to our customers'* is an example of the former whereas *'Check the card validity'* is an example of the latter. Therefore, it is natural for approaches to identify different levels of goal abstraction where high level goals represent business objectives and high level mandates and are refined in system goals (Anton et al., 2001; Anton and Potts, 1998) or system constraints (Lamsweerde and Letier, 2000a). Inspired by cognitive engineering, some goal driven RE approaches deal with means-end hierarchy abstractions, where each hierarchical level represents a different model of the same system. The information at any level acts as a goal (the end) with respect to the model at the next lower level (the means) (Leveson 2000; Rasmussen, 1990; Vicente and Rasmussen, 1992).

3.6. Types Of Goal And Goal Taxonomy

Several goal classifications have been proposed in the literature. In (Dardenne, 1993) a goal taxonomy composed of five types of goal, namely *Achieve, Cease, Maintain, Avoid* and *Optimise*, is introduced. Each type represents a certain type of behaviour formally defined using a temporal logic. *Achieve* and *Cease* goals generate system behaviours whereas *Maintain* and *Avoid* goals restrict behaviours, while *Optimise* goals compare behaviours. The same author classifies goals into *system goals* versus *private goals*. The former are application- specific goals that the system must achieve. The latter are agent- specific goals that the system might achieve as long as needed by the agent. System goals are themselves specialized into:

SatisfactionGoals, *InformationGoals*, *RobustnessGoals*, *ConsitencyGoals* and *SafetyGoals*.

(Yu, 1994) distinguishes between *functional* and *non-functional* goals. Functional goals express what the system must do (Thayer and Dorfman, 1990; Chung et al., 2000). Non-functional goals are quality goals expressing how the system shall fulfil its functional goals.

Annie Anton, (Anton, 1996a) defines types of goals according to their target condition and classifies goals into *Maintenance* Goals and *Achievement* Goals. *Achievement* goals address the actions that occur in the system whereas *maintenance* goals express actions or constraints that prevent things from occurring .

Prat proposes (Prat, 1997; Prat, 1999) a more complete hierarchical classification of goals based on a linguistic approach inspired by the Fillmore's case grammar. Every goal verb is defined by a verb frame constituted of parameters representing the semantic functions acceptable for this verb. The goal verb frames are organised in a hierarchical fashion using linguistic criteria. The first level of this hierarchy differentiate between *Maintenance* goals and *Evolution* goals, that is between static verbs (keep, remain, …) and dynamic verbs (become, achieve,…). Every level of the hierarchy constitutes a class of verbs characterized by a specific set of semantic functions.

Many goal classifications of *non-functional* requirements (NFR) have been proposed. Among the earliest approaches, (Boehm, 1976) introduces "the qualities" that a software must exhibit, and (Bowen et al., 1985) classify "software quality attributes" into consumer-oriented (software quality features) and technical oriented (software quality criteria). A survey is presented in (Sommerville, 1996; Sommerville and Sawyer, 1997). The notion of softgoal (Yu 94a) and the construction of an NFR framework for representing and analysing non-functional requirements can be found in (Mylopoulos, 1992; Chung, 2000).

A distinction between realisable and unrealisable goals is introduced by (Letier, 2002) who gives five pragmatic conditions of unrealisability. (Lamsweerde, 2000a) classifies the goals according to the category of requirements assigned to the associated agents. For evolving systems (Anton and Potts, 1998) proposes a classification depending of the type of target condition such as achievement of a state, preservation of a condition, avoidance of an undesired state. Another form of classification is based on the goal subject matter (Anton and Potts, 1998) which has some similarity with the notion of "problem frame" (Jackson, 1995). Finally in (Ramadour & Cauvet, 2001) an interesting classification of *generic* versus *reusable* domain goal is introduced. The former correspond to typical high level goals of a given domain whereas the latter are low level operational goal that might be reused in most applications over this domain.

3.7. Eliciting Goals

- The L'Ecritoire requirements elicitation process is organised around two main activities :
  goal discovery and,
  scenario authoring

In this process, *goal discovery* and *scenario authoring* are complementary activities, the former following the latter. As shown in Figure 4, these activities are repeated to incrementally populate the requirements chunk hierarchy.

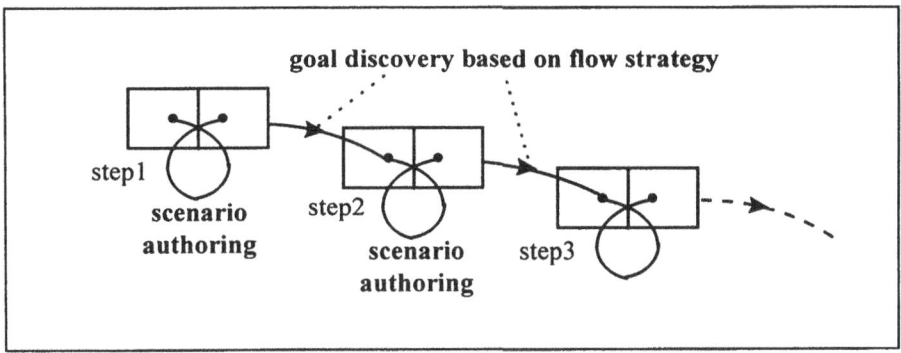

Figure 4. The L'Ecritoire requirements elicitation process

The requirements elicitation process can be viewed as a flow of steps: each step exploits the goal-scenario relationship in both, the forward and backward directions. A step starts with a goal and the goal-scenario relationship is then exploited in the forward direction to *author* a scenario which is a possible concretisation of this goal. Then, the goal-scenario relationship is exploited in the reverse direction to *discover* new goals based on an analysis of the scenario. In subsequent steps, starting from the goals of these new RCs, scenarios are authored and the requirements elicitation cycle (flow strategy) thus continues.

Each of the two main activities, goal discovery and scenario authoring, is supported by enactable rules, (1) *authoring rules* and (2) *discovery rules*. Authoring rules allow L'Ecritoire scenarios which are textual to be authored. Discovery rules are for discovering goals through the analysis of authored scenarios. We focus here on exemplifying the discovery rules. Detail about the authoring rules and the linguistic approach underlying them can be found in (Rolland and Ben Achour, 1997; Ben Achour, 1999).

Discovery rules in l'Ecritoire : Discovery rules guide the L'Ecritoire user in discovering new goals and therefore, eliciting new requirement chunks. The discovery is based on the analysis of scenarios through one of the three proposed discovery strategies, namely the *refinement, composition* and *alternative* strategies. These strategies correspond to the three types of relationships among RCs introduced above. Given a pair <G,Sc>:

- the composition strategy looks for goals Gi ANDed to G,
- the alternative strategy searches for goals Gj ORed to G,
- the refinement strategy aims at the discovery of goals Gk at a lower level of abstraction than G.

Therefore, *composition* (*alternative*) *rules* help in discovering ANDed (ORed) goals to G. These are found at the same level of abstraction as G. The <G,Sc> chunk is processed by the *refinement rules* to produce goals at a lower level of abstraction than G. This is done by considering (in a similar way to that suggested by (Cockburn, 1995)) each interaction in Sc as a goal. Thus as many goals are produced as there are interactions in Sc.

As shown in Figure 5, once a complete scenario has been authored, any of these three strategies can be followed. Thus, there is no imposed ordering on the flow of steps which instead, is dynamically defined.

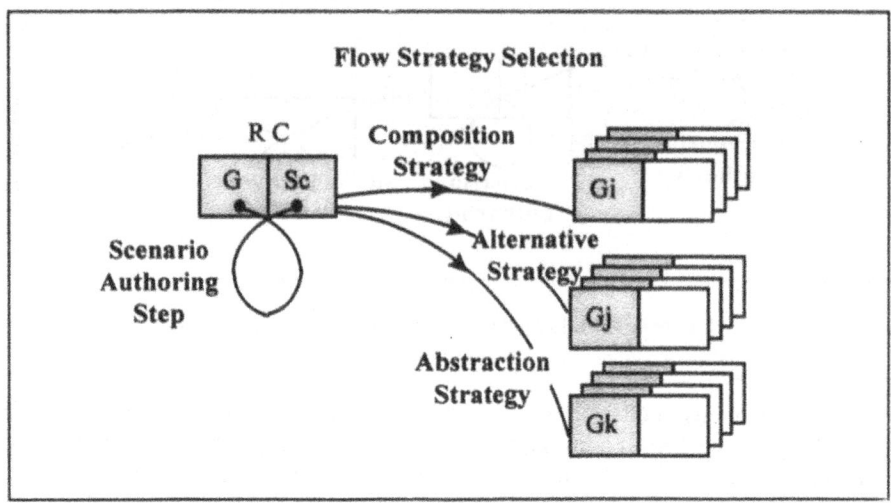

Figure 5. Selecting a discovery strategy in L'Ecritoire

L'Ecritoire uses six discovery rules, two for each strategy. Rules can be applied at any of the three levels of abstraction, contextual, functional and physical. A detail description of rules can be found in (Rolland et al., 1998; Tawbi, 2001, Rolland, 2002). As an example of a rule, we present the refinement rule R1 and exemplify it with the example of ATM system engineering.

Refinement guiding rule (R1) :

Goal :   Discover (from requirement chunk <G,Sc>)$_{So}$ (goals refined from G)$_{Res}$
*   (using every atomic action of Sc as a goal)$_{Man}$*

Body :

    1.   Associate a goal Gi to every atomic action Ai in Sc. Gi refines G

    2.   Complement Gi by the manner 'in a normal way'

    3.   User evaluates the proposed panel of goals Gi and selects the goals of interest

    4.   Requirement chunks corresponding to these selected goals are ANDed *to one another*

The guiding rule R1 aims at refining a given requirement chunk *(from RC<G,Sc>)$_{So}$* by suggesting new goals at a lower level of abstraction than G *(goals refined from G)$_{Res}$*.

The refinement mechanism underlying the rule looks to every interaction between two agents in the scenario Sc as a goal for the lower level of abstraction (step1). Let us take as an example the scenario of the requirement chunk RC presented below:

Goal G: Improve services to our customers by providing cash from the ATM

Scenario SC :

1. *If the bank customer gets a card from the bank,*
2. *Then, the bank customer withdraws cash from the ATM*
3. *and the ATM reports cash transactions to the bank.*

This scenario text corresponds to the structured textual form of the scenario as it results from the authoring step. The internal form is a set of semantic pattern instances which clearly identify three agents namely, the bank, the customer and the ATM as well as three interactions namely 'Get card', 'Withdraw cash' and 'Report cash transactions' corresponding to the three services involving the ATM. These services are proposed as goals of a finer grain than G, to be further made concrete by authoring scenarios for these goals.

We propose that these scenarios describe the normal course of actions. Thus, the manner parameter of every generated goal Gi is fixed to *'in a normal way'* (step2). This leads in the above example, to propose to the user the three following refined goals :

- *'Get card from the bank in a normal way'*
- *'Withdraw cash from ATM in a normal way'*
- *'Report cash transactions to the bank in a normal way'*

Assuming that the user accepts the three suggested goals (step3), the corresponding requirement chunks are ANDed to one another (step4).

❖ As illustrated above, L'Ecritoire develops a requirements/goal inductive elicitation technique based on the analysis of conceptualised scenarios. The conceptualisation of a scenario results of powerful analysis and transformation of textual scenarios using a linguistic approach based on a Case Grammar inspired by Fillmore's Case Theory (Fillmore, 1968) and its extensions (Dik, 1989; Schank, 1973). The pay-off of the scenario conceptualisation process is the ability to perform powerful induction on conceptualised scenarios. In (Lamweerde, 1998), a similar approach is developed that takes scenarios as examples and counter examples of the intended system behaviour and generates goals that cover positive scenarios and exclude the negative ones.

An obvious informal technique for finding goals is to systematically ask WHY and WHAT-IF questions (Potts et al, 1994), (Sutcliffe et al, 1998).

In L'Ecritoire the refinement strategy helps discovering goals at a lower level of abstraction. This is a way to support goal decomposition. Another obvious technique to perform decomposition is to ask the HOW question (Lamsweerde et al., 1995). A heuristic based decomposition technique has been developed in (Loucopoulos et al., 1997) and (Letier, 2001).

An attempt to retrieved cases from a repository of process cases was developed in (Le, 1999). The software tool captures traces of requirements engineering processes using the NATURE contextual model (Nature, 1999) and develops a case based technique to retrieve process cases similar to the situation at hand.

4. CONCLUSION

Goal-driven requirements engineering was introduced mainly to provide the rationale of the To-Be system. Beyond this objective, we have seen that there are some other advantages :

- goals bridge the gap between organisational strategies and system requirements thus providing a conceptual link between the system and its organisational context;
- goal decomposition graphs provide the pre-traceability between high level strategic concerns and low level technical constraints; therefore facilitating the propagation of business changes onto system features;
- ORed goals introduce explicitly design choices that can be discussed, negotiated and decided upon;
- AND links among goals support the refinement of high level goals onto lower level goals till operationalisable goals are found and associated to system requirements;
- Powerful goal elicitation techniques facilitate the discovery of goal/requirements;
- Relationships between goals and concepts such as objects, events, operations etc. traditionally used in conceptual design facilitates the mapping of goal graphs onto design specification.

There are other advantages which flow from issues which were not verified with in the paper and that we sketch here :

- Goal-based negotiation is one of them (Boehm and In H, 1996).
- Conflict resolution is another one. (Nuseibeh, 1994) explains how conflicts arise from multiple view points and concerns and in (Lamsweerde et al., 1998a) various forms of conflict have been studied.
- Goal validation is a third one. (Sutcliffe et al, 1998) use a scenario generation technique to validate goal/requirement and in (Heymans and Dubois et al., 1998) the validation is based on scenario animation.
- Qualitative reasoning about goals is provided by the NFR framework (Mylopoulos, 1992; Chung et al, 2000) and extended in (Kaiya et al., 2002). The process determines the extend to which a goal is satisfied/denied by its sub goals. Rules are provided to support a bottom-up propagation of positive/negative influences of sub-goal on their parent.

5. REFERENCES

Anton, A. I., 1996,Goal based requirements analysis. Proceedings of the 2nd International Conference on Requirements Engineering ICRE'96, pp. 136-144.

Anton, A. I, and Potts C., 1998,The use of goals to surface requirements for evolving systems, International Conference on Software Engineering (ICSE `98) , Kyoto, Japan, pp. 157-166, 19-25 April 1998.

Anton, A. I., Earp J.B., Potts C., and Alspaugh T.A., 2001,The role of policy and stakeholder privacy values in requirements engineering, IEEE 5th International Symposium on Requirements Engineering (RE'01), Toronto, Canada, pp. 138-145, 27-31 August 2001.

Ben Achour, C., 1999,Requirements extraction from textual scenarios. PhD Thesis, University Paris6 Jussieu, January 1999.

Boehm, B., 1976,Software engineering. IEEE Transactions on Computers, 25(12): 1226-1241.

Boehm, B., 1996,Identify Quality-requirements conflicts, 1996, Proceedings ICRE, Second International Conference on Requirements Engineering, April 15-18, 1996, Colorado spring, Colorado, 218.

Bowen, T. P., Wigle, G. B., Tsai, J. T., 1985,Specification of software quality attributes. Report of Rome Air Development Center.

Bubenko, J., Rolland, C., Loucopoulos, P., de Antonellis V., 1994,Facilitating 'fuzzy to formal' requirements modelling. IEEE 1st Conference on Requirements Engineering, ICRE'94 pp. 154-158.

Cockburn, A., 1995,Structuring use cases with goals. Technical report. Human and Technology, 7691 Dell Rd, Salt Lake City, UT 84121, HaT.TR.95.1, http://members.aol.com/acocburn/papers/usecases.htm.

CREWS Team, 1998,The crews glossary, CREWS report 98-1, http ://SUNSITE.informatik.rwth-aachen.de/CREWS/reports.htm.

Chung, K. L., Nixon B. A., and Yu E., Mylopoulos J., 2000,Non- Functional Requirements in Software Engineering. Kluwer Academic Publishers.. 440 p.

Dano, B., Briand, H., and Barbier, F., 1997, A use case driven requirements engineering process. Third IEEE International Symposium On Requirements Engineering RE'97, Antapolis, Maryland, IEEE Computer Society Press.

Dardenne, A., Lamsweerde, A. v., and Fickas, S., 1993,Goal-directed Requirements Acquisition, Science of Computer Programming, 20, Elsevier, pp.3-50.

Davis, A. .M., 1993, Software requirements :objects, functions and states. Prentice Hall.

Dik, S. C., 1989, The theory of functional grammar, part i : the structure of the clause. Functional Grammar Series, Fories Publications.

Dubois, E., Yu, E., and Pettot, M., 1998, "From early to late formal requirements: a process-control case study". *Proc. IWSSD'98 – 9th International Workshop on software Specification and design.* Isobe.IEEE CS Press. April 1998, 34-42.

ELEKTRA consortium, 1997,Electrical enterprise knowledge for transforming applications. ELEKTRA Project Reports.

ESI96, European Software Institute, 1996,"European User survey analysis", Report USV_EUR 2.1, ESPITI Project, January 1996.

Fillmore, C., 1968,The case for case. In ''Universals in linguistic theory'', Holt, Rinehart and Winston (eds.),Bach & Harms Publishing Company, pp. 1-90.

Gote, O., and Finkelstein A., 1994,Modelling the contribution structure underlying requirements, in Proc. First Int. Workshop on Requirements Engineering : Foundation of Software Quality, Utrech, Netherlands.

Haumer, P., Pohl K., and Weidenhaupt K., 1998,Requirements elicitation and validation with real world scenes. IEEE Transactions on Software Engineering, Special Issue on Scenario Management, M. Jarke, R. Kurki-Suonio (eds.), Vol.24, N°12, pp.11036-1054.

Heymans, P., and Dubois, E.,.1998, Scenario-based techniques for supporting the elaboration and the validation of formal requirements. Requirement Engineering Journal, P. Loucopoulos, C. Potts (eds.), Springer, CREWS Deliverable N°98-30, http:\\SUNSITE.informatik.rwth-aachen.de\ CREWS\.

Jacobson, I., 1995,The Use case construct in object-oriented software engineering. In Scenario-Based Design: Envisioning Work and Technology in System Development, J.M. Carroll (ed.), pp.309-336.

Jarke, M., and Pohl, K., 1993,Establishing visions in context: towards a model of requirements processes. Proc. 12th Intl. Conf. Information Systems, Orlando.

Kaindl, H., 2000, "A design process based on a model combining scenarios with goals and functions", IEEE Trans. on Systems, Man and Cybernetic, Vol. 30 No. 5, September 2000, 537-551.

Kardasis P., and Loucopoulos P., 1998,Aligning legacy information system to business processes. Submitted to CAiSE'98.

Kaiya, H., Horai, H., and Saeki, M., 2002,AGORA : Attributed goal-oriented requirements analysis method. IEEE Joint International, Requirements Engineering Conference. 10th Anniversary. (Accepted paper). Essen (Germany), September 09-13, 2002.

Lamsweerde, A. v., Dardenne, B., Delcourt, and F. Dubisy, 1991,"The KAOS project: knowledge acquisition in automated specification of software", Proc. AAAI Spring Symp. Series, Track: "Design of Composite Systems", Stanford University, March 1991, 59-62.

Lamsweerde, A. v., Dairmont, R., and Massonet, P., 1995,Goal directed elaboration of requirements for a meeting scheduler : *Problems and Lessons Learnt*, in Proc. Of RE'95 – 2nd Int. Symp. On Requirements Engineering, York, pp 194 –204.

Lamsweerde A. v., and Willemet, L., 1998, "Inferring declarative requirements specifications from operational scenarios". In: IEEE Transactions on Software Engineering, Special Issue on Scenario Management. Vol. 24, No. 12, Dec. 1998, 1089-1114.

Lamsweerde, A. v., Darimont, R., and Letier, E., 1998a, "Managing conflicts in goal-driven requirements engineering", IEEE Trans. on Software. Engineering, Special Issue on Inconsistency Management in Software Development, Vol. 24 No. 11, November 1998, 908-926.

Lamsweerde, A. v., 2000,Requirements engineering in the year 00: *A research perspective*. In Proceedings 22nd International Conference on Software Engineering, Invited Paper, ACM Press, June 2000.

Lamsweerde, A. v., and Letier, E., 2000a,"Handling obstacles in goal-oriented requirements engineering", IEEE Transactions on Software Engineering, Special Issue on Exception Handling, Vol. 26 No. 10, October 2000, pp. 978-1005.

Lamsweerde, A.v., 2001, "Goal-oriented requirements engineering: a guided tour". Invited minitutorial, Proc. RE'01 International Joint Conference on Requirements Engineering, Toronto, IEEE, August 2001, pp.249-263.

Le, T. .L., 1999,Guidage des processus d'ingénierie des besoins par un approche de réutilisation de cas, Master Thesis, CRI, Université Paris-1, Panthéon Sorbonne.

Lee, S. P., 1997, Issues in requirements engineering of object-oriented information system: a review, Malaysian Journal of computer Science, vol. 10, N° 2, December 1997.

Leite, J. C. .S., Rossi, G., Balaguer, F., Maiorana, A., Kaplan, G., Hadad, G., and Oliveros, A., 1997, Enhancing a requirements baseline with scenarios. In Third IEEE International Symposium On Requirements Engineering RE'97, Antapolis, Maryland, IEEE Computer Society Press, pp. 44-53.

Letier, E., 2001,Reasoning about agents in goal-oriented requirements engineering. Ph. D. Thesis, University of Louvain, May 2001; http://www.info.ucl.ac.be/people/eletier/thesis.html

Leveson, N. G., 2000,"Intent specifications: an approach to building human-centred specifications", IEEE Trans. Soft. Eng., vol. 26, pp. 15-35.

Loucopoulos, P, 1994, The f³ (from fuzzy to formal) view on requirements engineering. Ingénierie des systèmes d'information, Vol. 2 N° 6, pp. 639-655.

Loucopoulos, P., Kavakli, V., and Prakas, N., 1997,Using the EKD approach, the modelling component. ELEKTRA project internal report.

Mylopoulos, J.,. Chung K.L., and Nixon, B.A., 1992,Representing and using non- functional requirements: a process-oriented approach . IEEE Transactions on Software Engineering, Special Issue on Knowledge Representation and Reasoning in Software Development, Vol. 18, N° 6, June 1992, pp. 483-497.

Mylopoulos, J.,. Chung, K..L., and Yu, E., 1999,"From object-oriented to goal6oriented requirements analysis". Communications of the ACM. Vol 42 N° 1, January 1999, 31-37.

Mostow, J., 1985, "Towards better models of the design process". AI Magazine, Vol. 6, pp. 44-57.

Nature, 1999, The nature of requirements engineering . Shaker Verlag Gmbh. (Eds.) Jarke M., Rolland C., Sutcliffe A. and Dömges R., .Jily 1999.

Nuseibeh, B., Kramer, J., and Finkelstein, A., 1994,A framework for expressing the relationships between multiple views in requirements specification. In IEEE Transactions on Software Engineering, volume 20, pages 760-- 773. IEEE CS Press, October 1994.

Plihon, V., Ralyté, J., Benjamen, A., Maiden, N. A. M., Sutcliffe, A., Dubois, E., and Heymans, P., 1998, A reuse-oriented approach for the construction of scenario based methods. Proceedings of the International Software Process Associations 5th International Conference on Software Process (ICSP'98), Chicago.

Pohl K., 1996, Process centred requirements engineering, J. Wiley and Sons Ltd.

Pohl K., and Haumer, P., 1997, Modelling contextual information about scenarios. Proceedings of the Third International Workshop on Requirements Engineering: Foundations of Software Quality REFSQ'97, Barcelona, pp.187-204, June 1997.

Potts, C., Takahashi, K., and Anton, A. I., 1994,Inquiry-based requirements analysis. In IEEE Software 11(2), pp. 21-32.

Potts, C., 1995, "Using schematic scenarios to understand user needs", Proc. DIS'95 - ACM Symposium on Designing interactive Systems: Processes, Practices and Techniques, University of Michigan, August 1995.

Potts, C., 1997,Fitness for use : the system quality that matters most. Proceedings of the Third International Workshop on Requirements Engineering: Foundations of Software Quality REFSQ'97 , Barcelona, pp. 15-28, June 1997.

Prat, N., 1997,Goal formalisation and classification for requirements engineering. Proceedings of the Third International Workshop on Requirements Engineering: Foundations of Software Quality REFSQ'97, Barcelona, pp. 145-156, June 1997.

Ramesh, B., Powers, T., Stubbs, C., and Edwards, M., 1995, Implementing requirements traceability : a case study, in Proceedings of the 2nd Symposium on Requirements Engineering (RE'95), pp89-95, UK.

Rasmussen, J., 1990, Mental models and the control of action in complex environments. Mental Models and Human--Computer Interaction, D. Ackermann and M.J. Tauber , eds., North-Holland : Elsevier, pp. 41-69.

Rolland, C, and Ben Achour, C., 1997,Guiding the construction of textual use case specifications. Data & Knowledge Engineering Journal Vol. 25 N° 1, pp. 125-160, (ed. P. Chen, R.P. van de Riet) North Holland, Elsevier Science Publishers. March 1997.

Rolland, C., Grosz, G., and Nurcan, S., 1997a,Guiding the EKD process. ELEKTRA project report.

Rolland, C., Nurcan, S., and Grosz, G., 1997b,Guiding the participative design process. Association for Information Systems Americas Conference, Indianapolis, Indiana, pp. 922-924, August, 1997

Rolland, C., Souveyet, C., and Ben Achour, C., 1998,Guiding goal modelling using scenarios. IEEE Transactions on Software Engineering, Special Issue on Scenario Management, Vol. 24, No. 12, Dec. 1998.

Rolland, C., 2002,L'e-lyee: l'ecritoire and lyeeall, Information and software Technology 44 (2002) 185-194.

Rolland, C., Grosz, G., and Kla, R., 1999,Experience with goal-scenario coupling. in requirements engineering, Proceedings of the Fourth IEEE International Symposium on Requirements Engineering, Limerik, Ireland.

Ross, D. T., and Schoman, K. .E., 1977,Structured analysis for requirements definition. IEEE Transactions on Software Engineering , vol. 3, N° 1, , 6-15.

Robinson, W. N., 1989, "integrating multiple specifications using domain goals", Proc. IWSSD-5 - 5th Intl. Workshop on Software Specification and Design, IEEE, 1989, 219-225.

Schank, R. C., 1973,Identification of conceptualisations underlying natural language. In ''Computer models of thought and language'', R.C. Shank, K.M. Colby (Eds.), Freeman, San Francisco, pp. 187-247.

Sommerville I.,1996, Software Engineering. Addison Wesley.

Sommerville, I., and Sawyer, P., 1997, Requirements engineering. Worldwide Series in Computer Science, Wiley.

Standish Group, 1995,Chaos. Standish Group Internal Report, http://www.standishgroup.com/chaos.html.

Sutcliffe, A.G., Maiden, N. .A., Minocha, S., and Manuel D.., 1998,"Supporting scenario-based requirements engineering", IEEE Trans. Software Eng. vol. 24, no. 12, Dec.1998, 1072-1088.

Tawbi, M., 2001,Crews-L'Ecritoire : un guidage outillé du processus d'Ingénierie des Besoins. Ph.D. Thesis University of Paris 1, October 2001.

Thayer, R., Dorfman, M. (eds.), System and software requirements. IEEE Computer Society Press.1990.

Vicente, K. J., and Rasmussen, J., 1992, Ecological interface design: Theoretical foundations. IEEE Trans. on Systems, Man, and Cybernetics, vol. 22, No. 4, July/August 1992.

Yu, E., 1994,Modelling strategic relationships for process reengineering. Ph.D. Thesis, Dept. Computer Science, University of Toronto, Dec. 1994.

Yue, K., 1987,What does it mean to say that a specification is complete?, Proc. IWSSD-4. Four International Workshop on Software Specification and Design, Monterrey, 1987.

Zave P., and Jackson M., 1997, "Four dark corners of requirements engineering", ACM Transactions on Software Engineering and Methodology, 1-30. 1997.

TOWARDS CONTINUOUS DEVELOPMENT
A Dynamic Process Perspective

Darren Dalcher[*]

1. INTRODUCTION

The starting point for exploring software engineering often revolves around the process view of software development and the implications and limitations that come with it. Software development is life cycle driven: the traditional linear life cycle, where progress is represented by a discrete series of transformations (state changes) has had a defining influence over the discipline. In fact, many software development processes are either predicated directly on various forms of a rational model concept or are designed to overcome perceived problems with this process through increments, timeboxes, user participation and experimentation.

Many of the life cycle variations appear to share a concern with software development as a one-off activity. The process is thus viewed as discrete projects or incremental sets of micro-projects that lead to a final integrated artefact. Indeed, the life cycle notion represents a path from origin to completion of a venture where divisions into phases and increments enable engineers to monitor and control the activities in a disciplined, orderly and methodical way. The control focus however retains a short-term perspective. In practice, little attention is given: to global long-term considerations, to activities such as project management and configuration management that are continuous and on going, to the justification of the acquisition of tools and improvement strategies, to the need to maintain the value of the product and, more crucially, to the accumulation of knowledge, experience or wisdom.

The dynamic nature of knowledge and software evolution and usage present a pervasive challenge to system developers. Discrete attempts to create such systems often lead to a mismatch between system, expectation and a changing reality. This paper attempts to address the implications of continuity and highlight the mechanisms required to support it. The rationale for a Dynamic Feedback Model stems from the need to focus on a continuous and long-term perspective of development and growth in change-intensive environments. The paper makes the case for a learning and knowledge-driven view of software development and presents such a model in a way that accounts for the long-term survival, growth and evolution of software-intensive systems.

[*] Darren Dalcher, Software Forensics Centre, Middlesex University, Trent Park, London N14 4YZ, UK.

Information Systems Development: Advances in Methodologies, Components, and Management
Edited by Kirikova *et al.*, Kluwer Academic/Plenum Publishers, 2002

2. DEVELOPING A CONTINUOUS PERSPECTIVE

A life cycle identifies activities and events required to provide an idealised solution in the most cost-effective manner. No doubt many enacted life cycles are attempting an optimisation of determining a solution in 'one pass', typically stretching from conception through to implementation, but others explicitly accept the unlikelihood of such a solution and are therefore iterative, optimised over a single pass. The *singular* perspective can be represented as a black-box that takes in a given problem and resources as input and completes a transformation resulting in a delivered system that addresses the problem (see Figure 1). The fundamental implications are: that the input consists of essentially perfect knowledge pertaining to the goal and starting state; that the transformation is the straightforward act of management and control within the established constraints; and, that the quality and correctness of the derived product is the direct output of that process. In this singular mode, knowledge is assumed to be fully available in a complete, well-understood, absolute and specifiable format. In order for the initial (static) abstractions to maintain their validity throughout the development effort, to ensure consistency and relevance, such projects need to be short enough (and possibly simple enough) so as not to trigger the need to update or discard the initial abstractions and thereby invalidate the starting baseline.

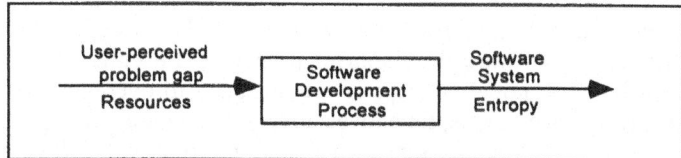

Figure 1. The Singular Software Development Process

However, many real projects are not about well-understood problems that can be analysed and optimised, but are to do with knowledge that is elusive, tacit, incomplete, ephemeral and ambiguous. Working in ill-structured environments calls for continuous and adaptive design rather than complete and optimal analysis. The act of design is assumed to be a creative, experimental, argumentative and negotiation-based, discovery effort. Rather than being complete before inception, such situations require on-going elicitation of knowledge (requirements) and also require adaptability so as to respond to change. The primary output of requirements therefore is understanding, discovery, design, and learning, so that the effort is concept-driven[1, 2].

Modern problem situations are characterised by high levels of uncertainty, ambiguity and ignorance requiring dynamic resolution approaches. Rather than expend a large proportion of work on the initial analysis of needs and expectations, such situations require openness to different domains and perceptions (rather than a search for optimality) and a recognition of the effect of time. As modern development focus moves from conventional commodities to a knowledge-based economy, intellectual capital becomes a primary resource. The main output of software development can be viewed as the accumulation of knowledge and skills embedded in the new emergent social understanding (see Figure 2). Rather than act as input (cf. Figure 1), knowledge emerges continuously from the process of development and learning. Continuous discovery and experimentation persist through the development, utilisation and maintenance phases. In addition to this primary output, the process may also create an artefact, which will represent a simplified, negotiated and constrained model of that understanding. Change combines with inputs of varying perspectives, and personal biases, attitudes, values and perceptions to offer the raw materials requiring constant compromise.

The resolution process will thus result in a continuous output of understanding and insight that can replenish organisational skill and asset values. In change-rich environments, this output plays a far more crucial role to the future of the organism than the partial artefact whose construction fuelled the enquiry process. In the long run, the intellectual asset defines the range of skills, resilience, competencies and options that will become available.

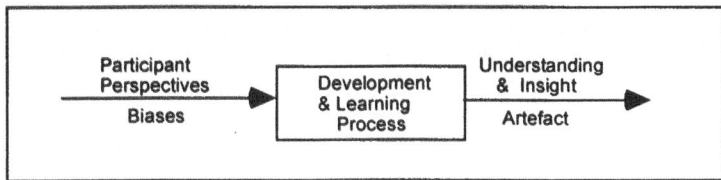

Figure 2. The Development and Learning Process

The process of Problem Solving leads to improved understanding of a situation through discovery, experimentation and interaction with the problem domain. As the act of software design is increasingly concerned with the generation of understanding and knowledge, the main transformation needs to reflect the need for on-going exploration. A focus shift from delivering to discovering allows continuous exploration rather than temporal targets. Changing perceptions, needs, new learning and continuous experimentation will thus involve designers beyond discrete products and fixed specifications. Continuously evolving customer perspective and satisfaction levels provide the on-going success measures for determining achievement and utility. In a changing environment, quality is a moving target. Change and the lack of a permanent and static specification permit the view of customer satisfaction as an evolving level that is allowed to grow and to improve alongside other assets. This stems from the recognition that satisfaction is neither a fixed nor a negotiated construct. The customer focus therefore extends beyond the product development view to incorporate usage and adaptation (of the product to the user rather than the other way round). Maintaining the evolving focus, value, relevance and satisfaction levels, requires a dynamic mechanism for conducting trade-offs as part of the sensemaking process. The process driver of knowledge-intensive development is the need to adjust to new discoveries and make trade-offs as part of the attempt to make sense.

3. TOWARDS CONTINUOUS SOFTWARE ENGINEERING

Figure 3 emphasises some of the differences between the assumptions and the operating modes of different domains. Software engineering, as traditionally perceived, appears to relate to the resolution of Domain I problems. The domain of application for this type of approach is characterised by well-defined, predictable, repeatable and stable situations, reasonably devoid of change that can be specified in full due to a low level of inherent uncertainty, and thus amenable to 'singular' resolution. Progression is carried out sequentially from the fixed initial definition of the start-state to the envisaged final outcome. The main emphasis revolves around the transformation and fixed initial parameters to ensure that the end product is directly derivable from the initial state. Traditional approaches to quality assurance likewise depend on the belief that a sequence of structured transformations would lead to a high quality result. Repeatable and well-understood problems will thus benefit from this approach, assuming little scope for risks and surprises.

The middle ground (Domain II) is occupied by situations characterised by higher levels of uncertainty that cannot be fully specified. Despite a reasonable knowledge of the domain, there is still a degree of uncertainty as to the solution mode, making them more amenable to incremental resolution. Quality is also created and maintained incrementally but requires on-going effort (and organisation).

Far from being a structured, algorithmic approach, as is often advocated, software engineering in Domain III, is a *continuous* act of social discovery, experimentation and negotiation performed in a dynamic, change-ridden and evolving environment. Domain III systems are closely entwined with their environment as they tend to co-evolve with other systems, and the environment itself. Such systems affect the humans that operate within them and the overall environment within which they survive, thereby, affecting themselves[3]. This continuity acknowledges the evolving nature of reality (in terms of meanings and perceptions) and the fact that absolute and final closure in Domain III is not attainable. However, discovery and the concern with knowledge, enable a new focus on the long-term implications of development in terms of assets, improvement and growth. This view underpins the new emphasis on the growth of organisational resources, including knowledge, experience, skills and competencies, and on the improvement that it promises. (The shift therefore is not only from product to process, but also from discrete to continuous-and-adaptive.)

Generally, the temporal and discrete focus of software engineering for Domain I does not justify a long-term view. Rather than create assets for the long-run, software engineering is concerned with optimising new undertakings that are viewed as single and discrete projects thus ignoring the long-term perspective. As a result, investment in quality and improvement is difficult to justify on a single project basis. Furthermore, no specific mechanisms are introduced to capture and record knowledge. Software engineering for Domain III should be concerned with the longer-term, where justification of resources extends beyond single projects. Investment in tools, improvement and new skills are therefore viewed against corporate targets of improvement and future performance levels which span multiple projects. Significant assets can thus be created in the form of knowledge, skills and competencies thereby serving the business and providing a competitive edge. Indeed, fostering a strategic view enables one to view all projects as strategic (as part of an asset portfolio). Maintenance becomes a part of normal protection, improvement and growth of assets, which also incorporates initial development.

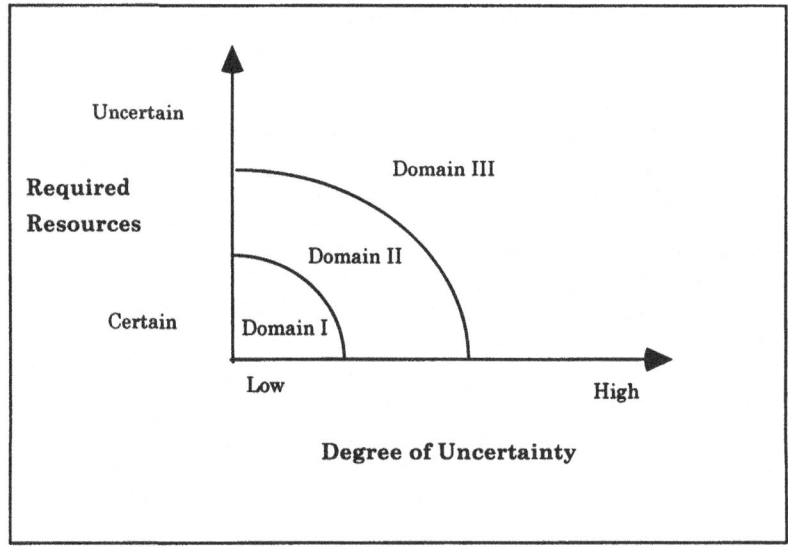

Figure 3. Selecting a Life Cycle Approach

A long-term perspective views design as on-going balance between concerns, constraints, challenges and opportunities. A unique feature of this perspective is the focus on the problem in terms of bi-directional adjustment of the starting position and expectations (rather than a uni-directional optimised process towards a static target). The main implication is that users need to be involved throughout as their needs are likely to evolve as a result of the interaction between plan, action and reality. Customer interaction becomes on-going throughout the evolution of the system and the growth of the resource. Indeed, participation itself can be viewed as a growing asset which offers competitive advantages.

The type of thinking required for Domain III situations relies on embracing a continuity dimension which enables long-term strategic planning. The general conceptual shift therefore is from discrete software development (in Domain I) to problem solving in a continuous sense (in Domain III). A long-term perspective can account for knowledge through the required infrastructure supported by the adoption of the growing-asset perspective, thus underpinning normal development and evolution of values, assets, and satisfaction levels (assets in themselves). This allows for the fostering and improvement of expertise, skills and competencies offering a competitive edge through value-driven and experience-based improvement.

The proposed interpretation of growth and improvement is thus concerned with:

Intellectual Assets: Focused primarily on generating a strategic view of development that accounts for long term investment in, and growth of, intellectual assets.

Continuous Learning: Grounded in a systematic and multi-disciplinary perspective that is responsive to the results of continuous learning to enable such improvement.

Satisfaction and Utility Trade-offs: Concerned with the trade-offs and their implication on the resulting outputs and the emerging satisfaction and utility.

Monitoring Feedback: Based on constant monitoring and feedback which play a part in driving the process and directing the shifting of perceptions and values.

4. MODELS TO DEPICT DEVELOPMENT

Models are used as tools to aid in simplifying and reducing the complexity of reality by abstracting and focusing on certain essential portions of it. They are thus utilised in explanation, demonstration, description or prediction. One important distinction in modelling is that between static and dynamic models. Static models typically focus on the representation of states, while dynamic models concentrate on the representation of processes. Dynamic models have the power to depict processes as a continuous description of relationships, interactions and feedback. Feedback systems are critical to understanding relationships, interactions and impacts. They link causes and effects in dense and often circular causal patterns[4, 5]. Feedbacks contribute to the ability to learn and improve and are therefore essential to understanding the growth of knowledge within structures and relationships. Information feedback plays a key part in connecting the different entities and in contributing to the growth and accumulation of knowledge resources. Utilising a model that uses feedback and is dynamic thus leads to a number of advantages:

Cross Discipline Perspective. First, an essential difficulty is in integrating knowledge from across disciplines in a way that supports growth and improvement. The traditional depiction of the process focuses on technical actions largely ignoring management processes and supporting activities. Indeed, while it is recognised that there are many interactions and interdependencies,[6] little attention has been paid to how the different functions and domains are integrated[7]. Rather than being mutually exclusive, the different sets of activities support and impact across disciplines. An integrated model of the process would guide managers and developers towards a systems thinking approach to process modelling and away from the current focus on addressing a single aspect, or perspective of the process[8]. The benefit is in seeing a more complete picture and addressing the long-term concerns in a more informed way. An integrated view of processes serves to reaffirm the importance of major stakeholders involved in the enterprise of designing software, however indirectly associated with the act of software production. Indeed, such a model would be explicitly directed at the identified need for communication between disciplines[9].

Exploring Complexity. Second, the creation of extended systems takes in additional perspectives inherent in the composite of people, skills, and organisational infrastructure[10]. This enables the use of feedback to highlight and explore the complex interactions and loops that are formed within the software environment[5, 6].

Viewing the Long-Term Horizon. Third, such a model enables reasoning about the long-term perspective. This facilitates the consideration of the long-term corporate objectives of enhancing quality, reducing maintenance, improving productivity, enhancing organisational assets and supporting growth which extend beyond the remit of single or discrete efforts. By looking beyond the delivered outputs it thus becomes possible to support the meeting of current and future objectives, including the acquisition and maintenance of tools, skills, knowledge, co-operation, communication, and competencies.

Wiser Trade-offs. Fourth, since, decisions are typically based on a single dimension or a single perspective view of the process prescribed by product-oriented or management activity models focusing on certain activities, a multi-perspective dynamic model can encourage intelligent trade-offs encompassing change and various concerns from the constituent disciplines. Furthermore, decision making and risk management assist in balancing long-term objectives and improvement with short-term pressures and deadlines.

Dealing with Knowledge. Fifth, the model provides the mechanism for reasoning about knowledge and a framework for the long-term growth of knowledge and assets. This enables reasoning about different perceptions and perspectives as well as encouraging a long-term approach to growth, improvement and support of organisational ability, knowledge, skills and competencies. The corporate perspective thus enabled, supports the growth and justification of resources that underpin an improving and evolving organisational asset.

Evolution and Continuous Learning. In addition, a dynamic model acknowledges the evolution of the system over time, thereby recognising the role of change and the need for continuous learning and adjustment (particularly prevalent in Domain III development). This goes to the core of the act of design, which involves learning and interaction with the various elements of the system in a constant effort to improve the fit in a rapidly changing environment. It thus underpins the notions of a long-term perspective, evolution and growth.

5. THE DYNAMIC FEEDBACK MODEL

This rest of the paper offers an example of a Dynamic Feedback Model (DFM) depicting the on-going inter-disciplinary links and trade-offs embedded in development. Domain III software development projects embrace a complex set of interacting organisational entities that can normally be divided into three or four basic groups (encompassing functional domains). The relationships between the different groups can be understood by studying the feedback cycles operating between them. Modelling the relationships in a non-linear fashion allows a continuous view of the development effort to emerge, which in turn facilitates a long-term view of the management and dynamics involved in such an effort. The discussion that follows focuses on four different functional domains that are intertwined throughout the act of designing systems. Each perspective offers a valid representation of the attempt to develop a system. While they are often depicted as mutually exclusive, they tend to interact and influence other domains as they essentially refer to the same development undertaking[8]. This recognises the fact that design requires a more dynamic perspective that is responsive to changes, adaptations and complications, as the more traditional perspective is ill-equipped to handle dynamic complexity. Design thus entails trade-offs between perspectives and disciplines including management and quality of the required artefact.

The *technical domain* extends beyond the traditional notion of development to encompass the design and evolution of artefacts continuing throughout their useful life span. Maintenance, from an asset perspective, is therefore part of this continuous cycle of improved value and utility. By extending the horizon of interest beyond the production stage, one becomes concerned with the useful life and persistence of artefacts, and in the continuous match between functionality, need and environment. The focus on the act of design as the creative activity, shifts the interest to issues of knowledge, learning and discovery of information. Experimentation, discovery and learning are essential to the act of design as they form the basis for pursuing intelligent negotiation, argumentation and agreement regarding needs and expectations. The continuous production of knowledge (viewed as development or design) facilitates the prospects of a long-term horizon[3, 11, 12]. This points to a continuous focus on adaptation which makes knowledge, and the resulting competencies and skills a corporate asset. Development and maintenance thus become a continuous activity of improving corporate organisational resources, through increased and orchestrated adaptability and responsiveness to change. The corresponding shift in perspective for software is from a product-centred focus to an asset-based perspective emphasising continuous usage, enhanced client emphasis and recognition of changing needs and perceptions.

The *management domain* is charged with the planning, control and management of the action that constitutes overall effort. It is therefore concerned with identifying mismatches, planning for action and assessing the degree of progress achieved, as well as with allocating resources to facilitate such progress. Trade-offs between levels of resources, attention and constraints will shape the resulting technical activity as the two are tightly bound. In the singular mode, management was perceived as a discrete and optimising activity guiding the transformation between the fixed input and the single output. However, management is primarily concerned with an on-going balance between objectives, benefits and obstacles,[13] as well as emerging opportunities and constraints[14]. Management action is carried out against the backdrop of continuous change and evolution[15, 16]. Once the position of knowledge as an on-going commodity attained through on-going probing and discovery is acknowledged, and technical development is perceived as a long-term activity, management also takes on continuous dimensions[5, 16]. Guiding exploration, adaptation and negotiation in search of a good fit, suggests a reduced role for absolute planning and a higher degree of responsiveness[14]. Furthermore, delivery is not limited to a single incident that only occurs once. The focus on delivering a product is thus replaced with the continuous need for generating and maintaining an on-going flow of knowledge requisite for the asset-based view of continuous development.

The *quality domain* is no longer optimised around the delivery of a product. The quality perspective in the singular 'one-pass' perspective was assumed to be a derived attribute of the product emerging from the process. This notion also entailed correctness and implied acceptance by users. In Domain III, quality is more difficult to capture and 'fix'. As discovery, learning and experimentation continue, and as adaptation rather than optimisation leads to a good fit, quality becomes an on-going concern. Experimentation and evolution lead to changes in expected functionality; and participation in experiments changes perceptions and expectations. Quality is thus perceived as a dynamic dimension, which continuously responds to perceived mismatches and opportunities reflected in the environment. It thus becomes more of a dynamic equilibrium[17], which requires constant adjustments and trade-offs. As development extends beyond the delivery stage, quality must also encompass the useful life beyond the release of an artefact. Satisfaction must cater to expectations and functional correctness leading to notions such as Total Customer Satisfaction[18] as the primary measurement of acceptability. In common with other domains, this perspective continues to evolve and grow, but can benefit from being viewed as an organisational resource or strength that needs to be maintained and enhanced.

The move from a target-based system towards an evolving fit entails a greater degree of uncertainty coupled with essential decision making and sensemaking. The ability to make decisions, especially in continuous and evolving contexts, and to trade-off multi-dimensional quantities, values and perspectives rests on the availability of a provision for making sense, taking decisions, and more generally managing opportunity and risk. In fact, the ability to make risky decisions and trade-offs, endows the design process with the sense necessary to balance and understand the interacting characteristics and their impact on the context, as well as the derived solution. Change, discovery and evolution also suggest a more open-ended environment seeking to balance and trade-off opportunities and risks. *Risk management* offers a domain for conducting argumentation, negotiation and trade-offs, while keeping the growth of the capital asset as a major objective. This enables skills and knowledge to benefit from the continuous search for fit, adaptation and evolution. Not only does risk oversee all other domains including knowledge, it also directs the evolution by continuously balancing potential with constraints while enabling re-adjustment as a result of the accumulation of new knowledge. Risk management is therefore approached from a continuous perspective which attempts to allow for the distinction between reality and possibility by acknowledging the role of opportunity. Well-being, or enhanced survival prospects can thus be viewed as an organisational resource that can be addressed through a conscious attempt to control evolution and prosperity by risk management. Risk management thus offers 'the glue' that unites the disciplines and concerns, thereby providing the capability for intelligent and 'enlightened' trade-offs.

6. THE RELATIONSHIPS

The discussion so far has established four general domains of interest (management, technical, quality and risk management). Generally, models fail to acknowledge the relationship between different domains such as management, engineering and product assurance primarily because they seem to emerge from one of the disciplines. The interactions between the different functions add to the overall complexity of the development effort and lead to emergent, yet unpredictable, behaviour dynamics[19]. This directly contributes to the need to focus on the ecology and interrelationships between the different components that make up the software environment.

The interactions between the four functions should allow the ultimate completion of feedback loops resulting in optimised performance of the system. These interactions are expressed in terms of (systems control) activities. Figure 4 links the four domains of interest in a dynamic feedback loop.

Project management can thus be viewed as a system incorporating inputs and outputs. The main output from project management is the act of planning (which must precede the activity of doing)[20]. Project management, and indeed the act of planning itself, are continuous and therefore require an on-going input of new and responsive information. The inputs to the process of project management include the visibility that emerges as a result of the act of doing[21, 22] and the knowledge that emerges from the act of checking performed as part of product assurance (in the form of reports). Effective project management requires dynamic updating of information which is reflected in altered plans[23]. Planning and project management are continuous activities thus requiring a continuous feedback loop that leads from the results of the planning, to new inputs of updated information[24].

Figure 4. The Dynamic Feedback Model (DFM)

Similarly, the output of *risk management* is the act of controlling which is used to implement and control the technical activity[25]. Control strategies are adjusted continuously on the basis of risk management activities designed to re-assess the process of doing in light of changes in plans, priorities and opportunities[14]. The input to risk management is provided through the channels of planning[22] and the results of monitoring[26]. The implication is that the control output is fed back (through a chain of activities described in the next section) as updated planning and monitoring inputs.

Technical development can also be viewed as a system. The result of the activity is in producing artefacts (in the form of established baselines)[27]. The second output is the enhanced visibility[28]. The input is the activity of controlling[22]. Indeed, visibility is the key to effective control and thus needs to be reflected back in the form of feedback loops connecting both outputs in a way that will allow adjustments to inputs (i.e. the act of controlling) based on observed results and changes in circumstances and perceptions.

Figure 4 also provides a basic system diagram of the *product assurance* entity. Once again, it is instructive to start with outputs to emphasise the fact that inputs are adjusted continuously as a result of observed outputs. The outputs emerging from this entity are primarily concerned with the results of quality evaluations. They comprise the act of monitoring and the additional visibility provided by management reports[24]. Product assurance is initiated through the input of baselines (artefacts)[29]. The outputs lead to re-work, which comes back to this system in the form of new baselines (and adjusted products).

The DFM appears to offer certain advantages in comparison with other attempts to depict the nature of the process and the effort required to manage it:

Dynamic: Unlike other representations of the life cycle[cf. 9, 30], the DFM concentrates on the interaction between processes rather than on the (temporal) representation of states.

Feedback: Rather than focus on a single pass,[see for example, 29, 31] the DFM takes account of feedback loops that affect inputs through circular and repeatable loops. Observation of effects and continuity provide an explicit opportunity to respond to feedback.

Risk Management Driven: Dynamic complexity and change are handled through the explicit use of risk management in handling trade-offs and complex decisions in light of feedback and requests for adjustments. Typical models ignore risk management or impose it on the technical or management stages[22, 30].

Dimensionality: Processes such as the waterfall and prototyping focus on a single-dimension micro-orientation perspective which is typically technical. Various attempts have introduced parallel processes, with explicit synchronisation points[see for example 29, 32]. However, these still do not allow explicit trade-offs or adaptation to changing circumstances. Similar attempts in project management to introduce product development cycles alongside project management cycles are subject to the same constraints[see 8]. The spiral model and subsequent work[30] went further than other models in tackling dynamic complexity by introducing risk management, however, this is done as part of the development process, rather than as an explicit and interacting discipline[33].

Continuity: Most processes adopt a single-project perspective (i.e. they are constructed and organised around a single,finite effort with little long-term perspective for evolving the system and, moreover, the processes involved). The DFM assumes an on going perspective enabling adjustments in time as part of the approach. The continuous project view accommodates growth and improvement as an on-going effort, thereby treating maintenance and evolution as a natural part of this process.

Knowledge: The view of continuity justifies learning, long-term improvement and investment in resources. This includes tools, skills, knowledge and competencies that can be developed on a continuous basis and improved to enable greater responsiveness and the ability to capitalise from opportunities and improvements.

Uncertainty: Various models have attempted to contain uncertainty through iterations and risk management (while more traditional approaches ignore this aspect completely). Incomplete and imperfect information, can be explicitly addressed through the risk management domain in the DFM. The long-term perspective and the explicit feedback underpin improvement as a responsive feature of observing impacts and continuously addressing uncertainty, ambiguity and ignorance.

7. FEEDBACK MECHANISMS

The dynamic model is in essence, a set of interactions and feedback loops governing and controlling the production of software from a continuous and dynamic perspective. It is also possible to analyse the model in terms of causal-loop diagrams. Space considerations prevent a detailed description of the loops which will only be briefly highlighted.

The basic loop in the dynamic system is the *planning-control-visibility loop* which enables the project manager to plan and control the production, evolution and growth of software by technical developers through the use of trade-offs and risk management (decision making). The loop enables the continuous generation of new information as well as feedback knowledge. The use of this knowledge is crucial in changing plans to adapt to reality and opportunities, modifying the decisions and re-examining the assumptions and the risk assessment completed previously. It also offers the framework for meeting scope, cost and interval targets. The recognition and establishment of visibility completes the basic feedback loop by providing a continuous process to ensure the system remains relevant with regard to its objectives.

Some form of product (or service) assurance is required in order to store, assess, and evaluate completed baselines at milestones and analyse the effectiveness of the processes producing them. In addition, the planning and controlling activities operate in concert with the risk management function. This is provided by the *configuration control loop*, linking control, baselines and monitoring, that enables the (closed-loop) link between monitoring and controlling. Monitoring provides feedback on the strategies implemented, thus enabling more effective control. Monitoring relies on quality assurance techniques[32] and feedback mechanisms to evaluate progress and quality. Monitoring is the missing link between risk management and quality as it makes the feedback system a closed-loop system. In this way, monitoring enables learning from past mistakes, improvement, the gradual accumulation of knowledge and the growth of competencies.

Reporting and evaluation are not represented in typical life cycle models or in decision making processes[23]. Their role is to provide the feedback (and long-term knowledge) that allows plans and decisions to be revised. Such revisions pertain to integrated technical plans as well as to objectives and strategic targets, in which case they can play a part in adjusting management plans. This is depicted by the *reporting-planning loop*.

8. THE DETAILED MODEL

A more detailed breakdown of the leading domains (entities) based on the key activities within each domain is useful in highlighting the key links between the activities that enable the various feedback loops. This is depicted in Figure 5, which emphasises internal physico-structural decomposition by function of each of the non-technical domains.

Project management is depicted in its twofold capacity. At the general level (corporate to strategic levels), management functions include providing direction, ensuring co-operation between projects and representing general management's concerns (i.e. representing a strategic long-term perspective). The project level itself is concerned with a specific project from a strategic, tactical and operations level. The main emphasis is on the assignment, allocation and control of all project resources expressed in terms of funding, time, personnel and equipment as embodied in project plans and status (progress) representations.

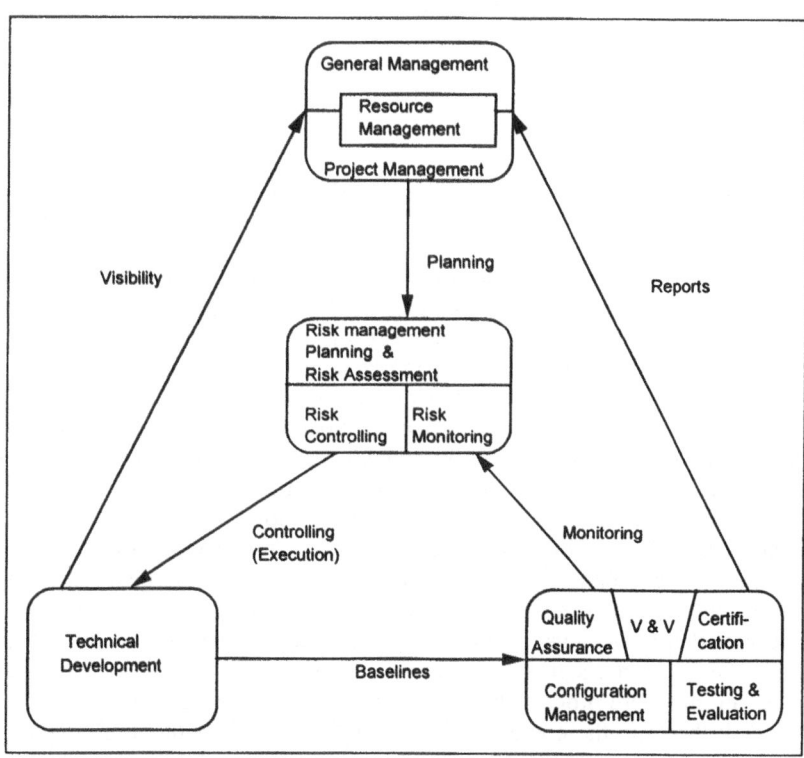

Figure 5. The Detailed Dynamic Feedback Model

Technical development deals with the creation and continuous improvement of the software system (and the resulting knowledge). This activity is dominated by the on-going focus on the transformations the system has to go through from inception to evolution and thus remains continuous. In other words, it is primarily concerned with maintaining the functionality, and utility of artefacts by periodically re-visiting the relationship and fit between function, form, context, expectations, emotions and limitations. Significantly, therefore, evolution and maintenance are treated as part and parcel of this activity as they represent the growth and development of the system over an unlimited time horizon.

Product assurance is concerned with assuring the quality of the product developed and the process used to develop it, providing external visibility for management and assessing the value of management decisions and estimates made in the course of the project. The individual disciplines called upon are: configuration management (CM), quality assurance (QA), validation and verification (V&V), testing and evaluation (T&E) and certification. The provision of independent verification and validation, as well as of independent testing and evaluation is explicitly recognised and included within the product assurance function (even though this is often a matter of either interpretation or politics).

The co-ordinating function at the centre of the dynamic model is responsible for directing and controlling the entire development effort. *Risk management* in this context, is used in its broader sense to incorporate the disciplines of risk assessment and risk control[25,26]. Risk management is an attempt to improve the quality of decisions based on intelligent and informed trade-offs and on knowledge obtained from previous iterations and projects. Figure 5 depicts a decomposition of the risk management entity that emphasises the planning, controlling and monitoring aspects of risk management. As can be seen, due to its iterative, continuous and feedback-centred focus, risk management continuously drives the processes taking place in the development environment.

9. EVALUATION

The DFM depicts relationships between the different organisational entities and can be used as a framework for understanding the dynamic nature of the interactions between entities in Domain III development. This model breaks away from linear thinking and the temporal project perspective to offer a continuous perspective for understanding and implementing the relationships and their long-term effects. This is done through the depiction of on-going feedbacks and interactions. The model therefore offers an alternative method of reasoning about development. A single project perspective ignores the dynamics that are established between functions thus obscuring the nature and critical importance of any interactions other than the prescribed linear ones. A long-term perspective enables developers to view the process and the organised structure as a framework for capturing knowledge and regarding it as an asset. This can lead to gradual continuous improvement.

The DFM highlights the fact that some consideration must be given to non-technical aspects of the development effort. Indeed, the purpose of the central subsystem is to integrate different perspectives and considerations and make sense of the emergent knowledge, thereby enabling intelligent decision making and balanced trade-offs. The establishment of four major domains suggests that effort must go into recognising and organising all four systems. Many organisations are characterised by defective management strategies, ineffective quality assurance and an immature process. Intelligent deployment depends on improvement to these areas, which will thrive in the context of holistic (if somewhat contention-inducing) decision making. In fact, creative contention exercised on a continuous basis, may result in richer understanding, wider choice options and greater utilisation of possibilities leading to increased prosperity.

The driving process is essentially, a simple feedback web that offers great diversity through complex interactions. The recognition of the process as a dynamic and on-going activity seems to lead to improved management perception which in turn, may result in more adequate planning and resource allocation. Better understanding of the process and the available knowledge deals with the essence of the problem, offering the long-term potential for reduction of associated problems such as maintenance (which can be viewed as an integral part of on-going value enhancing of corporate assets). The simplicity of the model, an inherent characteristic of complex dynamics[see for example, 34, 35] makes it an attractive and workable formalism for process improvement.

10. SUMMARY AND DISCUSSION

This paper focused on the on-going aspects of Domain III software development by introducing the DFM that encompasses the main functions concerned with design. Dynamic models rely on feedback which enables focus on the continuous nature of processes and their long-term survival, growth and evolution. The model highlights the need for continuous discovery of knowledge which essentially drives the feedback interactions leading to growth. The focus on knowledge as a key asset can be transported to software development to support the view of continuous growth and improvement as an essential corporate asset.

The DFM encourages thinking about software development in terms of the constituent parts and their interactions. The model addresses another fundamental deficiency in contemporary processes by offering a decision making perspective. This view of the process in terms of problem solving offers visibility into the mechanisms required to deal with the inherent uncertainties embedded in the environment. Most contemporary models miss this aspect of decision making[7] and subsequently cannot adequately address the hidden aspects of uncertainty, ambiguity, risk and conflicting objectives. The DFM uses risk management and the control framework driving it to direct the entire process. By taking a long-term perspective, it becomes possible to reason beyond the time horizon or resources of a single project to obtain an enlightened insight and structure that can improve performance and become a valued and growing corporate resource in the long-term.

The DFM is receptive to changes in the environment (characteristic of Domain III) and tackles them by feeding acquired knowledge back into the decision making subsystem. Thus, if, for example, a prototyping strategy is employed to reduce the inherent risk by buying information, the information can be used to improve the overall process and the knowledge resource. This approach to risk-containment and adjustment to new knowledge, can be adapted to any situation. The interest in risk, enables integration with any risk-driven approach and supports management in appreciating the organisational relationships, dynamic interactions and dependencies which affect and direct such risk assessments.

Activities such as configuration management are phase-independent and cannot be fitted into an individual stage in more traditional life cycles. Many such activities are not integrated into conventional models, as they represent a longer-term perspective, and are consequently considered to reside outside the remit of development. The DFM on the other hand, justifies the long-term investment in supporting facilities.

The model provides a unique possibility for accounting for and reasoning about long-term decisions such as creating a reusable software library, investing in tools or information, or accepting new and innovative processes. Moreover, the DFM seems to account for evolution, growth, change and maintenance (i.e. there is no need to re-organise the structure, change the functions, or create a new model). The software process can thus be perceived as a continuous growth of software assets over time. A corresponding improvement in shared skills and competencies across projects is also supported.

As multiple views of the process are integrated, a more complete picture of the effort can be constructed. However, as software development becomes more integrated in management practices and ever-growing organisational contexts, and as software becomes more integrated in other areas, the importance of continuous learning, knowledge, and skill acquisition as underpinned by this model will remain central to improved control, visibility and management. Therein lies the value of models such as the DFM: the availability of the model coupled with risk management and control principles, and adaptive frameworks for growth and knowledge could serve as a stepping-stone on the way to establishing an overall holistic and strategic project management framework, thereby supporting development as a continuous activity subject to long-term improvement.

Moreover, as software development becomes embedded in larger and more comprehensive processes, it becomes a more global concern that extends beyond the remit of software engineering and impacts the organisation at a number of levels. A Domain I short-term perspective appears incapable of supporting and addressing these aspects. A Domain III long-term perspective can account for knowledge through the required infra-structure supported by the adoption of the growing asset perspective, thus underpinning normal development and evolution of values, assets, and satisfaction levels (assets in themselves). This allows for the fostering and improvement of expertise, skills and competencies offering a competitive edge through value-driven and experience-based improvement.

Rather than bring into focus the accepted facets of the IS process, this paper reflects a different direction—one that considers a longer-term focus, while bringing interactions, feedbacks and accumulation of competencies into play. An important aspect is its focus on continuity and growth from a long-term perspective, *beyond the normal remit of development*, where the accumulation of intellectual assets represents continuous improvement in organisational capability and potential.

This notion appears to represent a distinct departure from the traditional Domain I focus on discrete and limited efforts viewed in isolation. It also re-emphasises human issues and attention to social and political constraints, issues and processes. Furthermore, a long-term view justifies the adoption of multiple perspectives, the reuse of knowledge and the utilisation of a dynamic perspective which underpin feedback and improvement. Taken together, this points to a move from a structured engineering to a growing design perspective; a move from discrete and project-based Domain I software engineering to a strategic, continuous and evolving Domain III discipline of software design.

This interpretation is not proposed as a final answer, but as an intermediary step—an opening of a new perspective on long-term development and improvement.

REFERENCES

1. Lawrence, B., *Designers Must Do the Modelling*. IEEE Software, 1998. 15(2): p. 30-33.
2. Reifer, D.J., *Requirements Management: The Search for Nirvana*. IEEE Software, 2000. 17(3): p. 45-47.
3. Simon, H.A., *Sciences of the Artificial*. 3 ed. 1996, Cambridge, Mass.: MIT Press.
4. Dalcher, D. *Feedback, Planning and Control- A Dynamic Relationship*. in *FEAST 2000*. 2000. Imperial College, London.
5. Weick, K.E., *The Social Psychology of Organising*. 2 ed. 1979, Reading, Mass.: Addison Wesley.
6. Lehman, M.M., *Software's Future: Managing Evolution*. IEEE Software, 1998. 15(1): p. 40-44.
7. Curtis, B., H. Krasner, and N. Iscoe, *A Field Study of the Software Design Process for Large Systems*. Communications of the ACM, 1988. 31(11): p. 1268-1287.
8. Dalcher, D., *Life Cycle Design and Management*, in *Project Management Pathways: A Practitioner's Guide*, e. al., Editor. 2002, APM Press: High Wycombe.
9. Yeh, R.T., J.D. Naumann, and R.T. Mittermir, *A Commonsense Management Model*. IEEE Software, 1991. 8(6): p. 23-33.
10. Senge, P.M., *The Fifth Discipline*. 1990, New York: Doubleday.
11. Alexander, C. and e. al., *A Pattern Language*. 1977, Oxford: Oxford University Press.
12. Dym, C.L. and P. Little, *Engineering Design: A Project Based Introduction*. 2000, New York: John Wiley.
13. Kerzner, H., *Project Management: A Systems Approach*. 7 ed. 2000: VNR.
14. Dalcher, D., *Safety, Risk and Danger: A New Dynamic Perspective*. Cutter IT Journal, 2002. 15(2): p. 23-27.
15. Gilbreath, R.D., *Working with Pulses, not Streams: Using Projects to Capture Opportunity*, in *Project Management Handbook*, W.R. King, Editor. 1988, Van Nostrand Reinhold, 1988: New York. p. 3-15.

16. Cleland, D.I., *Strategic Planning*, in *Field Guide to Project Management*, D.I. Cleland, Editor. 1998, Van Nostrand Reinhold. p. 3-12.
17. Turner, J.R., *The Handbook of Project-Based Management*. 1993, London: McGraw-Hill.
18. Ireland, L.R., *Total Customer Satisfaction*, in *Field Guide to Project Management*, D.I. Cleland, Editor. 1998, Van Nostrand Reinhold: New York. p. 351-359.
19. Lehman, M.M. *Rules and Tools for Software Evolution Planning and Management*. in *FEAST 2000*. 2000. Imperial College, London: Imperial College.
20. Gorog, M. and N.J. Smith, *Project Management for Managers*. 1999, Sylva, North Carolina: Project Management Institute.
21. Youll, D.P., *Making Software Development Visible: Effective Project Control*. 1990, Chichester: John Wiley.
22. Meredith, J.R. and S.J. Mantel, *Project Management: A Managerial Approach*. 1995, New York: John Wiley.
23. Adams, J.R. and M.E. Caldentey, *A Project-Management Model*, in *Field Guide to Project Management*, D.I. Cleland, Editor. 1998, Van Nostrand Reinhold: New York. p. 48-60.
24. Forsberg, K. and H. Mooz, *Visualising Project Management*. 1996, New York: John Wiley.
25. Chapman, C. and S. Ward, *Project Risk Management: Processes, Techniques and Insight*. 1997, Chichester: John Wiley.
26. Hall, E.M., *Managing Risk: Methods for Software Systems Development*. 1998, Reading, Mass.: Addison-Wesley.
27. Bersoff, E.H., V.D. Henderson, and S.G. Siegel, *Software Configuration Management*,. 1980, Englewood Cliffs: Prentice-Hall.
28. Cave, W.C.C. and A.B. Salisbury, *Controlling the Software Life Cycle-The Project Management Task*. IEEE Transactions on software Engineering, 1978. **SE-4**(4): p. 326-334.
29. Evans, M.W., *The Software Factory: A Fourth Generation Software Engineering Environment*. 1989, New York: John Wiley.
30. Boehm, B.W., *A Spiral Model of Software Development and Enhancement*. IEEE Computer, 1988: p. 61-72.
31. Charette, R.N., *Applications Strategies for Risk Analysis*. 1990, New York: Intertext/ McGraw-Hill.
32. Charette, R.N., *Software Engineering Environments, Concepts and Technology*. 1986, New York: Intertext/ McGraw-Hill.
33. Boehm, B.W., *Requirements that handle IKIWISI, COTS and Rapid Change*. IEEE Computer, 2000. **33**(7): p. 99-102.
34. Marion, R., *The Edge of Organisation: Chaos and Complexity Theories of formal Social Systems*. 1999, London: Sage.
35. Stacey, R.D., *Manging Chaos: Dynamic Business Strategies in an Unpredictable World*. 1992, London: Kogan Page.

DEVELOPING WEB-BASED EDUCATION USING INFORMATION SYSTEMS METHODOLOGIES

John Traxler [*]

1. INTRODUCTION

In creating useful or valuable artefacts, societies resort to a variety of paradigms (taken to mean very crudely, schools of thought, or models of acceptable intellectual or professional practice (Kuhn, 1970)) to describe the process of creation. Sometimes the model of experts (with apprentices) is used; sometimes the idea of automata or robots is used. In the first case, the nature of the creative process, for example the process of making a violin, can scarcely be articulated and it only passes from one generation to the next by osmosis. In the second case, the nature of the creative process, for example the assembling of volume cars, can not only be articulated but can also be delegated to machines (or to people acting like machines). (Many artefacts, for example novels, movies, ballets, haute cuisine, jazz, operas, gardens and Parliamentary statutes, seem to come into existence without any obvious paradigm for their creation, development or production, except trial-and-error, genius or experience.). There is apparently no paradigm to support the development of educational experiences. This paper develops the argument that information systems development methodologies can be adapted to underpin the rational and systematic development of web-based education.

---

[*] John Traxler, School of Computing and IT, University of Wolverhampton, Wolverhampton, WV1 1SB, UK

2. THE PRACTICAL NEED FOR A LEARNING TECHNOLOGY THEORY IN HIGHER EDUCATION

Historically and at the risk of caricature, education in the older English universities has worked with low student densities in relatively informal settings where students learnt from a simple linear thread of narrative and explanation provided and controlled by the lecturer, who could inspire, improvise and revise as necessary. Increasingly and especially in the new universities, this situation has given way to one involving much higher student densities in formally documented and modularised settings where lecturers provide and facilitate learning. They do this through a variety of resources, media and technologies that students can access randomly, concurrently, often unsupported at a distance and under their own control rather than that of the lecturer. (Traxler, 1993) Modules are delivered in a variety of modes; in particular, there is now a commercial and territorial imperative to investigate high technology modes. This change may be part of a wider trend, most noticeable in technology supported distance learning, of the "industrialisation" and "globalisation" of higher education of accompanied by a politically driven "massification" and "commodification" of higher education in the UK (Peters, 1998).

Many of the new media being used involve high technologies, needing large up-front investments of resources and efforts, which themselves need explicit quality assurance, increased staff development, resource control and project management. There are considerable economies of scale associated with many of these technologies but also considerable risks, increased visibility and new problems. There is now consequently a far greater need to articulate processes and techniques by which universities can develop and deploy cost-effective quality courses – hence the need for a theory, or perhaps a framework or paradigm.

3. THE ENGINEERING PARADIGM

In the area of traditional engineering - civil, mechanical, nautical *etc* - a vaguely defined set of practices, standards and tools have been fairly successful and engineering has superseded craft and cottage industries as a model for how industrialised societies meet many of their physical needs.

This vague notion of engineering has been turned into a much more specific agenda and adopted as the dominant paradigm for the development of computer software - hence, software engineering.

In the course of the software community's debate about its own identity and its attempts to develop and clarify a sustainable, explicit and viable production paradigm, an idealisation of engineering has emerged. Its characteristics, some overlapping, included

- the use of diagrams, models and notations
  - to elicit, elucidate and reason about what clients and customers require
  - to support communication amongst developers, possibly composed of large teams drawn from different disciplines

- to design and synthesise acceptable solutions composed of disparate technical elements that meet clients' requirements, perhaps functionality, delivery deadline and cost, whilst optimising other attributes, perhaps speed or efficiency
- the use of abstraction to deal with complexity and often to separate the function required from the means to deliver it
- the use of mathematics and formal reasoning to add rigour and quantity and thus reinforce rigour
- the use of prototypes, modules, components to facilitate re-use and maintenance and to reduce the commercial costs of going into production
- the use of teams, machines, tools and prototypes to increase productivity
- the use of project management techniques, especially of estimation and life cycle models to put production onto an economic footing;
- the development of quality assurance ideas and practices, such as traceability, validation, verification and testing, to confirm or enhance fitness-for-purpose
- the use of ergonomics and applied psychology, in the computing context called human-computer interaction, to include the user in the system definition
- the development of methodologies or methods to provide "tool-kits" for experienced practitioners and "cook-books" for novice ones.

These characteristics are routinely recited in any undergraduate textbook introducing software engineering (Pressman, 1994), (Somerville, 1992). Software engineers play up similarities between their activities and those of the idealised "real" engineers, whilst playing down discrepancies - such as the fact that software has no tangible raw material, and hence no worries about mass production costs and no worries about subsequent any rot or rust.

Later systems thinking, with its notions of purpose, synergy and components was incorporated into the engineering paradigm and at the same time the focus of computer professionals moved from the development of software to the engineering of information systems. In "conventional" engineering, the discipline of systems engineering eventually emerged explicitly to deal with large and complex technical systems.

4. COURSEWARE ENGINEERING

The idea of engineering as idealised above can provide a useful framework for developing self-contained episodes of learning.

The idea had been explored before in the guise of courseware engineering. However, systems thinking would have suggested that this concept was doomed from the outset. It drew the system boundary too narrowly (around a piece of software rather than the total learning experience) and committed to the "how" to teach before thinking about the "what" to teach. This was equivalent to allowing clerks to choose their individual computer systems before analysing the totality of their firm's business priorities and practices. The result was expensive multimedia software that was not integrated into courses (because of a lack of synergy with any of the other media and methods being used to teach) or was quickly discarded (because adaptation, maintenance and re-use were so difficult).

A better software development paradigm for educational multimedia software developed in isolation from its users and from other educational technologies might have been that model espoused but never clearly documented for large-scale shrink-wrapped retail software by, amongst others, Microsoft (Cusamano and Selby, 1996). This would at least have treated educational software packages as an autonomous entity, rather than a sub-system, and would have avoided the issue of attempting to define the customer where none may have actually existed in the classic project-based engineering sense.

5. LEARNING ENGINEERING

The proposed adaptation of the engineering paradigm to learning is more fundamental than the courseware engineering episode: the appropriate system to build is the total learning experience. In any formal setting such as higher education, the learning experience with the most clearly defined boundary is the course or module and its purpose or goal as a system is defined in its Aims and Objectives (or Learning Outcomes). Interestingly, software engineers would regard developing a large and complex system from a requirement composed of a paragraph of natural language as a risky undertaking, yet this is exactly what educators do when they deliver 150-hour courses on the basis of some 30 words of Aims and Objectives.

The customer for the course or module, namely the university or college hoping to take delivery of the learning experience on behalf of its students, will have markets, business practices and institutional priorities that determine how this learning experience is developed. They might want a course that can be delivered unchanged over many iterations, with a relatively high developmental effort and minimal maintenance effort. They might on the other hand want to develop the course over successive iterations with a steady input of perfective and adaptive effort. They might want a one-off prestige course developed rapidly to meet a market need. They might not have a very clearly defined course in mind at the outset of the development.

They might have the flexibility and resources to choose any of these options. They might find, on the other hand, that their freedom of choice is constrained by the curriculum area (for example, is it tracking a topic in constant flux or fixed on the eternal verities?) or by the university's liquidity (can the university afford big up-front investments of effort or do they work with the "rolling remake"?).

For the sake of this argument, we will look at larger institutions able to make a substantial investment in development and with a clear idea of the course, the "product", it wants to develop. In software engineering terms, the course is a candidate for a structured method of development. This term describes any software development that proceeds through a sequence of phases that start with the full elicitation of the customer's requirements and end with the implementation of the finished system. Classically these methods are applied to stable business data processing operations with clear-cut objectives. The absence of these conditions would not undermine the argument being made. If we chose a different scenario, it would merely be necessary to opt for the analogue of a different information systems development methodology, for example, one of the prototype-based Rapid Applications Development methods (Martin, 1991), one of the Soft Systems Methods (Patching, 1990) or one of the methods of incremental or exploratory development and delivery.

In this example, the proposed adaptation or framework avoids a premature and inappropriate rigid categorisation of courses. It avoids saying at the outset that a course should be completely web-based or that a module should be offered entirely in distance learning mode. In structured methods, these issues of delivery or implementation are routinely deferred. Instead, the framework tries to view them as quantitative issues of implementation in which a mix of technologies can be optimised, in terms of whatever parameters the customer has prioritised, possibly overall cost, once the educational purpose has been identified, articulated and refined.

5. THE ESSENCE OF AN EDUCATIONAL EXPERIENCE

One of the tenets of structured methods is that abstraction and modelling should be used to reason about the problem or the requirement before an attempt is made to solve it. If we are to proceed from analysis to design, this means reasoning about the "what" before the "how". If we can model accurately "what" the course's Aims and Objectives mean, we can check consistency and completeness. This would create an implementation-free Essential Model of the course (Yourdon, 1989).

For a new course, this essential model must be derived, in the traditions of structured methods, by some process of decomposition. This means breaking down the requirement into smaller and smaller fragments until some heuristic suggests stopping the decomposition. In the context of educational courses, perhaps these fragments are the facts, procedures, concepts and skills that are inherent in the course's Aims and Objectives. With an existing module, this essential model can be obtained by a process analogous to logicalisation in SSADM (Eva *et al,* 1994) that takes the current implementation of the module, removes any mention of sequence, schedule, and timing, and removes any references to media, technology and teaching methods. Whichever is the case, structured methods would attempt to represent the outcome of the decomposition diagrammatically. Structured methods would use the diagrams as the basis of validation and verification and then transform them into a design for the required system. In Yourdon's portrayal, the only difference between the essence of a system and the implementation of a system is that the essence can be delivered on a "perfect" and abstract technology, whilst the implementation must be delivered on a "real" and flawed technology. Developing an implementation or working on a design is the business of accepting and reconciling the various constraints and limitations of "real" technologies. In the case of developing an educational experience, this is still true. The technologies in question might include lectures, books, web sites, packages, handouts and assessments.

It would be useful in quality terms if these essential models could be explored for completeness, consistency and all the other attributes of a good functional specification. It might even be possible to generate metrics such as complexity, density, volume or coupling from this model. These would reveal something about the course's innate characteristics, for example its relative size or duration.

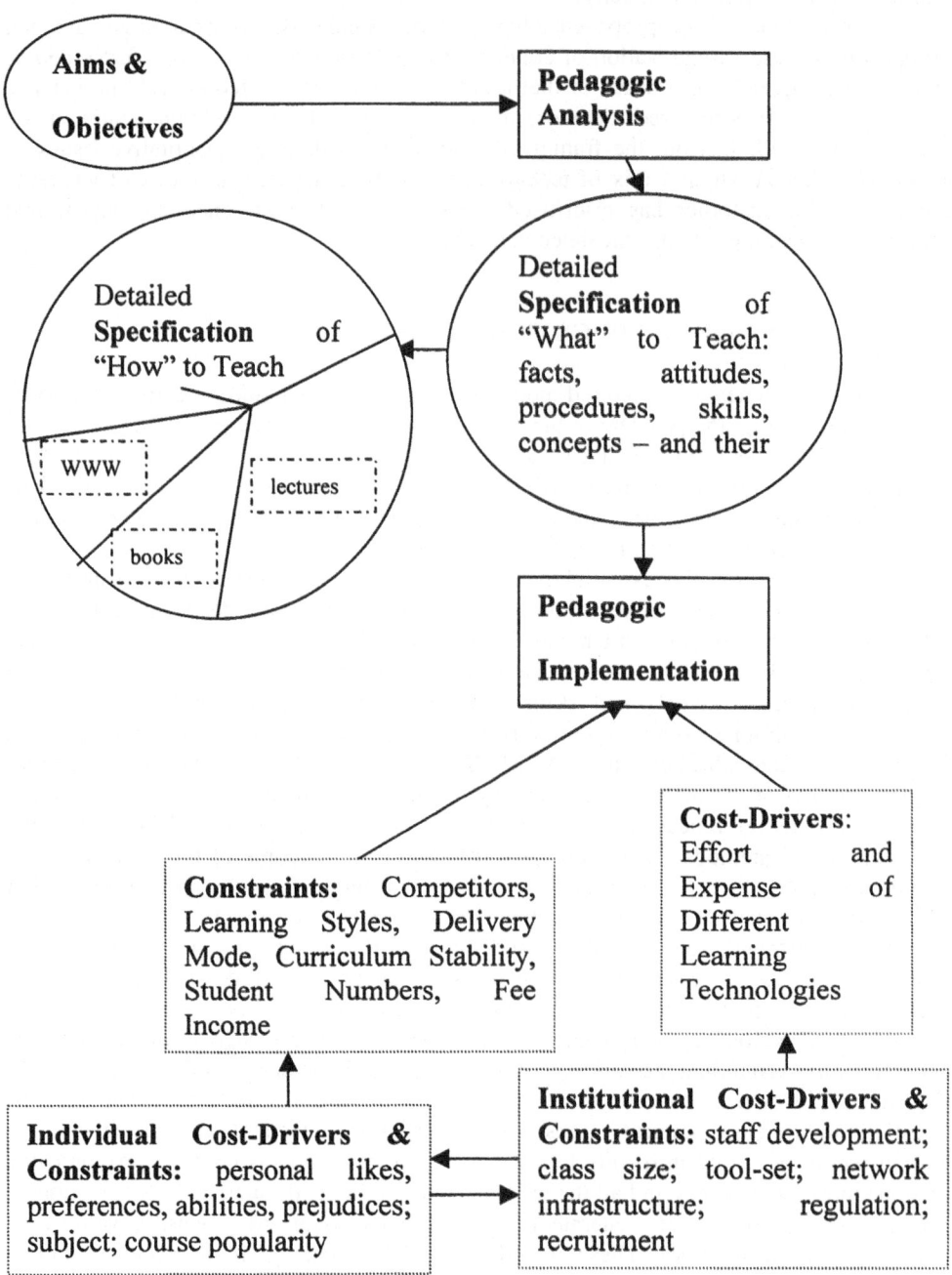

Figure 1. A Learning Development Methodology

6. THE IMPLEMENTATION OF AN EDUCATIONAL EXPERIENCE

If we proceed using Yourdon's techniques, a diagrammatic form of the essential model - namely the course decomposed into fragments of learning - acts as the basis of the implementation. The outcome of this phase is a partitioning of the essential model. In the original version, each partition becomes the responsibility of a different technology or "processor" – which could be a clerk or a computer – based on cost, reliability, performance and other non-functional requirements. In the proposed adaptation to education, the same principles could be used but the technologies will be those identified above and the criteria used will be comparable, including cost and efficiency. The stages are illustrated in Figure 1.

For each learning technology, however technically simple, efficiency must mean, in this context, the effort or cost to achieve a standard educational outcome, for example an hour's learning. Looking at each fragment of the course, for each learning technology there are associated costs. These are the costs, in terms of time and effort, to implement that particular fragment of learning. The technologies in question might include "home-grown" courseware, "home-grown" web-sites, presentations, "bought-in" courseware, books, "home-grown" notes, external web-sites and so on, and the efforts will include authoring and adaptation. The costs of teaching a fragment are not fixed for each technology. They will each depend on generic cost-drivers such as the level of interactivity, power of developmental tools, subject expertise, amount of structure, number of media, *etc.* They will also depend on the nature of the fragment (for example whether it is learning a fact, a concept, a skill or a procedure and whether it is exposition, example, rehearsal *etc*) and there may be also be cost-drivers specific to the different media. This part of the proposed framework builds on work on educational multimedia software cost-estimation and the cost-drivers associated with this form of software (Marshall 1994, 1995). There is however, a further division of cost-drivers into start-up costs (*i.e.* learning to use a particular technology to develop learning materials), authoring costs (*i.e.* actually developing the material) and maintenance costs (*i.e.* keeping the material up-to-date). This latter cost will depend on the durability or stability of the topic of each fragment.

This approach offers a more rational approach to the "media-mix" problem (that is the problem of choosing the media with which to teach a course) that has only been tackled before in a sporadic fashion (Reiser & Gagne, 1983). The proposed approach also offers a more principled approach to costs than previous work (Rumble, 1997) and consequently a more adaptable one in the face of evolving technologies.

If the operation of the various cost-drivers seems to produce too much freedom of choice in how to deliver courses or modules, there are always constraints on the design of the course. These will include designing it for all types of learner, different conceptions of teaching (Trigwell and Prosser, 1996), (Laurillard, 1993) and for specific market and institutional characteristics (stability of content, student cohort size, distance learning mode, commercial course and so on).

The constraint that the course must be accessible to different styles of learner could be applied in a variety of ways. It might mean checking that the fragments were connected in such as a way as to make it possible for learners to get from concepts to examples (if they were deductive learners) or from examples to concepts (if they were inductive learners). It might mean checking that learners could skim through summaries (if they were holists) or work on topics in depth (if they were serialists) (Richardson, 2000). Although these are only crude examples, they do suggest that hyper-linked material might have an innate superiority to print.

In practice in developing large data processing systems, Yourdon's method at this stage is used iteratively, with various data processing activities moved backwards and forwards between clerk and computer until what seems like an optimal design or perhaps a deadline is reached. In the case of education, if the various cost-drivers were ever known, a similar practice would have to be followed since design is necessarily a creative activity. There are too many implementation permutations to be eliminated systematically. If the implementation need not be "frozen" on delivery - possibly by the needs of mass-production - then this need not cause concern, since the maintenance and operation phase can include further tuning. If the course is to be tuned though, then metrics must be defined metrics by which to measure the improvements whilst delivering the course. Student retention, progression and pass-rates for the course are obvious candidate metrics.

Actually, when the proposed framework is examined critically, not only are the cost-drivers unknown (for most technologies) but they are also in constant flux as technologies improve. In addition, since they represent effort, they are personal, not global or institutional. If the proposed framework has the potential to put course development choices on more explicit and systematic basis, then the next step would be an investigation of the range of cost-drivers and constraints that individual lecturers work with.

7. CONCLUSIONS

Whilst it is interesting and productive to develop this framework further (and doing so certainly throws up many minor but intriguing insights), it is sensible at this point to conclude by discussing briefly some of the issues raised. These fall into three categories: those issues that focus on the relationship between teaching and learning on the one hand and the ideas of systems, engineering and products on the other; those that focus on the specifics of the example used here and those that clarify the potential for theory in learning technology.

Looking first at those issues that focus on the relationship between teaching and learning on the one hand and systems, engineering and products on the other, there are several reservations. There is the inference in using the engineering paradigm that learning is somehow bounded, in order to justify representing it as a system or product. Whilst learning in its most general sense is certainly not bounded, it is within most institutional settings and so this inference is a useful tool if used cautiously.

Moving onto those issues that focus on the specifics of structured methods in higher education, there was the assumption that learning could be decomposed or atomised. In some subjects, it certainly can be and this assumption would be consistent with some learning theories, particularly those favouring a Programmed Learning or an Instruction Design approach. The assumption that it could not be atomised would not invalidate the idea of engineering episodes of learning, it would necessitate using a development methodology base on a different premiss. The idea of decomposing or atomising learning might also raise the spectre of the Taylorism or "scientific management" of learning, or indeed the continued industrialisation of learning mentioned earlier, since these are based on the notion that workers' tasks can be analysed by observation and decomposition. Some current trends in English higher education might reinforce this worry but these trends are independent of theories of learning.

The extent to which the current ideas clarify the potential for theory in learning technology is limited. "Engineering" theories that prescribe best practice are judged by "market forces" - if they work, they get used – and in that sense, they may be loosely characterised as paradigms, in the sense of being schools of accepted professional practice, rather that theories in any other sense. Clearly at the most rigorous level, "engineering" is theory incapable of proof. So the question becomes, is it useful and worth refining?

8. REFERENCES

Cusamano, M. A. & Selby, R. W, *Microsoft Secrets*, HarperCollins, 0-00-255692-8, 1996

Eva, M., *SSADM Version 4: A User's Guide*, McGraw-Hill International, 0-07-707409-2, 1994

Halstead, M. H., *Elements of Software Science*, North Holland, 1977

Kuhn, T.S, *The Structure of Scientific Revolutions*, University of Chicago Press, 0-226-45804-0, 1970

Laurillard, D. *Rethinking University Teaching*, Routledge, London, 1993

Martin, J., *Rapid Application Development*, Macmillan, 0-02-946531-1, 1991

Marshall, I M, Sampson, W B, & Dugard, P I, Predicting the developmental effort of multimedia courseware, *Information and Software Technology*, 36, 5, 251 - 258, 1995

Marshall, I M, Sampson, W B, & Dugard, P I, Courseware - How much will it cost to develop?, Proceedings of 2nd All-Ireland Conference on the Teaching of Computing, Dublin, 175 - 184, 1994

Peters, O., *Learning and Teaching in Distance Education: Analyses and Interpretations from an International Perspective*, London: Kogan Page, 1998, 0-7494-2855-4

Patching, D., *Practical Soft Systems Analysis*, Pitman, 1990, 0-273-03237-2

Paulk, M. C., Curtis, B., Chrissis, M. B. & Weber, C. V., The Capability Maturity Model for Software in *Software Engineering*, eds Dorfman, M. & Thayer, R. H., IEEE Computer Press, 1997, 0-8186-7609-4

Pressman, R. S., *Software Engineering – A Practitioner's Approach*, McGraw-Hill, 1994, 0-07-707936-1

Reisner, R. A. & Gagne, R. M., *Selecting Media for Instruction*, Prentice Hall International, 1983

Richardson, J. T. E., *Researching Student Learning*, Open University Press, Buckingham, 2000, 0-335-20515-1

Rumble, G., *The Costs and Economics of Open and Distance Learning*, Kogan Page: London, 1997

Sommerville, I., *Software Engineering*, Addison-Wesley, 1992, 0-201-56529-3

Traxler, J., The Nature and Impact of Current Trends in Higher Education on the Teaching of Computer Science and Software Engineering, First All Ireland Conference on Curriculum and Computing, Dublin, 1993

Traxler, J., Courseware Engineering: a Paradigm for Courseware Development?, Technical Report 98/09, Centre for Informatics Education Research, Faculty of Maths and Computing, Open University, 1998

Trigwell, K. and Prosser, M. Congruence between intention and strategy in university science teachers' approaches to teaching, *Higher Education*, 32, 57-87

Yourdon, E, *Modern Structured Analysis*, Prentice Hall, 0-13-598632-X, 1989

TRENDS IN DEVELOPING WEB-BASED MULTIMEDIA INFORMATION SYSTEMS

Ingi Jonasson[1]

1. INTRODUCTION AND BACKGROUND

Information systems are getting more and more multimedia-based as well as network-based. Clear examples of this trend are various Internet applications for areas such as business, education and entertainment. In this paper, we present the result from an interview study directed towards trends in developing web-based multimedia applications. The focus of the study is trends in methodologies and competencies required for developing such systems. The result show that multimedia systems development is a multidisciplinary effort requiring cooperation between people with different backgrounds and specific competencies, methodologies and views of the world. Multimedia development comprises the parallel processes of software engineering, content development and project management.

Technically, interactivity and multimedia raise new challenges in several areas. For instance, new techniques are needed for storing and manipulating new data types (e.g. still pictures, videos, audio), not to mention transmission of huge amount of data over different networks. More expressive media opens up new opportunities to create effective visual applications, but building interesting multimedia applications is also a question of dramaturgy (Wiman, 2000; Laurel, 1993). Well-proven concepts from traditional storytelling, music, literature, drama, arts etc. are important as well, because multimedia presentations are fundamentally no different from any other type of human communication (Tannenbaum, 1998).

The creation of multimedia software is similar to software development in general, but particular skills are needed for the creation of the content part of the application (Tannenbaum, 1998). When using the term multimedia information system in this paper, we refer to information systems with a multimedia user interface.

---

[1] Ingi Jonasson (ingi@ida.his.se), Department of Computer Science, University of Skövde, P.O. Box 408, SE-541 28 Skövde, Sweden.

2.   INTERVIEW STUDY RESULTS

Below, we analyse the multidisciplinary dimensions of multimedia development projects and present a model for developing information systems for the web, which often are multimedia-based.

2.1 The Complexity of a Multimedia Development Project

The core idea of multimedia systems is to manage information that requires various types of media to be expressed and accessed in an efficient way. It is, for instance, reasonable to claim that one key factor in increased public Internet use is the multimedia interface offered in web-browsers. Compared to text-based systems, multimedia systems attract users by offering various types of media, such as graphics and sounds. It is common to create an interface representing an environment that the user has to investigate. The motivation to use the systems is often excitement and curiosity, even if the goal of using the systems may be learning or training. In order to gain these motivational effects in multimedia systems, substantial effort must be put into the creation of the systems' content. It may, for example, concern embedded scenarios in order to make the systems attractive and interesting to use.

In developing multimedia systems it has been common to create the major part of the content of a system before the system is taken into use. In order to create content, competencies are needed from various fields of practice. The content is often of a graphical, narrative or other aesthetic nature. Since software engineers perhaps are not the best creators of aesthetic content they have to seek content providers in museums, publishing houses, television and movie studios (Druin and Solomon, 1996). Recently, the term content industry has become common in media, referring to providers of aesthetic content of any kind.

During the interview study multimedia development appeared to be a composition of three main processes, software engineering, content development and overall management of the whole project (project management and marketing). The processes are illustrated in Figure 1.

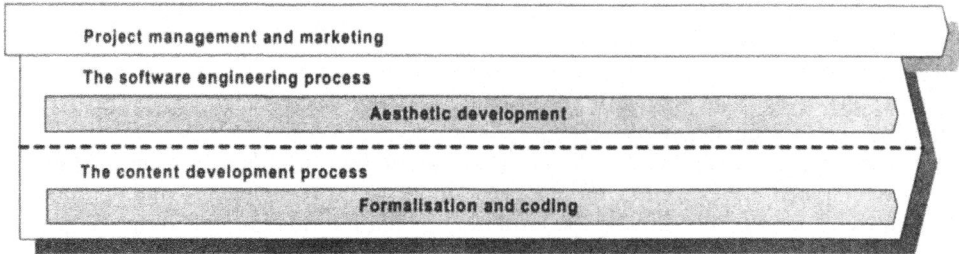

Figure 1: Developing multimedia information systems can be regarded as three parallel processes; a software engineering process and a content development process and an overall management process.

In the content development process, the content is created or retrieved from various sources and prepared for realisation. The software engineering process is about formalising and coding the content, as well as making it run as defined in the content process. Multimedia systems are based on computer software and hardware. From that point of view, their development does not differ from any other computerised information systems. Common software engineering techniques and development methods can be applied in the software process of a multimedia project (Tannenbaum, 1998). Several methods and techniques are used for management and control of software engineering processes, and the competencies needed are well defined, even if they differ slightly from methodology to methodology and from project to project. In content development, needed competencies vary as well. A large part of the content is of an aesthetic nature. The aesthetic content often exists in non-electronic forms, so the main work of the content producers in a multimedia project is often to convert the original content to an electronic form (Philips, 1997). Each competence of the aesthetic production belongs to a discipline that has its own methods [i1]. One major challenge is thus to coordinate all the involved disciplines and various ways of working into a homogeneous process.

Results of the aesthetic development can be design of graphics, audio, photos, animations, text etc. Stories or scenarios are also a matter of aesthetic development. These along with the characters in the stories can define or have great impact on the design of the system. The software engineering and content development processes can start simultaneously, which is quite common in the computer game industry [i1], but normally the content part is planned and partly developed before the software engineering process starts. The two activities normally run in parallel through implementation and testing, as the "content" process is often affected by different technical issues and vice versa.

The respondents agree on that the role of the project manager is critical for the success of a project involving multidisciplinary staff. [i6] and [i7] emphasise that it is extremely important that all parties involved truly understand both the mission of the project, and each others working conditions. These insights can only be obtained if it is made clear that the technical architecture only exists in order to make the creation of an application possible [i2]. To support this view, [i2] claims that it is critical to present clear IT and business strategies that are made clear to all parties involved.

[i3] points out the fact that programmers usually have totally different attitudes to problem-solving compared to the content staff. [i3] claims that the experienced technical staff contributes with structure, while the content staff have insights about problem solving that technicians lacks, such as introducing alternative working methods when the ordinary methods do not work.

[i4] has experience from html-based projects where the graphical design was a question of compromising from beginning to end. He finds it critical that those who develop content are capable of adapting their ideas to the technical possibilities at any time. [i7] has the same experience and even claims that it is common that programmers put restrictions on the content producers. These restrictions may be necessary because of technical considerations, but in some cases the restrictions continue even if the technical problems have been solved. In order to avoid this, [i7] has changed their approach from a technically oriented approach to a more content-oriented approach. Now, the content producers create and design solutions without reflecting on the technical solution. The technicians then do their best to implement the content. After changing their approach, the company has made the fastest technical progress ever.

Multimedia development is clearly a multidisciplinary issue. Several competencies are needed from the fields of engineering, fine arts and media as well as from project management and marketing. These issues are discussed and a list of required competencies is presented in Jonasson (2000).

2.2 A process model for developing web-systems

When it comes to methodology use, most respondents claim that there is no specific comprehensive methodology applied in their organisations. The lack of systematic ways of working is a general problem in several development projects [i3]; [i5]. [i6] and [i7] spontaneously answered that no methodology is used at all in their companies. As the interviews progressed, however, a notion of a structured way of working appeared. The structured work at the company of [i6] is inherited from the advertisement business. [i7] describes their working methodology as follows: "We do have our special way of working, even if we do not have any rigid framework for it." [i4] uses similar working methods as he used to do, while designing printed matters.

[i6] claims that as far as he knows, methodology use is not common among web-bureaus in general. He recognises the same Klondike atmosphere in the current web-business as in the advertisement business twenty years ago. He talks about how it was back then; "It was just to buy a Mac, rent a basement flat and start".

Both [i6] and [i7] point to tight time schedules as being the main problem in developing web-solutions. There is usually no time for any deep analysis.

At the time of the interview, the company of [i5] works on introducing a more methodological way of working. The work culture at the company has been very free. Fast results and creativity has been more important than structured way of working. In the long run, however, more structured working methods must be introduced in order to meet the demands of the customers, according to [i5].

Most of the respondents have described some kind of a structured framework, including various activities resulting in more or less well defined deliverables that are checked with the customer before the next activity starts, as illustrated in Figure 2. For every activity, suitable competencies, tools and techniques are picked from a 'toolbox'. The different competencies are of both technical and content nature. The outline of the model in Figure 2 is inspired by [i2], but it corresponds well to the working processes applied by the other respondents. There is of course a difference between the large Internet consultancy companies and the small web-bureaus owing to the supply of various competencies needed.

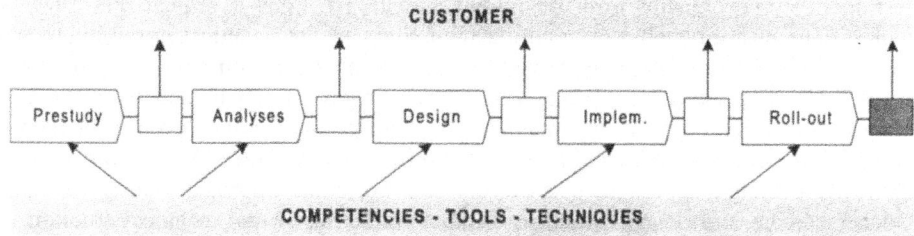

Figure 2: A simple process model for developing web-applications. Inspired by [i2]

The advantages of using a process model instead of a formal method are several. First of all, no two projects are identical, so it is usually necessary to customise the process for every new project [i2]. Developing multimedia web-systems requires competencies from several domains that use specific methods and techniques and normally, formalised methods do not address political and cultural structures, which may affect the development work. Here, specific techniques and competencies can be picked for each issue [i2] as the process model provides the possibilities to use the techniques and tools that best fit each situation. The process model also provides a good opportunity for shifting between a wide and a narrow perspective. Finally, the process model emphasises that continuous validations are carried out with the customer throughout the entire development process.

A process model such as the one described in Figure 2 gives opportunities for a flexible way of working, but there are also some disadvantages. Everything depends heavily on the project manager and his/her ability to coordinate all the required competencies and the adopted tools and techniques. Some of these tools and techniques may be incompatible. Even if the outcome works well, the question is how well it works if it must be integrated with other systems.

Both [i3] and [i5] emphasise the importance of structured working methods. If a process model as in Figure 2 is to work well, it is important to apply structured modelling and design principles, even if different working methods and techniques are mixed.

2.3 Discussion and summary of the interview study

The (r)evolution of web-based information systems has opened up for various new applications and services. Many actors have entered the scene and the late nineties were a Klondike for companies that specialised in developing web-applications. It was primarily not the established IT-consultants that provided the market with web-solutions, but new actors with new ideas and working methods, the so-called web-bureaus. The web-bureaus usually had the knowledge and skills to create good-looking web sites and sell their services. However, many of these companies failed owing to lack of knowledge regarding how to design and maintain complex information systems. It is one thing to design a spectacular web-site and another thing to integrate it with a complex back-office system.

To sum up the result of the interview study, there were some issues that seemed to be common for the respondents. All respondents referred to a lifecycle model for their products, but none of the companies where the respondents worked applied a comprehensive development methodology. Instead, process models as the one presented in Figure 2, where referred to as working models. All respondents referred to various competencies needed for developing multimedia web-systems and problems of coordinating these. The role of a project manager is pointed out as very critical for efficient coordination and cooperation between the different categories of competencies. Development lifecycles must be speeded up in order to deal with fast change rates. Finally, the contribution of multimedia technologies to the information systems domain is regarded as adequate, since visualisation and interaction makes information systems more interesting and easier to use.

The interview study referred to in this paper was carried out in autumn 2000, before the major dropout by the web-bureaus. It would be interesting to investigate what the different actors have learned from the mistakes made.

3. CONCLUDING REMARKS AND FUTURE WORK

The focus in the development of web-based information systems will in the future be more directed towards the purpose and the content of the systems. This development will result in a need for far more categories of competencies for developing future multimedia based information systems. As a consequence, development projects will be more multidisciplinary and thereby more complex, involving various models and techniques for facilitating the coordination and cooperation of the competencies involved, as well as use of domain specific methodologies.

An interesting future research issue would be to identify the factors that facilitate and hinder efficient cooperation in multimedia development projects, i.e. how should project management be carried out in these multidisciplinary development projects?

Another possible research area is to investigate if working methods from different disciplines can be integrated or harmonised in a single project. This might be related to the area of method engineering (Brinkkemper, 1996) or to the 'method in action' approach (Fitzgerald, 1998).

Finally a study of whether multimedia development projects should be primarily driven by technical or content issues is of interest.

4. REFERENCES

Brinkkemper, S., 1996, Method engineering: engineering of information systems development methods and tools, *Information and Software Technology*, 38:275-280.

Druin, A. and Solomon, C., 1996, *Designing multimedia environments for children*, Wiley, New York.

Fitzgerald, B., 1998. An empirical investigation into the adoption of systems development methodologies, *Information & Management*, **34(6)**:317-328.

Jonasson. I., 2000, *Developing the Information Systems of Tomorrow: competencies and methodologies*, M.S. thesis, Department of computer science, University of Skövde; Available from: www.ida.his.se/ida/htbin/exjobb/2000/HS-IDA-MD-00-016

Laurel, B., 1993, *Computers as theatre*. 2nd ed, Addison-Wesley, Reading.

Phillips, R., 1997 *The developer's handbook to interactive multimedia: a practical guide for educational applications*, Kogan Page, London.

Tannenbaum, R.S., 1998, *Theoretical foundations of multimedia*, Computer Science press, New York.

Wiman, B., 2000, *Att skriva manus för interaktiva medier* (in Swedish), Studentlitteratur, Lund.

APPENDIX: A BRIEF PRESENTATION OF THE INTERVIEW STUDY

The total number of interviews is seven. We have interviewed specialists representing various competencies in order to obtain different viewpoints and as a broad perspective on the development of multimedia web-solutions as possible. The interviews were carried out in the autumn of 2000. Below are the profiles of the respondents and the organisations they were working for. The identifications of the respondents are used as references in the text. More comprehensive presentation of the interview study is found in Jonasson (2000).

[i1] Graphical art director with international experience from three different companies in the computer game industry.

[i2] Strategy consultant with e-business as a special area of interest. Working for an Internet consultancy company with about 800 employees at the time of the interview.

[i3] Business consultant with decades of experience from different companies. Works for the same company as [i2].

[i4] Graphical designer with experience from designing books, records albums etc. He had worked on multimedia web-design for a period of two years at the time of the interview.

[i5] Project manager with a background as a programmer and database designer. Working for an Internet consultancy company with about 2500 employees at the time of the interview.

[i6] Managing director for a web-bureau with six employees.

[i7] The respondents in this interview where five and represented the entire staff at a web-bureau.

THE ORGANISATIONAL DEPLOYMENT OF SYSTEMS DEVELOPMENT METHODOLOGIES

Magda Huisman and Juhani Iivari[1]

1. INTRODUCTION

It is generally assumed that systems development methodologies (SDM) are used in practice (Saeki, 1998), and there exists a widespread belief that adherence to SDM is beneficial to an organisation (Fitzgerald, 1996; Hardy et al., 1995). Furthermore, organisations are facing a lot of pressure to use SDM (Fitzgerald, 1996). Despite the high investment in the development of SDM and the pressure to use it, their practical usefulness is still a controversial issue (Fitzgerald, 1996; Introna and Whitley, 1997; Nandhakumar and Avison, 1999). While many organisations claim that they do not use any methodologies (Hardy et al., 1995; Chatzoglou and Macaulay, 1996; Fitzgerald, 1998), others are using it with positive results (Chatzoglou and Macauly, 1996; Rahim et al., 1998). Apart from this, we do not know why SDM are used or not used, or what factors influence its use and effectiveness.

The purpose of this paper is to investigate what factors influence the organisational deployment of SDM. Researchers emphasise that a distinction must be made between the adoption and acquisition of technology at the organisational level and its adoption and implementation at the individual level (Fichman, 1992; McChesney and Glass, 1993; Senn and Wynekoop, 1995; Dietrich et al., 1997; Lai and Guynes, 1997). In terms of Rogers (1995), SDM are contingent innovations with organisations as primary adopting units and individuals as secondary adopting units. In this paper we will study the deployment of SDM at the organisational level. More specifically, we will study the deployment of SDM among primary adopters (IS departments). The deployment of SDM

---

[1] Magda Huisman, Department of Computer Science and Information Systems, Potchefstroom University for CHE, Private Bag X6001, Potchefstroom, 2531, South Africa, Fax: +27 18 2992557, Phone: +27 18 2992537, rkwhmh@puknet.puk.ac.za.
Juhani Iivari, Department of Information Processing Science, University of Oulu, P.O. Box 3000, 90014 Oulun yliopisto, Finland, Fax: +358 8 5531890, Phone: +358 8 5531922, juhani.iivari@oulu.fi.

Information Systems Development: Advances in Methodologies, Components, and Management
Edited by Kirikova et al., Kluwer Academic/Plenum Publishers, 2002

87

among secondary adopters (individual system developers) is reported in Huisman and Iivari, 2002.

2. CONCEPTUAL RESEARCH MODEL AND RESEARCH HYPOTHESES

2.1 Theoretical Background

Most of the previous research into SDM did not have any theoretical orientation but the idea had been just to report the state of use of SDM and techniques in purely descriptive terms, e.g. Hardy *et al.* (1995), and Chatzoglou and Macaulay (1996). In general terms the present work is influenced by the diffusion of innovations (DOI) theory (Rogers, 1995), which is becoming an increasingly popular reference theory for empirical studies of information technologies (Fichman, 1992; Prescott and Conger, 1995). More specifically, our work is based on the IS implementation model suggested by Kwon and Zmud (1987). They combined IS implementation research and the DOI theory. The conceptual model for our work is presented in Figure 1.

The DOI theory has been criticised. Fichman (1992) points out that it has mainly addressed individual adoption of relatively simple innovations. It is obvious that SDM are fairly complex innovations. They are technologies of Type 2, according to the IT diffusion research framework presented by Fichman (1992), which are characterised by a high knowledge burden or high user interdependencies. This means that our study tests the validity of DOI theory outside its major focus area. Therefore the detailed hypotheses concerning the deployment of SDM, derived from the classical DOI theory, are quite tentative.

As pointed above there is not much theoretically oriented empirical research into the adoption of SDM, on which we can draw in our discussion of detailed hypotheses. To compensate this we mainly use existing empirical research on the adoption of CASE technology. There are two reasons for this. Firstly, CASE tools represent relatively complex technologies that are contingent innovations just as SDM. Secondly, the methodology companionship of CASE tools (Vessey *et al.*, 1992) implies that their adoption includes a significant aspect of SDM.

2.2 The Innovation: Systems development methodologies

Trying to define SDM is no easy task. There is no universally accepted, rigorous and concise definition of SDM (Avison and Fitzgerald, 1995; Wynekoop and Russo, 1997; Iivari *et al.,* 1999). Avison and Fitzgerald (1995) argue that the term methodology is a wider concept than the term method, as it has certain characteristics that are not implied by method, i.e. the inclusion of a philosophical view. We use the term "methodology" to cover the totality of systems development approaches (e.g. structured approach, object-oriented approach), process models (e.g. linear life-cycle, spiral models), specific methods (e.g. IE, OMT, UML) and specific techniques.

Each innovation has the following characteristics: perceived relative advantage, compatibility, complexity, trialability and observability (Rogers, 1995). In this paper we study the influence of relative advantage, compatibility and complexity on the organisational deployment of SDM.

Relative advantage is the degree to which an innovation is perceived as being better than the idea it supersedes (Rogers, 1995). After a decade's intensive research on TAM (Davis *et al.*, 1989) in particular, there is significant empirical evidence that relative advantage or perceived usefulness is positively related to innovation use, even though Iivari (1996) discovered it not to have any significant relationship with CASE usage. This overwhelming evidence leads us to the following hypothesis:

- H1: There is a positive relationship between relative advantage and the organisational deployment of SDM.

Complexity is the degree to which an innovation is perceived as difficult to understand and use (Rogers, 1995). It is generally believed that the more complex an individual perceives an innovation to be before using it, the less likely it is that the innovation will be adopted and implemented. Although perceived complexity has generally been assumed to be negatively related to the adoption of innovations (Davis *et al.*, 1989; Moore and Benbasat, 1991; Rogers, 1995), the empirical results regarding the relationship between perceived ease of use (or perceived complexity) and use has been inconclusive (Gefen and Straub, 2000). This is also the case in the adoption of CASE tools (McChesney and Glass, 1993; Iivari, 1996). Despite the inconclusive empirical evidence, we postulate in accordance with the DOI theory (Rogers, 1995) and TAM (Davis *et al.*, 1989) the following:

- H2: There is a negative relationship between complexity and the organisational deployment of SDM.

Compatibility is the degree to which an innovation is perceived as being consistent with the existing values, past experiences and needs of potential adopters, and it is positively related to innovation use (Rogers, 1995). Compatibility is sometimes described as the "fit" between an innovation and a particular context, which implies that an innovation must match its context in order to be effective. McChesney and Glass (1993) remark that a detailed assessment should be made of the "fit" between CASE methodology and the systems development tasks it is designed to support, when studying the acceptance of CASE methodology. Iivari (1996) also found some evidence for the significance of compatibility for CASE usage. Following the DOI theory, we postulate the next hypothesis as follows:

- H3: There is a positive relationship between compatibility and the organisational deployment of SDM.

2.3 Innovation Diffusion Process

Our main focus is on the deployment of SDM, which is related to the implementation and confirmation stages of the innovation-decision process as described by Rogers (1995). Since deployment is part of the post-implementation stage, we use the description of McChesney and Glass (1993) in our conceptual research model. After implementation, deployment will follow, which in turn is followed by incorporation. We visualise deployment as two stages, namely use followed by acceptance. (See Fig.1)

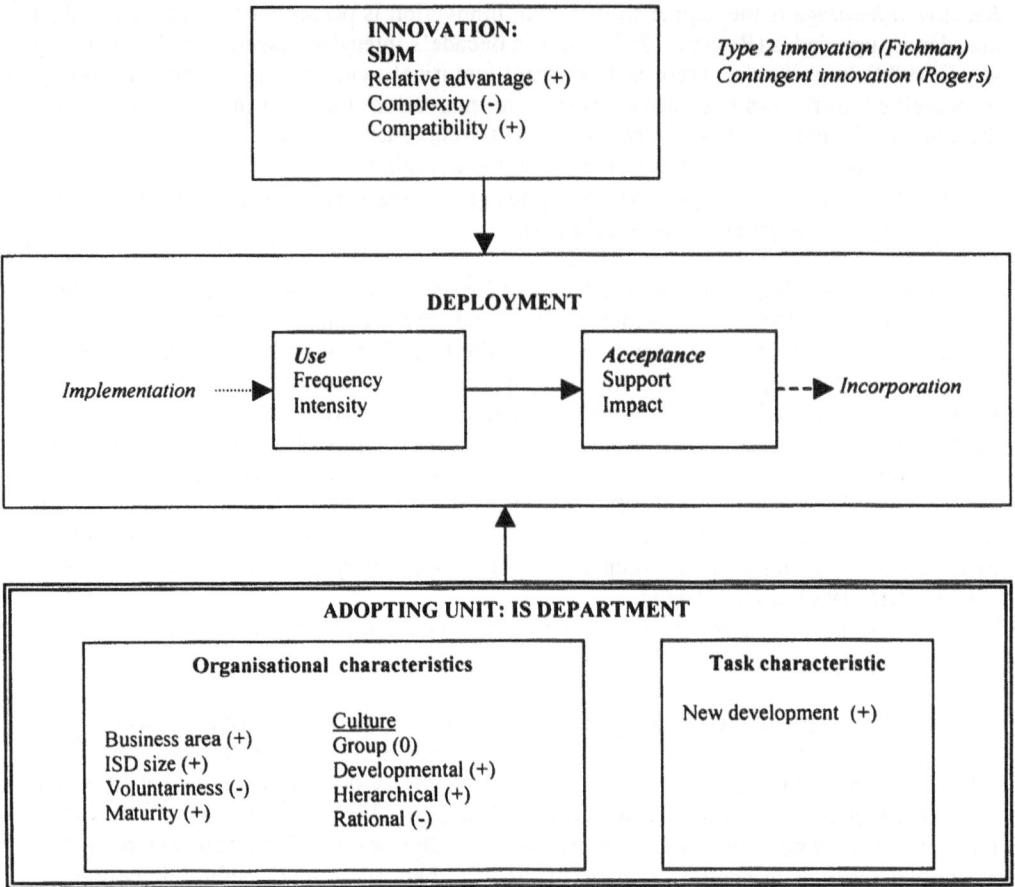

Figure1. Conceptual research model for the organisational deployment of systems development methodologies.

Use will be studied along two dimensions, namely frequency of use and intensity of use (McChesney and Glass, 1993). The acceptance of SDM will be studied from two perspectives, namely their impact on systems development and the perceived support it provides (McChesney and Glass, 1993). When studying the impact of SDM, we will focus on their impact on both the developed system and the development process (Heineman *et al.*, 1994; Wynekoop and Russo, 1997). The support that SDM provide will be studied along three dimensions, namely the perceived support as production technology, control technology, and cognitive/co-operative technology (Henderson and Cooprider, 1990).

2.4 Organisational Characteristics

Business area: Currie (1996) argues that business types have considerable influence on the structure of the IS activities in an organisation. This view is supported by Premkumar

(1992) who claims that certain types of organisations, particularly in the finance sector, have a greater stake in IS operations. These organisations view IS as a strategic device and major investments in IS are made. This is confirmed by Rahim *et al.* (1998), who found that the nature of the business sector was significantly related to methodology use. They found that the majority of organisations that used SDM were government agencies and were primarily engaged in the administrative services and finance sectors. Therefore we can expect that the deployment of SDM will differ among business areas. This leads to our next hypothesis:

- H4: The organisational deployment of SDM differs by business area.

Size of the ISD: Larger organisations are more likely to be adopters of technological innovations (Ettlie and Rubenstein, 1987; Tornatzky and Fleischer, 1990; Damanpour, 1992; Rogers, 1995). Size is probably a surrogate measure of several factors that influence the diffusion of innovations like total resources, slack resources and organisational structure (Rogers, 1995). Research indicates that organisational size is a significant predictor of CASE diffusion (Huff, 1992; Rowe, 1993; Rai, 1995). Grover *et al.* (1997) found strong support for the hypothesis postulated by Swanson (1994) that early adoption of Type 1 innovations is positively related to the size of the IS unit. Rai and Howard (1994) found a positive relationship between the natural logarithm of size of an IS department and the degree of CASE usage. The size of the IS department is positively related to CASE usage up to a point, after which the rate of increase diminishes. On the other hand, very large firms may find it difficult to implement new innovations due to the high investment in existing systems development practices. This leads to our next hypothesis:

- H5: There is a positive relationship between the size of an IS department and the organisational deployment of SDM.

Voluntariness is the degree to which an innovation is perceived as being voluntary or of free will (Moore and Benbasat, 1991). When we consider contingent innovations, the secondary adopters rarely have complete autonomy regarding their adoption and use in the workplace. Furthermore, SDM are complex innovations, and unless management declares their use mandatory, systems developers will have difficulty to fit them into their tight schedule. Iivari (1996) found strong support for the negative influence of voluntariness on CASE usage, and Green and Hevner (1999) report a strong negative influence of voluntariness on the use of the Personal Software Process innovation. This leads to our next hypothesis:

- H6: There is a negative relationship between voluntariness and the organisational deployment of SDM.

Maturity if ISD: Humphrey (1989) notes that when an organisation is at the initial level of maturity, it typically operates without formalised procedures, cost estimates, and project plans. He also claims that when an organisation does not have an effective project planning system, it may be difficult or even impossible to introduce advanced methods and technology. Much of the later literature has echoed these views. Curtis (1992) and McChesney and Glass (1993), for example, contend that when organisations in their earlier stages of maturity deploy software engineering technologies, the acceptance, use and impact of these technologies is very low. Sumner (1995) notes that the moderate to low success with CASE technology overall can be attributed to the relatively low process maturity of the software development organisations that implemented it. These organisations are not sufficiently mature to implement and use the technology effectively. Furthermore, Huisman and Iivari (2000) found a significant positive relationship between system development methodology use and the maturity stage of Information system departments. Although not statistically significant, they also found indications that as organisations mature, they perceive SDM to become more useful as production, control and cognitive/co-operation technology, and that the impact of SDM on the developed system and the development process increases as maturity increases. We can therefore expect that organisations with high process maturity will implement and use software-engineering technologies more effectively. This leads to our next hypothesis:

- H7: There is a positive relationship between the maturity stage of an IS department and the organisational deployment of SDM.

Organisational culture: The competing values framework is based on two distinctions: change vs. stability and internal focus vs. external focus. Change emphasises flexibility and spontaneity, whereas stability focuses on control, continuity and order. Internal focus underlines integration and maintenance of the socio-technical system, whereas the external focus emphasises competition and interaction with the organisational environment (Denison and Spreitzer, 1991). The opposite ends of each of these dimensions pose competing and conflicting demands on the organisation.

Based on the two dimensions, one can distinguish between four culture types. The *group culture* (change and internal focus) is primary concerned with human relations and flexibility. Belonging, trust and participation are core values. Effectiveness criteria include the development of human potential and member commitment. The *developmental culture* (change and external focus) is future-oriented, considering what might be. The effectiveness criteria emphasise growth, resource acquisition, creativity and adaptation to the external environment. The *rational culture* (stability and external focus) is very achievement-oriented, focusing on productivity, efficiency and goal achievement. The *hierarchical culture* (stability and internal focus) is oriented toward security, order and routinisation. It emphasises control, stability and efficiency through following regulations (Quinn and Kimberly, 1984; Denison and Spreitzer, 1991).

Iivari and Huisman (2001) used the competing values framework to study the relationship between organisational culture and the deployment of SDM. They found that the hierarchical culture was most consistently associated with the deployment of SDM, when assessed by developers. They also found the developmental culture to have some positive associations with methodology deployment, although not systematically. Furthermore, they identified negative associations between the rational culture and methodology support and impact, when assessed by managers. No relation was found between the group culture and the deployment of SDM. This leads to the following hypotheses:

- H8: There is no relationship between the perceptions of group culture and the organisational deployment of SDM.
- H9: There is a positive relationship between the perceptions of developmental culture and the organisational deployment of SDM.
- H10: There is a positive relationship between the perceptions of hierarchical culture and the organisational deployment of SDM.
- H11: There is a negative relationship between the perceptions of rational culture and the organisational deployment of SDM.

2.5 Task Characteristics

SDM are primarily used in the development of new systems from scratch, and are often not as helpful in the enhancement of operational systems (Peacham, 1985). Rowe (1993) found that departments using CASE tools tend to spend more time on the development of new systems than those departments without any CASE tools. SDM are most effective for analysing the functionality of a newly developed system (Isoda *et al.*, 1995). This leads to the formulation of the following two hypotheses:

- H12: There is a positive relationship between the amount of time an IS department spends on the development of new systems and the organisational deployment of SDM.

3. RESEARCH DESIGN

3.1 Survey

This study is part of a larger survey on the deployment of SDM in South Africa, which was conducted between July and October 1999. The 1999 IT Users Handbook (the most comprehensive reference guide to the IT industry in South Africa) was used and the 443 listed organisations were contacted via telephone to determine if they were willing to participate in the study. 213 organisations agreed to take part. A package of questionnaires was sent to a contact person in each organisation who distributed it. This package consisted of one questionnaire to be answered by the IS manager, and a number of questionnaires to be answered by individual systems developers in the organisation.[2] The response rate of the survey was as follows: 83 organisations (39%), 234 developers (26%) and 73 managers (34%) responded. The responses came from organisations representing a variety of business areas, manufacturing (33%) and finance/banking/insurance (15%) as the major ones. At the individual level the respondents reported considerable experience in SD, 22% between 3 and 5 years, 23% between 5-19 years and 38% more than 10 years.

3.2 Measurement of Dependent Variable: Deployment

In order to measure how frequently SDM are used in an IS department, managers were asked to indicate the proportion of projects that are developed in the IS department by applying systems development methodology knowledge, and the proportion of people in the IS department that apply systems development knowledge regularly. Frequency of use was calculated as the average of the above two items. Reliability analysis was performed and the resulting Cronbach alpha was 0.89.

In order to measure how intensively SDM are used in the IS department, maximum intensity of method use was calculated. The developer and manager data were merged (i.e. each respondent from an organisation had an equal weight), and the average of the responses that indicate that a certain method was in use in that specific organisation was

---

[2] The questionnaires are available from the first author on request.

calculated. The maximum of all the indicated methods, both commercial and in-house developed, was selected.

The perceived methodology support as production, control and cognitive/co-operative technology, were based on the average of the manager perceptions and the aggregated developer perceptions. This was also the case for the measurement of the perceived impact of methodologies on the quality of the system, and the quality and productivity of the development process (Iivari and Huisman, 2001).

One might also ask what independent variables explain organisational deployment in total. In order to describe deployment with the minimum number of variables, we performed factor analysis on the seven aspects of deployment. The factor analysis resulted in two factors, which correspond to the use (reliability = 0.65) and acceptance (reliability = 0.95) stages of deployment as illustrated in our conceptual research model. The values for the two factors were computed as the average of the items included in the factor.

3.3 Measurement of Independent Variables

The measurement of the independent variables is summarised in Table 1.

Table 1. Measurement of independent variables

Type	Characteristic	Respondent	Measurement	Reliability
Innovation Characteristics	Relative advantage	Developers	5 items (adapted from Moore and Benbasat, 1991)	0.94
	Complexity	Developers	3 items (adapted from Moore and Benbasat, 1991)	0.88
	Compatibility	Developers	3 items (adapted from Moore and Benbasat, 1991)	0.91
Task Characteristics	Time spent on the development of new applications	Manager	%	-
Organisational Characteristics	Business area	Manager	List presented	-
	ISD size	Manager	Total number of people employed in the organisation's ISD at all locations	-
	Voluntariness	Developers	2 items (adapted from Moore and Benbasat, 1991)	0.82
	Maturity	Manager	38 items (Humphrey and Sweet, 1987)	-
	Group culture	Developers	3 items (Yeung et al., 1991)	0.68
	Developmental culture	Developers	3 items (Yeung et al., 1991)	0.69
	Hierarchical culture	Developers	2 items (Yeung et al., 1991)	0.71
	Rational culture	Developers	2 items (Yeung et al., 1991)	0.71

3.3.1   Maturity of IS Department

The 38 asterisked items and the algorithm in the instrument developed by Humphrey and Sweet (1987) were used to calculate maturity levels 1 and 2 (items 1-9, 11, 18 and 24), maturity level 3 (items 10, 12-17, 19-23, 33 and 34) and maturity level 4 (items 25-

32 and 35-38). Dekleva and Drehmer (1997) found that in practice organisations do follow the prescribed practices of the CMM, but not precisely in the optimal manner as described by Humphrey. They found that heterogeneous items are grouped within a maturity level and that the different maturity levels usually overlap. Therefore, besides the maturity level, the percentage of key practices marked by the respondents at each of the maturity levels was also calculated. This was used to classify organisations into different stages of maturity.

The vast majority (92%) of organisations was at the initial maturity level. K-means cluster analysis was conducted, using the percentages of the key practices marked at the different maturity levels as clustering variables. Experimenting with an alternative number of clusters (3-5), a four-cluster solution turned out as the easiest to interpret. This gave an indication of the different maturity stages of the organisations. The results are given in Table 2.

The results show that all the clusters roughly follow the hierarchy of levels so that they exercise a higher percentage of key practices at the lower level than at the higher levels (cluster 4 is the only exception where the percentage of key practices at the level of 4 is higher than at the level of 3). At the same time, the results confirm the findings of Dekleva and Drehmer (1997) that organisations do not follow the CMM prescriptions precisely. Instead of devoting their efforts to achieve a certain maturity level, they skip some key practices of a certain maturity level and proceed with practices of the next maturity levels in their own fashion. There is nevertheless considerable variation in the degree to which they follow the key practices. Organisations in cluster 3, although categorised at maturity level 1 according to CMM, on the average exercise more than 60% of the key practices of maturity levels 2 and 3. Even more interestingly, some organisations in cluster 4, also categorised at maturity level 1 according to CMM, perform more than 70% of the key practices of maturity level 3, and more than 80% of the key practices of maturity levels 2 and 4. In the analysis that follows we used the stages of maturity as independent variables.

Table 2. Results of cluster analysis

	Cluster 1 (n=14)	Cluster2 (n=29)	Cluster 3 (n=17)	Cluster4 (n=9)
% of level 2 key practices	15.5	56.3	65.7	84.3
% of level 3 key practices	12.8	30.8	61.8	71.4
% of level 4 key practices	3.6	20.7	39.2	82.4
Interpretation of clusters (maturity stages)	Low maturity	Medium level 2	Medium level 3	High maturity

3.4   Data Analysis

Data analysis was performed using Statistica (version 6) software. In order to test whether organisational deployment of SDM differs by business area, ANOVA/MANOVA analysis was performed. In the second analysis the seven different aspects of deployment were treated separately as the dependent variables. To identify the most important independent variables that explain the dependent variables, best subset multiple regression analysis was performed. A third analysis was conducted, where the two stages of deployment (resulting from the factor analysis), namely use and acceptance were treated separately as the dependent variables.

Multiple regression analysis assumes interval or ratio scale measurement, linearity, homoscedasticity, i.e. the constancy of the residual terms across the values of the predictor variables, independence of residuals, normality of residuals, and no multicollinearity (Hair *et al.*, 1992). These assumptions were tested and no violations were detected.

4. RESULTS

4.1 ANOVA/MANOVA analysis

In order to determine if systems development methodology use differs significantly between IS department in different business areas, ANOVA/MANOVA analysis was performed. Software houses/Software consulting firms report the highest values for frequency of use and intensity of methodology use. They are followed by IS departments in the Retail/Wholesale section, who reported the second highest values. The results indicate that methodology use differs significantly between IS departments in different business areas. In order to use business area in the regression analysis, it was ranked on an ordinal scale from 1 (lowest use) to 5 (highest use) as follows: Administrative services (1), Manufacturing (2), Finance/Banking/Insurance (3), Retail/Wholesale (4), and Software house/Software manufacturing (5).

4.2 Regression Analysis with the seven different aspects of deployment treated as dependent variables

The results of the regression analysis with the seven different aspects of deployment treated separately as the dependent variable are presented in Table 3.

4.3 Regression Analysis with Use and Acceptance as the dependent variables

The results of the regression analysis with use and acceptance as the dependent variables are presented in Table 4.

5. DISCUSSION AND FINAL COMMENTS

In this paper our purpose was to determine the factors that explain the organisational deployment of SDM. To analyse deployment in total, we performed factor analysis on the various aspects of deployment. This resulted in two factors that correspond with our description of deployment in our conceptual research model, namely "use" and "acceptance".

Table 3. Results of regression analysis

N=58	Frequency of use	Intensity of use	Support: Production technology	Support: Control technology	Support: Cognitive/Co-operation technology	Impact: Developed System	Impact: Development process
Relative advantage			0.59**	0.87***	0.67**	0.56*	0.78***
Complexity							
Compatibility		-0.18	-0.38'	-0.67**	-0.34	-0.31	-0.46*
Business area	0.39***		0.25*	0.24*			
IS department size	-0.19		-0.34*	-0.22'		-0.37**	-0.21
Voluntariness	-0.33**	-0.25*					-0.22'
Maturity	0.30**	0.26*		-0.15			
Group culture	0.19						
Development culture		0.25	0.16				0.20
Hierarchical culture		0.22	0.53***	0.45**	0.40**	0.34*	
Rational culture	-0.39**	-0.66***	-0.39*	-0.17	-0.25'	-0.17	-0.24
Time: New development	0.35**	0.20	0.25*	0.15	0.24'	0.14	
R	0.75	0.67	0.74	0.72	0.59	0.62	0.65
R²	0.56	0.45	0.55	0.51	0.35	0.38	0.42
Adjusted R²	0.50	0.37	0.47	0.43	0.28	0.30	0.35
F	8.83***	5.62***	6.98***	6.02***	5.18***	4.88***	5.75***

' $p \le 0.10$ * $p \le 0.05$ ** $p \le 0.01$ *** $p \le 0.001$

Table 4. Results of regression analysis

N=58	Use	Acceptance
Relative advantage		0.66**
Complexity		
Compatibility		-0.38'
Business area	0.26**	0.20
IS department size		-0.31*
Voluntariness	-0.30**	
Maturity	0.29**	
Group culture	0.20'	
Development culture		
Hierarchical culture	0.16	0.43**
Rational culture	-0.59***	-0.27*
Time: New development	0.36**	0.21
R	0.77	0.67
R²	0.59	0.45
Adjusted R²	0.53	0.36
F	9.86***	5.19***

' $p \le 0.10$ * $p \le 0.05$ ** $p \le 0.01$ *** $p \le 0.001$

One of our aims was to test the validity of DOI theory outside its major focus area. The results are very interesting. There is a strong positive relationship between relative advantage and the acceptance of SDM, but no relationship with its use. One explanation might be that relative advantage reflects the perceived support provided by SDM and their impact on the quality of the development system and the development process. Contrary to our hypotheses, no relation was found between complexity and the deployment of SDM. Very surprisingly, the relationship between compatibility and acceptance was negative (at the level of $p < 0.1$), while we expected it to be positive. An explanation may be that the respondents perceive SDM to be compatible with their current way of working; they neither do see them to provide essential support nor to lead to positive impacts.

The results indicate that the business area of an IS department is related to the frequency of use of SDM at the organisational level. More specifically, deployment of SDM is highest in IS departments in Software houses and Retail/Wholesale companies.

In contrast to our hypothesis where we postulated that there is a positive relationship between the size of an IS department and the deployment of SDM, the results indicate the opposite. The size of an IS department was not related to the use of SDM, but it was negatively related to the acceptance of SDM. This finding is quite surprising. Often SDM are perceived as mechanisms devised especially for large organisations. Our results suggest that they are used in large organisations as they are in smaller ones, but their perceived support as production and control technology, and their impact on the quality of the developed system is perceived significantly lower in large than in small organisations.

The results indicate that voluntariness is negatively associated with methodology use, but not with acceptance.

The maturity of an IS department is significantly related to systems methodology use but not with acceptance. This is consistent with the findings of Huisman and Iivari, 2000.

Regarding the culture of an IS department, the results indicate that hierarchical culture has a very significant positive relationship with systems development methodology acceptance, and that rational culture is negatively related to both systems development methodology use and acceptance. These findings are also consistent with the results of Iivari and Huisman (2001). To some extent contrary to the findings of Iivari and Huisman (2001), the developmental culture was not discovered to be associated with methodology deployment.

Furthermore, the time that an IS department spends on the development of new applications is positively related to the use of SDM.

6.   REFERENCES

Avison, D.E., and Fitzgerald, G., 1995, *Information Systems Development: Methodologies, Techniques and Tools*, McGraw-Hill Publishing Company, Berkshire, England.

Chatzoglou, P.D., and Macaullay, L.A., 1996, Requirements capture and IS methodologies, *Information Systems Journal*, Vol. 6, pp. 209-225.

Currie, W.L., 1996, Organisational structure and the use of information technology: preliminary findings of a survey in the private and public sector, *International Journal of Information Management*, Vol. 16, No. 1, pp. 51-64.

Curtis, B., 1992, The Case for Process, in: *The impact of computer supported technologies on Information Systems Development*, Kendall K.E. et al. (eds.), Elsevier Science Publishers B.V., North Holland, pp. 333-343.

Damapour, F., 1992, Organizational size and innovation, *Organizational Studies*, Vol. 13, pp. 375-402.

Davis, F.D., Bagozzi, R. P., and Warshaw, P.R., 1989, User acceptance of computer technology: A comparison of two theoretical models, *Management Science*, Vol. 35, No. 8, pp. 982-1003.

Dekleva, S., and Drehmer, D., 1997, Measuring Software Engineering Evolution: A Rasch Calibration, *Information Systems Research*, Vol. 8, No. 1, pp. 95-104.

Denison, D.R., and Spreitzer, G.M., 1991, Organizational culture and organizational development: A competing values approach, in: *Research In Organizational Change and Development*, Woodman, R.W. and Pasmore, W.A (eds.), Vol. 5, JAI Press Inc, Greenwich, CT, pp. 1-21.

Dietrich, G.B., Walz, D.B., and Wynekoop, J.L., 1997, The failure of SDT Diffusion: A Case for Mass Customization, *IEEE Transactions on Engineering Management*, Vol. 44, No. 4, pp. 390-398.

Ettlie, J.E., and Rubenstein, A.H., 1987, Firm size and product innovation, *Journal of Product Innovation*, Vol. 7, pp. 89-108.

Fichman, R.G., 1992, Information Technology Diffusion: A review of empirical research, in: *Proceedings of the Thirteenth International Conference on Information Systems*, DeGross, J.I., Becker, J.D. and Elam, J.J. (eds.), Dallas, TX, pp. 195-206.

Fitzgerald, B., 1996, Formalized systems development methodologies: a critical perspective, *Information Systems Journal*, Vol. 6, pp. 3-23.

Fitzgerald, B., 1998, An empirical investigation into the adoption of systems development methodologies, *Information & Management*, Vol. 34 , pp. 317-328.

Gefen, D., and Straub, D., 2000, The relative importance of perceived ease of use in IS Adoption: A study of E-Commerce adoption, *Journal of the Association for Information Systems*, Vol. 1, http://jais.isworld.org/articles/1-8/article.htm.

Green, G.C., and Hevner, A.R., 1999, Perceived Control of Software Developers and Its Impact on the Successful Diffusion of Information Technology, SEI Technical Report CMU/SEI-98-SR-013, Carnegie Mellon, Software Engineering Institute, Pittsburgh, PA.

Grover, V., Fiedler, K., and Teng, J., 1997, Empirical Evidence on Swanson's Tri-Core Model of Information Systems Innovation, *Information Systems Research*, Vol. 8, No. 3, pp. 273-287.

Hair, J.F., Anderson, R.E., Tatham, R.L., and Black, W.C., 1992, *Multivariate Data Analysis with Readings*, Macmillan, New York.

Hardy, C.J., Thompson, J.B., and Edwards, H.M., 1995, The use, limitations and customization of structured systems development methods in the United Kingdom, *Information and Software Technology*, Vol. 37, No. 9, pp. 467-477.

Heineman, G.T., Botsford, J.E., Caldiera, G., Kaiser, G.E., Kellner, M.I., and Madhavji, N.H., 1994, Emerging technologies that support a software process life cycle, *IBM Systems Journal*, Vol. 33, No. 3, pp. 501-529.

Henderson, J.C., and Cooprider, J.G., 1990, Dimensions of I/S planning and design aids: A functional model of CASE technology, *Information Systems Research*, Vol. 1, No. 3, pp. 227-254.

Huff, C.C., 1992, Elements of a realistic CASE tool adoption budget, *Communications of the ACM*, Vol.35, No.4, pp. 45-54.

Huisman, M., and Iivari, J., 2000, Perceived maturity of IS departments and the deployment of systems development methodologies, in: *Proceedings of Conference on Software: Theory and Practice*, Feng, Y., Notkin, D., and Gaudel, M. (eds.), 16th IFIP World Computer Conference, 21-25 August, 2000, Beijing, China, pp. 135-144.

Huisman, M., and Iivari, J., 2002, The individual deployment of systems development methodologies, *Lecture Notes in Computer Science*, Springer, Vol. 2348, pp.134-150.

Humphrey, W.S., 1989, *Managing the Software Process*, Addison-Wesley, Reading, Massachusetts.

Humphrey, W.S., and Sweet, W.L., 1987, A method for assessing the software engineering capability of contractors, SEI Technical Report CMU/SEI-78-TR-23, Carnegie Mellon University.

Iivari J., 1996, Why are CASE Tools not used?, *Communications of the ACM*, Vol.39, No. 10, pp. 94-103.

Iivari, J., Hirscheim, R., and Klein, H.K., 1999, Beyond Methodologies: Keeping up with Information Systems Development Approaches through Dynamic Classification, in: *Proceedings of the 32nd Hawain International Conference on Systems Sciences*, pp.1-10.

Iivari, J. and Huisman, M., 2001, The relationship between organisational culture and the deployment of systems development methodologies, *Lecture Notes in Computer Science*, Vol. 2068, pp. 234-250.

Introna, L.D., and Whitley, E.A., 1997, Against methodism: Exploring the limits of methods, *Information Technology & People*, Vol.10, No. 1, pp. 31-45.

Isoda, S., Yamamoto, S., Kuroki, H., and Oka, A., 1995, Evaluation and Introduction of the Structured Methodology and a CASE Tool, *Journal of Systems Software*, Vol.28, No.1, pp. 49-58.

Kwon, T.H., and Zmud, R.W., 1987, Unifying the Fragmented Models of Information Systems Implementation, in: *Critical Issues in Information Systems Research*, Boland, R.J. and Hirschheim, R.A. (eds.), John Wiley & Sons, New York, pp. 227-251.

Lai, V.S., and Guynes, J.L., 1997, An assessment of the influence of organizational characteristics on Information Technology Adoption Decision: A discriminative approach, *IEEE Transactions on Engineering Management*, Vol. 44, No. 2, pp.146-157.

McChesney, I.R., and Glass, D., 1993, Post-implementation management of CASE methodology, *European Journal of Information Systems*, Vol. 2, No. 3, pp. 201-209.

Moore, G.C., and Benbasat, I., 1991, Development of an instrument to measure the perceptions of adopting an Information Technology innovation, *Information Systems Research*, Vol. 2, No. 3, pp. 192-222.

Nandhakumar, J., and Avison, D.E., 1999, The fiction of methodological development: A field study of information systems development, *Information Technology & People*, Vol. 12, No. 2, pp. 176-191.

Peacham, D., 1985, Structured methods – ten questions you should ask, *Data Processing*, Vol. 27, No. 9, pp. 28-30.

Premkumar, G., 1992, An empirical study of IS planning characteristics at common industries, *OMEGA*, Vol. 20, pp. 611-629.

Prescott, M.B., and Conger, S.A., 1995, Information Technology Innovations: A Classification by IT Locus of Impact and Research Approach, *DATABASE*, Vol. 26, No. 2/3, pp. 20- 42.

Quinn, R.E. and Kimberly, J.R., 1984, Paradox, planning, and perseverance: Guidelines for managerial practice, in: *New Futures: The Challenge of Managing Organizational Transitions*, Kimberly, J.R. and Quinn, R.E. (eds.), Dow Jones-Irwin, Homewood, ILL, pp. 295-313.

Rahim, M., Seyal, A.H., Raham, N., 1998, Use of software systems development methods. An empirical study in Brunei Darussalam, *Information and Software Technology*, Vol. 39, p. 949-963.

Rai, A., 1995, External information source and channel effectiveness and the diffusion of CASE innovations: an empirical study, *European Journal of Information Systems*, Vol.4, pp. 93-102.

Rai, A., and Howard, G.S., 1994, Propagating CASE usage for Software Development: An Empirical Investigation of Key Organizational Correlates, *OMEGA*, Vol. 22, No. 2, pp. 133-147.

Rogers, E.M., 1995, *Diffusion of Innovations*, Fourth edition, The Free Press, New York.

Rowe, J.M., 1993, Can enforced standardization affect CASE usage?, *Journal of Systems Management*, Vol. 44, No. 3, pp. 29-33.

Saeki, M., 1998, A meta-model for method integration, *Information and Software Technology*, Vol. 39, pp. 925-932.

Senn, J.A., and Wynekoop, J.L., 1995, The other side of CASE implementation, *Information Systems Management*, Vol. 12, No. 4, pp. 7-14.

Sumner, M., 1995, Factors influencing the success of Computer-Assisted Software Engineering, *Information Resources Management Journal*, Vol. 8, No. 2, pp. 25-31.

Swanson, E.B., 1994, Information Systems Innovation Among Organisations, *Management Science*, Vol. 40, No. 9, pp. 1069-1092.

Tornatzky, L., and Fleischer, M., 1990, *The Processes of Technological Innovation*, Lexington Books, Lexington, MA.

Vessey, I., Jarvenpaa, S.K., and Tractinsky, N., 1992, Evaluation of vendor products: CASE tools as methodology companions, *Communications of the ACM*, Vol. 35, No. 4, pp. 90-105.

Wynekoop, J.L., and Russo, N.L., 1997, Studying system development methodologies: an examination of research methods, *Information Systems Journal*, Vol. 7, pp. 47-65.

Yeung, A.K.O., Brockbank, J.W. and Ulrich, D.O., 1991, Organizational culture and human resource practices; An empirical assessment, in: *Research In Organizational Change and Development*, Woodman, R.W. and Pasmore, W.A. (eds.), JAI Press Inc, Greenwich, CT, Vol. 5, pp. 59-81.

THE RATIONALIZATION OF ORGANIZATIONAL LIFE: THE ROLE OF INFORMATION SYSTEMS

Dubravka Cecez-Kecmanovic, and Marius Janson*

1. INTRODUCTION

This paper focuses on the relationship between ISs and organisational processes from a perspective of *rationality* of actors, processes and organisations. Actors in organisational processes are considered rational to the degree to which their actions contribute to the achievement of their goals. Furthermore, organisational processes governed by rational actions are considered rational. More generally, the increase in rationality that characterises modern organisations and society is called *rationalisation*. The major role of many ISs has been to assist actors in selecting the best actions (e.g., production schedule) to achieve a predefined goal (maximise throughput or minimise waiting times). Given a particular criterion (e.g., minimise cost, maximise margins), such ISs automate the generation of alternative actions and the selection of the best or optimal action, thereby achieving optimal control and ultimate rationalisation of these processes.

Rationality of organisational processes achieved through IS has often been approached from a mechanistic and decision theoretic perspective. For instance, according to a control-oriented view, ISs are conceived as instruments for achieving greater control and surveillance over the employees (Mowshowitz, 1976; Bjorn-Andersen and Eason, 1980). The role ISs play in organisations was understood as monitoring and scanning internal and external environments so as to inform decision units and then transmit the decision to implementation units. The use of IS to automate production and other organisational processes and optimise decision processes are an ultimate contribution to rational decision-making (Jeske, 1982).

Rationality and rationalisation in the context of IS are considered highly problematic (Nurminen, 1986; Van Tocht, 1994). For example, Nurminen (1986) elucidates the

---

* Dubravka Cecez-Kecmanovic, School of Information Systems, Technology and Management, Faculty of Commerce and Economics, UNSW, Sydney NSW 2052, Australia, Ph (61-2) 9385 4735, ubravka@unsw.edu.au. Marius Janson, Department of Information Systems, University of Missouri-St. Louis, St. Louis, Missouri, US, Ph: (314) 516 5846, janson@umsl.edu.

implications of alternative views of rationality by considering IS and the human beings it should serve. In particular he mentions how rationality without countervailing emotionality leads to a weakening of the role of the individual. These authors investigated the dangers inherent in applying rationality with a purely mechanistic perspective. We suggest that these concerns can be addressed by developing a richer theory of IS' role in rationalization of organizations, not limited to mechanistic and decision-theoretic approaches.

If it were possible to consider developing new ISs from more than one type of rationality perspective, then many problems arising during the development and implementation of the new systems could be avoided. Assessment of organizational benefit and value becomes relative and will change with the rationality criteria. Systems fully justified under one rationality type could be of dubious value seen from another rationality point of view. Similarly, the use to which systems are put could change from one rationality consideration to another. The choice of an inappropriate type of rationality leads to the development of systems of low value to the organization. Lack of agreement within the development team as to which rationality applies in a particular process could be another source of trouble. System objectives, the approach to implementation, and the use to which the IS is put, are all aspects likely to vary from one rationality view to another. If confusion reigns as to which rationality is relevant then the development stage will be fraught with conflict almost certainly reducing the final value of the system to the target organization.

In our view, organizations that make a conscious choice of the desired rationality increase to be achieved by an IS and which seek to obtain agreement of all participants to this choice stand a better chance of obtaining a high value from their new system. The aim of this paper is to develop a rationality framework and by applying this framework to several case examples of ISs to illustrate the impact of the choice of rationality. Our framework is based on previous ideas of rationality and we fully expect this to be developed further. This paper argues that as long as more than one rationality exists, the choice between available options will be an important factor in IS success.

In the sections that follow we first present different views of rationality and rationalisation in modern organisations and briefly describe the proposed rationality framework. We then explain our research methodology and its use to study IS in the retail company (section 3). Next, we illustrate how our framework helped us identify different IS's roles and explain their specific social consequences, including both substantial benefits and risks (section 4). In the concluding section we discuss potential contributions this approach may have to the understanding of social and organisational consequences of IS.

2. THE RATIONALISATION OF ORGANISATIONS – A THEORETICAL FRAMEWORK

While developing our rationality framework, we drew on the ideas of Klein and Hirschheim (1991). Their framework was developed with a view toward assessing how far each IS development methodology could meet the needs of each of four rationality types. Our framework is more focused on the process of identifying system purpose and objectives, and the final use of ISs. We propose a framework based on two dimensions –

one captures alternative ways of viewing organizations, whereas the second differentiates the individual from the collective perspective.

We begin with two basic conceptualisations of organisations distinguished by different ontological assumptions. One is organisation as a system, which conceives of organisations as concrete facticities such as aggregations of actors, physical artefacts (machinery, buildings, technology), process and structures integrated to achieve certain goals. Accordingly, managing is defined as the activity of intervening by actors with formal status and legitimate authority (Gephart, et al., 1996). Systems, such as production system, administrative system, decision-making process, financial system, and the like are defined in terms of objects, processes, states and events about which we claim that they exist, have happened or are likely to happen. In other words, organisations are defined as part of the *objective world* [1].

Alternatively, organisations are conceived as both a *system and lifeworld* of its members based on the assumptions that besides the world of facticities (within which we define systems), there is the *social world* of values and norms, and *subjective worlds* of many individuals[2]. Unlike the objective and social worlds that organisational members share, each individual has his/her own experiences, desires and feeling, that is their inner subjective worlds, which they may or may not disclose to others. The lifeworld is the symbolically created, taken-for-granted universe of daily social activities of organisational members, which involves knowledge related to all three worlds. Whatever happens in an organisation or whatever organisational members may raise and talk about belongs to these three worlds. The lifeworld is permanently (re)created by its members in contextually embedded social discourse.

Two conceptualisations of organisations, based on two sets of ontological assumptions, determine what is considered to be subject to rationalisation: systems in the first, and both systems and lifeworld in the second conception. We use the ontological assumptions (and two concepts of organisation) as one classification dimension to formulate basic types of rationality and rationalisation of organisations. The second dimension is determined by different approaches to reason and rationality.

There exist two fundamentally different and mutually opposing approaches to reason and rationality. One is *subject-centered* reason concerned with self-assertive individual interests that determine the goodness of goals and means to achieve them. Subject-centered reason is behind the individual perspective of rationality. The other is reason *situated in social interaction* exemplified by intersubjectivity of mutual understanding of the participants that denotes the collective perspective of rationality. The individual and the collective perspective of rationality coupled with two views of organisation (as a system or as both a system and lifeworld) form a framework for exploration of different types of rationality of actors and their actions (Table 1).

From an individual perspective, assuming the view of organisation as a system, rational actors pursue their interests and make decisions so as to intervene in a system and achieve pre-defined ends. This type of rationality, following Weber (1978), will be called formal rationality. Formal rationality is 'a matter of fact' and refers to efficacy of the means to intervene in the objective world and achieve a desired state of affairs (e.g., in production or administrative systems). Using Habermas's (1984) categorisation, it is further differentiated as instrumental rationality and strategic rationality. Instrumentally rational actors calculate means based on technical knowledge to achieve given ends disregarding other human beings involved. Strategically rational actors follow rules of rational choice and achieve given ends by influencing other actors, perceived as rational opponents. The more accurate an actor's knowledge of the target system, the more effective his/her intervention in the system, and therefore the more instrumentally rational the actor. Similarly, the better an actor's knowledge of other actors (opponents) and their likely counter-actions, the more effective his/her influence on these actors and therefore the more strategically rational the actor.

When we change the ontological assumptions and include all three world, while still looking from an individual perspective, the nature of rationality changes as actors are oriented to achieving ends not only related to systems (in the objective world) but also those referring to norms and values, justice and fairness, political or ideological affiliations, etc. (related to their shared social world and their inner subjective worlds). Such rationality, which Weber calls substantive, is 'a matter of value' and refers to substantive ends, beliefs and values. The issue here is that different actors pursuing their (different) interest, driven by their (different) substantive ends and values, will usually disagree in their judgement of rational action. As irreconcilable conflict of interests and values is endemic in modern organisations, Weber maintains, substantive rationality is inherently limited (1964, 1978). Klein and Hirschheim (1991) outline the key assumption behind effective application of substantive rationality, that individual actors can and do share a common set of values. Each is 'held accountable for the degree to which his actions are consistent with an ultimate value ideal'. Clearly the potential for conflict arises when actors hold differing values about either or both of their shared objective and social worlds. Conflict of this nature is particularly difficult to handle in situations where the lack of agreement over values is hidden and there is no mechanism for identifying it.

An alternative, collective perspective of rationality, becomes of great significance when viewing the organisation as both systems and lifeworld, is communicative rationality, the third type in our framework. In contrast to the subject-centered reason, Habermas (1984, 1987) proposed reason situated in social interaction and intersubjectivity of mutual understanding of actors. Instead of rationality defined in relation to a self-interested individual, Habermas defined communicative rationality in relation to individuals as social actors that interact to coordinate their activities. Communicatively rational individuals use language to develop intersubjective understanding of a situation, as a basis for a rationally motivated agreement and coordination of their action plans (aimed at achieving their, in principle different ends). It is via communicative rationality that the hidden disagreements of substantive rationality can be identified and, possibly resolved.

Communicative rationality connotes argumentative speech free from any force or constraint. The key assumption here is that participants in communication understand the internal relationship between the raising of intersubjective *validity claims* and the commitment to give and be receptive to arguments. Communicative rationality in essence "signifies a mode of *dealing with* (raising and accepting) validity claims" (Wellmer, 1994, p. 53). Communicative rationality can thus be said to express a reflexive conception of human speech, which means that all validity claims can only be redeemed in human discourse and can only be justified through argumentation. This also implies that the validity claims are not limited to the objective world of facts (as in instrumental and strategic rationality) but can also refer to the social world of values and norms, as well as to the subjective world of individual experiences, desires and feelings. Consequently, communicative rationality implies restructuring of the lifeworld, that is cultural reproduction, social integration and socialisation (Habermas, 1984, p. 341).

Klein and Hirschheim (1991) proposed a further refinement of communicative rationality, that of emancipatory rationality. When communication works to create an effective shared understanding of all significant elements of a situation, it may emerge that differences of opinion among the actors are extreme enough to prevent 'consensually orientated action'. Emancipatory rationality is proposed as a way of dealing with such conflicts. Habermas gave a detailed set of conditions for emancipatory communication by which agreement could be reached. However a number of authors have taken the view that his conditions could not be met in any practical organisational situation (Wilson, 1997). Perhaps the best that we can do is to consider the emancipatory potential of communicative rationality as an *ideal* that can never be fully achieved.

The processes of reaching understanding and communicatively achieved agreement may be limited by competing interests, underlying power asymmetry, different levels of communicative competence among actors and unequal access to knowledge and resources. For instance, actors in power positions or with privileged access to knowledge may unintentionally exert influence on others while believing to be oriented to understanding. In another scenario, they may pretend to be oriented to understanding while in fact being oriented to success, thus intentionally deceiving others. In both cases communicative rationality is distorted: unconsciously in the former and consciously in the latter. Distorted communicative rationality (paradoxically) assumes a collective perspective in order to preserve the appearance of communicative rationality and thus enable covert strategic acting. However, a practice of distorted communicative rationality does not genuinely take into account or refer to the lifeworld of participants but rather remains concerned with systems aspects.

Table 1 shows a taxonomy of rationality along two dimensions: 1) ontological assumptions about the world underlying two different conceptions of organisations, and 2) the individual perspective based on subject-centered reason versus a collective perspective, based on reason situated in intersubjectivity.

The rationality framework presented here suggests several lines of IS inquiry. First, it illustrates the rationality potential of IS-Organisation relationships in relation to the four rationality types. Second, for each type of rationality, it helps understand the meaning of rationalisation (to be) achieved by an IS and the resulting benefits and risks. Moreover, the framework also provides a conceptual foundation for analysis and classification of different types of ISs and the development of standards for their evaluation. It is a valuable tool for ISD because the choice of rationality affects systems objectives, the way ISs are used, and whether completed ISs are successful. In the next section we illustrate these lines of inquiry by drawing on the examples of IS from the field study.

3. RESEARCH METHOD

In this paper we draw from a field study conducted in a discount food chain Colruyt, Belgium's third largest food retail company. The Company evolved from a one-store enterprise in 1960s to a highly profitable food retail chain, currently comprising 120 stores located throughout Belgium. The Company's success is attributed, among other things, to its innovative use of Information Technology (IT) and its integration with Company's management philosophy regarding worker empowerment and participation in decision-making. Namely, as the late Jo Colruyt, founder and Company Board Chairman,

Table 1. The Rationality Framework

Perspective	Ontological assumptions	
	Organisations understood as systems (objective world)	Organisations understood as both systems and lifeworld of their members (thus involving the objective, social and subjective worlds)
Individual perspective (subject centered reason)	Formal rationality • Instrumental rationality • Strategic rationality	Substantive rationality
Collective perspective (reason situated in intersubjectivity)	Distorted communicative rationality or Covert strategic rationality	Communicative rationality

explained in his interview (1993), from its very beginning the Company used IT to explore innovative organisational structures and to enable and support open and inclusive management practices that stimulated employees' initiative, responsibility and risk taking.

The field study started in 1992 and continues to this day. Initially it was an interpretive field study conducted by non-participant observers (Janson et al., 1997a, 1997b). Gradually, as we became concerned with assumptions behind the application of IT and with the ways in which ISs are used to achieve improvements in work processes and decision-making, we added a critical dimension to our study. Namely, on one hand we experienced the Company's attempts to build genuine participative decision-making and empower employees in which the use of IS played an important role. On the other hand, we saw union accusations that Company management had hidden agendas and used IS to mask their pure commercial objectives. As a result, we adopted a critical orientation that aimed to interpret and explain but also to inform and change practice (Cecez-Kecmanovic and Janson, 1999; Cecez-Kecmanovic, 2001). Consequently, informed by critical social theory, our interpretation and analysis turned the study into a critical field inquiry (Lyytinen, and Klein, 1985; Lyytinen, 1992; Klein, 1999). In this paper we report how we adopted the rationality framework to explore underlying assumptions behind several Colruyt's IS failures and successes.

In our field-based study we used document analysis, in-dept interviews and non-participant observation research techniques developed for interpretive field studies (Walsham, 1993, 1995). However, by setting a particular research agenda (rationalisation of organisational processes), focusing on specific explanatory substantive problems (such as assumed rationality of actors; intended and achieved rationalisation due to IS use; manipulation and control of employees versus emancipation and participation), and adopting a historic perspective, our study became a critical inquiry (Cecez-Kecmanovic, 2001).

We collected and analysed over thirty Company and Union documents (both hard copy and electronic ones). We conducted and analysed eighteen in-depth semi-structured interviews (five with the company's founder and high level managers, and thirteen with shop managers and clerks) (e.g., Colruyt, 1993; Lengeler, 1993, 2000). From these sources we reconstructed stories about Company information systems, including the purpose and history of their development, assumptions about the context in which they were developed and implemented, types of rationality addressed and rationalisation aimed and achieved, as well as other intended and experienced effects, risks and dangers. We have select three IS cases to illustrate how our rationality framework assisted us in understanding the roles and social effects these systems had and what contributed to their failure or success.

4. INTERPRETATION OF IS INFORMED BY THE RATIONALITY FRAMEWORK

In this section we briefly present four IS. In order to make the presentations more compact and economical, we include both some relevant empirical evidence collected about each IS and then our interpretation of issues informed by the rationality framework.

4.1. IS For Product Distribution

In the early 1960s the company was a wholesale food distributor supplying small neighbourhood stores. Colruyt salesmen would periodically visit a store, collect inventory replenishment data, and instruct warehouse personnel to send the store a food shipment. This process was time consuming, unreliable, and resource-intensive. The company conceived of an IS that would automate this replenishment process. The goals of the IS were to reduce the time spent by the salesman collecting reordering data, increasing data accuracy, and making the ordering process more reliable. For each reordering cycle storeowners were to fill out keypunch cards indicating the quantities needed on an item-by-item basis. Next, Colruyt salesmen would collect these cards, and submit them to the IS department for processing. The result of this operation was a warehouse picking list for each store that formed the basis for food shipments. This IS was a resounding failure.

During our interview Mr. Lengeler (1993, 2000), former Colruyt salesman, explained that after the IS was introduced Colruyt salesmen found the keypunch cards not filled out when they visited stores during subsequent reorder cycle. This left Colruyt salesmen no other choice but to continue collecting data manually. As Mr. Lengeler (1993, 2000) explained, IS designers never sufficiently took the role of storeowners into consideration. In Mr. Lengeler's (1993, 2000) words:

"After a long day in the store, owners have something else on their mind than filling out keypunch cards." This meant that they postponed this task to the next day but, in fact, it was never completed.

It seems that the key problem was the inadequate assumptions concerning rationality of actors involved in the process. The IS designers considered storeowners inanimate elements of the 'objective world' and consequently modelled them as 'object-origins of data'. The IS design reflected the company's the then current view that the reordering process was inherently instrumental and that the IS should achieve optimally and rational distribution, based on cost minimization and shortening cycle times.

Designers did not recognize storeowners as people and actors with their own interests. Designers failed to understand that reordering was not governed by instrumental rationality but, instead, by strategic rationality. Designers failed to understand that storeowners are rational actors as well, with their own strategic intents. As rational actors, storeowners also undertook actions based on information concerning other players, including the Colruyt Company, in order to achieve their objectives. Consequently, the IS designed to optimise the distribution process and therefore increase its instrumental rationality, failed. Disregard for strategic rationality of storeowners in the distribution process was a fatal failure.

However, by itself the need and wish to increase the degree of rationalization of stock replenishment is not fated to end in failure. In the early 1980s a manager commented on computer-based restocking of store shelves:

> "Automatically computer-based counting of a store's daily sales on an item-by-item basis enables restocking the shelves the next morning." (Luc Rogge, 1984)

On the other hand, the 1960s and 1980s systems differ in an important way. The 1960s restocking IS ignored the human aspects involved in data collection. The 1980s restocking IS collected sales data automatically at the cash register and thus eliminated the human factor. The 1960s IS failed whereas the 1980s IS succeeded.

4.2 IS for Increasing Client Service and Supporting Checkout Clerks.

Corporate life at Colruyt has since the company's inception in the 1960s consistently focused on customers who are 'the raison d'être of the company's existence.' Thus, for example, in the late 1970s the company's founder stated:

> "Not a single customer is inclined to even pay us a single franc that we don't earn by delivering value-added services" Colruyt (November 1979, pp.35-37)

On a similar note a clerk commented:

> "[Less enjoyable] for those who come in contact with the client is the concept of the "client as king." Thus any client-clerk conflict is always resolved in the client's favor. However, that is the way it should be." (Matterne, November 1982, pp.168-169)

The client-focused mentality permeates the entire organization and forms the driving force behind all processes, initiatives, and activities. More importantly, a client-focus is accepted as normal, legitimate, and the way things should be throughout the company. The Union, however, is concerned about indoctrination of Company personnel by top management. We found the following opinion expressed in a Union publication:

> "The thoughts of the clerk concerning customers are not open for discussion. His personality is subjugated to demand and supply with little room for personal creativity." (Dossier Colruyt, 1984, p.11)

Colruyt managers and company documents frequently stress that employees' suggestions are listen to, taken seriously, and acted on. However during an interview with one of the authors a Union respresentative (2001) stated:

> "[Management] listens when [workers] suggest ways a checkout procedure can increase effectiveness but pays no attention to employee complaints about back pain arising from checkout work."

In light of the aforementioned comments we evaluate the development and implementation of a new customer checkout system. During the early 2000s the server involved in the store customer checkout system had aged to such a degree that replacement became necessary. The company's IT department considered this an opportune moment to not only install new servers with the necessary application software but to also take a fresh look at the entire customer checkout application process.

Even though the existing customer checkout application software was robust, reliable, and in general worked well, clerks had experienced and reported numerous complaints over a period of years. First, checkout clerks complained about physical discomfort caused by standing for long periods of time in the same place. Second, occasionally customers would purchase groceries for themselves and a family member. Then, while checkout had already started the customer would suddenly inform the clerk that two receipts were needed. Because receipts were printed while checking out each individual item, the unfortunate clerk would have to start the process all over again. Finally, because Belgium is a bi-lingual country receipts have to be printed in French or Flemish. The checkout clerk picks up whether the receipt is to be in French of Flemish because customers usually speak several sentences before the checkout process begins. However, some customers are sullen and don't say anything which leaves the clerk no choice but to guess which language should appear on the receipt. Yet, even a sullen person is apt to say something while checkout is in progress. If the clerk had assumed the incorrect language the checkout process had to start anew using the correct language.

The new checkout system solved these and many other problems. User information needs and system specifications were determined in a for-this-purpose created prototype room. Potential solutions to problems of the existing checkout system were tested in this prototyping environment that allowed different physical and software arrangements. First, the solution to physical discomfort constituted separating the checkout system into separate data entry and printer stations. During checkout the clerk and the customer would start at the data entry desk and then walk toward the printer station. This physical configuration provided the clerk movement and allowed an efficient customer traffic flow. Savings accrued because one printer station served several data entry stations, thus reducing the number of printers. The problems arising from customers wanting two separate tickets and customers not readily disclosing their native language were solved by enabling the checkout application software to make adjustments on the fly.

The new system achieved several goals. First, the new checkout system supported the checkout clerk's motivation to provide the customer with efficient, effective, friendly, value-added, and human-oriented service. Second, the new checkout system made possible for the store's manager to be more efficient, and increase revenue per store employee. Finally, despite earlier Union statements, the new checkout system did indeed solve the clerk's back pain complaints. As the system's goals are determined from individual perspectives (implying the subject-centered reason) and are defined in relation to the objective, social, and subjective worlds of actors (organization as lifeworld), the IS contributes to substantive rationality (Table 2). Note here that the IS contribution to individual goals would not be sufficient to ensure its success. It is the fact that the IS's goals were mutually agreed upon by all involved actors, that explains how it succeeded by increasing substantive rationality. Understanding IS's impact on a rationality type (in this case substantive rationality) and conditions of sustaining that impact, that is remaining committed to substantial rationality, is an important contributor to systems success.

4.3. Groupware: ISID

In keeping with the idea that information should be available to anyone, the Colruyt Company developed an interactive system for information dissemination (ISID). The system was designed to meet the company's objectives for open, public, effective, and efficient communication. Company policy ensured that information about decisions, actions, and events as well as inter-office correspondence, outbound and inbound communication, and minutes of meetings were captured by ISID. An important system feature was its wide accessibility (80% of information is accessible to all company members and union stewards, 20% is confidential with access limited to authorized individuals).

The key role of ISID is to assist employees to engage in problem identification and problem resolution and to become genuine actors in decision-making processes. Any employee can raise a problem via ISID and initiate its resolution. Other employees may respond (via ISID) with relevant information or, perhaps, a ready-made solution. If no immediate solution exists a team of self-nominated individuals is created to explore the problem further and to propose possible courses of action. The team chooses a moderator, based on self-nominations or nominations by others. Next, team members establish a common understanding of the problem situation and develop one or more potential solutions to the problem at hand. This is then communicated via ISID so that other company employees with an interest in the problem or its solution get promptly informed and can participate in the problem solving. Once publicly announced on ISID, the problem definition and its potential solutions are open to questioning, criticism, and counter proposals. New inputs to the problem definition and its solution may trigger reassessment by team members and this process continuous until, ideally, an agreement is reached. However, this is not always feasible due to time limitations or deep-seated personal differences. In this case, the team moderator weighs all arguments, comments, and counter proposals, and makes a final decision and communicates it to all employees via ISID. The decision for which the moderator carries ultimate responsibility is then implemented. While the entire decision making process is lengthy, the democratically assigned rights of the moderator ensure that the process stays within time limits that are tolerable for the retail industry.

ISID is an example of an IS designed to increase communicative rationality of all Company members. Evidence from its two decade long history, indicate that its stated objectives – wide access; open, public and efficient communication across the company, participation in problem identification and solving by all relevant employees; democratisation of the work place – have been achieved. ISID is therefore considered a great success. For example, a Union representative (2001) stated:

> "There exists communication within the company at all levels – from the lower to the upper levels and vice versa. One will never be adversely affected by asking questions. ISID is essential to the company [it enables] communication between all levels. It encourages communication within the company."

The IS success however has to be situated in a larger social context. Namely, the Colruyt Company has for years built a participative and democratic culture (Janson, et al. 1997a; 1997b). As part of it, the company has an extensive range of in-house courses that focus on employee self-knowledge, emancipation, assertiveness, company values, company policies, job skills, inter-personal skills, and communication skills. Employees attend these courses at their own discretion and during company time. Employees so trained share a common perspective and participate in company affairs significantly less constrained than is normally the case. ISID is an integral part of the Company's cooperative culture and participative decision-making. ISID provides a technologically assisted environment that enables communicative rationality and makes communicative actions possible. By providing easy access to knowledge, an ability to ask questions and test validity claims, and thereby construct shared understanding, ISID assists in creating the basis for rationally motivated agreement.

However, ISID carries with it the danger of being misused. Several incidents were discovered and publicly discussed. For example, two individuals at the same level in the managerial hierarchy ran into conflict that they tried to resolve using ISID. One individual would inform his opponent posting the following ISID message:

> "You did this wrong, you did that wrong, et cetera, while all the time realizing that his messages would be read by many colleagues. This then would call for a rejoinder by the aggrieved person. It would turn into a soap opera. So these types of communications were discontinued." (Lengeler, 2000)

Furthermore, given certain conditions, actors can disguise strategic actions by appearing to act communicatively. Members of top management can systematically distort communication by restricting lower level employees access to certain pieces of information (Table 2). As a result, norms and rules regarding the use of and working with ISID are permanently revisited and re-negotiated.

5. CONCLUSION

Our rationality framework enables a categorical exploration of the wide-ranging impacts of ISs on rationalisation of organisations beyond the narrow view of instrumental rationality. Our taxonomy of rationality in organisational context is based on 1) organisation ontology, that is to say 'organisation as system' versus 'organisation as both system and lifeworld,' as one dimension, and 2) generic perspectives and location of reason, that is to say, the 'individual' versus 'collective' as the second dimension. Our framework systematises concepts of rationality originating in social theory that are relevant for examining the role and impact of IS on organisations. We contend that the basic rationality types: formal (instrumental and strategic), substantive, communicative and distorted communicative rationality type—with their well established theoretical foundations are useful constructs that contribute to a better understanding and prediction of risks and benefits of IS in organisations. The presented IS examples illustrate how our rationality framework is used to examine and explain failures or successes of systems in light of their rationality.

The rationality framework enabled us to understand designers' assumptions about users and processes and to categorize ISs in relation to the rationality types they represented. An explicit analysis of the rationality potential and likely implications (i.e., benefits and risks) of an IS is essential to designers and users. It leads them to consider alternative rationality types that need to be supported/enabled by an IS which may otherwise be overlooked. It suggests that our rationality framework may assist users to define requirements and express these together with expected rationality benefits as well as risks.

Table 2. Impacts of Colruyt's IS on Rationality

	Intended IS effects	Observed IS use and its effects	Risks and challenges
1960 IS for product restocking	Increased instrumental rationality and optimisation of the reordering process	IS failed due to focus on instrumental rationality of the reordering process and neglect of strategic rationality of actors	Disregard for strategic rationality of actors affected by the IS prevented planned functioning of the IS; storeowners' actions prevented increased rationalisation of reordering process
1980s IS for product restocking	Increased instrumental rationality and optimisation of the reordering process	IS succeeded due to a focus on instrumental rationality of the reordering process	The IS counts sales at the cash register. Hence, the IS does not take into account shrinkage caused by spoilage and theft. Product shelve contents need to be reconciled weekly to keep IS restocking successful.
IS for improving checkout process	Increased substantive rationality—achievement of mutually agreed goals related to customer service	Increased efficiency and improved customer service Improved work environment and decreased physical stress.	There is a risk that managers and supervisors misuse the IS and obtain detailed customer processing data and use these data against individual clerks. Introduction of clear policies to prevent IS misuse and nurturing shared values and norms regarding employees' rights (through training) should be considered key to achieving intended goals
ISID (Interactive System for Information Dissemination)	Increased communicative rationality—increased mutual understanding of issues, enabling cooperative interpretation of problems, assisting members in reaching agreement and consensus	Generally improved communication: open, public and efficient Company-wide communication Raised awareness of Company problems and increased workers' participation in problem-solving and decision-making	Individuals can deceive others by pretending to act communicatively while in fact acting strategically The challenge is to train Company members to be communicatively competent and capable of detecting misuse of ISID and potential deception. A further challenge is to ensure access to as wide a range of information as possible.

A combination of the evidence from our case study and other published case studies suggests that our rationality framework provides a starting point for the further exploration of rationality in IS-enabled and supported organisations. Additional explorations are needed to address less obvious and hidden consequences of IS on rationalisation of organisations. For example, our rationality framework helps investigate ISs role in the increasing formal rationality, bureaucratisation, employee subordination, increasing levels of depersonalisation of working relationships, control, and alienation.

An essential contribution of our rationality framework is raising awareness and building knowledge about social and organisational consequences of rationalisation enabled and supported by ISs. We hasten to add that it is not our intention to replace but rather to complement other theoretical perspectives that inform our understanding of inherently contradictory rationalisation processes arising from IS/organisation interaction in contemporary society. Our rationality framework also serves to analyse published studies of IS and their organisational implications, thus potentially leading to new learning and further theory building. Finally, the rationality framework of IS and organisations should be tested against other theoretical developments, in particular postmodernist thinking in organisation theory.

REFERENCES

Bjorn-Andersen, N. and K. Eason, (1980) Myths and Realities of Information Systems Contributing to Organisational Rationality. In A. Mowshowitz (Ed.) *Human Choice and Computers*, North Holland, Amsterdam.

Cecez-Kecmanovic, D. (2001). Doing Critical IS Research: the Question of Methodology. In *Qualitative Research in Information Systems: Issues and Trends* (E. Trauth, Ed.), p. 142-163, Idea Group Publishing, US.

Cecez-Kecmanovic, D. and M. Janson (1999). Communicative Action Theory: An Approach to Understanding the Application of Information Systems. 10th *Australasian Conference on Information Systems ACIS'99*, Wellington, New Zealand, 183-195.

Colruyt, J. (November, 1979) Social-Economic Models, In *There are no Gentlemen Here, Sir*, (T. Penneman, Ed.) (in Flemish), Druco, Halle, pp.35-37.

Colruyt, J. (May 1993). *Interview*, Halle.

Colruyt, J. (April 1984). What is Different at Colruyt? In *There are no Gentlemen Here, Sir*, (T. Penneman, Ed.) (in Flemish), Druco, Halle.

Gephart, R.P.Jr., Boje, D.M. and T.J Thatchenkery (1996). Postmodern Management and the Coming Crises of Organisational Analysis. In *Postmodern Management and Organization Theory* (D.M. Boje, R.P.Jr Gephart and T.J Thatchenkery, Eds.), p. 1., SAGE, London.

Habermas, J. (1984). *The Theory of Communicative Action – Reason and the Rationalisation of Society* (Vol I). Beacon Press, Boston, MA.

Habermas, J. (1987). *The theory of Communicative Action – The Critique of Functionalist Reason*. (Vol II). Beacon Press, Boston, MA.

Janson, M., Brown, A.P., and T. Taillieu (1997a). Colruyt: An Organization Committed to Communication. *Information Systems Journal*, 7, 175-199.

Janson, M., Guimaraes, T. Brown, A. and T. Taillieu (1997b). Exploring a Chairman of the Board's Construction of Organisational Reality: The Colruyt Case. In *Information Systems and Qualitative Research* (Lee, A., Liebenau, J. and Degross, J.I., Eds.), p. 303, IFIP, Chapman and Hall, London.

Jeske, J. (1982) Designing A Decision Support System For A Changing Bell System. In *Decision Support Systems* (M. J. Ginzberg, W. Reitman, and E. Stohr, Eds.) p. 133, New York, NY.

Klein, H.K. (1999). Knowledge and Methods in IS Research: From Beginnings to the Future. In *New Information Technologies in Organization Processes—Field Studies and Theoretical Reflections on the Future of Work* (O. Ngwenyama, L. Introna, M.D. Myers, and J.I. DeGross, Eds.), p.13. IFIP, Kluwer Academic Publishers, Boston.

Klein, H. and R. Hirschheim (1991). Rationality Concepts in Information System Development. *Accounting, Management and Information Technology*, 1,2, 157-187.

Koningsveld, H., and J. Mertens (1992). *Communicative and Strategic Action*, Murderer, Continuo, (in Dutch), Netherlands.

Lengeler, M. (1993). Interview transcript, Brussels, Belgium.

Lengeler, M. (2000). Interview transcript, Brussels, Belgium.

Lyytinen, K. (1992). Information Systems and Critical Theory. In *Critical Management Studies* (Alvesson, M. and H. Willmott, Eds.), p. 159, SAGE, London.

Lyytinen, K. and H. Klein (1985). The Critical Theory of Jurgen Habermas as a Basis for a Theory of Information Systems. In *Research Methods In Information Systems* (Mumford, E., Hirschheim, R., Fitzgerald, G. and T. Wood-Harper, Eds.), p. 219, Elsevier Science Publishers (North Holland), Amsterdam.

Matterne, M. (1985) In *There are no Gentlemen Here, Sir*, (T. Penneman, Ed.) (in Flemish), Druco, Halle, pp.168-169.

Mowshowitz, A. (1976) The Conquest of Will: Information Processing in Human Affairs, Addison Wesley, New York.

Rogge, L. (1984) Automatic Inventory Replenishment, in *There Are No Gentelmen Here Sir*, (T. Penneman, Ed.) (in Flemish), Druco, Halle.

Union Representative (June, 2001), Interview with one of the Authors, Brussels, Belgium.

Van Tocht, J. (March 1994). Professionalisation of the Informaticus. *Information*, 1-10.

Walsham, G. (1993). *Interpreting Information Systems in Organisations*. Wiley, Chicester.

Walsham, G. (1995). The Emergence of Interpretivism in IS Research. *Information Systems Research*, 6(4), 376-394.

Weber, M. (1958). *The Protestant Ethic and the Spirit of Capitalism* (Trans. T. Parsons). Scribner's, New York.

Weber, M. (1964). (Winckelmann, J. ed.) *Wirtschaft und Gesellschaft, Studienausgabe*. 4 Edition, German, 2 Vols., Kiepenheurer and Witsch, Koln.

Weber, M. (1978). (Roth, G. and Wittich, C., eds.) *Economy and Society*, 2 Vols. University of California Press, Berkeley.

Wellmer, A. (1994). Reason, Utopia, and the Dialectic of Enlightenment. In *Habermas and Modernity* (J. R.. Bernstein, Ed.), p. 35, The MIT Press, Cambridge, MA.

Wilson, FA (1997) The truth is out there: the search for emancipatory principles in information systems design, *Information, Technology and People*, 10, 3, pp. 187-204

---

[1] We adopt here Habermas's definition of the *objective world* as "the totality of states of affairs that either obtain or could arise or could be brought about by purposeful intervention"   (Habermas, 1984, p.87).

[2] Habermas defines the *social world* as a "normative context that lays down which interactions belong to legitimate interpersonal relations." (Habermas, 1984, p.88). The social world embodies moral practical knowledge in the form of norms, rules, and values. Complementary to the objective and social worlds, which are external to an actor, Habermas defines and internal or *subjective world*, which is defined "as the totality of subjective experiences to which the actor has privileged access." (Habermas, 1984, p.100).

INFORMATION SYSTEMS DEVELOPMENT
IN EMERGENT ORGANIZATIONS
Empirical findings

Toni Alatalo, Harri Oinas-Kukkonen, Virpi Kurkela, Mikko Siponen[*]

1. INTRODUCTION

Traditional ISD methods have been criticized for placing too much emphasis, at least implicitly, on stability of IS development activities (Truex and Baskerville, 1998). Advocates of so-called lightweight methods (Beck, 2000; Cockburn, 2000) have presented similar thoughts. Discussions on IS development in emergent organizations (Truex et al., 1999) provide recent profound critique of ISD methods (Truex et al., 2000). In fact, advocates of IS development in emergent organizations have set new goals for developing ISs (Truex et al., 1999).

Emergent organizations are organizations that constantly try to adapt to their changing environments, but they never achieve the stability they are striving for. The idea of emergent organizations as it bears on IS development stems from (Truex et al., 1999; Truex and Baskerville, 1998). They see that emergent organizations have a constantly changing environment in which they ongoingly try to adapt to meet the evolving requirements. In fact, "ideal" emergent organizations are always in change — they will never achieve any stability (Truex et al., 1999).

As can be seen from above, emergent organizations are the very opposite to stabile organizations. Due to fundamentally different assumptions on the nature of reality and IS development, the process for designing stabile and emergent ISs are seen different (Truex et al., 1999). In fact, the process in these two cases follow goals contradictory to each other: Ideal ISD in stabile organizations follow five goals, namely: 1) economic advantages of lengthy analysis, 2) user satisfaction, 3) abstract requirements, 4) complete and unambiguous specifications and 5) new system projects as achievements (Truex et

[*] HYTEC Research Lab, University of Oulu, Department of Information Processing Science.
http://hytec.oulu.fi/

Information Systems Development: Advances in Methodologies, Components, and Management
Edited by Kirikova *et al.*, Kluwer Academic/Plenum Publishers, 2002

115

al., 1999). As to IS development in emergent organizations, Truex et al. propose the following four goals instead: 1) always analysis, 2) dynamic requirements negotiations, 3) incomplete, usefully ambiguous specifications and 4) continuous redevelopment (Truex et al., 1999).

The existing work on ISD methods in the context of emergent organizations is mostly conceptual-theoretical research (Truex et al., 1999; Truex and Baskerville, 1998; Truex et al., 2000). There is work where emergent technologies have been studied empirically in field studies of intranet and web development (Bansler et al., 2000; Balasubramanian and Bashian, 1998), but they do not address question of emergent organizations, and call for further empirical studies, reaching over longer time periods (Bansler et al., 2000). As a move towards addressing this gap in literature, we seek to evaluate the goals of (Truex et al., 1999) empirically. With the specific organization in this case study, the still ongoing series of development projects, where we have participated as researchers and designers, began already in 1998. Here the focus is on a subproject that took place during year 2001.

This paper is organized as follows: Section 2 describes the research setting. Section 3 describes the research results. Section 4 discusses the results in light of the research problems and related research. The work is concluded and future research questions addressed in section 5.

2. RESEARCH SETTING: THE MEDIA CENTRE POEM

To investigate how IS development occurs in an emergent organization, we have studied a particular development process, with the aforementioned proposed goals in mind. The organization in question is POEM, the Northern Film and Media Centre, which is the first regional film and media resource centre in Finland, situated in the city of Oulu. It's main goals include increasing skills, knowledge and production resources and establishing an infrastructure for the regional film and media industry in Northern Ostrobotnia. POEM is a member of the European network of regional and film and media centres, and promotes the distribution of short and documentary film exploring the sources of digital film production and distribution. To advance towards these goals, POEM decided to develop an information system to be used by the different parties over the Internet, mainly using Web browsers — hence we have conceptualized it as a Web IS (WIS) (Isakowitz et al., 1998).

Early ideas about the POEM WIS were discussed already in 1998, when the organization itself did not even exist yet but was in formation. As cooperation with the local university, the first author of this article was involved in this early planning. The ideas included a location database, with information about e.g. skilful people and suitable locations for film and media production. Also the possibilities of digital media distribution and new forms of content were discussed from early on. The organization was officially launched in autumn 1999, and during spring 2000 two of us worked on requirements analysis and design specification for the IS as a part of the research at the university. Already at this early phase the emergent nature of POEM became apparent. Things changed at a fast pace and the tools that were used to model the WIS were not flexible enough to support change making process. After the preliminary design phase POEM started to search for financing and contractors for the actual implementation of the

WIS. During spring 2001 the WIS development project was launched officially. By then the financing and contractors were set.

There were four main participators in the project during year 2001: POEM, a software house, an advertising agency and our research project OWLA. POEM produced all the content for the WIS and connected the database / business logic and the user interface to each other by server page programming. The software house was responsible for database design and business object programming, whereas the advertising agency planned and built the visual look and the user interface. The OWLA project performed usability evaluations utilizing questionnaires, heuristic evaluation and usability testing, and supported the other participants when needed.

The project was divided into four iteration cycles. This paper tackles the first cycle, that was completed before the end of year 2001. Currently, in 2002, the systems exists and several new applications are developed to provide services around it.

To be able to answer specific questions about the ISD process, two of us interviewed at the end of October 2001 the eight key people involved in the development. They were from POEM and the two contractors. The interviews were semi-structured and non-leading, i.e. basically only concepts that had already been introduced by the interviewee during that session were brought up by us. However, we had specified research questions and related interview themes beforehand. None of these questions were shown to the interviewees nor directly asked, but they did speak about them in detail in their own terms. One of the goals during the interviews was to find out how the development process had proceeded from the perspectives of the different parties. Based on these results, we aimed at finding out how the process that had occurred reflected the aforementioned goals of ISD in emergent organizations. The other two themes where: 1. the role of conceptual modelling (Wand et al., 1995; Mylopoulous, 1998) in collaboration between the participating organizations 2. the usefulness of usability evaluations. These other themes will be discussed here aside the main question, as they are also related to the goals: firstly, (Truex et al., 1999) suggest that conceptual models (e.g. diagrams) help to redesign changes to the systems, and secondly, usability evaluations can be seen as a form of analysis, related to the goal "Always analysis".

POEM as an emergent organization is summarized as follows: POEM is a young organization where things change fast all the time. This builds up pressure and stress in the minds of the personnel. The publicly (mostly EU) funded organization is one of its kind in Northern Finland, with a lot of expectations and requirements from the outside. However, POEM has a strong vision of it's mission. This vision guides the WIS development process and helps to keep the ultimate goal clear in the minds of developers and other people who are involved with the project. Also the potential users (stakeholders) expect a lot from the WIS. POEM tries to satisfy these needs. In the regular meetings with domestic and international (potential) collaborators new requirements constantly emerge. POEM also collaborates closely with the vibrant IT industry in the area.

3. RESULTS

In this section we describe our findings related to the four goals of ISD in emergent organizations as stated in (Truex et al., 1999).

3.1. Always Analysis

Plans, models and other work done were analyzed all along the whole project. This was carried out, for example, during initial background research, requirements analysis, design specification, when specifications and propositions were given to POEM, when needs for changes arose, when server page programming was done, during inspections / evaluations and when risks and time use was analyzed.

The whole project was built on continuous interaction between creating and commenting on proposals. The latter took place either immediately after the change proposals had risen and been presented, for example in a meeting, or afterwards through e.g. email or phone.

At POEM, the technical person was the one spending time reading the models the software house had made (UML Class Diagrams). He described them as "handy". The person that is in charge of the content creation in POEM said that the models were "easy and quite logical to interpret" and that it was easy to detect the relationships between different entities and groups. During this interpretation process, needs for changes arose.

Analyzing was also done, when POEM's technical person connected the user interface and the database together with so-called server page programming. Indeed, he found some points from plans and the code that needed to be fixed or implemented in a different way. And again, there was the question about addressing the issue during the current iteration cycle or later. Moreover, as a by-product, some new ideas rose during everyday working and in meetings.

In addition, the OWLA project evaluated WIS usability. The techniques used included usability testing, heuristic evaluation and questionnaires. With usability evaluations we aimed at detecting problems, e.g. functions that were hard to use, learn and/or remember. POEM manager told us that, in his opinion, usability tests created a sense of security that the project is going to the right direction. Another interviewee told us that performed tests made her think about new things.

3.2. Dynamic Requirements Negotiations

Requirements specifications are often used in contracting between the user organization and software supplier(s) to define the work to be done beforehand. The notion of dynamic requirements negotiations obviously makes this difficult. In this case there was a sort of an open contract between POEM and the software house, where the cost and roughly the amount of work was predefined, but the actual requirements were constructed and prioritized dynamically as described. Before this was possible, mutual trust had to be established — that was a subtle process in itself.

During the actual working period, the software house and the advertising agency implemented change propositions if they were seen as reasonable and technically possible. Sometimes change propositions were moved to next iteration cycle or further, if they came up too late during the particular iteration cycle, if the change propositions were too complicated to implement or if they were still too abstract.

Most of the change propositions came from the person who is in charge of content creation at POEM and from POEM's manager. POEM's technical person discussed with the software house representatives how difficult these changes would be to implement and when it would be best done. In addition, the person who is in charge of content creation in POEM discussed these same things with the advertising agency regarding the user interface. Change propositions that were presented by the software house and the advertising agency were again reviewed by POEM. Also the actual changes done to the database / business logic and user interface were commented on by POEM..

3.3 Incomplete, Usefully Ambiguous Specifications

In this case, design specifications where mostly in-house tools at the software house. They were, however, also means of communication from the software house to the people working on the project at POEM, which supported constant analysis as described earlier.

The database and business logic specifications were UML diagrams. When needed, associations, objects, attributes and components were easy to change, delete or add. This is in line with the statement of Truex et al. (1999, p. 122): "easily maintainable specifications, like object-oriented designs, make it easier and cheaper to re-specify IT systems when change is needed". However, the changes were easy to do into the database / business logic specifications also due to the fact that they were mostly quite small. As more major changes to the system are on the way, this point can be re-evaluated later.

The specifications were incomplete indeed, i.e. not everything was prescribed in them. Related to this, one major misunderstanding occurred, namely the work on the server page programming. To put it simply, everybody thought that somebody else would do it. Finally, POEM's technical person did the programming. This mix-up occurred probably because there are so many participants in this project. Therefore, the possibility of human error is high. But undeniably it was an ill-defined issue — something to bear in mind while evaluating the usefulness of ambiguous specifications.

3.4 Continuous Redevelopment

The project was divided into four iteration cycles. Requirements were divided between the different iteration cycles based on how critical each requirement was. The first cycle naturally included most of the basic functionality. The software house made a suggestion about the division and then it was discussed and agreed upon in one of the steering group meetings. This division changed somewhat during the project.

Change needs occurred all the time and most of them were carried out: new features were added and old features were improved. Changes were almost entirely small ones, which however caused hours of work. There was only one major change that had to be carried out, about colours and photographs. Generally, change needs have been about changing, deleting and adding attributes and associations. Based on the interviews, one could expect substantial change needs to rise in the near future and during the WIS' whole life cycle. They can have a huge effect on the WIS. For example, there will be a need for new databases, services and features (currently user registration and personalization are on the way) plus different user interfaces (e.g. mobile use, possibly digital television).

The user interface is much harder to change than the database and the business logic because of the way the user interface has been implemented. For example, every link in the link list consists of two pictures. Therefore, even small changes to the links would require a lot of work. Indeed, the updater has to be good at graphics processing, hypertext mark-up and client page programming in order to be able to update the user interface when needed. However, WIS actual content is easy to update.

Prototyping was an important aspect of the development process. The schedule for it was tight, but the work was finished in time. At the interview, one of the participators said it was a good idea to construct a demo with a certain deadline. This brought structure to the process and required everyone to be in contact with other participators. In addition, the demo was part of the first usability evaluation. Through this, POEM had a chance to find out early in the development process how the potential users felt about the WIS, and some problems were discovered that had not been noticed before. Truex et al. (1999, p. 122) emphasize the importance of prototyping. The experiences received during the development of POEM WIS support this statement.

3.5 An Overview of the Findings

Overall the first iteration cycle of the POEM WIS development proceeded along the goals presented in (Truex et al., 1999). This may not be surprising, as the process described resembles the immature ad-hoc way software is often made. Interestingly, however, it became clear during the investigations that the development process of POEM WIS differed from the software houses' other projects with respect to process' formality. Usually the software house performs very formal inspection sessions a few times during the project. Between these sessions, the software house concentrates on programming. In this case, the software house did conduct some formal in-house inspections, but also quite frequently went through the specifications and checked for change needs quite informally with POEM. However, despite this informality, the software house tried to follow its own process model as much as possible, but, possibly because of POEM's emergent nature, realized that they could not perform the project as formally as they were accustomed to. The project manager at the software house thought the reason was that POEM's employees are artists, and not used to software business (unlike many other customers of the software house, such as hardware manufacturers and heavy industry).

Also, when the advertising agency starts a project, they try to plan completely in detail how the project will proceed. Therefore, the plan is not too flexible when change needs rise. Then again, the software house was paid for making the changes, whereas for the advertising agency it was not so. This explains partly why the software house was much more flexible with regard to change making than the advertising agency was.

The questions a) whether the process was productive and b) how it was and could be supported, remain. Of the ways of supporting discussed in (Truex et al. 1999, p. 122), we see here that easily maintainable specifications and prototyping were essential in the process. Whether the architecture resulted as an open one that can be easily utilized in di erent settings is put to test in near future. So far, there has not been the kind of end-user development nor back-channel communications suggested in (Truex et al., 1999), but these might indeed be good improvements. In this case there is not an internal IT department but a myriad of changing partnerships. Finally, the importance of a proper rewards system, that values adaptation, was demonstrated by a) how it was crucial for POEM and the software house to gain mutual trust to form an open enough relationship that allowed emergence b) the understandable reluctance of the advertising agency to make all the changes to the user interface that had not been properly agreed upon. Of the productivity we only know that the shared understanding at the time of the interviews was that all the major objectives had been reached on schedule —— that the project was considered a success so far.

4. DISCUSSION

The major limitation of this paper is the poor generalization of the results. As here only one organization has been studied — furthermore quite a special one — this may not tell much about other organizations. However, it may be that the issues related to emergence are quite common. In the words of the software house's responsible manager in this project:

> "People have realized that there will always be changes that need to be made into systems. Experienced designers take this into consideration in almost every case, and they try to design the systems so that they can be changed whenever needed."

Shortly after this the the interviewee referred to the ideology of extreme Programming, that he was interested in — albeit in this case it was not actually in use, but an attempt at a more traditional iterative software engineering process took place instead.

Of the interactions between the people in the project and the relationships of their organizations it is difficult to pin-point the most significant events and their root causes. Also the decision to look at the process so strongly in the light of (Truex et al., 1999) has coloured the interpretation. With this modest account we, however, wish to contribute to advancing the understanding of these sometimes overwhelming issues.

5. CONCLUSION

Goals of information systems development (ISD) in emergent organizations were presented from the literature, and analyzed based on a case study with such an organization. The process under study did proceed according to the principles for ISD in emergent organizations, even though those principles are in contradiction with the traditional ISD goals, and the company doing the actual software development for the organization is a rigorous one (i.e. tries to manage projects along the lines of the software engineering doctrine). The reasons for this were threefold: 1. software professionals have learned to embrace change in any case, 2. the emergent organization's nature and actions encouraged the software company to adopt practices stated in the goals, such as continuous redevelopment and dynamic requirements negotiations, and 3. the participating academics were studying the phenomena of emergent organizations and therefore intervened in the process in ways that did not enforce a rigid process model. A major dilemma, that the theory did not address, was found in contracting / outsourcing. As for continuing research, the ways of supporting emergence need to be studied more closely and developed further.

6. ACKNOWLEDGEMENTS

We wish to thank POEM and the interviewees — without them this study would not have been possible. We are also grateful for the anonymous reviewers who helped us to see what this work is actually about.

7. REFERENCES

Balasubramanian, V. and A. Bashian, 1998, *Document Management and Web Technologies: Alice Marries the Mad Hatter.*, Communications of the ACM 41(7).

Bansler, J., J. Damsgaard, E. Havn, J. Thommesen, and R. Scheepers, 2000, *Corporate Intranet Implementation: Managing emergent technologies and organizational practices*, Journal of the AIS 1, 1–39.

Beck, K., 2000, *Emergent Control in Extreme Programming*, Cutter IT Journal 13(11), 22–24.

Cockburn, A, 2000, *Balancing Lightness with Sufficiency*, Cutter IT Journal 13(11), 26–33.

Isakowitz, T., M. Bieber, and F. Vitali, 1998, *Web Information Systems*, Communications of the ACM 41(7), 78-80.

Mylopoulous, J., 1998, *Information Modeling in the Time of the Revolution*, Information Systems 23(3/4), 127–155.

Truex, D., R. Baskerville, and J. Travis, 1998, *Amethodical systems development: the deferred meaning of systems development methods* Accounting, management and information technologies 10, 53–79.

Truex, D. P. and R. Baskerville, 1998, *Deep Structure or Emergence Theory: Contrasting Theoretical Foundations for Information Systems Development*, Communications of the ACM 42(8), 117–123.

Truex, D. P., R. Baskerville, and H. Klein, 1999, *Growing systems in emergent organizations*, Information Systems Journal.

Wand, Y., D. Monarchi, and J. Parsons, 1995, *'Theoretical Foundations for Conceptual Modelling in Information Systems Development*, Decision Support Systems 15(4), 285–304.

SUCCESS FACTORS FOR OUTSOURCED INFORMATION SYSTEM DEVELOPMENT

Murray E. Jennex and Olayele Adelakun[1]

1. INTRODUCTION

Offshore outsourcing is the transference of an Information Technology, IT, function, from a company to a supplier organization located outside the borders of the parent company's country. It is a commonly used strategy among leading companies in the United States, US, and Western Europe. Companies typically invest in offshore outsourcing with the expectation of lower costs, economies of scale, access to specialized resources, and/or new business ventures, Aubert, et. al. (1998). Contributing to the rise of offshore outsourcing was the shortage of IT professionals in the US in the late 1990s.

The late 1990s saw an increase in the outsourcing of software development, and in particular, offshore software development. Rajkumar and Mani (2001) point out that Year 2000, Y2K, and converting systems to accommodate the European change in currency to the Euro have stressed organizations' ability to keep up with necessary development leading to outsourcing more development projects to offshore developers. Contributing to the increase in offshore development are advances in telecommunications technology and personal computers that have increased the ability of companies outside of the US to provide development services. Currently, the high demand for e-business and the Internet based solutions is continuing the drive for offshore development.

Traditional offshore software development is primarily application development. These applications tend to be highly structured requiring little or no changes to the requirement specifications. These projects require less client interaction and project management from the client. They are ideal for outsourcing as deliverables and bids are understandable and predictable and risks are better understood. Current offshore development includes e-business and web application development, and "follow-the-sun" or "round the clock" application development. These projects tend to be less structured in nature and need more client contact and project management than traditional offshore

[1] Murray E. Jennex, Ph.D., P.E., San Diego State University and Olayele Adelakun, Ph.D, DePaul University

Information Systems Development: Advances in Methodologies, Components, and Management
Edited by Kirikova *et al.*, Kluwer Academic/Plenum Publishers, 2002

123

development projects. They are less ideal for outsourcing as deliverables, costs, and risks are less predictable.

Offshore software development offers an opportunity to significantly reduce the cost of application development. However, given the change in the types of projects being outsourced, there is a need for organizations to assess the likelihood of developer companies to successfully complete the project. Conversely, given the ease of starting a company that supplies development services, it is important for startups to understand what it takes to be a successful outsource developer. This paper proposes a set of success factors for startup offshore development companies. These factors were determined through detailed study of two successful startup software developers. Research has shown that the major player in the offshore software development outsourcing is India (Rajkumar and Mani, 2001), with approximately $6 billion in software and services in 2000 (Smetannikov, 2001). Some research has been done with respect to what has made Indian companies successful. This paper expands this research by looking at companies located in Western and Eastern Europe. The question to be answered is what causes some offshore developers to succeed while others are less successful.

2.   BACKGROUND

2.1.   Offshore Outsourcing

Offshore outsourcing is when a company providing an outsourced service is located outside the country of the company contracting for the service. Offshore IT outsourcing is primarily concerned with the development of IT applications, mostly in geographically low-cost regions abroad. The major driver for offshore outsourcing is reducing internal operational cost. Many US and Western European companies are located in high-wage geographical areas such as Silicon Valley, Los Angeles, New York, Boston, London, and Paris, driving them to seek the low-cost programmers found in low-cost areas.

2.2.   Risks from Offshore Software Development

Before determining success factors it is important to understand failure by understanding the risks. Generally, the risk of any software development project is that the developed software does not fulfill the software requirements or expectations of the users. The degree of this risk varies from total failure where the developed system cannot be used to marginal failures where the developed system can be used with some difficulty. These difficulties are overcome by applying additional effort or resources or by operating at less than optimal efficiency or effectiveness. Risk of failure is minimized through the application of project management and software engineering techniques and practices, examples of which are specified by the Software Engineering Institute's, SEI, Capability Maturity Model, CMM, or the International Standard Organization's, ISO, 9000 certification processes; and the certification of personnel developing the software.

Other risks are the viability and ability of the supplier to meet its commitments. To successfully develop software the supplier must remain in business long enough to complete the project. Additionally, the supplier must have the resources needed to perform the development and to deliver the final product. Finally, the supplier and the client must have the ability to transfer funds and to conduct business within the existing regulatory environment.

2.3.   The Need to Identify Offshore Development Success Factors

As the market for offshore IT outsourcing grows and companies are formed to meet the demand, it is critical to understand what factors will contribute to their success. Empirical studies that systematically study these factors are few.

Rajkumar and Mani (2001) discuss why India's software developers are successful. Reasons for this success are an abundant supply of highly educated but low paid software engineers, English is the language of education and business in India, and the Indian government's development of infrastructure and tax and financial incentives. They further discuss the need for the business to be organized to support the client through interfacing, project management, and contract management. Ultimately they list four categories of success factors: Management, Customer, Project, and Staff. The management factor focuses on the leadership of an organization. Managers are expected to guide the organization and ensure adequate resources are allocated through business plans and strategies. Also, management must ensure that facilities and staff are developed to support the target market. The customer factor focuses on developing a relationship with the client. It includes setting up communications with the client, visiting the client, learning from the client, and integrating practices when feasible with the client. The key is to understand the client so that the developer can add value and develop a long-term relationship. The project factor focuses on project management. The software developer must be able to estimate and manage the project. Care has to be taken to ensure project scope is clearly understood as well as who is authorized to change it. Two key concerns are avoiding research and development projects and ensuring that functional test requirements are clearly communicated by the client. The staff factor focuses on hiring and retaining technical talent. Career paths must be established and care taken to ensure that management is meeting the needs of their staff. Support for travel to client sites in the form of funds and visa support. Finally, flexibility in hiring needs to be established to allow for clients who wish to hire developers of their systems.

Smetannikov (2001) discusses problems facing Russian software developers rather than success factors. Chief among these problems are travel difficulties, cultural differences, real and perceived, between Russian and American programmers, and an unstable Russian business climate. Russian companies are impatient to get business, they feel it is only a matter of time before they will be credible competitors with India.

Raval (1999) discusses his secrets of successful offshore software development from the viewpoint of the client. These secrets or factors are having a strategy for offshore development, understanding the countries you are outsourcing too and the risk associated with outsourcing, preparing the organization and the offshore developer to work with each other, delegating offshore administration to local expertise, and not letting cultural and language issues affect the project.

3.   METHODOLOGY

An action research and mini case study approach was used to perform a detailed study of two companies. The subject companies were selected because: 1) they had ties to the authors, 2) they were newly formed, 3) they provided software development services, and 4) they were experiencing success. Each company was visited and their principals interviewed. Operations were observed and processes reviewed. Also, since this was action research, the researcher provided recommendations to the companies. From this data a list of success factors was generated. The list was then administered as a survey to the leaders of each company who rated the importance of each attribute. The scale used was 1-Critical, 2-Very Important, 3-Important, 4-Useful, and 5-Not Important. The completed surveys supported the initial list. To further validate the survey, the survey was given to the graduate system analysis and design classes of one of the authors. These students were familiar with software development principles, most had business experience, and were mixed international and US. Nineteen of the students had experience with outsourcing. Results of the survey were analyzed with respect to the origin of the respondent and in total. Results of the analysis were used to create the final form of the survey. The survey was then administered to personnel from outsource and client companies. Distribution of the survey was through meetings, personal contacts, and email. Surveys were analyzed with respect to the origin of the respondent and as an outsourcer or a client. Means and standard deviations were calculated with attributes that scored less than 2.0 determined to be critical success factors.

4. THE COMPANIES

4.1.   International Business Solutions

International Business Solutions, IBS, was founded in March 2000, by an American expatriate and is located in Kyiv, Ukraine. The company has an affiliated US company, Energy Solutions, also formed in March 2000. IBS is a small enterprise with three full time and five part time employees, all Ukrainian, in addition to its founder. To minimize costs and increase flexibility, IBS utilizes an extensive network of independent contractors to provide its services. IBS has three areas in which it provides services of which one, web development, is of interest for this study.

Web development services are offered at the suggestion of the author who had discovered during a Y2K risk assessment that there is a large web development talent pool available in Kyiv that is very reasonably priced and motivated to work. Given the weak business climate in Ukraine these services are targeted to United States and other developed countries' companies. Originally IBS offered fixed fee contracts for development projects. Problems with managing projects remotely and incomplete user requirements drove IBS to switch to providing developers on an hourly basis working under direct control of the contracting client. IBS is able to offer developers at a very attractive rate, approximately $15.00 per hour United States Dollars, USD. The new approach has been more successful and IBS has several developers under contract.

IBS markets itself through word of mouth and its web site. The web site provides a good overview of IBS services, examples of the skills of the IBS web developers, and a means of contacting IBS. The sites are hosted by a US based Internet Service Provider, ISP. A US based ISP was chosen because they were considerably less expensive than Ukrainian ISPs, approximately $7.00 USD versus approximately $40.00 USD. Another issue was bandwidth, 48 kbps is becoming common in Kyiv while DSL is being introduced but is very expensive, greater than $100.00 USD per month. Reliability is still an issue with Ukrainian ISPs as power quality is poor and phone lines are degraded.

4.2. IT Business Solutions

IT Business Solutions, ITBS, is located in Castle Franco, Italy. It consists of a German database expert who recently completed his Masters of Science in Information Systems in the US, and an Italian expert in computer aided manufacturing, web mastering, and web interfaces. ITBS was formed in 2000 to provide web development services for vacation rentals after getting its start by building a web site for the villa resort owned by the parents of one of the principals. A US vacation rental company has contracted with ITBS to build a web portal and database. This project is succeeding.

ITBS has no formal organization, processes, or office. The principals work from their own computers on their own parts of the project. They get together to set the design framework and to integrate their portions of the system. Each principal is responsible for their own computer and network connections. Local ISPs are used as costs and service are not issues. There is no central repository. One principal serves as the main client liaison and does most of the interfacing in the US.

Ultimately, the success of ITBS is based on the expertise of its principals. ITBS does not advertise. Both are young, technically gifted and talented professionals. The continued success of ITBS will depend upon the principals developing business savvy and building the business side of ITBS. The lead author has provided several recommendations on how ITBS can develop its business side.

5. ANALYSIS OF THE COMPANIES

5.1. Non-IT Issues

It has taken IBS over eighteen months to become operational. This is primarily due to the business climate in Ukraine. Things take time to complete. Paying expedited fees is normal as it is rare to pay the standard fee and have something done. Americans pay more for everything, up to 10 times what a Ukrainian would pay. IBS avoids this by the founder living in Kyiv and by having Ukrainian directors and partners. Regulations and tax requirements are difficult to understand and comply with. Getting paid is difficult. Transfer payments are expensive, an approximate $350.00 USD fee is applied, and take a long time, approximately 5 weeks. IBS is set up to accept off shore payments through a non-Ukrainian bank. This allows IBS to accept credit card payments. Since Ukraine is a cash economy, checks aren't accepted. Non-Ukrainian customers must pay in cash or via credit card. This is difficult for many customers to understand and work with.

Networking within the business community is vital. Business within Ukraine is done face-to-face and usually over meals or social gatherings. Reputation is important and whom you know is vital. Doing a good job supports continued business but does not get the initial job. IBS discovered this after several months. The founder now spends a majority of his time mixing and meeting people in Kyiv.

Learning to work within the Ukrainian culture has been difficult. Ukraine has 33 holidays compared to approximately 10 in the US. Punctuality is not a standard practice. Break times are whenever. Language has been an ongoing issue since the founder did not speak Russian when he arrived in Kyiv and still is not fluent. Finally, the founder is from California and was used to California smoking laws. Ukraine has not adopted those laws and smoking is prevalent. Learning to deal with smokers has been a challenge.

Most of the issues ITBS faces are non-IT with communication with the client as the main issue. The owner of the US based company is the point of contact for ITBS. Neither the owner nor his employees are computer literate. This lead to misunderstandings as the client does not understand software requirements and limitations and ITBS has difficulties in understanding client needs. Fortunately ITBS has a background in vacation rentals and so understood the basic business of the client. Additionally, English is not the principle language for ITBS. One principal speaks English with an accent. The client owner also has an accent. The principals admit to several occasions where language barriers led to misunderstandings and disagreements.

The contract is an issue. ITBS did not use a lawyer to negotiate the contract. The contract provides for payment when the project is complete except that expenses are paid as they are incurred. This has required the principals to rely on sources of income outside of ITBS during the project. Cash flow has not affected the project as both principals had outside sources of income but it is not recommended that startup companies use this approach. The lead author has recommended ITBS use a lawyer on many occasions.

Project management has also been an issue. No formal project scope was established resulting in disagreements as to what should and should not be included. Additionally, scope and requirements creep have extended the project schedule. Finally, a loss of version control event was observed that took a day to correct. All these issues could be corrected through implementation of project management practices and techniques.

ITBS has little business infrastructure. There is no formal billing or accounting system, organization, legal representation, or advertising. ITBS has no plan for growth or for what it wants to do. Its success has been due to falling into a project and relying on its technical expertise to do a good job. Doing a good job is essential, but having a business plan is important for continued success. ITBS needs to avoid the trap of believing that doing a good job is all that is necessary and that the business will run itself.

A final issue that has arisen since the terrorist attack on the US on September 11, 2001, is physical access to the client. The client interface principal originally traveled to the client offices while completing his degree program and resident in the US under a student visa. Obtaining student visas has become difficult. The one principal has lived in the US for the last 2 years and has been able to interface with the client regularly. However, he cannot obtain a resident visa easily. He can obtain a business visa but this will require him to return to Italy as a permanent place of residence.

5.2.   IT Issues

Phone communications for calls outside Ukraine are very expensive. Calling cards are available that significantly reduce costs, to about $0.45 USD per minute, but require accessing special numbers and switches. File transfers are slow. Bandwidth is not readily available due to degraded communication lines. When lines are new or have been upgraded, bandwidth is available for a reasonable cost from a US perspective. One of the authors found an Internet café that offered 128 kbps connections for a cost of about $4.00 USD per hour. This is considered expensive in Ukraine as confirmed by the clientele being primarily non-Ukrainians. Another negative is the poor condition of the telecommunications infrastructure. Jennex, et. al. (1999) found no digital/IT equipment used for energy management and plant communications. Analog phone switches and equipment were normal. Telecommunication service between dispatch centers and plants or other dispatch centers were very unreliable. Ukraine has a wide area network for monitoring the power system that is based on SM1420 and SM2 computers (these are DEC1000 and PDP11 clones). This system was observed to be frequently out of service requiring system operators to rely on voice communications for dispatch functions. These also were frequently out of service requiring system and plant operators to use analog radios or to simply load follow. Load following is a technique used for system control only when communications are out of service, it results in poor power quality with little frequency stability with the previously discussed effects on IT. Observations of phone lines in hotels catering to westerners found that dial up connections of greater than 9800 bps were difficult to impossible to sustain for more than a few minutes due to line noise and errors. Ultimately this reduces the effectiveness of internet, email, and fax processes, raising the cost for these services. The Central Intelligence Agency's World Fact Page confirms the antiquated state of the system.

Jennex (2001) looked at IT in the energy sector of Ukraine. In general, the energy sector was found to be at a 1960s or 70s technological level. Poor reliability and power quality was observed with frequency oscillations of 0.5 hertz or more routine and power outages common. Most critical building and hotels, as well as many residences, keep and maintain backup generators. North American standards have frequency oscillations controlled to 0.05 hertz or less. Digital equipment does not function well nor last long with the large observed frequency swings. Ukraine has an average load of approximately 0.375 watts per person while Southern California has approximately 1.375 watts per person. This translates to the average house or office in Ukraine having a 45 amp fuse box while the average house in Southern California has 150-200 amps. What this means is that the electrical infrastructure in Ukraine does not readily support a modern office's IT electrical needs. Large companies compensate by installing their own power equipment. Small companies make do with what they have with the result that they have less reliable IT. An example is the failure of the lead author's laptop a few weeks after returning from Ukraine. While it couldn't be proved that the weeks of poor power quality caused the failure, it is highly coincidental and was at least a contributing factor.

Availability of hardware and software can be issues. Leading edge hardware such as personal computers, digital cameras, printers, and communications equipment are very expensive and hard to get. Additionally, some companies differentiate between hardware sold in Europe and that sold in the US. As an example the coding used on printer cartridges can be different for models sold outside the US than for those sold in the US. Another issue is incompatible character sets. Ukraine uses the Cyrillic alphabet. The character set used to display this alphabet on computers in Ukraine makes the generated files unreadable on computers running English character sets.

Software is readily available. The issue is pirated software. Virtually any software package can be purchased in the local markets for approximately $2.00 USD per compact disc, CD. Authentic software is available but costs as much or more than it would in the US. This makes buying and using authentic software unattractive. IBS uses only authentic software. Competing companies in Kyiv do not have this constraint so this gives IBS an unattractive cost differential. Fortunately this isn't a significant issue as IBS does not provide software to its developer contractors and only needs software for its own business use. Software costs can be a significant issue to other companies in Ukraine, especially if they are trying to use "legal" versions and the competition does not.

Jennex (2001) reported availability of technical talent in Ukraine. Olearchyk (2001) in the Kyiv Post English language newspaper reports that there is a growing shortage of talent. It is stated that approximately 2500 IT specialists are leaving Ukraine each year. Additionally, the schools are not producing usable IT professionals due to their focus on theory and not practical education. The net impact is that Ukrainian software developers are turning down contracts due to lack of work force. Two issues exist. The first is that IBS will not be able to keep web developers under contract for potential work. The second issue is the upward pressure on wages. As developers become shorter in supply, companies will have to pay more to retain them. This will force IBS to pay more, resulting in higher prices that IBS must charge and making IBS less competitive.

The only IT issues that existed were with the client. The client did not have adequate connections and bandwidth requiring ITBS to upgrade the client. Additionally, the client did not have IT literate employees. ITBS assisted the client in hiring and training an IT specialist. Both of these issues were critical as ITBS needed to have adequate support within the client for maintaining and managing the system being developed.

6. FINDINGS

The main finding is a list of success factors for companies providing software development services. There are five main success factors: People factors, Technical Infrastructure, Client Interface, Business Infrastructure, and Regulatory Interface. Figure 1 is a model of how these factors affect the relationship between client and provider. Attributes of each factor are listed in Table 1.

The Outsourcer Success Factor Model shows the relationships between the five main factors and the participants in an outsourced development project. The model shows that both participants operate within a regulatory environment. This environment provides the legal framework in which both entities must operate. The environment can hinder the ability of the participants to perform the project, or it may encourage it. India is an example of a regulatory environment that encourages outsourcing while Ukraine is an example of one that does not. Participants also operate within an external technical infrastructure comprised of each country's telecommunications and electrical systems, technical education systems, and availability of modern software and hardware. The reliability and availability of these infrastructure components directly impact the ability of the participants to perform the system development project.

Internal to the participants are the client interface, internal technical infrastructure, business infrastructure, and people factors. The client interface is the defined communications process between the participants. This directly impacts the transference of requirements and knowledge and guides the participants in the resolution of conflicts. The business processes of the outsourcer determine the likelihood that the outsourcer will remain viable for an extended partnership. The internal technical infrastructure ensures the outsourcer has the ability to develop systems. Finally, people factors ensure the outsourcer has the ability to understand the context in which the client operates.

Table 2 lists the means and standard deviations of the survey importance rating. Table 2 lists values for clients, outsourcers, and overall, additionally, outsourcers are broken down into US, non-US, and overall ratings. Critical attributes are discussed below.

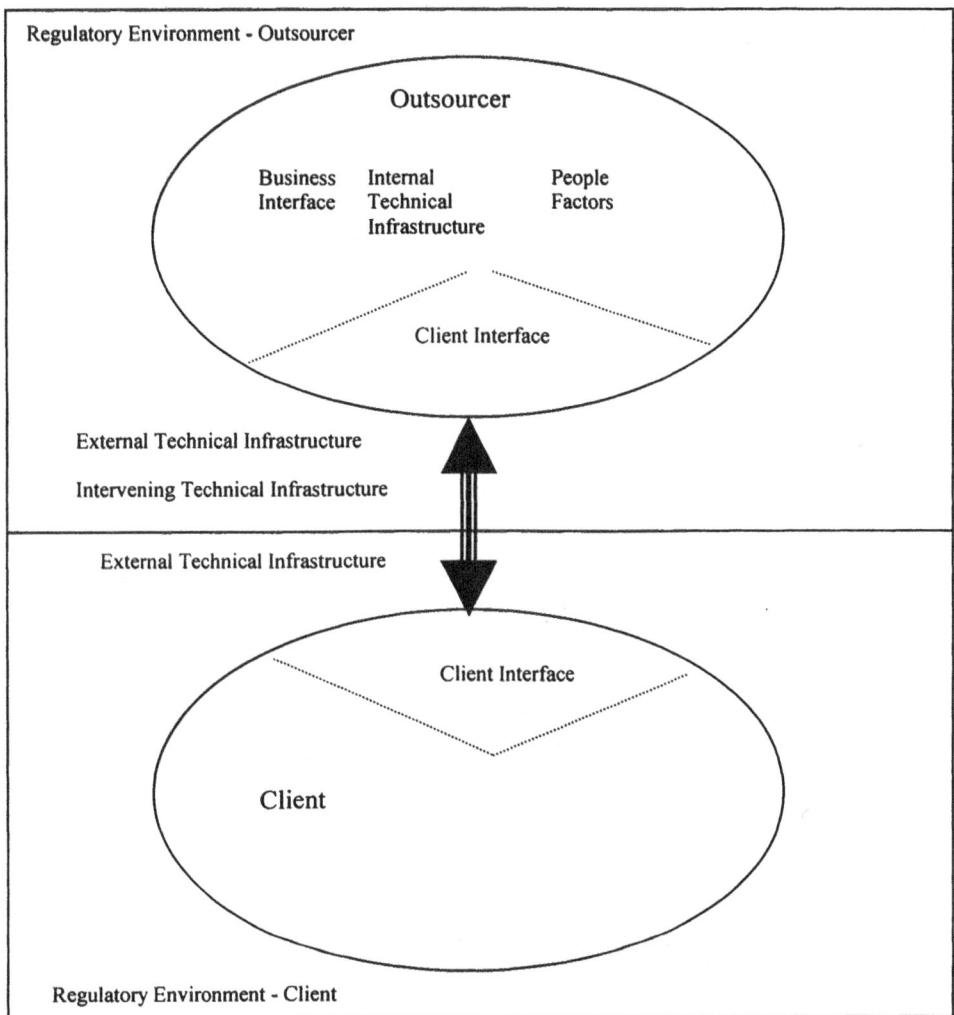

Figure 1. Outsourcer Success Factor Model

People factors attributes PF1 (knowledge skills) and PF4 (project management skills) are considered critical and PF2, PF3, and PF5 are very important. Most interesting is cost not being critical. This indicates awareness that low cost without project performance is not a bargain.

Only Technical Infrastructure attribute TI3, outsource workers having up to date technical skills, is considered critical. However, all other Technical Infrastructure attributes are considered very important.

Table 1. Factor Attributes

Factor	ID	Attribute
People Factors	PF1	Knowledge Skills of outsource workers (read, write well, good overall knowledge of processes and work) support ability of company to do work
	PF2	Language Skills of outsource workers (know the language of the client) support ability of company to do work with other nations
	PF3	Cultural Awareness of outsource workers (understand the culture of the client and how it differs from yours)
	PF4	Project Management People Skills (managers know how to manage workers and users)
	PF5	Cost of outsource workers is low enough to give cost benefits for competing with other outsource companies
Technical Infra-structure	TI1	Telecommunications infrastructure is reliable and supports large file transfers at reasonable speeds
	TI2	Outsource country location has availability of up to date PCs, other computer hardware, and software
	TI3	Outsource workers have up to date technical skills
	TI4	Outsource Managers have software development Project Management skills and tools
	TI5	Technical control processes such as version control, configuration control, etc. are in place
Client Interface	CI1	Client contact point is knowledgeable of client needs and knows technology well enough to communicate the requirements
	CI2	Trust exists between client and outsource company
	CI3	Client contact can communicate in a language the outsource company understands
	CI4	Problem resolution process is in place with the client
	CI5	Time difference between client and outsource locations allows for overlap of work day
	CI6	Travel between client and outsource locations is relatively easy, fast, and inexpensive
Business Infra-structure	BI1	Business Plan in place and guiding company
	BI2	Business Organization (corporation, LLC, etc.) in place
	BI3	Business Process such as accounting, billing, etc. are in place
	BI4	Cost/Cash Control Processes exist with sufficient cash flow/reserves to cover delayed payments
	BI5	Advertising such as a web site, etc. is in place and effective in communicating company abilities
	BI6	Potential client contact methods such as web site, email, phone, fax, etc. are in place and effective
	BI7	Payment Processes are in place and support client payments to company
	BI8	Legal Representation/Support is available for contract review/other
Regulatory Interface	RI1	Intellectual Property Right Protection Laws in place
	RI2	Tax laws allow/favor export/overseas work
	RI3	Banking/Wire Transfer laws support overseas payments
	RI4	Customs/Import/Export Laws support overseas work
	RI5	Exchange Rules/Rates favorable
	RI6	Travel Restrictions (Visa rules) favor business travel
	RI7	Telecom Regulations favor business

The Client attributes CI1 (knowledgeable client contact point), CI2 (trust), and CI3 (common language) are considered critical, CI4 is very important, and CI5 and CI6 are important.

No Business Interface attributes are critical but all are very important. An interesting observation is that clients do not consider advertising by the outsourcer very important, while as expected, outsourcers find it very important. Additionally, there is more deviation between respondents than with the previous factors. This is expected to be due to differing levels of business experience among the respondents.

Table 2. Importance of Factors Attributes

ID	Total		Outsourcers						Clients	
			Total		US		Non-US			
	Mean	Std Dev	Mean	Std Dev	Mean	Std Dev	Mean	Std Dev	Mean	Std Dev
PF1	1.68	0.69	1.61	0.67	1.68	0.69	1.50	0.63	1.75	0.72
PF2	2.21	0.96	2.13	0.99	2.24	1.05	1.93	0.88	2.30	0.93
PF3	2.59	1.03	2.56	0.98	2.68	1.03	2.38	0.89	2.61	1.08
PF4	1.96	0.83	1.98	0.80	1.92	0.81	2.07	0.80	1.95	0.86
PF5	2.58	0.83	2.61	0.92	2.60	0.87	2.63	1.02	2.56	0.76
TI1	2.04	0.88	2.02	0.96	1.68	0.69	2.56	1.09	2.05	0.81
TI2	2.32	0.93	2.33	1.07	2.13	1.08	2.63	1.02	2.32	0.80
TI3	1.69	0.76	1.63	0.77	1.64	0.70	1.63	0.89	1.75	0.75
TI4	2.26	0.91	2.22	0.91	2.12	0.97	2.38	0.97	2.30	0.93
TI5	2.13	0.98	2.07	1.03	1.88	1.05	2.38	0.96	2.18	0.90
CI1	1.79	0.88	1.68	0.76	1.60	0.65	1.81	0.91	1.89	0.98
CI2	1.75	0.89	1.56	0.74	1.56	0.65	1.56	0.89	1.93	0.97
CI3	1.98	0.77	1.95	0.74	2.08	0.70	1.75	0.77	2.00	0.81
CI4	2.28	0.86	2.00	0.83	1.92	0.76	2.14	0.95	2.52	0.82
CI5	3.27	1.04	3.02	1.04	2.80	1.00	3.38	1.02	3.50	1.00
CI6	3.21	1.07	3.12	1.10	2.88	1.09	3.50	1.03	3.30	1.05
BI1	2.40	0.94	2.44	0.92	2.24	0.83	2.75	1.00	2.36	0.96
BI2	2.48	0.97	2.37	0.92	2.28	0.89	2.50	0.97	2.59	1.02
BI3	2.40	1.00	2.29	0.98	2.08	1.00	2.63	0.89	2.50	1.02
BI4	2.62	1.05	2.44	1.00	2.28	0.84	2.69	1.20	2.80	1.07
BI5	2.96	1.03	2.59	0.92	2.68	0.89	2.44	0.81	3.32	1.01
BI6	2.02	0.89	1.88	0.84	1.88	0.88	1.88	0.81	2.16	0.91
BI7	2.35	0.83	2.22	0.79	2.20	0.82	2.25	0.77	2.48	0.85
BI8	2.48	1.14	2.44	1.00	2.32	1.03	2.63	0.96	2.52	1.27
RI1	2.14	1.15	2.05	1.05	1.96	1.06	2.19	1.05	2.23	1.24
RI2	2.48	1.14	2.35	1.14	2.38	1.21	2.31	1.08	2.60	1.14
RI3	2.47	0.99	2.45	0.99	2.54	1.10	2.31	0.79	2.49	1.01
RI4	2.36	1.12	2.35	1.06	2.50	1.18	2.13	0.89	2.38	1.17
RI5	2.75	1.06	2.60	1.03	2.75	1.07	2.38	0.96	2.88	1.12
RI6	2.79	1.04	2.73	1.11	2.58	1.18	2.94	1.00	2.84	0.99
RI7	2.50	1.02	2.33	1.13	2.39	1.16	2.25	1.13	2.65	0.90

No Regulatory Environment attributes are considered critical but all are very important. An interesting observation is that US outsourcers considered intellectual property protection critical while everyone else considered it very important. Additionally, these attributes had the least degree of agreement between respondents. This is because these attributes are very country specific. The lack of agreement between respondents is not considered an issue; rather, it indicates global applicability as the list is not tailored to specific countries or regions.

7. CONCLUSIONS

The list of success factors and attributes presented in this paper can be considered complete and a reasonable indicator of potential success for an outsourcer development company. The list of attributes is considered to be reasonably independent of regional or country biases. Companies from Eastern and Western Europe where looked at while studies from India were used. The strength of these success factors lies in its incorporation of published studies, action research with actual offshore development companies, and validation by a mix of US and Non-US respondents.

As expected worker's skills, communications, and the client interface including trust were the most critical factors. A mild surprise is the strong showing of protection for intellectual property. This can be taken as an indicator that there is more global awareness with respect to the issues of software piracy and knowledge capital. Almost as critical are telecommunications and project management skills. This also was expected. Finally, the strong ratings for the regulatory environment show awareness of the impact governments have on business.

What was unexpected was the relatively low rating for cost given that the premise of this paper was that cost drivers are the primary reason for offshore outsourcing. It indicates that awareness is growing that it takes more than low cost programmers to ensure successful projects, which is ultimately the finding of this paper.

Ultimately, the value of this list is to startup companies as it provides a good listing of what it will take to build a successful company. There is also value to companies that outsource development as it provides a listing of criteria that should be looked at when selecting an offshore or inshore development company.

REFERENCES

Aubert, B. A., Patry, M., and Rivard, S., *Assessing the Risk of IT Outsourcing*, IEEE, 1998.

Central Intelligence Agency World Fact Page for Ukraine located at: http://www.odci.gov/cia/publications/factbook/geos/up.html

Central Intelligence Agency World Fact Page for Italy located at: http://www.odci.gov/cia/publications/factbook/geos/it.html

Jennex, M.E., *IT in the Energy Sectors of Ukraine, Armenia, and Georgia*, Global Information Technology Management Conference, GITMA, June 2001.

Jennex, M. E., Sears, D., Furumasu, B., Severn, S., Olson, J., Khoeler, R., Bogardt, W., and Holliday, R., *Ukraine Y2K Risk Assessment Final Report*, U.S. Aid Project, October, 1999.

Olearchyk, R., *Software Development Workforce Slipping*, Kyiv Post, November 29, 2001, found at http://www.kpnews.com/main/10154

Rajkumar, T. M. and Mani, R. V. S., *Offshore Software Development, the View from Indian Suppliers*, Information Systems Management, Spring 2001, Volume 18, Issue 2.

Raval, V., Seven Secrets of Successful Offshore Software Development, *Information Strategy: The Executive's Journal*, Summer 1999, Volume 15, Issue 4.

Smetannikov, M., The New Russian Revolution, *Inter@ctive Week*, June 4, 2001, Volume 8, Issue 22.

ACTABLE INFORMATION SYSTEMS
- Quality ideals put into practice

Stefan Cronholm, Göran Goldkuhl[*]

1. INTRODUCTION

The problem we are approaching in this paper is that the actions offered by information systems (IS) often seem to disharmonise with the actions performed in the work practice. Several researchers report lacks in IS use. For example, Hägerfors (1994) claims that there is a lot of IS which is not fully usable in the context wherein they exist. Henderson & Kyng (1994) claims that there is a discrepancy between creation of IS and work situations. Bannon (1994) claims that there is need for a better understanding among researchers and system designers about users and their work settings. We need to understand people as actors with a set of skills and shared practices based on work experiences (ibid.)

There are many different philosophies, methodologies or checklists aiming at supporting the information systems development (ISD) process. One of the most popular methodologies today is the object-oriented approach Rational Unified Process (RUP), (e.g. Kruchten, 1999). Another familiar methodology is Structure Analysis and Structure Design (SASD) (Yourdon, 1989). The tradition of participatory design has paid a lot of attention to user influence in the ISD process. Followers argue for a broad and genuine participation aiming at agreement of IS and work (Hägerfors, 1994). Carlshamre (1994) claims that participatory design is more of a philosophy than a methodology. Vonk (1990) discusses a prototyping approach and claims that this approach will put more attention to the user interface then traditional ISD methodologies do. In the Human Computer Interaction (HCI) area usability and IS are focused. In this area we can find checklists such as Nielsen's (1994) ten usability heuristics and usability models such as

---

[*] Stefan Cronholm, Dept. of Computer and Information Science, Linköping University, SE-581 83 Linköping, Sweden. Göran Goldkuhl, Centre for Studies on Human, Technology and Organisation (CMTO) Linköping University, SE-581 83 Linköping, Sweden

Information Systems Development: Advances in Methodologies, Components, and Management
Edited by Kirikova *et al.*, Kluwer Academic/Plenum Publishers, 2002

135

Shackel (1984) and Nielsen (1993). What they all seems to miss or at least not have in focus is the action character of the IS.

The purpose of the paper is to propose actable quality ideals for ISD. Following this purpose we have formulated our research question as: How can more actable IS be achieved? In order to answer our research question we have used an action-oriented perspective as a basis (see section 3). This perspective has been operationalised into actable quality ideals that support the ISD process (see section 5). In section 6 we show some examples where the quality ideals are put into practice.

2. THEORETICAL BASE: INFORMATION SYSTEMS ACTABILITY

There are many different approaches to support ISD. Even though they all aim at building good IS, they can differ in several ways. These differences often depend on underlying values and perspectives. When designing IS there is always one (or several) perspectives applied (e.g. Nurminen, 1988; Löwgren 1995). Nurminen (1988) claims that a perspective is defined through how to look at a phenomenon and how to think of a phenomenon. Perspectives contain principles, values, ideas, experiences, categorisations and definitions. Further more, perspectives can exist on an unconscious level, not reflected, unclear and not sufficiently articulated. It is obvious that the choice of perspective has implications for how we design IS.

We adopt an action perspective on information systems in this paper. Actions are humans' intentional way of changing the world. Humans intervene in the external world. These intervening actions are overt actions, which can be communicative or material. Actions can also be covert. In such a situation a human tries to make sense of something external. He performs an interpretative action. He is not changing something externally as in intervening actions. He is instead trying to change his inner world, his knowledge of the external world. Besides interpretative actions, there are other covert actions. When a human is intentionally trying to solve a problem mentally through reflection this can be seen as a covert action. This action view is inspired by American pragmatism (e.g. Mead, 1938), social phenomenology (e.g Schutz, 1962) and language action theories (e.g. Austin, 1962). Confer also Goldkuhl (2001) and Goldkuhl & Ågerfalk (2002).

From an action-oriented perspective, IS are viewed as communication systems, as distinct from strict representational views of information. A representational view of information means that designers try to create an 'image' of the reality in order to have the analysed piece of reality properly represented in the systems database. This strict representational view can be challenged, which an action perspective certainly does (e.g. Goldkuhl & Lyytinen, 1982; Winograd & Flores, 1986). In the action-oriented perspective, IS are not considered as "containers of facts" or "instruments for information transmission" (Goldkuhl & Ågerfalk, 2001). The action-oriented perspective emphasises what users do while communicating through an IS (*ibid.*). IS are systems for action in work practices, and such actions are the means by which work practice relations are created.

IS have an action ability. We call this IS actability. We define actability as an IS ability to perform actions and to permit, promote and facilitate users to perform their actions both through the system and based on messages from the system, in some work practice context (Ågerfalk, 1999; Goldkuhl & Ågerfalk, 2001).

Within the actability perspective the notion of IS can be defined in the following way (ibid): An IS consists of

- an action potential (a predefined and regulated repertoire of actions)
- actions performed interactively by the user and the system and/or automatically by the system.
- action memory (a memory of earlier actions and including other prerequisites for action)
- documents (as action conditions, action media, action results)
- a contained structured work practice language (giving frames for actions, action memory and documents)

Designing an IS means suggesting and establishing an action potential. An action potential both enables and delimits actions. It entails a repertoire of actions and a related vocabulary. The vocabulary consists of concepts related to the work practice language. An IS must also offer a record of actions performed. Information about these performed actions can normally be found in the IS database. We call it an action memory, which is part of an organisational memory.

In order to design actable IS three types of IS use situations have been identified in Ågerfalk (1999). First, there is an interactive use situation (a user interactively using an IS, for example, registering some information). Secondly, there is an automatic use situation (an IS perform an action independently by the user, but of course according to some predefined rules). Finally, there are consequential use situations (information from the IS that will be used in other situations). The generated quality ideals are of the interactive use situation type.

3. RESEARCH APPROACH

In this study we have used an action-oriented perspective (see section 2). Contemporary approaches do not seem to give the necessary support for specifying IS in a way that sufficiently emphasises the action character of information and IS needed to achieve the work practice goals (Ågerfalk, 1999). In several IS it is not clear which actions are possible to be performed in a specific use situation. One reason for applying an action-oriented perspective is that there is a lot of communication needed in the work practice studied.

IS Actability is articulated as a perspective (see Ågerfalk, 1999; Goldkuhl & Ågerfalk, 2001) and as a requirement engineering method (ibid; Ågerfalk et al. 2001). The notion of an actable information system is described in a general way. There is however not in this literature an explicit list of quality ideals for IS actability, which can govern the design of such systems. When reflecting upon this perspective and this Requirement Engineering (RE) method we obtained some preliminary ideas of how to formulate quality ideals for IS design. Having these preliminary ideas in mind we carried out an ISD-project (see section 4). The ISD-project has contributed to a refinement of these ideas. This means that quality ideals were developed continuously during the ISD-project, which was performed in an action research manner. The developed quality ideals have guided the design process and they have been implemented in a prototype and been tested on a small scale (see figure 1).

The research approach has been performed in a dialectal way where both theoretical claims and empirical findings from the ISD-project have influenced the quality ideals. In order to structure our quality ideals we have used a paradigm model suggested by Strauss & Corbin (1998). The model has acted as a means for categorising the ideals. The reason for choosing this model is that it is action oriented and the categories (ideals) identified are either considered as conditions for or as consequences of actions/interactions. This model also harmonises with our general pragmatic view on research and IS.

The quality ideals have been tested by user reviews of prototypes during the ISD process and limited user test of executable prototypes. For full empirical grounding there is a future need for studying users full utilization of implemented IS in their work.

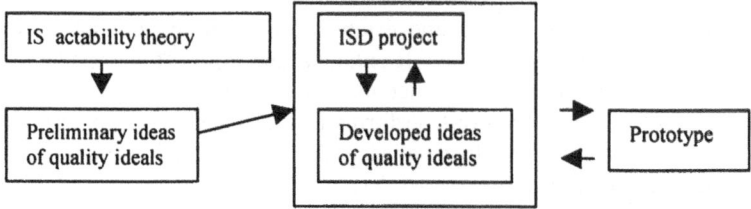

Figure 1. The overall research model

4. DESCRIPTION OF THE WORK PRACTICE AND THE ISD PROJECT

The ISD project was performed in a municipal home care unit for serving elder people. The major tasks of the home care are to help the elders with daily hygiene, simple medical tasks, cleaning, doing laundry, shopping etc. The personnel consists of two home care managers who are responsible for the home care unit and a number of home care assistants. The home care assistants are responsible for the daily work with the elders.

The home care assistants are well qualified and experienced. Their work can be characterised as flexible and responsive to the different needs of the elders. This kind of flexibility is also characterising the administrative work at the home care unit. The home care assistants are governed in their work by much tacit knowledge. Documentation routines have evolved gradually. There are many types of documents; a number of self-made as well as pre-printed forms (e.g. journals, diaries, note pads, schedules etc.). These documents are used for communication about clients, assignments, measures and work procedures. Many documents, especially the self-made forms, are unclear. There are no exact rules for what to write in different documents. The terminology is rather fluid. Many documents lack a clear rubric and after intervewing the staff it became obvious that some documents lack a common name. From an information systems perspective it is easy to be critical towards this fluid and vague communication and document treatment.

There are programs for improved quality assurance in the home care service. There are initiatives made to have a more ensured home care service. The home care routines should be designed in ways making it possible even for inexperienced substitutes to perform work in a proper way. This necessitated a redesign of several work documents and the introduction of prescriptive routine descriptions. It necessitated the development of IT-based information systems. In the ISD project four researchers and two home care assistants and two home care managers participated.

One main objective for the home care service is the individualisation of the home care. To perform home care is not a standardised service. The home care unit strives for maximum individualisation. The elder clients should live their lives in their own desired ways. The home care assistants should support the clients to live in their own ways. In order to do this there is great need for knowledge. The home care assistants must have a good understanding of every person, about their personal life history, their current social and medical situation and their habits and needs. This partially changing knowledge must be transferable to all members of the home care team since there is not one single assistant who takes care of a particular elder. One objective of the IS to be developed was to contribute to this knowledge sharing (Goldkuhl & Braf, 2001).

5. ARTICULATION OF QUALITY IDEALS

Earlier literature on IS actability (e.g. Goldkuhl & Ågerfalk, 2001; Ågerfalk, 1999; Ågerfalk et al, 2001 is not, as said above in sec 3, explicit concerning quality ideals. In order to develop and refine quality ideals we have based our work on an elementary interaction loop (EIAL) model (Ågerfalk, 1999). The EIAL model describes in a generic sense the interaction between a user and a computerised IS. The interaction is divided into three phases within the loop: 1) user action, 2) IS action and 3) interpretation (by the user). When working with the quality ideals we have refined this model.

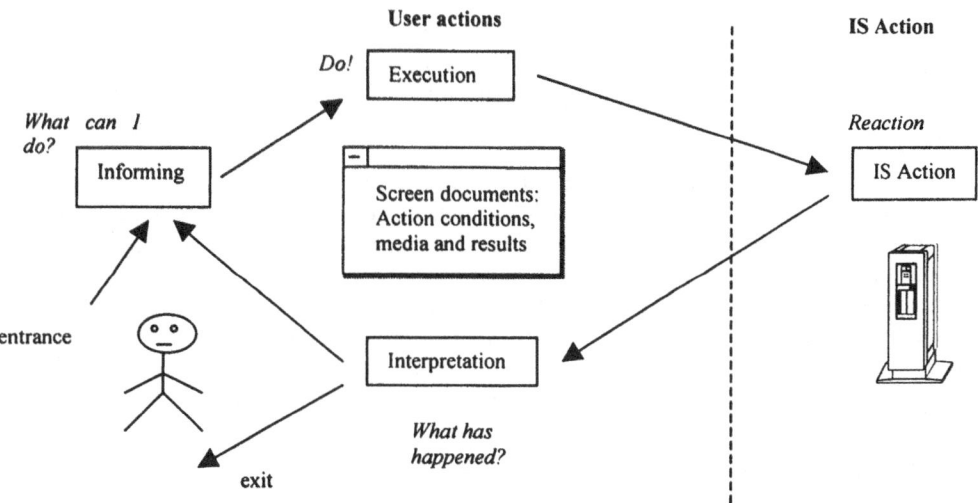

Figure 2. The elementary interaction loop (EIAL) – refined model

The first phase (user action) has been divided into two phases: Informing and execution. This means that our refined model consists of four phases: Informing, execution, IS action and interpretation (see figure 2). It describes in a generic sense the interaction between a user and a computer-based IS. In the informing phase the user has to be informed from the screen document about what can be done. He/she must have knowledge about which possible actions can be carried out. After being informed, the user executes an action (for example by clicking on a button on the screen document). The IS reacts by performing its corresponding IS action. When the IS action is performed the user interprets what the IS has done.

The screen document plays important roles in the interaction. One can say that the screen document is multifunctional. It contains *information about the action possibilities and other action conditions*. In this sense it used in the informing phase by the user when he is reading the screen figuring out what to do. The screen document functions as an *action media* in the execution phase when the user for example clicks on some button when performing his action. The user can also (in the execution phase) type some information in a field and the screen document consists in this sense of *action results* of the user execution action. The IS action can result in changes of the screen document (as a feed-back to the user). This means that it contains of *results of the IS action*.

The earlier (three steps) EIAL model has been compared to the action model of Norman (1988) in Goldkuhl & Ågerfalk (2001). Performance (= user action) and assessment (= interpretation) are explicit steps in Norman´s model. The reaction part (= IS action) is however missing which is noticed (ibid). The refined EIAL model can be compared to the classical action model developed by Mead (1938). Mead's model consists of four stages of an act: 1) impulse, 2) perception, 3) manipulation, 4) consummation. Our first phase (informing) corresponds to Mead´s two first perceptual stages. Before one can act, one must perceive the action environment and become informed about action possibilities. The actor assigns meaning to the situation in accordance to his pre-understanding. This pre-understanding is "action-penetrated", i.e. the world is understood as a world to act in. Our second phase (execution) corresponds to the manipulation stage of Mead, and our last phase (interpretation) corresponds the last stage of Mead (consummation). Mead defines an action consisting of all these four stages. We define each of the four phases in our interaction loop as separate actions. This follows the principal division of intervening and receiving actions made by Goldkuhl (2001).

Our analysis has resulted in two modes of interaction: navigation and performance. We distinguish between navigation and performance. In order to navigate, the user must first inform him-/herself about: where am I? and in what direction am I heading? We call this part of navigation for orientation. The answer to these questions makes it possible to make a move in the IS. Navigation consists of both orientation and movement. From the navigation mode the user can reach a performance mode. In the performance mode the users perform work practice actions. We have used the model in figure 2 for descriptions of quality ideals both for the navigation mode and the performance mode.

5.1. Navigation mode

5.1.1. Informing

5.1.1a. Orientation. Orientation should not be confused with navigation. Need for orientation arises when a use situation is approached. The users need to locate the current use situation in relation to other use situations (Nygren, 1996). Orientation is to understand where I am and to decide where I want to move.

5.1.1b. Focusing in contexts. As mentioned above, the studied work practice can be characterised as complex with a high degree of communication needed. This implies that there is a need for understanding a specific task in a context. In order to fulfil or solve a specific task you need information from the IS (the context). The context acts like a background and the specific task as the foreground (se figure 3). This means that the context work as conditions for action. This is in line with hermeneutic theory of interpretation (e.g. Bleicher, 1980).

5.1.1c. Supporting understanding of navigation principles. There are several types of navigation. First, there is hierarchical navigation which for example take place when you access the next higher or a lower level in an IS. Second, there is sequential navigation. Sequential navigation means that you access an adjacent use situation within the same level. Finally, there is direct navigation. Direct navigation means that it is possible to access a use situation anywhere in the IS. We claim that the type of navigation offered by the IS must be clear in order to support the users mental model of the IS.

5.1.1d. Informing about action mode possibilities. The action mode of the use situation should be explicit. Is it a read, an update or a write situation? Confusion about what mode of action offered should be eliminated. We claim that text, in for example buttons or other screen elements used for navigating, should be built up with the name of the action combined with the name of the actual object (i.e. planning tasks, register information about clients).

5.1.2. Execution

Actions should be easy to execute. This means that the way of how the execution is performed should be easily managed. The selection of the desired destination must be done easily and without hesitation. The IS should support execution alternatives for both novices and experts (e.g. Nielsen, 1993).

5.1.3. IS-action

The IS should be able to execute what is asked for; i.e. perform the correct movement in the document space.

5.1.4. Interpretation

A feedback should be given to report whether that the intended navigation has succeeded. A way to do this is to clearly label each use situation. The user needs to interpret what the IS has done (see figure 2). He/she must recognise that the action has succeeded.

5.2 Performance mode

5.2.1. Informing

5.2.1a. Understanding of the screen document. The contents of the screen document should offer good conditions for performing actions both within the IS and outside the IS. This means that information presented must be easily interpreted, actions must be easily accessible and understandable. The user shall understand the consequences of offered actions. Relations between actions performed within the IS must be visualised in a way that the users easily understand if there is a specific order among the offered actions.

5.2.1b. Action memory – easily accessible. Earlier stored information should be easy to access. This means that information about previous actions should be easily accessible. The action memory can consist of both historical information (actions that have been performed and other action conditions) and expected actions (actions that should be performed).

5.2.1c. Action memory – personalised. In our case study there was a lot of communication between home care assistants. The home care assistants communicated through messages, both spoken and written. Often when written messages were used the receiver of the message felt that there was a need to know more than that actually has been written. There was a need for contacting the sender of the message. We claim that it should be clear who is responsible for the content of the message. Information about "who has said what" should be stored in the IS as part of the action memory. This quality ideal can be seen as an exhortation to avoid anonymity in information systems.

5.2.2. Execution

Actions should be easy to execute. This means that the way the execution is performed should be easily managed. The IS should support execution alternatives for both novices and experts (e.g. Nielsen, 1993).

5.2.3. IS-action

The IS should be able to execute what is asked for. IS execution can consist of changing the action memory (data base) and/or sending some message outside the system (Ågerfalk, 1999). The IS can also change the contents of the currently used screen document as a feed-back to the user (see below).

5.2.4. Interpretation

The IS should always give an understandable response to a performed action (see figure 2). The response should consist of a description of the IS action performed (and eventually the actions that will be performed). The user should understand the consequences of chosen interactive actions.

5.2.5. General quality ideal

5.2.5a. Work practice language. The IS language should harmonise with the work practice and the users' language. There should be no confusion of the meaning of the concepts used. The IS should offer explanations of all concepts and a description of the actions that could be performed through the IS.

6. THE QUALITY IDEALS PUT INTO PRACTICE

Below, there are two screen documents presented (see figure 4 and 5). The purpose of showing the screen documents is to illustrate how some of the quality ideals discussed above can be put into practice. As you see in figure 4, we have labelled the buttons with name of the action combined with the name of the actual object. In figure 5, the title of the screen document is clearly labelled with "Perform tasks". This is the same name as appears in the menu (see figure 4). This is an easy way to support navigation and feedback.

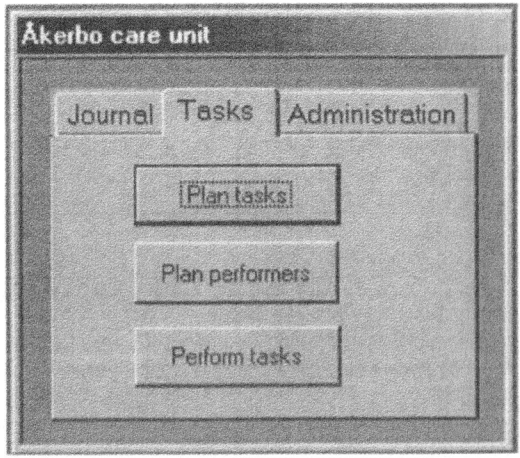

Figure 4. Screen document: Main menu

In the upper part of the screen document "Perform tasks" (figure 5) there are several parameters that can be adjusted according to what should be shown in the grid below. All the parameters help the user to make a selection of the tasks he/she wants to look at. Together with the buttons below they form the action repertoire. The language and concepts that appear are the same as are used in the work practice. Using a familiar language and clear labels helps the user to more easily understand the content of the screen document and which actions that can be performed.

It is also easy to access earlier stored information. If the user wants to check if all the tasks for yesterday were performed he/she simply changes the date and the grid will show the status for yesterday tasks.

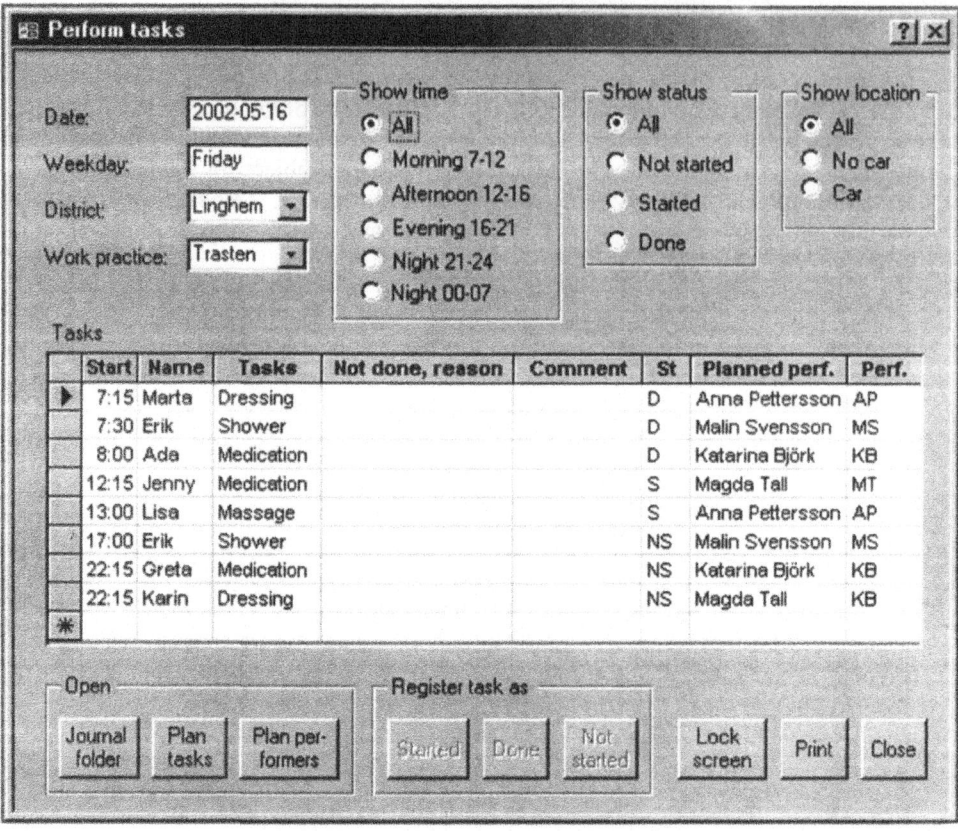

Figure 5. Screen document: Perform tasks

7. SUMMARY AND DISCUSSION

In this paper we have argued for the importance of viewing IS as an action system. This study can be criticised for using a perspective that is too limited. Of course, when designing IS you need to take care of different aspects such as organisational, cognitive, economical etc. All perspectives have their limitations and opportunities. Perspectives tell us what to view, they are not telling us what we are missing to view. We claim that the action-oriented perspective has brought forward several important quality ideals for designing IS.

When we make a brief comparison between the action perspective quality ideals to other perspectives mentioned in section 1 we can see that there are both differences and similarities. The object-oriented approaches have an obvious focus on objects and relations. However, they are not neglecting the action part (see i.e. Behaviour Diagrams in Unified Modelling Language (Object Management Group, 2002)) but it is not as well articulated as in the actability approach. SASD is a traditional ISD method that focuses on data. Well-known diagrams in this method are Dataflow Diagrams, Entity-Relationship Diagrams and State-Transition Diagrams. In these diagrams there is no way to describe user actions. The perspective in SASD method is data oriented. We have not compared the quality ideals with participative design since this perspective is more of a philosophy than a methodology.

In the HCI area we can find Nielsen's 10 heuristics (1994). Some of these heuristics are similar to quality ideals. For example Nielsen (ibid.) claims that "The system should speak the users' language" and "The system should always keep users informed about what is going on, through appropriate feedback within reasonable time". The first quote has relations to what we have formulated as "The IS language should harmonise with the work practice" (see section 5 "Work practice language"). The second quote are related to what we are calling "feedback". There are also differences. It is clear that Nielsen's 10 heuristics are more oriented towards the user-tool relation in Shackel's model (see figure 6). The heuristic that is closest to the task component is "match between system and the real world". This heuristic discusses the importance of that the system should use business concepts that are familiar to the user. The heuristics can be classified as a checklist for user interface design. The heuristics are not confronting the task component to the same extension as the user and tool. Another difference is that the heuristics are more oriented towards cognitive aspects of the user than the quality ideals.

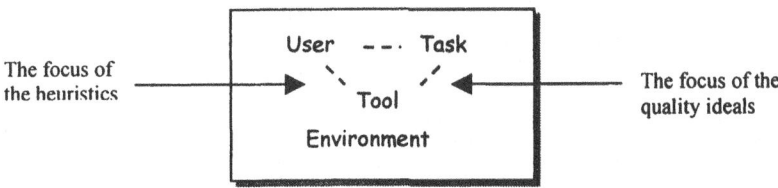

Figure 6. Overview of focus (utilising the usability model from Shackel, 1984, as a base)

A lack in Nielsen (1994) is that he presents the heuristics in a sequential list without any order. We have categorised our quality ideals according to two use situations and into conditions for and consequences of action.

We claim that if one follows the quality ideals described in this paper, one will have a high degree of probability to reach an actable information system. Of course there are other ways to arrive at actable systems. Following other approaches does not exclude the possibility to create an actable system. For example following object-oriented approaches or Nielsen's usability heuristics might well lead to actable systems although those approaches do not contain explicit criteria for actability design. In such cases actable systems are created *by chance*. In the case of using the actability quality ideals, actable information systems are created by *intentional and conscious design*.

Our results are based on one case study. Despite this, we think that it is possible to make some generalisations. We think that our approach should be considerate in every context where a high degree of communication exists. We also think that several of the quality ideals have more of a general character that they should be valid for the most IS.

We can see two interesting future research directions. One direction is to study the effects in the work practice from using the prototype. Will the actions offered by the prototype harmonize with the actions performed in the work practice? The other direction is to more thoroughly compare our quality ideals with the usability heuristics in Nielsen (1994) and to other ISD approaches.

REFERENCES

Austin J L (1962). *How to DoThings With Words*. Oxford University press

Bannon L (1994). From Human Factors to Human Actors, in *Design at Work* (Greenbaum J & Kyng M eds), Lawrence Erlbaum Associates. Hillsdale, New Jersey.

Bleicher J (1980). *Contemporary hermeneutics*. Hermeneutics as method philosophy and critique. Routledge & Kegan Paul, London

Carlshamre P (1994). *A Collaborative Approach to Usability Engineering*, Licentiate Thesis. Dept of Computer and Information Science, Linköping University

Goldkuhl G (2001). Communicative vs material actions: Instrumentality, sociality and comprehensibility, in Schoop M, Taylor J (Eds, 2001), *Proocedings of the 6th Int Workshop on the Language Action Perspective (LAP 2001)*. RWTH, Aachen.

Goldkuhl G, Braf E (2001). Contextual knowledge analysis - understanding knowledge and its relations to action and communication, in proceedings of *the 2nd European Conference on Knowledge Management*. IEDC-Bled School of Management, Slovenia

Goldkuhl G & Lyytinen K (1982). A language action view of information systems, *in proceedings of third Intl. Conference on Information systems*, Ginzberg & Ross (eds). Ann Arbor

Goldkuhl G & Ågerfalk P (2001). Actability: A way to understand information systems pragmatics, *in Coordination and Communication Using Signs: Studies in Organisational Semiotics - 2*, Liu, K. et al. (eds.). Kluwer Academic Publishers, Boston, to appear in December 2001.

Henderson A & Kyng M (1994). There's No Place Like Home, in *Design at Work* (Greenbaum J & Kyng M eds), Lawrence Erlbaum Associates. Hillsdale, New Jersey.

Hägerfors A (1994). *Co-learning in Participative Systems Design*, PhD thesis. Lund University

Kruchten P (1999). *The Rational Unified Process: an Introduction*. Addison Wesley Inc. Reading, MA.

Löwgren J (1995). *Perspectives on Usability*, Dept. of Computer and Information Systems. Linköping University

Mead G H (1938). *Philosophy of the act*. The University of Chicago Press

Nielsen J (1993). *Usability Engineering*. San Diego, California, Academic Press

Nielsen J (1994). *How to Conduct a Heuristic Evaluation*. http://useit.com/papers/heuristic/heuristic_evaluation.html

Norman D A (1938). *The psychology of everyday things*. Basic Books, New York

Nurminen M (1988). People or Computers: Three Ways of looking at Information Systems. Studentlitteratur, Lund

Nygren E (1996). *From Paper to Computer Screen*, Human information processing and user interface design, PhD thesis. Dept. of Technology, Uppsala University, Sweden

Schutz A (1962) *Collected papers I*, Martinus Nijhoff, Haag

Searle J R (1969). *Speech Acts. An essay in the philosophy of language*. Cambridge University press, London

Shackel B (1984). The Concept of Usability, in *Visual Display Terminals: Usabiblity Issues and Health Concerns* (Bennet J, Case D, Sandelin J and Smith M eds), Prentice Hall, Englewood Cliffs, NJ

Strauss A, Corbin J (1998). Basics of Qualitative Research, Techniques and Procedures for Developing Grounded Theory. Sage Publications, Beverly Hills, California

Winograd T & Flores F (1986). Understanding Computers and Cognition: A new foundation for design. Ablex Publishing Corporation, N J

Vonk R (1990). *Prototyping*. Prentice Hall, Englewood Cliffs, New Jersey

Yourdon E (1989). *Modern structured analysis*. Prentice Hall, Englewood Cliffs, New Jersey

Ågerfalk P J.(1999). Pragmatization of Information Systems – A Theoretical and Methodological Outline, Licentiate thesis. IDA, Linköping University, Sweden

Ågerfalk P, Goldkuhl G, Cronholm S (2001). *Actability Design*. Dept of Economics, Statistics and Computer Science, Örebro University. Sweden

Object Management Group (2002). UML Resource Page. http://www.omg.org/uml/. Site accessed 2002-02-20

MANAGEMENT SUPPORT METHODS RESEARCH FOR INFORMATION SYSTEMS DEVELOPMENT

Malgorzata Pankowska*

1. INFORMATION SYSTEMS DISCIPLINE

Information systems are considered as an academic discipline at universities. It is defined as the effective analysis, design and implementation of information systems and information technology in social organizations. Information systems domain is multidisciplinary and includes strategic, tactic and operational activities connected with collecting, processing, storing and disseminating, usage of information. These activities constitute information management process. Information systems as a research discipline is focusing on information as well as on information technology and software development. Information processing and communications are known from the beginning of human activity and people always had the same problems of appropriate information selection and collection for the decision-making.

Although the key to a critical understanding of the system idea can be found in Kant, in this paper system is perceived functionalistically and cybernetically. Contemporary systems science has failed to understand the critical significance of the systems idea in Kant (Ulrich, 1983). Whereas contemporary systems science understands the systems concept functionalistically, as referring to a set of variables to be controlled in a context of instrumental action, Kant understands it as referring to the totality of relevant conditions on which theoretical or practical judgments depend, including basic metaphysical ethical, political and ideological a priori judgments. Much of present-day systems science relies on a cybernetic systems concept, not necessarily on the mechanistic systems concept of early cybernetics of Wiener, or Ashby, but rather on the so-called organic systems concept. The organic paradigm claims to overcome the narrowness of the machine paradigm by taking account of the intrinsic capability of complexity absorption, self-regulation and self-organization (organic control) that is characteristic of biological, ecological and social systems (Morgan, 1996). The organic systems concept of cybernetics appears to be oriented toward the idea of intrinsic control rather than intrinsic motivation. Intrinsic control means the capability of a system,

* Malgorzata Pankowska, Information Systems Department, University of Economics, 40 226 Katowice, Poland.

Information Systems Development: Advances in Methodologies, Components, and Management
Edited by Kirikova *et al.*, Kluwer Academic/Plenum Publishers, 2002

independently of an external controller i.e. by means of internal complexity control, to maintain its stability (e.g. its structure, its boundaries, and particularly a given goal state) across a range of environmental or internal variations. The idea is that the sources of control are spread through the architecture of the system; the controller and the controlled are inseparable. System's source of purposefulness (source of motivation) when this source cannot be localized within an external decision maker but is distributed throughout the system itself. Intrinsic control does not imply intrinsic motivation; just because a system is capable of maintaining a given goal state by means of intrinsic homeostatic control, it is not necessarily self-responsible with regard to the purposes it serves. Cybernetically rational management of complexity is then what helps a system to generate requisite control variety or else what reduces its need for control variety by destroying environmental perturbation variety.

From the point of view of practical reason the system planner must reflect upon his activity as an effort in ISs design rather than merely methodology design. Methodology design is a problem-solving attitude that takes problems and purposes as given and refers decision on them to rational acts. Methodology design relies on a decisionistic model of the relation of technical and social sciences on a pragmatist model. The general methodology has its status because of its success in use, but an every future use of the methodology is the opportunity to adapt, test and evaluate with a view to improvement.

Four elements are present within any problem-solving context: the problem situation, the problem solver (developer), the problem solving process (methodology) and the evaluation. In any given context in the problem solving process the problem solvers will utilize their mental constructs to influence their thinking processes and actions (Hudges, 1999). These mental constructs influence their own sense-making and decision making activities and thus explain the ways in which a methodology is used differently by one person as compared with another and indeed as compared with the original proponents of the methodology.

2. INFORMATION SYSTEMS DEVELOPMENT

Information systems development literature mainly includes the often perceived view that the system development process is technical and predominantly rational. This prevalent position is coupled with a view of information systems methodology as providing common standards of practice and documentation. System development is intentional; to the extent it reflects a planned change. It is based on developers' intentions to change object systems towards desirable ends. Information systems development is an organized collection of concepts, beliefs and normative principles supported by material resources. The system methodology purpose is to help a development group successfully change object systems that is to perceive, generate, assess, control, and to carry out change action of them.

All methodologies include concepts and beliefs that enable developers to identify and order phenomena. All of them suggest a pool of methods, languages and techniques (i.e. normative principles) for going about representing, selecting and implementing the change. Methodologies act like perceptual filters, which identify certain phenomena at the cost of neglecting other. Each information system can be analyzed in three contexts: the technology context, the language context, and the organizational context (Boland & Hirschheim, 1987).

A technology context means a viewpoint, which confines object systems to a view of how to efficiently process and store signs (data) in some material carrier. Information system in this context consists of work-processes that are integrated user-machine systems. During systems development there is a need to select a technology context, because in the final stage every information system is introduced as a technical system consisting of hardware, software, database, models, and manual procedures. Phenomena in this context are mainly deterministic. The goal emphasize the effectiveness of technological change: to minimize cost or maximize monetary benefits over the life span of the information system. This can be expressed by several measures, which represent systems quality attributes such as robustness, portability, and efficiency of operation, fault-tolerance. In this context quantitative methods are widely applied like statistical analysis of faults and failures.

The language context means a viewpoint which confines object systems to the use, nature, content, context and form of signs included into the information system. The language context is necessary in systems development, because an information system always has a symbolic function. In general there are no universal desirable properties of languages. They depend on the chosen concept structure and its preferred view of the linguistic phenomena.

The organizational context means a viewpoint, which confines information systems to the origin, nature, purpose and form of systematic relationships and interactions between people. The organizational context is necessary in systems development, because the information system must serve the organizational mission. Information system is to achieve a particular objective, some measures of that degree of achievement must be defined and activities included in the model that make use of the measure to take control action to improve that degree of achievement. This is defined as a measure of performance and information collected according to that measure would be used by some decision-taking procedure to take control action through control mechanisms. Thus if the system objective is defined as the satisfaction of a perceived market need, the measure of performance must be related to how well the particular sector of the market is satisfied. i.e. in terms of market share or customer complaints or some combinations of the two. Based upon information collected in these categories action can be taken to improve the product or improve market definition or selling activities. So development of information systems involves three categories of factors: human factors (motivation, ergonomics, and aesthetics) technical factors (function, mechanism, structure) and business factors (production, economics, presentation, support). This interdependences defines the need to utilize management methods to support information systems development

3. MANAGEMENT SUPPORT METHODS FOR INFORMATION SYSTEMS DEVELOPMENT

Scientific management is an organizational ideology and a set of techniques conceived to deal with such problems as control, supervision, order or disorder, managerial arbitrariness, greed, power. As an ideology scientific management assumes that all actors are rational. Humans are rooted in the excessive mechanization and individual industrialization achieved by scientific management. Workers are seen as people with group identification and emotional dependences, driven by psychosocial norms and needs. Managers are not essentially different from workers, but their superior ability and skills in dealing with human relationships allows them to exercise self-control. Management is a social activity focusing on materials, information and people as well. So it can be seen as the process of acquiring and combining human, financial, and physical resources to attain the organizational primary goal of producing a product or service desired by client.

Economic organizations have a wide range of methods supporting effective and efficient management for cost reduction, service improvement, and product quality increase (Euske & Player, 1996). Management Support Methods (MSMs) include as following:

- Quality-based methods - Quality Function Deployment, Total Quality Management (TQM)
- Activity-based methods - Activity-Based Costing (ABC), Activity Based Management (ABM)
- Personnel-based methods - Individual and Organizational Learning Development, Outsourcing, Total Employee Involvement (TEI)
- Time-based methods - Kanban, Just-In-Time (JIT), Total Productive Maintenance, Overall Equipment Effectiveness, Time Critical Manufacturing, Time Compression Management (TCM)
- Technology - based methods - MRP, MRPII, Distribution Resource Planning, CASE tools methods (like Critical Path Method CPM)
- Process- based methods - Business Process Reengineering (BPR), Process Mapping, Benchmarking (see Figure 1).

These methods enables multidimensional improvement of organizations and also their information system. Some selected methods can constitute a methods' portfolio optimized from the general value perspective. Market globalization creates new risks and opportunities. Practices and management philosophies, customers' expectations, product standards, production processes, market approaches change frequently. Companies supporting by the last management information systems can effective produce according to MRPII method and can deliver products just in time (JIT). Also information systems development changes frequently, similarly as for many other products. Mentioned above methods are applied to support information systems development. Business environment dynamics demand to constantly renew information systems (upgrade implementation) to come up changes business condition in enterprises. Although Computer Aided Software Engineering (CASE) packages causes reducing the work input for information system maintenance as well as easily modified application development but still there is the pressure concerning development of new alternative technologies.

Correct and appropriate development of information systems depends on implementation of management methods to support information systems similarly as any other product launch. Reduction of information system life cycle means immediate improvement of other business functions within company. The enterprise can get competitive advantage through the decrease of time necessary for conceptualization, design, realization and implementation of information systems. Information System Life Cycle reduction is the subject of studies supported by Time Compression Management (TCM) and Quality Management (Cordata, 1995). Total Quality Management (TQM) is the collection of strategic practices that enable implementation of constant improvement initiatives. TQM strategies form the point of view of information systems development include the following imperatives: constant improvements, zero failures and mistakes, correct performance at first, performer's duty of task correction (Grant et al., 1994).

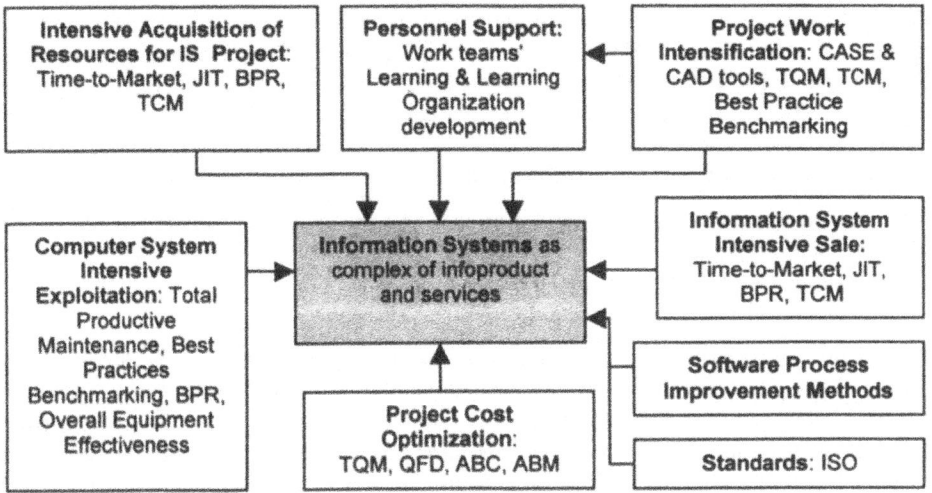

Figure 1. Management Support Methods for Information Systems Development

Activity Based Costing (ABC) method refines costing information system development by focusing on individual activities as the fundamental cost objects. An activity is an event, task, or unit of work with a specified purpose for example designing information system. ABC systems calculate the costs of individual activities and assign costs to cost objects such as products and services on the basis of the activities undertaken to produce each product or service (Cooper & Kaplan, 1999). Activity-Based Management (ABM) describes management decisions that use activity-based costing information to satisfy customers and improve profitability. ABM is defined to include pricing, and product-mix decisions, cost reduction and process improvement decisions and product design decisions (Horngren et al., 2000). Organizational Learning approach encourage to continuous improvement of activities and controlling the people within the whole company staff working on information systems. Process Mapping and Benchmarking focus on the implementation of the best practices for ISs development.

Activity speed is important factor in economic venture cycle. The decision concerning clients, suppliers, production lines, technological processes must be correct and immediate. So TCM is an approach to simplify the operation and integration of information flow within software company from supplier to client. That means reduction and simplification of business processes (BPR). The key concept in TCM method is an intensification of the whole organization and reduction of Total Cycle Time (Stone, 1997). This approach suggests redesign of business process (BPR), higher product quality, deeper understanding of clients' requests. Cycle Time reduction is defined as decreasing time for work because of the automation. So the time is treated as company's strategic resource. Some classes of time resources can be specified for example time for concept's generation, time for decision-making, time for gathering the necessary resources for tasks, time for tasks' learning, time for tasks' performance.

Although for the purpose of this paper the Management Support Methods are the most important for the information systems development. It should be noticed that software process improvement methods like Software Engineering Institute (SEI) Capability Maturity Model (CMM), BOOTSTRATP, Personal Software Process, Software Process Improvement for Capability dEtermination (SPICE) have the influence on the eventual form of software that is the main part of information systems. Information systems developers, as for example software companies or network service providers in their practice utilize standards. So ISO 9241-11, ISO/IEC 13236, ISO/IEC/TR 13243, ISO/IEC 14598, PN-EN 28402:1991, ISO 8402: 1986, PN-ISO 10011 can be treated as primers of information systems development and maintenance control.

MSMs are complementary and as a collection of methods support development of information systems. Each of the methods has:

1. Selected essential base, general purposes and spectrum of problems to solve
2. Special language
3. Tools for actual situation analysis
4. Tools for changes' implementation.

Understanding of these four components is the base of the methods' application and *ex ante* and *ex post* success evaluation. Management Support Methods are based on beliefs that clients (precisely information system users) should be satisfied and other suppliers in information system life cycle are partners, not competitors and collaboration with them is necessary for IS project success and sustain competitive advantage. Project teams are unique and as such they must be treated because they are the competitive advantage source. In each IS development case the corporate executives must select the portfolio of methods to ensure that the expected benefits will be bigger than assumed costs. Management Support Methods implementation strategy requires as following:

1. Executives' acquaintance with methods and establishing the purposes of method application
2. Method acceptance and executives' engagement
3. Selection of project leader
4. Elaborating strategy of method implementation and identification of method application field
5. Elaborating the detail plan of implementation and project feasibility study
6. Consistent realization of implementation plan, including users' training
7. Regular post-implementation reviews in management support method exploitation environment.

Knowledge about MSMs lets to combine them into subgroups of complementary components. Analysis of relations among the methods helps organization to preserve flexibility in solving problems. New implementation of methods as well as invention of a new management support method is based on the post-implementation experiences taking into account corporate strategy, work organization, culture and management styles.

4. MANAGEMENT RESEARCH METHODOLOGIES

The research term is used for a range of activities, from desk research (ascertaining information from public sources) through action research, to research for a PhD or post-doctoral work in a management school. It may be conducted to solve an immediate problem (applied research) to assess the performance or impact of an action or policy of a person, group or organization, or to develop a test some theory with varying degrees of abstraction (Oulton, 1995). Many researchers in the management field combine methods drawn from two traditions: positivism or phenomenology. Development of the research project an extremely interactive process at three levels:

1. At the research design level, since each phase (theoretical frame development, purpose definition, modeling, operationalising, empirical research design and data analysis instrument construction) did not follow a linearly planned flow but constantly interacted with each other and required reassessment and returns to previous one.
2. At the interpersonal level, given the number of minds simultaneously involved in each phase
3. At the theoretical framework level - formulation of theoretical conclusions (Stenbacka, 2001).

Conducting research will make demands on intellectual abilities and also on management skills and competences. Securing resources, managing time, negotiating a brief, and in all probability doing two jobs at one is not something everybody can manage. General outline of research includes exploration, explanation and validation. To proceed along this trajectory a researcher may choose from several research strategies (also called designs) and within each strategy from various data collection and analysis methods. When researchers conduct empirical research they establish methods that help them observe what is, not what they think should be. Their methods help them to accept approaches to observe occurrences objectively and record findings accurately. Generally there are two approaches to research: quantitative and qualitative. Quantitative research is concerned primarily with acquiring and analyzing relatively small amounts of data from large numbers of subjects. Its tools include mass interview studies, self-completion questionnaires, and statistical techniques. Qualitative research is concerned with acquiring and analyzing relatively large amounts of data from a small number of subjects, to investigate experiences, attitudes and opinions. Qualitative research is a simulated activity that locates the observer in the world. It consists of a set of interpretative, material practices that make the world visible. These practices transform the world. They turn the world into a series of representations, including field notes, interviews, conversations, photographs, recordings and memos to the self. At this level qualitative research involves an interpretative, naturalistic approach to the world. This means that qualitative researchers study things in their natural settings, attempting to make sense of or to interpret, phenomena in terms of the meanings people bring to them. Case studies, ethnography research, action research and grounded theory are the most frequently applied qualitative methods. Qualitative research is an interdisciplinary, transdiscplinary field. It embraces two tensions at the same time. On the one hand it is drawn to a broad, interpretive, post experimental, and critical sensibility. On the other hand it is shaped to more narrowly defined positivist, and naturalistic conceptions of human experience and its analysis (Denzin & Lincoln, 2000).

5. INFORMATION SYSTEM DEVELOPMENT AS A RESEARCH ACT

As each academic discipline information systems requires research methods development. Analysis of the applied methods can lead to the conclusion on how research works are carried out, what results are received, what problems are ignored, and eventually what new methods should or could be applied. A lot of information systems researchers concentrate on methods of analysis and design of information systems. This focusing leads to method engineering (Brinkemper et al., 1995). Methods engineering premises are as following:

- Need of description of different approaches to information systems development and their comparisons

- Need to improve these methods using Computer Aided Method Engineering (CAME) tools.

Information systems draw attention researchers from many disciplines like mathematics, computer science, management, economics, and psychology. This multidisciplinary base ensure multiplicity of concepts, axioms, opinions, experiences to understand development, usage and impact of information systems. Generally informatin systems research can be classified as positivist if there is evidence of formal propositions, quantifiable measures of variables, hypothesis testing, and the drawing of inferences about a phenomenon from a representative sample. Information systems research can be classified as critical if the main task is seen as being one of social critique, whereby the restrictive and alienating conditions of the status quo are brought to light. Critical researchers assume that people can consciously act to change their social and economic conditions. Information systems research can be classified as interpretative if it is assumed that our knowledge of reality is gained only through social constructions such a language, consciousness, shared meanings, documents, tools, and other artifacts. Interpretative research does not predefine dependent and independent variables, but focuses on the complexity of human sense making as the situation emerges It attempts to understand phenomena through the meanings that people assign to them. Interpretative methods of research in information systems are aimed at producing an understanding of the context of the information system and the process whereby the information system influences and is influenced by the context (Klein & Myers, 1999). For information systems research both qualitative (case research, grounded theory, ethnography, action research) and qualitative methods (survey, experimental study) are included. A common approach to researching information systems is to focus on the construction and technical aspect of a system and to treat the management as the context in which the development and adoption take place. This paper suggest certain re-orientation and proposal to research of information systems in context of management methods, which have influence on the results and processes of information systems development. Working in this way there is a great opportunity to directly explain management impact on information systems development. On the other side information systems development lets verify and evaluate application and quality of management methods.

6. CONCLUSION

In information systems there is a very close relationship between research for systems development and research into systems development. Management Research Methods provide first a basis for understanding and improving practice and secondly insight for researchers. The páper has explored the use of Management Support Methods in system development methodologies. It has focused on the use of methods as the approaches to support system development. The paper emphasizes the value of management for project and for information systems development, and the necessity or place of qualitative methods. The qualitative systems could be applied for the study of the human side of information systems development. The proposed approach seems to be useful in a number of ways. It has been used successfully as a basis for teaching undergraduate computing students about the relationships between the management methods and the relationship between management methods and information systems research methods. It provides an easily understandable basis on which to explain how a given Management

Support Method may be used in a variety of context within system development, particularly for information system project management.

7. REFERENCES

Avison D., Lau F., Myers M., Nielsen P.A (1999) Action Research, *Communication of the ACM*, January, vol 42, No 1., pp.94-97.

Boland R.J. Hirschheim R.A (1987) Critical issues in information systems research, Wiley Series in Information Systems, Chichester

Brinkkemper S., Lyytinen K., Welke R.J. (1995) Method Engineering, Principles of method construction and tool support, Chapman & Hall, London.

Carvalho M. Hudson J. (2000) Grand Theory and Grounded Theory, The Research Centre for the Built and Human Environment, The University of Salford.

Chand D.R., (1989) Some observations on information systems as an academic discipline, Journal of IS Education on-line, 3/89, vol 1, no 3.

Cooper R., Kaplan R.S. (1999) The Design of Cost Management Systems, Upper Saddle River, NJ, Prentice Hall

Corbitt B.J., (2000) Developing intraorganizational electronic commerce strategy: an ethnographic study, Journal of Information Technology , 15, pp. 119-130.

Cortada J.W. (1995) TQM for Information Systems Management, Quality Practives for Continuous Improvement, McGrawHill, Inc. New York.

Correia Z., Wilson T.D. (2000) Scanning The Business Environment For Information: A Grounded Theory Approach, National Institute for Engineering and Industrial Technology, Information Research Vol.2, No.4, Lisbon, Portugal.

Cyber-ethnography webring (2000) http:// www. pitt.edu/ ~ gajjala /define.html

Denzin N.K., Lincoln Y.S. (2000) Handbook of Qualitative Research, Sage Publications Inc., London

Euske K.J, Player R.S (1996) Leveraging Management Improvement Techniques, Sloan Management Review, Fall, 69-79.

Gordon, S., (1994) Benchmarking The Information Systems Function, Working Paper Series 94-08, Center for Information Management Studies (CIMS) Babson College, Babson Park, Massachusetts 02157-0310

Hansen, G.A. (1994) Automating Business Process Reengineering, Prentice Hall, Englewood Cliffs.

Huber, G. (1991) Organizational Learning: The Contributing Processes and Literature, Organization Science 2 88-115.

Hughes J (1999) An Empirical Model of the Information Systems Development Process: A Case Study of an Automotive Manufacturer, Proc.10th Australasian Conference on Information Systems http://www.ac.nz/acis99/

Hughes J., Wood-Harper T (1999) Systems development as a research act, Journal of Information Technology, 4, 83-94

Horngren Ch.T., Foster G., Datar S.M. (2000) Cost Accounting A Managerial Emphasis, Upper Saddle River, NJ, Prentice Hall

Information Technology - Quality of Service - Guide to Methods and Mechanisms - Technical Report Type III, ISO/IEC TR 13236 (Editor's draft 1.0) 1997.

ISO 9241-11, Ergonomic Requirements for Office Work with Visual Display Terminals, Part 11, Guidance on Specifying and measuring usability, Draft International Standard, 1995.

ISO/IEC 13236 IT Quality of service, 1998.

ISO/IEC /TR 13243 IT Quality of service Guide to methods and mechanisms, 1999.

ISO/IEC 14598-1 Information Technology -Software product evaluation. General overview, 1999.

ISO/IEC 14598-2 Information Technology - Software product evaluation. Planning and Management, 1999.

ISO/IEC 14598-3 Software engineering -product evaluation (for developers), 1999.

ISO/IEC 14598-4 Software engineering - product evaluation (for acquirers), 1999.

ISO/IEC 14598-5 Software engineering -product evaluation (for evaluators), 1999.

Klein H.K., Myers M.D. (1999) A Set of Principles for Conducting and Evaluating Interpretative Field Studies in Information Systems, MIS Quarterly, March, Vol 23 No.1. pp. 67-94.

Knapp C.A (2000) A Grounded Theory Study of Successful Organizational Integrated CASE Technology Implementation, Pace University New York.

Lee, A. S., J. Liebenau, and J. I. DeGross (eds.) (1997) Information Systems and Qualitative Research, London: Chapman and Hall.

McQueen R., Kock N. (1997) A field study of the effects of asynchronous groupware support on process improvement groups, Journal of Information Technology, 12, 245-259.

Morgan G (1996) Images of Organization, SAGE Publications, London.

Myers M.D. (1999) Investigating Information Systems with Ethnographic Research, Communications of the Association for Information Systems, Vol 2, Article 23, Dec.

Myers M.D. (2000) Qualitative Research in Information Systems, http://www.auckland.ac.nz/msis/ sworld/

Oulton T (1995) Management research for information, Management Decision, Vol.33, no.5

Parker C.M., Wafula, E.N., Swatman P.M.C., Swatman P.A. (1994) Information Systems Research Methods: The Technology Transfer Problem, ACIS'94, 5th Australasian Conference on Information Systems, http: //www. ac. nz./acis99/

PN-EN 28402:1991, ISO Standard 8402: 1986, Jakość - terminologia (Quality-Vocabulary), Warszawa, Polski Komitet Normalizacji, Miar i Jakości, 1993.

PN-ISO 10011-1 Guidelines for auditing quality systems Part 1 Auditing, Wytyczne do auditowania systemów jakości. Auditowanie, Warszawa, PKN, 1994.

PN-ISO 10011-2 Guidelines for auditing quality systems Part 2 Qualification for auditing quality systems. Qualification criteria for quality systems auditors, Wytyczne do auditowania systemów jakości. Kryteria kwalifikowania audytorów systemów jakości. Warszawa, PKN, 1994.

PN-ISO 10011-3 Guidelines for auditing quality systems Part 3 Guidelines for auditing quality systems. Management of audit programmes, Wytyczne do auditowania systemów jakości. Zarządzanie programami auditów, Warszawa, PKN, 1994

Remenyi D., Williams B. (1995) Some aspects of methodology for research in information systems, Journal of Information Technology, 10, pp. 191-201

Senge, P.M. (1990) The leader's new work: Building learning organizations. Sloan Management Review, Fall: 7-23.

Senn J. (1998) The Challenge of Relating IS Research to Practice, Information Resources Management Journal, Winter, Vol 11, No 1., pp.23-28.

Smart G., (1998) Mapping Conceptual Worlds: Using Interpretative Ethnography to Explore Knowledge-Making in a Professional Community, The Journal of Business Communication, Vol 35, Number 1, January, pp.111-127

Stenbacka C (2001) Qualitative research requires quality concepts of its own, Management Decision, 39(7)

Stone J.A. (1997) Developing Software Applications in a Changing IT Environment, McGraw Hill, New York.

Susman G.I., Evered R.D., (1978) An assessment of the scientific merits of action research, Administrative Science Quarterly, 23, pp. 582-603.

Swatman, P.A. and Swatman, P.M.C. (1992) Formal Specifications - An Analytic Tool for Management Information Systems, Journal of Information Systems, 2, 121-160.

Trauth E.M (2001) Qualitative Research in IS: Issues and Trends, IGP, Hershey

Ulrich W (1983) Critical Heuristics of Social Planning, A New Approach to Practical Philosophy, Verlag Stuttgart

Ward K.J., (1999) Cyber-ethnography and the emergence of the virtually new community, Journal of Information Technology, 14, pp. 95-105.

PANEL ON CHANGE MANAGEMENT AND INFORMATION SYSTEMS

G. Harindranath, Björn Lundell, Robert Moreton, and Wita Wojtkowski

1. INTRODUCTION

This panel will constitute a series of paper presentations examining key issues in the area of information systems and organisational change. In particular, it will address the themes of process re-engineering, change management, evaluation frameworks for change management and impact of groupware implementation using case studies from a number of countries such as the UK, Sweden, USA and Germany. A brief outline of the contribution from each of the panellists is given below.

2. IMPLEMENTING GROUPWARE IN A LARGE GERMAN PHARMACEUTICAL COMPANY[1]

This case concerns the introduction of groupware into a large German pharmaceutical company. The company's Electronic Document Management & Information System (MEDIS) provides the basis of the discussion. As the case indicates, the main problem in using groupware efficiently is the need for organizational change (reengineering or redesign, change of management responsibility and authority, training and motivation). It is not enough to have a wide-ranging knowledge of software engineering methods and technologies. Rather, knowledge of the human factors and communication skills have to be taken into account. That means that besides technical and methodological knowledge a capacity for understanding human behaviour is needed by organisations developing groupware systems. Not least, this requires managers to set project goals and to support these goals by setting a good example in terms of their own contribution.

---

[1] Robert Moreton, University of Wolverhampton, UK (Panel Chair), r.moreton@wlv.ac.uk

Information Systems Development: Advances in Methodologies, Components, and Management
Edited by Kirikova *et al.*, Kluwer Academic/Plenum Publishers, 2002

157

3. IS AND PROCESS RE-ENGINEERING IN THE UK HEALTH SECTOR[2]

This paper presents a case study of an experimental project to establish an ambulatory care and diagnostic centre within the context of a highly politicised National Health Service Hospital (NHS) Trust in the United Kingdom (UK). This project involved large scale business process re-engineering, including the redesign of long-established healthcare procedures and the development of sophisticated new information systems through a unique partnership between the public sector (the UK's NHS) and a number of private sector companies. This paper examines some of the change management challenges faced by the project team from a technical, human and organisational perspective.

4. CHANGE MANAGEMENT: BUT WHAT HAS CHANGED?[3]

Managing change in Information Systems development presupposes that we have an understanding of how an organisation's practices are changing. This paper concerns an approach to documenting change in the working practices of an Information Systems development organisation. The original goal of the work was to facilitate tool selection to better support working practice. Such IT development tools are complex, and a selection process is likewise complex. Complexity comes not only from the product itself, but involves the social and human issues related to product usage in specific IS development contexts. We believe it is necessary to use a systematic approach in such analysis. One aspect of this is an organisation's ability to analyse and comprehend their own longer and shorter-term needs for IT-products. By means of a number of field-studies, a systematic method has been developed to address the critical task of evaluation framework development sensitive to a specific organisational context. Evaluation frameworks give detailed, but snapshot insight into a specific development context. If short-term, pragmatic issues can be identified and ignored, then comparing consecutive snapshots has the potential to give insights into organisational change. In this panel we will draw on two case-studies with the method, to illustrate our view and consider its potential usage in longitudinal studies of change.

5. CHANGE: COMPARING THE IMPACT OF TECHNOLOGICAL REVOLUTIONS OF THE PAST AND CURRENT INFORMATION TECHNOLOGIES[4]

To round out the discussion on information systems and organisational change, we will focus on the nature of change and ponder whether the New Economy (powered by the developments in the microchip and the Internet industries) can be treated as a fundamental industrial revolution as great in importance as the concurrence of inventions such as electricity and the internal combustion engine. We will also consider certain

---

[2] G. Harindranath, Royal Holloway, University of London, UK, g.harindranath@rhul.ac.uk

[3] Björn Lundell, Department of Computer Science, University of Skövde, Sweden, bjorn@ida.his.se

[4] Wita Wojtkowski, Networking, Operations and Information Systems, Boise State University, Boise, Idaho, USA, wwojtkow@boisestate.edu

inventions in chemical industry. We will discuss the effect of radio and movie industries which transformed everyday life between 1910s and 1970s and compare these effects to changes brought about by the inventions of the New Economy. The presentation will then explore, in very general terms, organizational change. We will close with a brief on why companies fail.

THE ROLE OF LEADERSHIP IN
VIRTUAL PROJECT MANAGEMENT

Janis Grevins and Voldemars Innus[1]

1. INTRODUCTION

With the dramatic advancement in information technology, organizations and institutions worldwide can now make much more effective use of virtual teams that employ distributed work patterns, rely on technology-mediated communications and take advantage of "around the clock" activity. For example, according to research conducted by the Gartner Group 137 million workers worldwide will be involved in some form of remote electronic work by 2003. By the year 2010 it is projected that employees will spend 40% of their time working with others who are in a different place and different time zone (Salamon 2001). In fact, in the project management arena 60% of all projects are already considered virtual (Guss 1998). However research on the effect of virtuality on project management processes and subsequent success is practically nonexistent. This paper concentrates on the role of leadership in virtual project management and focuses primarily on information systems (IS) projects.

The leadership skills of project managers have always been recognized as a critical project success factor (Crawford 2000). The project manager needs to facilitate the development of a stimulating teamwork environment fostering team member involvement and commitment (Thamhain 2000). These leadership skills need to be melded with the classical project manager's tasks, such as project planning and progress tracking. Project virtuality also adds complexity by introducing new social and task related dynamics to project team (Kayworth and Leidner 2001/2002). These dynamics affect the impact of leadership on project outcome.

---

[1] Janis Grevins, University at Buffalo, State University of New York, Buffalo, New York, 14260, USA.
Voldemars Innus, University at Buffalo, State University of New York, Buffalo, New York, 14260, USA.

2. PRIOR RESEARCH

2.1. Project Evaluation

While there is general agreement that project success is the ultimate goal in project management, there is a considerable difference in opinion in what constitutes project success and the criteria to judge success (Crawford 2000; Pinto & Slevin 1988 b). For example, some projects exceed budget and time and are still considered successful, while other projects accomplished within time, budget, and quality are judged unsuccessful. Project management literature addresses this issue by distinguishing between project outcomes and project success (Murphy, Baker & Fisher 1974; Lim & Mohamed 1999). Project outcomes are measured in terms of cost, time and quality objectives, and how well the project management process is followed. In contrast project success is determined by how "it meets the technical performance specifications and/or mission to be performed, and if there is a high level of satisfaction concerning the project outcomes among the key people on the project team, and key user or clientele of the project effort" (Murphy, Baker & Fisher 1974).

2.2. Virtuality

The concept of virtual teams, virtual organizations and virtual enterprises was first introduced in business literature during the 1990's (Lipnack & Stamps 1997). Researchers have identified differing virtual characteristics or dimensions – time, space and structure (Skryme 1998), people, purpose and links (Lipnack & Stamps 1997), boundary crossing, resource sharing, geographic dispersion, participant equality, electronic communications (Jagers, Jansen & Steenbakkers 1998), and modularity, heterogeneity, time and space distribution (Wassenaar 1999).

Virtual team research is still in an early phase and is dominated by controlled experiments (primarily student teams) and/or case studies with very few examples of field research (Maznevski and Chudoba 2000). In experimental research the teams are usually strictly divided in virtual and non-virtual, however in practice they differ based on the degree of virtuality (O'Leary & Cummings 2002).

2.3. Project Management and Leadership

A competent project manager or successful project champion has always been recognized as a prerequisite for project success (Crawford 2000; Pinto & Slevin 1989). One of the characteristics that a successful project manager exhibits is leadership, which is defined as "the ability to identify work that has to be done and to select the people who are best able to tackle it." (CRMP 1999). These people posses and exhibit behaviors that facilitate "getting the job done" and combine the art and science of project management.

However, with the emergence of self-managed teams, this understanding of a single leader's role should be reevaluated. Horner (1997) suggested that leadership in such teams often comes from within the team, where team members rotate or share leadership roles, thus the influence of single leader diminishes. In virtual teams, leadership behaviors need to become more complex as team members become more geographically distributed and use computer-mediated communication tools (Kayworthy & Leidner 2001/2002).

3. THEORETICAL MODEL

The frequent use of virtual teams in project management (Guss 1998) and preliminary field discussions suggest that virtuality is a continuous rather than dichotomous team characteristic with teams ranging in their degree of virtuality from conventional to virtual teams (O'Leary & Cummings 2002).

Based on previous research, this paper recognizes four dimensions that distinguish conventional and virtual teams: temporariness, physical distance, cross boundary distribution, and communication means (Table 1) (Grevins 2002) Purely conventional teams are permanent, collocated, work in the same organizational unit (all members reporting to the same supervisor) and only use face-to-face communications. In contrast, purely virtual teams assemble only for the duration of the task and are distributed in different places of the world (preventing convenient use of simultaneous communications, due to time differences). Virtual team members represent different organizations and most likely have never met or seen each other.

Table 1. Team virtuality boundaries

Dimension	Extreme state for	
	Conventional team	Virtual team
Temporariness (longitudinal virtuality)	Permanent team	Short term *ad hoc* project team (no history and no future)
Physical distance (space virtuality)	Collocated team	Space (and time) distributed team
Cross boundary distribution (boundary virtuality)	Team in single organizational unit	Cross-organizational team
Communication means (communication virtuality)	Only face-to-face communications	Only computer mediated communications

A project manager's leadership can be equally important in virtual projects and conventional projects. However to achieve an equal impact in virtual projects a much more targeted and sophisticated use of leadership skills is necessary.

Communication means differ in their ability to transmit different types (e.g. task related, social) of information. Face-to-face communications can transmit all types of information at the fastest rate, thereby contributing to effectiveness of leadership. Technologically mediated communications are slower and perceived to be less capable to transmit social information, however they still contribute to the impression building of one person about the other (Walther 1993). Therefore of the four dimensions of virtuality, physical distance and technology-mediated communications have the greatest negative impact on the ability of the leader to influence project success. It is important, therefore, for the project manager to focus effort on these particular negative influences. The project manager needs to target limited face-to-face experiences of team members to ensure alignment of the team. The other modes of communication used by virtual teams must reinforce the alignment process as well.

4. RESEARCH METHODOLOGY

To test the described model the study used a web[2] survey to collect data from team members of various information systems projects. Out of total 215 surveys distributed 102 were completed yielding the response rate of 47%. The surveys were distributed to organizations representing education, banking and accounting service industries.

The scales for the survey instrument were adapted from relevant previous research. Project success ($\alpha = .92$) was measured using an 11-item instrument adapted from Pinto, Pinto and Prescott (1988). Project outcomes were measured using a single question for each outcome (e.g. schedule, task and budget). Project manager's leadership ($\alpha = .97$) was measured using a scale from Jiang, Klein and Shepherd (2001). Expected future interaction ($\alpha = .78$) scale was adapted from Walther (1994). All these items were measured on a 5-point Likert's scale (anchored between "Strongly agree" and "Strongly disagree", or "Never" and "Very Frequently"). Frequency of face-to-face meetings was measured using a 5-point Likert's scale, and information exchange through e-mail was measured as percentage of project related information exchanged through e-mail. Tenure was measured as the average time of reported team member relationships with each other. Data on organizational association and physical location of the team members was collected separately of the survey. As suggested by Pelled, Eisenhardt & Xin (1999) and O'Leary & Cummings (2002), the organizational diversity and physical distance index was calculated using Teachman index (Teachman 1980).

5. RESULTS

Although the multi-item measures exhibited good psychometric properties in previous studies, the reliability and validity was checked before further analysis. The Cronbach alpha scores, used to assess measure reliabilities, for the multi-item constructs ranged from .78 to .97; good for commonly used scales (Carmines & Zeller 1979, p. 51). Convergent validity was assessed using factor analysis with principal axis factoring. For all constructs only one factor emerged with factor loadings between .51 and .96.

The analysis of the correlation table revealed that independent variables were strongly correlated with their respective interaction terms. Consequently all independent variables were centered, to minimize the effects of the high correlations onto regression analysis (Neter et al., 1996).

To test the proposed model the Ordinary Least Squares regression analysis was conducted in two stages for every dependent (project performance) variable (Table 2). In the first stage only leadership and all virtuality constructs were entered in the analysis. In the second stage interaction terms between leadership and virtuality variables were entered. The change in R^2 between the stages was analyzed to see whether interaction terms significantly contribute to the explanatory power of the model. The change was significant only for project task outcomes.

Overall the analysis reveals that leadership has a significant effect on project performance and organizational diversity has a strong direct effect on project performance. Frequency of face-to-face communications and geographical diversity were

---

[2] The survey was primarily distributed thought the Internet, however respondents had an option to print and complete the survey in a hard copy.

found to have direct effects on some project performance measures (e.g. task outcome and project success). Expected future interaction has a moderating effect on leadership – project schedule and task outcomes. This result suggests that leadership has a less significant effect on long-term teams than on short-term teams (not expecting future interaction).

Table 2. Regression results

	Project Outcomes						Project Success	
	Schedule		Task		Budget			
Leadership	0.32*	0.15	0.57***	0.62**	0.34*	0.37	0.50***	0.28
Geographical diversity	-0.11	-0.17	-0.26	-0.39*	-0.41*	-0.36	-0.14	-0.13
Organizational diversity	1.00***	1.05***	0.79***	0.88***	0.58***	0.53***	0.61***	0.61***
Tenure	0.03	0.06	0.03	0.09	-0.07	-0.11	-0.08	-0.09
Expected future interaction	-0.09	-0.19	-0.12	-0.26	-0.18	-0.26	0.21	0.2
Face-to-face meetings (frequency)	-0.1	-0.13	-0.19*	-0.27**	0	-0.03	-0.19**	-0.16*
E-mail (information exchange)	-0.05	-0.21	-0.29	-0.63	-0.1	-0.4	-0.13	0
Leadership interaction with								
x Geographical diversity		-0.36		0.02		-0.11		-0.39
x Organizational diversity		0.06		-0.04		0.07		0.09
x Tenure		-0.03		-0.14		-0.2		-0.01
x Expected future interaction		-0.45*		-0.80***		-0.19		-0.07
x Face-to-face meetings		0.18		0.25		-0.37		0.01
x E-mail		-1.25		-2.06		-1.14		0.71
R^2	**0.526**	**0.581**	**0.446**	**0.552**	**0.298**	**0.36**	**0.491**	**0.518**
ΔR^2		**0.06**		**.11*****		**.06**		**.04**

*** - $p<0.01$; ** - $p<0.05$; * - $p<0.1$

6. CONCLUSION

It is clear from this research that leadership is an important determinant of successful project performance, irrespective of project virtuality, and that project managers modify behavior in response to virtuality to enhance project success. The study also found that the influence of leadership on teams decreases when team members expect to interact with each other in the future. This could suggest that team members actively seek alignment when they expect future collaboration and that they move towards self-direction (Horner 1997). Further, the positive direct effect of organizational diversity on project performance provides strong support for the increasing trend of using cross-functional team in project management settings. The negative effect of face-to-face meetings on project task outcomes and project success seems somewhat counter intuitive. This may be the result of the groupthink effect on teams that engage in very frequent face-to face interactions (Janis 1982).

In summary, this research confirms the importance of the project manager's leadership, irrespective of project virtuality, in achieving successful project performance.

Also it supports the contention that project virtuality significantly affects project performance.

REFERENCES

Carmines, G., Zeller, A., 1979, *Reliability and Validity Assessment,* SAGE Publications, Beverly Hills.

Centre for Research in the Management of Projects (CRMP), University of Manchester, UK, 1999; http://www.UMIST.ac.uk/CRMP (the leadership citation is from Wideman Comparative Glossary of Common Project Management Terms V. 2.0: http://www.pmforum.org)

Crawford, L., 2000, Profiling the competent project manager, in: *Proceedings of PMI Research Conference 2000 Paris, France,* Project Management Institute, Newtown Square, pp. 3-15.

Guss, C. L., 1998. The virtual process environment and success - research and results, *Virtual-organization.net Newsletter.* 2(3): 22; http://virtual-organization.net/

Grevins, J., 2002. *Project Virtuality Effects on Project Team Processes and Project Success,* Doctoral Dissertation, University at Buffalo, State University of New York.

Horner, M., 1997, Leadership theory: past, present and future. *Team Performance Management.* 3(4):270

Jagers, H., Jansen, W., Steenbakkers, W., 1998, Characteristics of virtual organizations, in: *Organizational virtualness – proceedings of the virtual-organization.net workshop,* P. Sieber and J. Griese, eds., Institute of Information Systems, University of Bern, Bern; http://virtual-organization.net

Janis, I., 1982, *Groupthink,* Houghton Mifflin, Boston.

Jiang, J. J., Klein, G., Shepherd, M., 2001, The materiality of information system planning maturity to project performance, *Journal of the Association for Information Systems.* 2(5):1.

Kayworth, T. R., Leidner, D. E., 2001/2002, Leadership effectiveness in global virtual teams, *Journal of Management Information Systems.* 18(3): 7.

Lim, C. S., Mohamed, M. Z., 1999, Criteria of project success: An exploratory re-examination, *International Journal of Project Management.* 17(4):243.

Lipnack, J., Stamps, J., 1997, *Virtual Teams: Reaching Across Space, Time and Organizations with the Technology,* John Wiley & Sons Inc., New York, NY.

Maznevski, M. L., Chudoba, K. M., 2000, Bridging space over time: Global virtual team dynamics and effectiveness, *Organization Science.* 11(5):473.

Murphy, D. C., Baker, B, N., Fisher, D., 1974, *Determinants of Project Success.* Report for NASA. Management Institute, School of Management, Boston College, Chestnut Hill, Massachusetts.

Neter, J., Kutner, M. H., Nachtsheim, C. J., Wasserman, W., 1996, *Applied Linear Statistical Models,* Irwin, Chicago.

Pelled, L. H., Eisenhardt, K. M., Xin, K. R., 1999, Exploring the black box: An analysis of work group diversity, conflict, and performance. *Administrative Science Quarterly.* 44(1):1

Pinto, J. K., Slevin, D. P., 1989, The project champion: key to implementation success, *Project Management Journal.* 20(4):15.

Pinto, J. K., Slevin, D. P., 1988, Project success: definitions and measurement techniques, *Project Management Journal.* 19(1):67.

Pinto, M. B., Pinto, J. K., Prescott, J. E., 1993, Antecedents and consequences of project team cross-functional cooperation. *Management Science.* 39(10):1281.

Skyrme, D. J., 1998, The realities of virtuality, in: *Organizational virtualness – proceedings of the virtual-organization.net workshop,* P. Sieber and J. Griese, ed., Institute of Information Systems, University of Bern, Bern; http://virtual-organization.net

Solomon, C. M., 2001, Managing virtual teams, *Workforce.* 80(6):60.

Teachman, J. D. 1980. Analysis of population diversity: measures of quantitative variation. *Sociological Methods and Research.* 8(3):341.

Thamhain, H. J. 2000. Criteria for effective leadership in technology-oriented project teams, in: *Proceedings of PMI Research Conference 2000 Paris, France.* Project Management Institute: Newtown Square, PA. pp. 115-121

Walther, J. B. 1993. Impression development in computer-mediated interaction, *Western Journal of Communication.* 57(4):381.

Walther, J. B. 1994. Anticipated ongoing interaction versus channel effects on relational communication in computer-mediated interaction, *Human Communication Research.* 20(4):473.

Wassenaar, A. 1999. Understanding and designing virtual organization form, *virtual-organization.net Newsletter.* 3(1):6; http://virtual-organization.net

MANAGING KNOWLEDGE IN A NETWORKED CONTEXT

Per Backlund and Mattias Strand[1]

1. INTRODUCTION

The aim of this paper is to describe an information service within a virtual organisation and to provide a set of lessons learned concerning the implementation of information services in virtual organisations. The study was performed within a joint project between a major Swedish industrial research institute, a number of international companies within the electronic industry and the Department of Computer Science at the University of Skövde. A main goal for the project was to provide designers with information about manufacturability. This may be described as knowledge sharing in an inter-organisational network (Franke, 1999). In order for organisations to survive, it is of vital interest to acquire knowledge about the actors in the organisational environment e.g. customers, suppliers, and business partners (Alavi and Leidner, 1999; Hackathorn, 1999). Carlsson (2001) stresses the need for further research on the use of information and communication technology (ICT) in network-based knowledge processes. In our view an information system serves as a means for managing knowledge. Hence the terms information system and knowledge management system will be used interchangeably.

The case description and lessons learned will be presented using three levels: the organisational level, the service level, and the solution level. Furthermore, the three-levels will be elaborated in order to describe and classify some different types of problems, which have to be dealt with in developing support for knowledge sharing within virtual organisations. The elaborated framework is empirically anchored in the case study undertaken.

---

[1] Per Backlund Dept. of Computer Science, University of Skövde, P.O. Box 408, SE-541 28 Skövde, Sweden.
Mattias Strand Dept. of Computer Science, University of Skövde, P.O. Box 408, SE-541 28 Skövde, Sweden.

Information Systems Development: Advances in Methodologies, Components, and Management
Edited by Kirikova *et al.*, Kluwer Academic/Plenum Publishers, 2002

2. BACKGROUND

In this section we will describe the three levels of description, which were introduced above. Castells (1996) describes the paradigm of Information Technology as a combination of coherent technological, organisational, and business economical innovations, which together form a dynamic constellation. These aspects have influenced the three-level framework.

2.1. The Organisational Level

This new paradigm of Information Technology is characterised by the following: Information is raw material and technology is used to act on information; The new technologies pervade our society since information is an integrated part of all human activity; Networks may be introduced into all types of processes and organisations by using information technology; Processes and organisations become more and more flexible, and; Different information technologies are converging into one integrated system (Castells, 1996).

The items presented above form a basis for our ideas about web-based knowledge management system for virtual organisations. Knowledge is dispersed in different organisations, which have to cooperate in order to reach their goals. Shao et al. (1998) identify three generic types of virtual organisations: Organisations that outsource some of their business activities and thus form virtual alliances; Conceptual organisations only existing in the minds of those forming them; Organisations that are built up by virtual links through the use of IT.

Ahuja and Carley (1999) define a virtual organisation as a geographically distributed organisation whose members are bound by a long-term common interest or goal, and who communicate and coordinate their work through information technology. We define a virtual organisation as an organisation, which is formed to achieve some organisational purpose using IT as a means of doing so. The members typically belong to other organisations at the same time as they participate in the virtual organisation. In this sense the members have their loyalty in more than one organisation. We stress the use of IT since it provides the possibilities to work together over time and space.

Franke (1999) identifies three types of inter-organisational networks: internal networks, stable networks, and dynamic networks. The notion of virtual organisations described above may be complemented by this classification in order to provide a better understanding. The dynamic network (Fig. 1) is characterized by firms relying on core skills such as manufacturing, R&D/design, design/assembly, or pure brokering. The companies in a dynamic network may join or leave the network according to internal and/or external requirements. The dynamic network needs to be managed by a network broker who decides who is in and who is out.

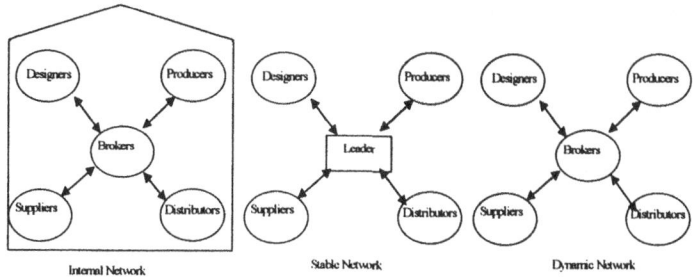

Figure 1. Common network types. From Franke (1999), p.205

2.2. The Service Level

Information systems of today are closely integrated with organisations and they support organisations at all levels. Some systems are focused on supporting the operational day-to-day work whereas others are focused on e.g. supporting top-level management. There are also other types of information systems, specifically aimed to support communication within and between organisations. Making an unambiguous categorization of information systems and their roles in organisations is not trivial, as the roles are changing over time.

Another problematic issue is that information systems overlap and change, as applications combine new capabilities with old ones (Alter, 1999). However, common for the different types of systems, independent of the organisational level or function they are aimed at supporting, is that they offer support for acquiring, storing, managing, and accessing data sources. Those data sources may, in turn, be of different kinds, ranging from operational systems internal to the organisation, to external systems such as the Internet. Furthermore, the data used may be of different types, such as numbers, text, audio, and video. In addition, the data may also be raw, such as data from the operational systems or it may be cleaned and consolidated, e.g. from a data warehouse system.

Since the service level description is focused on the service offered, via a Web-based information system, for virtual organisations, the focus on data needed will naturally be laid on supplying each participant with relevant external data. With external data we refer to data that is stored in systems residing outside the acquiring organisation. That data may still be internal to the virtual organisation and one may also claim that the core of the services offered via a Web-based information system for a virtual organisation is to *internalize* the total amount of relevant data in the whole virtual organisation to each individual participant. The increased need for external data and the increased interest in participating in virtual organisations and other types of network depends on the ever-changing environments, most organisations are facing. Therefore, it is of vital interest to acquire data about the actors in the organisational environment e.g. customers, suppliers, and business partners (Alavi and Leidner, 1999; Hackathorn, 1999).

2.3. The Solution Level

The technical solution must facilitate knowledge sharing within a distributed community of knowledge workers (Tiwana and Bush, 2001). The Internet provides an infrastructure for information sharing between organisations throughout the world. Hence, this infrastructure may be used to support the type of information sharing described above. Many Web sites contain more information of a dynamic nature. Maintaining such information in both a database and in separate HTML files can be an enormous task, and difficult to keep synchronized (Connolly et al., 1999; Florescu et al., 1998). For these reasons databases being accessed directly from the web are becoming of interest. We can identify two types of web pages: *Static web pages*: the content does not change unless the file itself is changed. *Dynamic web pages*: the page is generated each time it is accessed

Dynamic web pages have two major features: they can respond to user input from the browser (e.g. returning data requested by the completion of a form or the results of a database query) and they can be customized by and for different users. Dynamic pages such as those resulting from queries to databases, needs to be generated by the servers.

3. THE CASE – MANAGING KNOWLEDGE IN A VIRTUAL ORGANISATION

The project described implements a web-based information system for a virtual organisation consisting of members from various companies within the electronic components industry. The case will be described on the basis of the three levels presented above.

The work done in the joint project, AIS 5[2] may be categorised as an action research project in which the authors took on the roles of systems analysts and part time project manager (one of the authors). The use of action research in the field of information systems is discussed in, e.g. (Stowell et al., 1997). Action research provides a means of eliciting knowledge about the development process at the same time as domain knowledge is elicited within the project. The approach taken provides a convenient sample. The data presented was collected during project meetings, work sessions, project documentation, and by using unstructured interviews. The participants, representing the different companies, were designers and manufacturing experts with the strategic aim to create a knowledge management system, which allows for knowledge sharing between organisations within the electronic industry. The means of collecting data may constitute a risk for bias. However, we have been aware of these risks and tried to balance them in ongoing discussions about the project.

---

[2] "Implementering av BGA, CSP och Flip-Chip - AIS 5 is a joint project between the Swedish Industrial Research and Development Corporation (IVF), the department of Computer Science at the University of Skövde, and a number of companies within the electronic circuit industry: Parker Hannifin AB, Elektronikpartner AB, Digital Vision Sweden AB, Norrtelje Elektronik AB, Flextronics International Systems AB, and PartnerTech AB. The project is financed by NUTEK.

3.1. The Organisational Level

Our focus is on a virtual organisation for developing and manufacturing electronic components. The manufacturing process covers stages from initial product development to large volume manufacturing. Different companies typically perform different stages in the process. We describe this as horizontal cooperation and vertical competition (Figure 2).

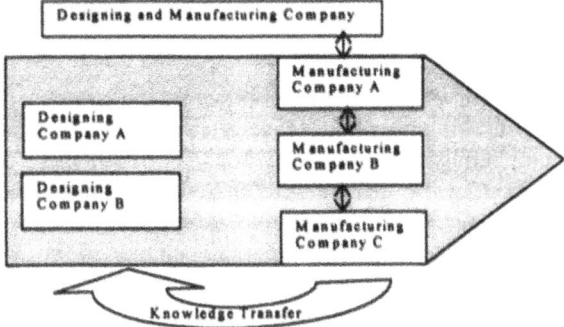

Figure 2. Horizontal cooperation in the production process. Vertical competition between producers.

The designing companies represent different application areas; hence there is no direct competition between them. We would like to point out that this is not necessarily the normal situation, even though it is the case here. The designing companies may engage one or more of the manufacturing companies depending on the type of production that is required. The manufacturers may, to some extent, compete for orders from the designing companies. However, they have different strengths and may therefore cooperate in some areas. Finally, there is a large, world wide manufacturing company that also does some design work, i.e. they span the whole production process. As can be seen by the description above we deal with an inter-organisational network.

The companies in a dynamic network may join or leave the network according to internal and/or external requirements. The dynamic network needs to be managed by a network broker who decides who is in and who is out. In our case a major Swedish research institute plays this role. We use Franke's typology to describe the type of virtual organisation we are dealing with, i.e. a dynamic virtual organisation. However, there are no distributors involved in this network (Fig. 1). The broker's role is to mediate knowledge between the participants of the network. Hence, the broker's role in this case is not focused on sales, which may be the normal interpretation.

Lessons Learned on the Organisational Level

Participating in a virtual organisation means sharing information and allocating time and resources. Thus, there is a need to evaluate the participants with respect to their willingness and potential to participate with respect to a number of issues.

Evaluate the organisations individual resource level: It is important to evaluate the different participants in the virtual organisation, since they may have different resources. That may, in turn, affect their ability to contribute to the virtual organisation. The different resource levels were clear in the case study in which one participant had significantly more resources than the others, being a multinational company with a large amount of know-how within the organisation. This resulted in an unbalanced division of the work conducted.

Evaluate relations between participants: The virtual organisation is established by a group of individuals and companies that benefit by pooling their skills. The individual participants can achieve things, which are hard on their own by participating in the organisation. There may be relations between participants hindering the pooling. We encountered some problems related to the fact that there is a situation of competition between some of the participants. We found it difficult to distinguish between data that is sharable and not. The problem is that data is not regarded shareable in all situations. That is, there are situations in which the participants are competing and other situations in which there is no competition. Some of the participants were hesitating to share information due to this issue. One example of this is that one representative described some of the participating companies as 'controversial' in the sense that they competed for the same customers.

Identify hidden agendas: With a variety of organisations participating hidden agendas are likely to appear. It may also be the fact that these hidden agendas are the main reason for some organisations to participate. This may constitute a problem if the hidden motives are more important then the goal of the virtual organisation. One such situation may be the desire to establish business contacts without any plans to contribute to the overall goal of the organisation. A situation like this may paralyse further work within the organisation. One example of this was found during a meeting when two of the participants tried to promote a bi-lateral project. When these plans failed some of their interest in the overall project was lost.

Identify clashing cultures: Since the virtual organisation may be constituted of organisations, with different cultures, it is important to identify if the organisations share the same culture. If not, the discrepancies may influence the success of the virtual organisation in a negative manner. All organisations do not have a culture of sharing knowledge in inter-organisational settings. There may be restrictions on the type of external cooperation and knowledge sharing allowed. We have encountered some problems of this nature. For example, one of the companies has a policy, which does not allow technical development other than in certain product development centres.

Develop a knowledge sharing culture: If a knowledge sharing culture is present, problems are less likely to arise. The project managed to develop a knowledge sharing culture to some extent. This was mainly done by meetings and visits to different plants, which allowed for knowledge sharing between the participants. Naturally there were restrictions concerning critical competitive knowledge, e.g. the capacity of plants etc. However, more work is needed in order to engage all participants in knowledge sharing activities.

Establish and clarify the goals of the virtual organisation: Common and clear goals are vital for success. It is of uttermost importance that there is a common goal that is known to and shared by all participants. Clear goals may also be beneficial in the establishment of the virtual organisation, since any organisation not willing to live up to the goals, may feel free to leave. We found that the hidden agendas constituted a problem in establishing the goals for the organisation. Goal modelling showed that there were a number of different types of goals, which to some extent were in conflict. For example: the broker stressed the issue of reliability connected to different ways of manufacturing as being the most important goal. This goal turned out to be incompatible with what the manufacturing organisations were willing to contribute. One possible explanation is that manufacturers have this information of their own and do not consider it shareable. Another explanation may be that it was considered to costly to elicit this type of data within the frames of the organisation.

Balance wagers and benefits: The wagers and the benefits have to be balanced for all contributors. We found that the manufacturers were supposed to provide a large amount of data before they could benefit from the better knowledge about manufacturability among the designers. This means that the amount of resources spent is not balanced against the immediate benefits and this constitutes a problem when motivating the participation within the own organisation.

3.2. The Service Level

The intention of the Web-based IS in this case was mainly to provide decision support for designers in their choice of components. The general idea was to form a feedback loop from the manufacturers, concerning experiences from mounting components, back to the designer. The intended user is a designer of electronic equipment who needs additional information about components for an intended design. The component vendors provide standard product information. However, this type of information needs to be complemented by information about reliability and manufacturability (Rowland, 2000). This information is provided by the system in terms of calculated values, estimated values, and text descriptions of experiences concerning the actual component. There is also a need to keep information about combinations of components, i.e. specific printed circuit assemblies since reliability is a function of combinations of printed circuit boards and the components attached to them. The service must provide data about components in order to provide designers with knowledge about manufacturing aspects as early as possible in the design process. Designers and manufacturers reside in different companies, as described in Fig. 3, which means that there is a need to transfer knowledge between organisations.

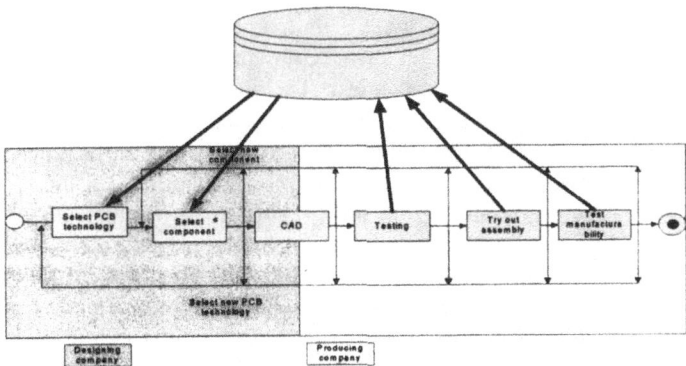

Figure 3. The production process, which the system aims to support

The providers of information in our information system are typically the manufacturers of the virtual organisation, whereas the users are representing the earlier phases of the product design process.

Lessons Learned on the Service Level

The core of the service supported is to internalize the data in the virtual organisation and to distribute that data, to the right person at the right time.

Establish routines for selecting relevant data. Since the participants may have different needs and thereby be interested in different data, it is important to establish routines on how to select the data to be pooled. To support this activity, the goal of the virtual organisation may be used as a guideline. In the case it was clearly shown that different participants had an interest in different data. The result was that the work on designing the underlying information system was delayed and much work previously conducted was in vain. One explanation to this may be that the goals were open for interpretations.

Establish routines for acquiring data. Since the data is located in different systems, among the participants, it is important to establish routines for how the data is to be acquired to the virtual organisation. The acquiring process may be performed manually or automatically, but the importance is on an agreement on how it is to be carried out. Furthermore, it is important that the participants agree on the amount of data they are expected to contribute with. In the case in was early decided that the initial data transfer was to be manually performed, but with a possibility for automatic data acquisition later on. In addition, it was also decided that every participant should contribute with the same amount of data. However, despite the early agreement on the manual acquiring of data the project was delayed, since the agreement on contributions was not fulfilled.

Establish responsibilities for maintaining data. When loaded into the underlying system, there is a need for maintenance. Otherwise, the data may be incorrect or polluted and this may lead to the end of the whole cooperation. For the certain project described, it was clearly stated that the broker was the stakeholder, which should maintain the data. Since this responsibility was clearly stated early in the project and agreed upon by the other participants, this never became a problem.

Establish routines for accessing data. In order to be able to use the data, the users must be able to access it. Therefore it is important to establish routines for how the data is to be accessed. In addition, it is also important to clarify who may be granted access to the data and to what extent. In the particular case described, such routines were established early on. The following types of users were identified: administrator, supervisor, user, and guest. The four types were also given different access- and system rights. The reason for having one and only one supervisor in each organisation was to simplify for each organisation to give colleagues access to the system, while restricting the organisational responsibility of the system to one person. The supervisor is also to be the organisation's contact person towards the broker.

Establish metrics. It is important to establish metrics to measure the benefits of the whole virtual organisation; each stakeholder group, and; each individual participant. In the case described, no metrics were established and that is to consider as a major disadvantage, since it became more difficult to motivate the organisations to participate in the cooperation. One participant stated that he had to motivate the resources spent within his own company. There was an unequal distribution of value between the groups. The designers were able to identify advantages of their participation early in the project, whereas the manufacturers were much less able to do so. Considering the general goal of the project this was rather natural, since the manufacturers benefit of the network only may come if the printed circuit assemblies were better designed.

Establish routines for evaluating, selecting, and implementing new, services. Since the participants may have different needs and thereby be interested in different services of the system, it is important to establish routines for how to evaluate new, proposed services. To support this activity, the goal of the virtual organisation may be used as a guideline. Since the participants in the virtual organisation may have limited resources to put into the cooperation, it is not always enough to evaluate the relevance of new, proposed services. It may also be necessary to select some of the new evaluated and approved services Finally, after evaluating new proposed services and selecting some or all of them for implementation, it is important to have routines for implementing new services. In the case described a number of new services were proposed. However it was unclear how they should be implemented and who should be responsible for doing so.

3.3. The Solution Level

In this paper we only deal with the part of the IS that has connection to the component database. The database may be queried by different types of users. A login system with four levels of access ensures that the user has the appropriate authorities. The *administrator* plays the role of the broker. For example the administrator can add new users and modify users rights. *Supervisors* are the key persons in the participating companies and thus have the right to add or modify the *users* from their own company. The *users* are entitled to search the database and to add components and experiences to the system. Finally, *guests* are entitled to search the database and view parts of the information.

The database is accessed via a web interface in order to support the organisational level and the service level as described above. An overall view of the system is shown in Fig. 4.

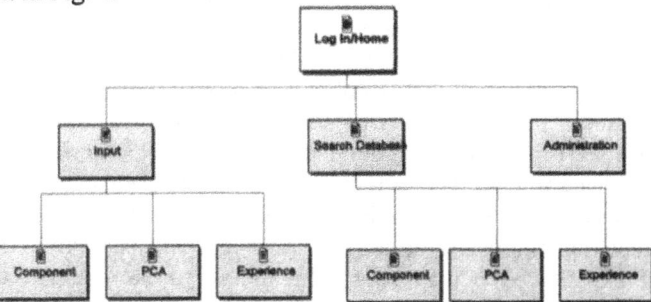

Figure 4. An overview of the web pages providing the database access.

The database is accessed via Java Server Pages (JSP), see for example Goodwill (2000). JSPs are used to generate dynamic HTML in order to separate content generation from content presentation. We use a three-tier architecture consisting of a database layer, a web server layer (Apache 1.3.9 and Tomcat 3.2.2[3]), and a browser layer to access the component database.

The database stores data about electronic components, electronic assemblies, and experiences connected to those. We have catered for the possibility to store data about assembly methods and IPC classes[4]. The conceptual schema of the database is described in Fig. 5. The functionality of the prototype, as until now, is limited to the possibilities to add users to the system, to add data about components, and to search and browse the database.

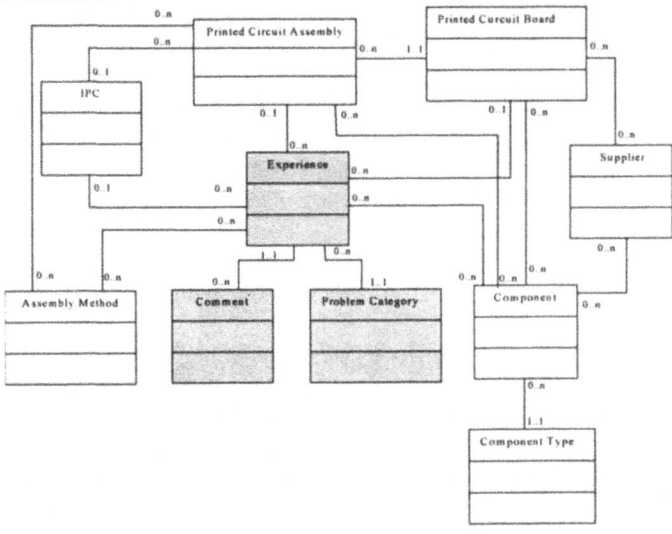

Figure 5. The conceptual schema of the experience database. The grey entities are the entities primarily allowing for knowledge sharing.

The *Experience* entity constitutes a means to store information about individual components, assembly methods, IPC classes, printed circuit assemblies, and printed circuit boards. It is also possible to relate experiences about different entities to each other. These features cater for the possibility to store additional information about components, i.e. information that is not provided from the supplier but has influence on manufacturability and reliability as described above. An experience is organized like a pattern, see for example Coplien and Schmidt (1995), consisting of a problem, a context, and a solution. The aim is to identify problem categories and make it possible to classify problems according to them. Furthermore, an experience may be complemented by comments added by other users. Concerning the storing and retrieval of experiences it may be fruitful to take a case-based reasoning (CBR)

---

3 http://www.apache.org/ and http://jakarta.apache.org/tomcat/
4 IPC: The Institute for Interconnecting and Packaging Electronic Circuits. IPC provide guidelines for reliability testing of surface mount solder attachments. For more details see IPC-SM-785 Guidelines for Accelerated Reliability Testing of Surface Mount Solder Attachments.

approach (Kolodner, 1997), since the context of an experience map well to the notion of indexing in CBR.

Lessons Learned on the Solution Level

The functionality of the prototype is limited to the possibilities to add users to the system, to add data about components, and to search and browse the database. We have identified the following aspects to consider, based on the work of Connolly et al. (1999) and Florescu et al. (1998).

Cater for platform independence: One reason for creating web-based access to a database application is that browsers are mostly platform independent. This advantage is disappearing since browser vendors are abandoning established standards when providing new browser specific features. We have identified this type of problems concerning the GUI look and feel in different browsers.

Deal with security issues and accessibility: User authentication and secure data transmission are critical since we are dealing with data that has competitive value. Thus security is of great concern for a virtual organisation that makes a database accessible on the web. We have implemented a system of different user authorities in order to ensure that users only access the data they are entitled to. Furthermore, in order to keep the data reliable it is important to see to that only authorized users are allowed to enter data. However, there is more to be done in security concerning for example virus protection and other types of unauthorized use.

Establish routines for maintenance: In order to keep a system of this type running and up to date there is a need to cater not only for the services provided, but also for the maintenance of the underlying technology. The broker must cater for the technological level since it constitutes an important foundation for the entire service. In fact it could be stated that the broker must be able to provide this technology, otherwise it would be a question of sharing data on a company-to-company basis.

Use well-known technology: Since the underlying technology is provided by the broker it is important that it is reliable and stable, or else the value of the service may be at risk. In this case we have chosen to implement a database in order to get access to advantages of database technology e.g. security, data independence, and query processing. Web technology is, to some extent, slow and unreliable but these issues are not considered as critical since we are not dealing with a time-critical application. Hence, the advantages of web technology concerning accessibility and platform independence outweigh the disadvantages.

4. CONCLUDING REMARKS

We have shown how a three level framework may be used to describe and evaluate an information service for a virtual organisation. All three levels have to be catered for and there are a number of important issues to be considered at each level. If problems arise in at any level there will be consequences for the entire project, which was the case in the case described. We have presented some issues that arose and caused problems during the project. The most important ones are summarised below.

The Organisational Level: Evaluate relations between participants; Identify hidden agendas; Identify clashing cultures; Establish and clarify the goals of the virtual organisation; Balance wagers and benefits.

The Service Level: Establish routines for selecting relevant data; Establish routines for acquiring data; Measure the benefits of the whole virtual organisation, stakeholder groups, and individual participants.

The Solution Level: Use well-known technology; Deal with security issues and accessibility; Cater for platform independence.

We found that there was an interest on behalf of the manufacturers to provide designers with knowledge of manufacturability in order to reduce resources spent on try out assemblies. There was also an interest on behalf of the designers to acquire such knowledge. However, much of the manufacturability knowledge is company specific, i.e. a potential competitive advantage, and therefore not considered sharable with potential competitors. Furthermore, this type of knowledge sharing is pursued in the bi-lateral contacts between designers and manufacturers.

We would like to point out that it is very important to be aware of the dynamics of the network. There are a number of influential factors within the participating companies, which are out of control for the network. These issues have to be dealt with as well, but they are not covered in this framework.

5. ACKNOWLEDGEMENTS

We would like to thank Ali Karimi and Gudmundur Hallgrimsson for their work on the technical solution; Per Johander for his valuable advice on the electronic circuit industry; and Benkt Wangler for reading and commenting the paper. This research is funded by the Swedish Knowledge Foundation (KK-stiftelsen).

6. REFERENCES

Ahuja, M. and Carley, K. (1999) Network Structure in Virtual Organizations *Organization Science*, **10**, 741-747.

Alavi, M. and Leidner, D. (1999) Knowledge Management Systems: Emerging Views and Practices from the Field. In *32nd Hawaii International Conference on System Sciences (HICSS)*Maui.

Alter, S. (1999) *Information Systems a management perspective*, Addison-Wesley, Reading, Massachusetts.

Carlsson, S. A. (2001) Knowledge Management in Network Contexts In *Ninth European Conference on Information Systems*Bled, Slovenia.

Castells, M. (1996) *Nätverkssamhällets Framväxt*, Daidalos, Göteborg.

Connolly, T., Begg, C. and Strachan, A. (1999) *Database Systems, A practical approach to design, implementation and management, 2nd Edition*, Addison-Wesley.

Coplien, J. O. and Schmidt, D. C. (Eds.) (1995) *Pattern Languages of Program Design*, Adison-Wesley Publishing Company, Reading, Massachussets.

Florescu, D., Levy, A. and Mendelzon, A. (1998) Database Techniques for the World-Wide Web: A Survey *SIGMOD Record*, **27**, 59-74.

Franke, U. J. (1999) The virtual web as a new entrepeneurial approach to network organizations *Entrepeneurship & Regional Development*, , 203-229.

Goodwill, J. (2000) *Pure JSP Java Server Pages*, Sams Publishing.

Hackathorn, R. D. (1999) *Web farming for the DW – exploiting business intelligence and knowledge management*, Morgan Kaufmann Publishers, San Francisco.

Kolodner, J. (1997) *Case-Based Reasoning*, Morgan Kaufmann Publishers, Inc., San Mateo.

Rowland (2000) DFM Rating Index *SMT - Magazine - Surface Mount Assembly of Printed Circuit Boards*, November.

Shao, Y. P., Liao, S. Y. and Wang, H. Q. (1998) A model of virtual organisations *Journal of Information Science,* **24,** 305-312.

Stowell, F., West, D. and Stansfield, M. (1997) Action Research as a Framework for IS Research, In *Information Systems: An Emerging Discipline?,* (Eds, Mingers, J. and Stowell, F.) McGraw-Hill, London, pp. 159-200.

Tiwana, A. and Bush, A. (2001) A social exchange architecture for distributed Web communities *Journal of Knowledge Management,* **5,** 242-248.

CREATING AN ORGANISATIONAL MEMORY THROUGH INTEGRATION OF ENTERPRISE MODELLING, PATTERNS AND HYPERMEDIA: THE HYPERKNOWLEDGE APPROACH

Anne Persson and Janis Stirna[1]

1. INTRODUCTION AND BACKGROUND

Modern organisations are expected to maintain a high level of innovation in their business and products. This means that they need to flexibly adapt to rapid change in their business environment. Among the main driving forces in this process are people and their knowledge. Organisations need to utilise this knowledge in the most efficient way since, in essence, it is part of their competitive advantage. It is therefore that managing experience, competence, knowledge about business processes, organisational practices, and best business practices are so important. The main goal of the Framework 5 project no IST-2000-28401 Hypermedia and Pattern Based Knowledge Management for Smart Organisations (HyperKnowledge) is to develop and test a novel approach to Knowledge Management (KM). The two cornerstones of this approach are the EKP method and the RETH tool (Bubenko, Persson and Stirna, 2001). During the course of the project two trial applications of the approach will be carried out – one involving a commercial company and one involving a public organisation.

The aim of this paper is to present the HyperKnowledge KM approach and to discuss the implications of introducing it in organisations. The remainder of the paper is organised as follows. The EKP method and the RETH tool are introduced in Sections 1.1 and 1.2. Section 2 provides an overview of the approach as well as explains its main activities. The implications and risks of introducing the approach in organisations are discussed in Section 3 while Section 4 gives some concluding remarks and discusses future outlook.

[1] Anne Persson, Dept. of Computer Science, University of Skövde, P.O. Box 408, SE-541 28 Skövde, Sweden. Janis Stirna, Dept. of Computer and Systems Sciences, Royal Institute of Technology and Stockholm University, Electrum 230, SE-16440, Kista, Sweden.

1.1 The EKP method

For the last decade the Royal Institute of Technology (KTH), Sweden has been active in building up an enterprise knowledge development methodology. Under the name of Enterprise Modelling (EM) an approach has been developed, first in the Framework 3 ESPRIT Project F3 (F-Cube-*"From Fuzzy to Formal"*[2] No 6621 and during the last 4 years in the Framework 4 projects ELEKTRA – *"Electrical Enterprise Knowledge for TRansforming Applications"*[3] No 22927 and HyperBank *"High Performance Banking"*[4] No 22693 under the name EKD. The approach employed in the HyperKnowledge project, Enterprise Knowledge Patterns (EKP), includes three components:

(1) A set of structured, goal/problem - driven models to be used for structuring and representing organisational knowledge. This modelling approach is called EKD - Enterprise Knowledge Development (Bubenko, Stirna and Brash, 1998; Loucopoulos et.al., 1997). Versions of this approach have been successfully applied in a number of European companies.

(2) A set of guidelines for conducting the knowledge acquisition and representation process. The basic assumption is that knowledge acquisition is strongly participatory, i.e. all involved actor and stakeholder types in an organisation are assumed to actively contribute. Within the ELEKTRA project two web-based tools for guidance of the EKD process were developed – the EKD Road Map (Nurcan and Rolland, 1999), and the EKD FAQ system (Sneiders, 1999).

(3) Support for reusing existing knowledge, business designs and enterprise models in the form of *organisational patterns*. Organisational patterns are *generic and abstract organisational design proposals that can be easily adapted and reused* that represent solutions to specific problems within an organisation. Each pattern couples a problem with a solution, reflecting the context and the way in which the pattern can be applied (ELEKTRA Consortium, 1999; Rolland et. al., 2000; Zaharova and Stirna, 2000). The pattern approach developed in the ELEKTRA project has been applied in Vattenfall AB (Sweden) and Public Power Corporation (Greece).

1.2 The RETH tool

Siemens AG Österreich, Austria, has developed and successfully used a method and a tool (RETH) for software requirements engineering. The tool represents textual objects and their relationships, including attributes, and makes use of hypertext functionality and multimedia objects, such as sound and video. The integration of hypertext and multimedia is mostly referred to as *hypermedia*. In contrast to comparable tools on the market, the RETH tool actively helps in the modelling effort. For example, it has the useful feature of semi-automatic generation of hyperlinks (Kaindl, Kramer and Diallo, 1999), which supports the RETH user in relating concepts that are defined using natural language. Based on the generation of hyperlinks, the RETH tool can semi-automatically generate object-oriented associations / E-R relations in the course of enterprise modelling (Kaindl, 1996). In the context of the HyperKnowledge project a significant capability of

---

[2] Information available on http://www.dsv.su.se/research/syslab/Projects/Archives/F3/f3.html
[3] Information available on http://www.dsv.su.se/~js/elektra.html
[4] Information available on http://www.dsv.su.se/research/syslab/Projects/HYPERBANK/hyperbank.html

the tool is that its knowledge contents can be automatically exported into a Web representation and thus be made available for a larger knowledge networking community. After the export, the complete repository, including all multimedia files attached, is available in the Web representation. This means that users who only want to access the knowledge repository can do so by using their web browser.

2. THE HYPERKNOWLEDGE APPROACH TO KNOWLEDGE MANAGEMENT

The HyperKnowledge Knowledge Management (KM) process (Figure 1) covers the whole lifecycle of knowledge in an organisation (Nonaka and Takeuchi, 1995). The cycle is similar to the spiral of organisational knowledge creation as presented by Nonaka and Takeuchi (1995).

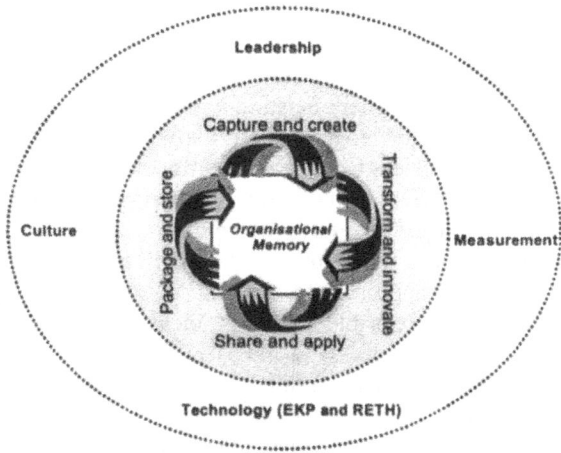

Figure 1. Overview of the HyperKnowledge KM approach

Creating knowledge can be done in many different ways – running day-to-day business operations, improving existing work routines, restructuring the organisation, planning organisational strategies for the future, etc. Often the creators of knowledge are not aware of this and valuable knowledge may therefore be lost. To prevent this, the knowledge needs to be captured in one way or another. This might require thinking in abstract terms, building models/mind maps, or simply writing down the experiences. Most often this should be done in a participative and collaborative way. Once knowledge is captured, the organisation and its employees are aware of its existence. The next step is to package and store the knowledge so that it is available and can be used by everyone in the organisation. The key element here is to make the specific knowledge *useful*. This usually requires some degree of generalisation of knowledge. Furthermore it also requires envisioning how each knowledge chunk will be used. The knowledge that is written down in some form usually resides in repositories, manuals, an intranet, etc. However, not everything can be written down. Most often the tacit knowledge is the most important knowledge. In this case we can only write down who knows what, where the knowledge sources are, and how to access it. This also becomes an important part of the corporate knowledge repository. After knowledge is properly documented and stored, it needs to be shared and applied. This is probably the most important task in KM. Knowledge sharing cannot be done mechanistically. It is not enough to install and fill a knowledge repository and expect the organisation to suddenly start sharing knowledge. Therefore, the main attention should be paid to building a knowledge sharing culture in the organisation. Technology can only play a supporting role in knowledge sharing and application – it can make knowledge sharing easier and more effective. Successful and effective knowledge sharing and application also stimulates innovation - improvement of existing knowledge and creation of new knowledge. This essentially closes the knowledge cycle. In the following we outline these activities from the point of view of the HyperKnowledge project and the EKP approach.

2.1 Capture and create knowledge

In order for knowledge to be shared between different stakeholders it has to be discovered. This is an activity where existing knowledge is *captured* and new knowledge is *created*. Creation of knowledge during knowledge discovery is an effect from integration of existing knowledge. In other words, the sum is more than the individual parts put together. This effect is particularly observable if the knowledge discovery process is *open, collaborative* and *participative*. Therefore, the HyperKnowledge approach to KM will use the EKD enterprise modelling method in this stage of the KM process. Its participatory approach to solving ill-structured problems will stimulate discovery of best practices, reusable solutions, ideas, suggestions, etc. The discovery of knowledge is an ongoing process in the learning organisation. New "chunks" of knowledge are of different size and complexity. E.g. adding details to existing knowledge is a smaller effort compared to generating knowledge for a new type of business problem. Employing the EKD method consumes resources. This means that the amount of knowledge generated in each case of the knowledge discovery process should be large enough to award that such resources are invested. Therefore, we perceive that the process of capturing and creating knowledge will use the EKD method when this is awarded and simple knowledge discovery techniques such as interviewing when this is found to be more suitable.

One of the main success factors for modelling projects is availability of skilled people who can support the project and drive it forward. If the organisation decides to incorporate modelling as part of the KM process it should establish a modelling support team. Without such a team the individual employees can engage only in very small modelling tasks. It is important to understand that establishing the team will take time. Typically a modelling support team includes modelling facilitators, model documenters and maintainers, as well as methodology experts.

2.2 Package and store knowledge

The main focus of this step is to package knowledge in a form that makes it reusable at some later point in time. Business knowledge in the form of enterprise models, ideas, suggestions, discussions, best practices, documents, and reports is described in the form of *organisational patterns* following the Enterprise Knowledge Patterns (EKP) approach.

Organisational patterns are linked to each other, thus essentially forming a pattern language. They are also linked to their knowledge sources, explanations, comments, as well as real life examples.

Pattern based approaches have been established in software programming, software design, data modelling, and in systems analysis (see e.g. Gamma et. al., 1995; Fowler, 1997; Coplien and Schmidt, 1995). The notions of pattern from these areas share two main ideas – (1) a pattern relates a recurring problem to its solution, and (2) each problem has its unique characteristics that distinguish it from other problems. In the context of the HyperKnowledge project we will use the term "organisational pattern". Organisational patterns are viewed as *generic and abstract organisational design proposals* (ELEKTRA Consortium, 1999; Prekas et. al., 1999; Rolland et. al., 2000). Patterns encapsulate organisational knowledge in a way that facilitates its reuse.

In the area of business development, patterns are relatively new and untested. Therefore there are no standard components for their use in the business development world. Patterns in the organisational setting can be used to describe anything that presents regular, repeatable characteristics: formal organisational and contractual relationships, informal relationships, responsibilities, work practices, etc. Well-known examples of Web-based pattern repositories are *Organizational Patterns* (Coplien et. al., 2000) and the *Portland Patterns Repository* (2000). These patterns cover solutions to a wide range of business environment issues. Within the ELEKTRA project a Web-based library of organisational patterns was developed for the purpose of sharing business practices in the European electricity supply industry. Despite the overall good evaluation results for ELEKTRA patterns (ELEKTRA Consortium, 1999; Rolland et. al., 2000), a number of challenges were discovered indicating that more research should be devoted to application of patterns for communicating organisational knowledge and expertise (Zaharova and Stirna, 2000). The HyperKnowledge project will further advance the results and experiences of the ELEKTRA project.

The community of design patterns emphasises that patterns should always be *well-proven* solutions to reoccurring problems. Whereas it is possible to argue that some of the HyperKnowledge patterns are designs that are tried and tested throughout their organisations, some others will only be proposals for organisational solutions, whose degree of applicability will need to be tested. Indeed we will not claim that all the design proposals put forward in this project are tested in a sufficient number of real cases so that they can be characterised as tested solutions. Therefore, we put the emphasis on the fact that organisational patterns *are generic and abstract design proposals, solutions to recurring problems* within the sector of interest that can be easily adapted and reused. An organisational pattern is more than just a description of "something" in the world. A pattern should also be a "rule" defining when and how to create that something. It should be both a description of the artefact and a description of the process that will generate the artefact.

In order to facilitate the reuse process we have structured the pattern template in two main components – the *knowledge perspective* and the *usage perspective* (Figure 2). The former is the part of the knowledge that is effectively reused whereas the latter aims to provide sufficient information about the pattern as well as to describe its intended context of reuse and how it should be used. The knowledge perspective defines the problem, that the pattern solves and the solution to that problem. The usage perspective answers questions such as, when can the pattern be reused, how can the pattern be reused, what are the consequences of reusing the pattern, where has the pattern been reused, etc.

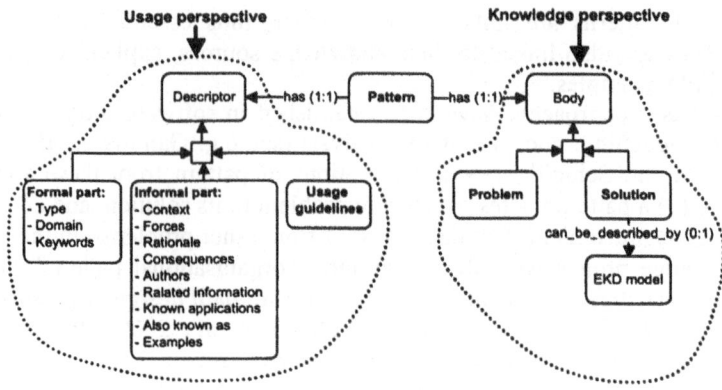

Figure 2. Conceptual view of the pattern template

In practice the various pattern components have been represented using *natural language* (e.g. see Coplien, 1995; Coplien et. al., 2000) or *conceptual models* (Fowler, 1997). In the HyperKnowledge approach free natural language will be used for the attributes of the informal part and for the usage guidelines. In the case where no other representation for the solution of the pattern is available, free natural language can also be used to represent the solution. Formalised natural language will be used to represent the formal part of the descriptor. Diagrammatic descriptions will be used to represent the pattern solution through the EKD modelling components (such as goal models, business process models, concepts models, etc.). The pattern solution can also be represented as a video, a drawing, etc. Furthermore, attributes of the informal part can also be linked to various multimedia sources. See http://www.dsv.su.se/~danny/patternlibrary/main.html for examples. In existing work on organisational patterns much emphasis has been placed on organising patterns using a meaningful and coherent mechanism. The mechanism preferred in the ELEKTRA project was that of a hierarchy (see ELEKTRA Consortium, 1999; Prekas et. al., 1999). In the HyperKnowledge project the hierarchy of patterns is mainly built by defining hyperlinks between patterns. Pattern relationships supported by the hyperlink mechanism of the RETH tool directly contributes to building a pattern language.

The main benefit of pattern development is *reusability*. This makes patterns an ideal concept in human endeavours that involve design and the kind of decisions that come with it. The process of reuse is classically divided into two major and complementary activities (Figure 3). The first one deals with the process of defining reusable components (*design for reuse*) while the second is dedicated to the effective usage of reusable components (*design by reuse*). The process of *design for reuse* aims to produce a description of reusable components. These can be organisational designs, solutions to common business problems, implementation guidelines for organisational policies, as well as designs of information system components. Identifying potential reusable knowledge and constructing the reusable component embedding this knowledge is a main target. Within the HyperKnowledge project the RETH tool will maintain the knowledge repository. Complementary, the process of *design by reuse* concerns, on the one hand, procedures for searching for and retrieving reusable knowledge components, and, on the other hand, procedures for customising and integrating the solution advocated by the component in the product under development. The feedback learning loop is also provided here since experiences from applying the proposed solution serves as a source for improving the reuse repository. The process of design by reuse involves search and retrieval of knowledge in the repository. This is supported by the web export functionality of the RETH tool.

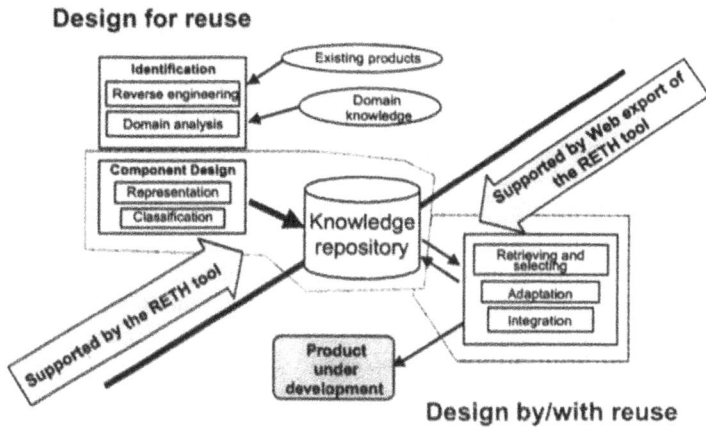

Figure 3. Two aspects of the reuse process

2.3 Share and apply knowledge

In knowledge sharing and application the main attention should be paid to building a knowledge sharing culture within the organisation. This primarily requires leadership that actively supports KM. Technology can only make knowledge sharing easier and more effective. Within the scope of the HyperKnowledge project the main technology will be the RETH tool. It will contain the core of the knowledge of the user organisations. Knowledge reuse will be supported by the web export functionality of the tool and the knowledge repository will be located on the intranets of both user organisations. As a result, employees will be able to access it via their web browsers. The tool itself will be used only for entering knowledge chunks, e.g. patterns, experiences, business practices, and job instructions in the repository. From a technology point of view, knowledge sharing implies that the employees will browse the knowledge in the repository and then apply it in their work. They will also use electronic discussion groups to discuss and comment various issues in the knowledge base. The results of these discussions will serve as input to the next step – knowledge transformation and creation of new knowledge. The repository will only work if the organisation has established other KM activities and infrastructures, such as expert groups, networks, knowledge-based performance appraisal systems, etc. However important, these activities are considered to be beyond the scope of the HyperKnowledge approach.

Organisational patterns are usually arranged in structures or hierarchies. These may or may not have sub-hierarchies and they might be linked to other hierarchies. Most likely hierarchies will deal with different organisational problems. They will have different owners and possibly different people contributing to them. In tightly integrated organisations such knowledge hierarchies will be more closely integrated than in large and distributed organisations. Therefore, large organisations should devote additional effort to integrating distributed knowledge hierarchies and to developing cross-references between knowledge chunks in the repository. The HyperKnowledge project will use the RETH tool for automatically finding and establishing references between entries in the repository. For example, in Figure 4 we can see parts of two knowledge hierarchies, one containing organisational patterns regarding competency management and the other one containing electronic job instructions. We can see that knowledge chunks are cross-referenced within a hierarchy as well as between different hierarchies. References are implemented as hyperlinks, which the RETH tool found and generated automatically. In practice, automatic finding of hyperlinks considerably improves the integration between different domains in the knowledge repository. This is particularly useful if people in different organisational units maintain knowledge hierarchies. To increase the possibility that the RETH tool will automatically find cross-references we recommend defining synonyms to terms and concepts commonly used within the domain. We recommend that synonyms be chosen from a predefined taxonomy.

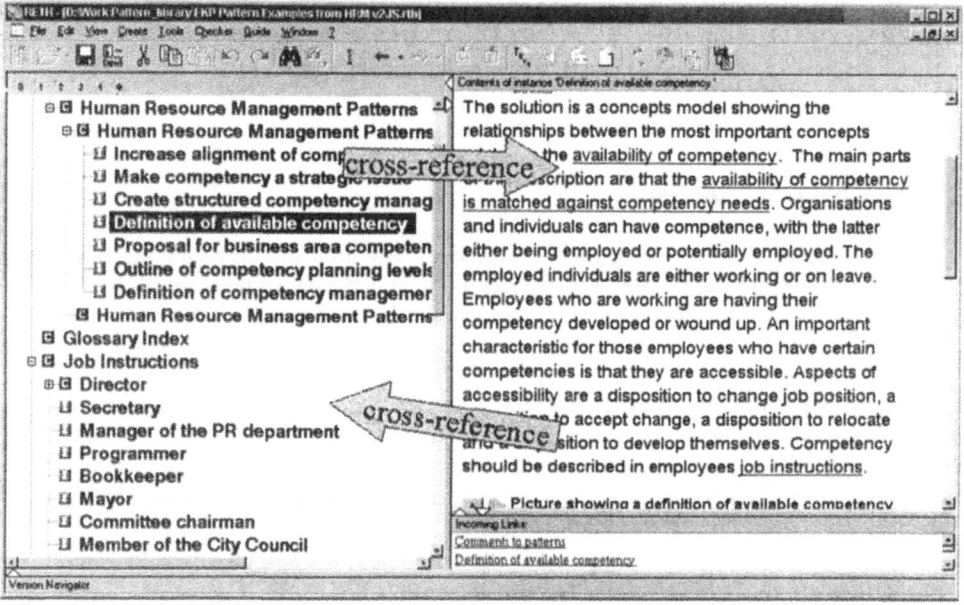

Figure 4. Knowledge entries in different hierarchies are cross-referenced in the RETH tool

Employees will use their web browsers to locate and retrieve knowledge. Browsing the knowledge hierarchies is one approach. This requires considerable time if the user is not familiar with the overall structure and contents of the repository. Another alternative is to search using keywords. The RETH tool offers simple text searching facilities. Since the knowledge repository will be located on the corporate intranet more powerful search and retrieval mechanisms can be used. For instance, once the knowledge repository grows sufficiently large and the number of similar queries asked to the system grows the organisation may want to install a Frequently Asked Question answering system like the EKD FAQ system suggested by Sneiders (1998).

Knowledge application is a creative human process and therefore it is difficult to predict and support. When it comes to applying organisational patterns we can foresee the following approaches:

- A pattern or a collection of patterns is applied in order to solve the problem that the pattern addresses. In this case the pattern user follows the usage guidelines given in the pattern template. In essence patterns are used as *proven solutions* to recurring problems.
- A pattern user browses the pattern library and on the basis of the patterns found creates a new solution to the problem at hand. This can be done by either modifying an existing pattern, merging several patterns, or starting completely from scratch. In either of these cases patterns in the repository are used as *sources of ideas* or organisational design proposals.

Other types of knowledge in the repository can be used in different ways. For example, the organisational structure and job descriptions can be linked to patterns. The taxonomy of the organisation's business objects can be used to cross-reference knowledge chunks, etc. No matter how the knowledge in the repository is applied, it is important to gather experiences and lessons learned from this application. We call this the learning loop. The process of gathering comments and experiences is also supported by the RETH tool. Once comments are entered, the tool finds corresponding hyperlinks.

2.4 Transform and innovate knowledge

Each application of the contents of the knowledge repository will generate new experiences and lessons learned which, in turn, will be captured, entered in the repository, and linked to the piece of knowledge from which they originated. On the basis of the gathered experience the repository users will be able to improve their new best practices and develop new organisational patterns. We perceive that the HyperKnowledge approach will contribute to the knowledge transforming and innovation process in the user organisations, enabling them to become learning organisations. The main focus would be on knowledge synthesis, which involves both explicit and tacit knowledge. Furthermore, patterns are not the only nodes in the knowledge repository – there are also success stories, lessons learned, suggestions, etc. Transformations and innovations of knowledge cover the following dimensions:

From specific to generic: Generic best practice descriptions and new business solutions are developed from many specific organisational problem cases. Analysing and learning from experiences made is an activity integrated in the learning organisation. In order for the organisation to collectively capitalise on such a learning process specific experiences are generalised so that they can be reused in similar situations. The EKD modelling process using a participative approach is a suitable vehicle for the generalisation process.

From "tacit" to explicit: We expect that some of the knowledge about a specific problem or topic should be considered as tacit. However, parts of that knowledge may in fact be possible to extract and represent, which would be an important contribution to building the corporate memory. The EKD modelling method using a participative approach is useful in attempting to extract knowledge that is considered tacit. The RETH tool is then a means to disseminate the represented knowledge.

From distributed to integrated: The integration of organisational knowledge from distributed sources provides a *"one stop shop"* of reusable knowledge chunks. Knowledge is distributed in different knowledge containers, such as documents, systems, the minds of business stakeholders etc. An understanding for how this knowledge can be integrated is developed through modelling the relationships between different types of knowledge using the EKD modelling method. The integration is then visualised through the RETH tool.

From operational to strategic: Knowledge created in the operational business processes will be extracted, analysed and transformed into strategic knowledge in order to support the *decision-making needs* of the organisation. The collected knowledge is used for various purposes, operational as well as strategic. On the strategic level the knowledge is a basis for formulating new business goals and for re-evaluating the situation at hand. It is therefore necessary that the representation of knowledge supports aggregation. In essence, this leads us to the first stage in the knowledge cycle creation and capturing once again. The main difference, however, is that we have created and captured new knowledge based on our previous knowledge – in other words, we have learned or completed one cycle in the learning loop.

3. INTRODUCTION OF THE HYPERKNOWLEDGE APPROACH IN ORGANISATIONS

The successful introduction of an explicit KM process in an organisation must be based on a carefully devised strategy in order to motivate employees to capitalise on the potential benefits of KM. Looking at the HyperKnowledge approach to KM we identify two aspects of concern regarding its introduction: one aspect regarding KM in *general* and one regarding how to introduce the *specific* approach that we propose.

Firstly it is essential that efforts be made to create a knowledge-sharing culture in the organisation and to introduce the principles of organisational learning. Introducing KM in an organisation that is unfamiliar with the concept of organisational learning requires that small and concrete steps be taken in the desired direction. Introduced KM activities should be perceived to make substantial improvements on the working situation of employees, i.e. they should be useful. If so, they will motivate employees to change their thinking and ways of working. A challenge is also how to motivate employees to keep contributing their knowledge to the system over time. Again this concerns the perceived usefulness of the activities and the system. Identifying common domains of interest is therefore important. A unified terminology is then required, which facilitates the sharing of knowledge between different parts of the organisation.

Introducing KM in an organisation that is not familiar with the concept involves a risk. KM is, therefore, a leadership issue. Without the proper leadership, KM activities have little or no chance of penetrating the organisation. Top management in the organisation should actively drive the implementation of KM. Their main responsibility is to motivate employees through different means and to support building of the KM infrastructure. It is also essential to develop indicators to assess KM performance at the organisational and individual levels.

Although supporting technology is an important aspect of KM, it should be emphasised that it is insufficient to install a single KM technology package. Instead, KM technology consists of numerous building blocks supporting various KM activities. In fact, the main potential problems and risks related to the implementation of KM are not concerned with technology in. Technology supports KM, but the *actual* KM is carried out by people. We perceive the greatest challenge for effectively implement the KM vision to be the changing of attitudes at all organisational levels towards knowledge sharing and organisational learning. This emphasises again the importance of culture and leadership.

The approach includes two main parts: the EKP methodology for capturing and representing reusable knowledge and the RETH tool for disseminating reusable knowledge. Applying the approach for the first time in an organisation requires that substantial resources be invested into setting up a functioning KM system. Setting up such a system involves working with the business stakeholders to discover the knowledge to be included. This requires that facilitators and analysts work together with business stakeholders according to the EKP approach. It also requires that the RETH tool be properly set up. However, the greatest challenge in applying the HyperKnowledge approach to KM is to maintain the KM system over time. This requires the organisation to obtain or develop the needed competency to continuously discover, represent and disseminate new knowledge to be used in the KM system. It is suggested, therefore, that specific roles are designed with an explicit responsibility to maintain the system. It is also suggested that the KM process is systematically integrated into existing and future business processes.

4. CONCLUDING REMARKS AND FUTURE OUTLOOK

In this paper we have presented a KM approach (EKP) that integrates an enterprise modelling method (EKD) with organisational patterns. This method addresses all four stages of the organisational knowledge cycle: create and capture knowledge; package and store knowledge; share and apply knowledge; as well as transform and innovate knowledge. Within the HyperKnowledge project the EKP approach will be used to create and maintain organisational memory of both user organisations – Verbundplan GmbH, Austria and Riga City Council, Latvia. The resulting knowledge repositories will be supported by the RETH tool. Preliminary results of applying the EKP approach allows us to make the following observations:

- The concept of knowledge reuse is much appreciated in the user organisations as they currently lack tools for gathering and sharing experiences, business practices, standard solutions, etc.
- Using participative enterprise modelling for knowledge discovery seems to be a natural and easy way of working for business stakeholders, which facilitates the process of making some of their tacit knowledge more explicit.
- Application of participative modelling and organisation patterns directly contributes to building a learning organisation.

- The tool support for KM is important, but not decisive. In the process of building an organisational knowledge repository many cultural, organisational, managerial issues should be resolved first.

In conclusion, integration of enterprise modelling and organisational patterns supported by a hypermedia-based tool has shown to be useful. More research efforts should be directed towards further customising the EKP method and the RETH tool according to the specifics of knowledge management applications. In addition more efficient use of video and audio media to capture some of the tacit knowledge should also be investigated in order to create a true hypermedia-based organisational memory.

5. REFERENCES

Brash D. and Stirna J. (1999). Describing best business practices: A pattern-based approach for knowledge sharing, In *Proceedings of SIGCPR 1999 Conference on Managing Organizational Knowledge for Strategic Advantage: The Key Role of Information Technology and Personnel*, New Orleans, USA.

Bubenko J.A. jr, Stirna J. and Brash D. (1998). EKD User Guide, Dept. of Computer and Systems Sciences, Royal Institute of Technology, Stockholm, Sweden.

Bubenko J. A. jr, Persson, A. and Stirna J. (2001). User Guide of the Knowledge Management Approach Using Enterprise Knowledge Patterns, HyperKnowledge project deliverable, project no IST-2000-28401, Dept. of Computer and Systems Sciences, Royal Institute of Technology, Stockholm, Sweden.

Coplien J., and Schmidt D. (Eds.) (1995). *Pattern Languages of Program Design*, Addison Wesley, Reading, MA.

Coplien, J., Berczuk, S., Cunningam, W., Kaul, U., Hanes Perry, B., Devos, M. and Harrison, N. (Eds.) (2000). Organizational Patterns, as is January 2000, http://www.bell-labs.com/cgi-user/OrgPatterns/OrgPatterns?FrontPage.

ELEKTRA Consortium (1999). Molière: The ESI Knowledge Base Specification, ELEKTRA Project Deliverable Document, ESPRIT Project No. 22927.

Fowler M. (1997). *Analysis Patterns: Reusable Object Models*, Addison-Wesley.

Gamma E., Helm R., Johnson R. and Vlissides, J. (1995). *Design Patterns: Elements of Reusable Object-Oriented Software*, Addison Wesley, Reading, MA.

Kaindl, H. (1996). How to Identify Binary Relations for Domain Models, in *Proceedings of. Eighteenth International Conference on Software Engineering (ICSE-18)*, Berlin, Germany, IEEE.

Kaindl, H., Kramer, S., and Diallo, P.S.N. (1999). Semiautomatic Generation of Glossary Links: A Practical Solution, in *Proceedings of Tenth ACM Conference on Hypertext and Hypermedia (Hypertext '99)*, Darmstadt, Germany.

Nonaka, I. And Takeuchi, H. (1995). The Knowledge-Creating Company, Oxford University Press.

Nurcan, S. and Rolland, C. (1999). Using EKD-CMM electronic guide book for managing change in organisations, in *Proceedings of the 9th European-Japanese Conference on Information Modelling and Knowledge Bases*, Iwate, Japan.

Portland Pattern Repository (2000). As is February 9, 2000, http://c2.com/cgi/wiki?PortlandPatternRepository.

Prekas N., Loucopoulos P., Rolland C., Grosz G., Semmak F. and Brash D. (1999). Developing patterns as a mechanism for assisting the management of knowledge in the context of conducting organisational change, In *Proceedings of 10th International Conference and Workshop on Database and Expert Systems Applications (DEXA'99)*, Springer.

Rolland C., Stirna J., Prekas N., Loucopoulos P., Grosz G. and Persson A. (2000). Evaluating a Pattern Approach as an Aid for the Development of Organisational Knowledge: An Empirical Study, In *Proceedings of Conference on Advanced Information System Engineering*, (WANGLER B. AND BERGMAN, L., Eds.), Springer.

Sneiders E. (1999). Automated FAQ *Answering on WWW Using Shallow Language Understanding*, Licentiate Thesis, Dept. of Computer and Systems Science, The Royal Institute of Technology and Stockholm University, Sweden, 1999, ISSN 1101-8526.

Zaharova S. and Stirna J. (2000). Using Organisational Patterns as a Technique for Knowledge Management, 5th CAiSE/IFIP8.1 International Workshop on Evaluation of Modeling Methods in Systems Analysis and Design (SIAU, K., Ed.).

APPLICATION DOMAIN KNOWLEDGE MODELLING USING CONCEPTUAL GRAPHS

Irma Valatkaite and Olegas Vasilecas[*]

1. INTRODUCTION

Recently a lot of efforts were directed towards finding a conceptual modelling language that would be simple to learn and use in information systems development process, but expressive enough in order to capture all aspects of domain knowledge. To model knowledge explicitly has become an important issue because information systems were started to be used not only for data processing activities but also for management purposes and long-term organisation development. The knowledge considered usually is related with different aspects of organisation – its processes of production, marketing, management, processes constraints, policies and guidelines, customers and suppliers. The knowledge of such domain considered in information systems should capture all aspects of the domain – its structure, the behaviour of objects in the domain, and the constraints that govern the behaviour. Thus, the complete description of the domain is achieved.

In order to represent all these aspects of organisation knowledge, traditional conceptual modelling languages lack semantic expressiveness (Mineau et al., 2000). Today UML is the most widely used modelling language in the industry (Booch et al., 2000). UML allows the designer to draw a set of objects and the way they interact, but the behaviour must be entered as OCL (Object Constraint Language) code, which is not far away from the programming code itself (Rational, 2002). Even though OCL aims to be used and understood by people that are not necessarily mathematicians, the first order logic knowledge is required to be able to express domain constraints.

The aim of research community is to achieve the conceptual modelling language that would be simple to learn and use, had graphical notation, and would be able to express all aspects of domain knowledge in a uniform manner. Furthermore, the knowledge specification must be precise and unambiguous. Conceptual graphs were proposed as

---

[*] Information Systems Laboratory, Vilnius Gediminas Technical University, Vilnius, Lithuania.

such a language. The original theory of conceptual graphs was introduced by Sowa (Sowa, 1984). A conceptual graph is a finite, connected, bipartite graph. It includes notions of concepts, relations, and actors. In the models, concept nodes represent entities, attributes, states, and events, and relation nodes show how phenomena classes are interconnected (Krogstie and Ivberg, 2000).

Later a number of extensions to conceptual graphs were proposed to overcome the inability of conceptual graphs to represent dynamic behaviour. Delugach proposed a new conceptual graph node called *demon* and argued that demon as a fundamental extension to conceptual graphs is needed to capture the idea of a process or transformation (Delugach, 1991). Later Mineau introduced processes (Mineau, 1998) that are based on Delugach work. In addition, it was argued that the original theory without any extensions is enough to model and directly execute conceptual graphs (Cyre, 1998). The comparative study of proposed approaches to dynamic conceptual graphs might be found in (Lucose and Mineau, 1998).

Despite all the arguments that support conceptual graphs as conceptual modelling language, those attractive modelling ideas remain the subject of research papers, not the business practice. Conceptual graphs have had a very limited influence on conceptual modelling practices and the development and maintenance of information systems in most organisations, even it has received much attention within computer science research (Krogstie and Ivberg, 2000). To overcome limitations of conceptual graphs usage for information systems modelling some work has been done. A mapping from ER model to conceptual graphs was proposed by Creasy and Ellis (1993). The implementation of the mapping rules was presented by Valatkaite and Vasilecas (2002). A relational database querying mechanism was developed by Haemmerle and Carbonneil (1996). Although the latter is more targeted to the usage of already implemented business systems, not for conceptual modelling, but the authors have shown how conceptual graphs may be applied to relational database design and querying. In addition, a number of research projects were carried out that proposed executable systems based on conceptual graphs, e.g. MODEL-ECS (Lukose, 1995), PROLOG+CG (Kabbaj and Lanta-Polczynski, 2000), pCG (Benn and Corbet, 2001).

However, the lack of effective techniques and tools based on conceptual graphs suitable for use by business practitioners is obvious, as noted in the recent publications (Mineau et al., 2000), (Krogstie and Ivberg, 2000). Such techniques should offer the convenient way to incorporate the conceptual graphs into the knowledge modelling process of some real business domain for which information system is to be built. In this paper, a step towards such a tool is proposed: a methodology for information systems development where conceptual graphs are used extensively as the basic language of knowledge modelling. The approach incorporates the means to aid the implementation of domain business rules into a relational DBVS as a target system. We believe that the proposed methodology may facilitate the integration of the conceptual graphs formalism into the market place. Consequently, in Section 2 we present the methodology, Section 3 outlines how this methodology may be used in practice. Finally, the discussion and conclusions are drawn in Section 4.

2. THE METHODOLOGY BASED ON CONCEPTUAL GRAPHS AS MODELLING LANGUAGE

The methodology proposed in this paper is based on the conceptual graphs as the basic modelling language. The striving to have one language suitable for knowledge modelling no matter it is structural or behavioural has been there for some time. Moreover, the knowledge model must be formal yet readable by domain experts. In research on knowledge systems development methodologies such as CommonKADS (Schreiber et al., 1999), MIKE (Angele et al., 1998) several languages were proposed to be used as modelling languages for knowledge model. The goal was the formal executable language, understandable by domain experts and able to represent complete domain knowledge. In CommonKADS, two languages are used: semi-formal language CML for conceptual modelling and formal language $(ML)^2$ for formal knowledge model. MIKE has KARL and its extension DesignKARL. Conceptual graphs were proposed to be used in CommonKADS as language for expertise model in (Moeller, 1995) because they fulfil the contradicting requirements that stem from the research on knowledge modelling languages:

1. Knowledge model must be in a formal language because then it captures domain knowledge precisely and unambiguously; also given a formal knowledge model it is possible to translate it to some executable knowledge system while preserving the original knowledge structure.
2. Knowledge model in informal language allows using it as the communication basis for the analyst and domain experts thus achieving active involvement of domain experts in the knowledge modelling process.

The conceptual graphs formalism being a graphical interface to a system of logic may be used in an informal way with their graphical notation, and in a formal way when translated to CGIF (conceptual graphs interchange format) or KIF (knowledge interchange format) (Sowa et al., 2000).

The difference between the proposed methodology and already established methodologies (e.g. CommonKADS which is *de facto* standard for knowledge systems development) or knowledge modelling methods with conceptual graphs (e.g. MODEL_ECS, pCG) (Lukose and Mineau, 1998) is the knowledge storage and execution system. Methodology proposed in this paper is targeted not to some specific knowledge execution environment (e.g. MODEL-ECS, PROLOG+CG), but to the relational database management system. The information system development process is tailored so because majority of business systems currently rely on relational database systems as the main business data and knowledge storage and processing unit. One may question the phrase "database system as knowledge storage unit". However, knowledge was always a part of any information system although it was not explicitly identified so. An example of knowledge stored in relational database system is triggers that enforce domain business rules. shows how the development process may be viewed using conceptual graphs as modelling language and the relational database system as the target system. There may be two approaches considered, those eventually produce the same result.

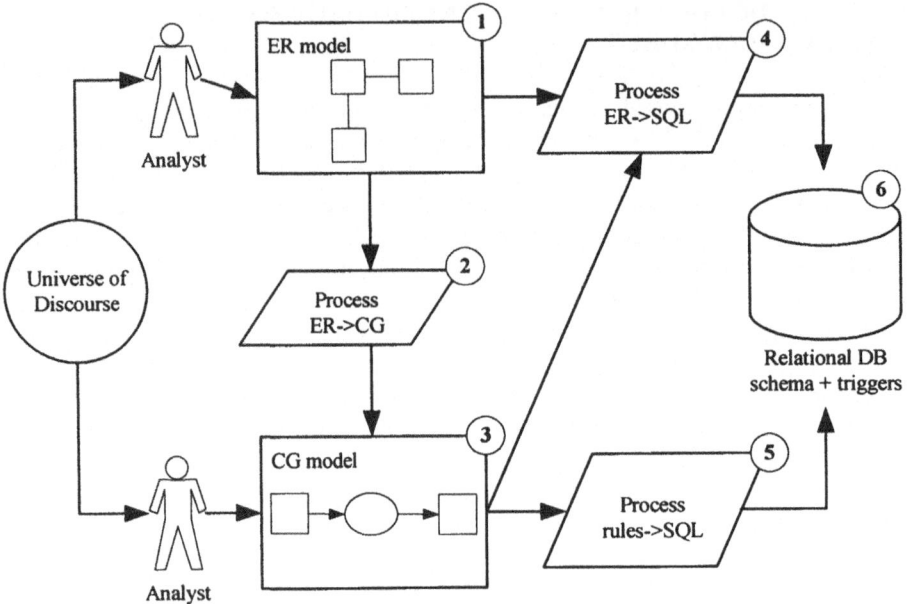

Figure 1. Methodology based on conceptual graphs as modelling language

The first approach would comprise the following phases:

1. Domain structural knowledge captured by analyst is represented in entity
    relationship (ER) language. ER model is used because it is the widely accepted
    language for conceptual modelling, taught in universities and therefore known
    by majority of systems analysts. In the proposed modelling approach, ER
    language is recommended to use for the purpose of easier integration of known
    and new modelling languages, i.e. ER language and conceptual graphs. The
    result is structural domain knowledge in ER language (marked as (1) in Figure
    1).
2. The ER model of structural knowledge is translated to conceptual graphs model.
    The translation is based on work proposed in (Creasy and Ellis, 1993) and
    extended in (Valatkaite and Vasilecas, 2002) where translation rules are
    formulated and implemented (process is marked as (2) in Figure 1). The result is
    structural domain knowledge in conceptual graphs language (marked as (3) in
    Figure 1).
3. The conceptual graphs model is augmented with dynamic (processes or
    behaviour) knowledge and the constraints by analyst. The dynamic knowledge
    may be represented using Delugach demons or Mineau processes as the means
    to model domain objects behaviour. The result is the full domain knowledge
    model in conceptual graphs language (marked as (3) in Figure 1).
4. The domain knowledge model in conceptual graphs language is used to generate
    triggers for a relational database system as the business rules implementation.
    The result is the business rules implemented in a database system.

5. The structural part of domain knowledge contained in ER model is used to generate database schema in a relational database system, the majority of database modelling tools support the generation of SQL scripts to create a database schema.

The second approach would comprise the same steps, although ER modelling is omitted:

1. Domain knowledge captured by analyst is represented in conceptual graphs language. In this approach all domain knowledge is considered, the analyst does not have to make knowledge analysis regarding knowledge type. The dynamic knowledge may be represented using Delugach demons or Mineau processes as the means to model domain objects behaviour. The result is the full domain knowledge model in conceptual graphs language (marked as (3) in Figure 1).
2. The domain knowledge model in conceptual graphs language is used to generate: (a) database triggers for a relational database system as the business rules implementation, the result is the business rules implemented in a DBVS; (b) the structural part of domain knowledge is used to generate database schema in a relational database system.

Both approaches may be used. However, the second one is the approach that utilises conceptual graphs as a uniform language to model all knowledge of domain therefore is more desirable as a pure conceptual graphs modelling approach.

Using the proposed methodology, the small experiment is presented to show how the modelling ideas and conceptual graphs language may be used in business practice.

3. APPLICATION OF THE PROPOSED METHODOLOGY

The domain used for the experiment is taken from the real business environment. However, for the purposes of simplicity and clarity, only excerpt from the real business environment is considered while presenting the modelling ideas. The second approach is used in order to utilise conceptual graphs fully.

3.1. Domain Knowledge Model

The structural domain knowledge model is presented in Figure 2. In short, the domain may be described as follows:

The business is run in terms of works. Each work has a number of technologies that must be accomplished before the work is considered as completed. Each technology defines the operation that must be performed. The number of hours spent on each technology is tracked.

A work is identified by a name, also must be in some state and have a completion date.

A technology is identified by some numeric identifier, also must be in some state, have a completion date, and number of hours spent by a worker on this technology.

A operation is identified by a name.

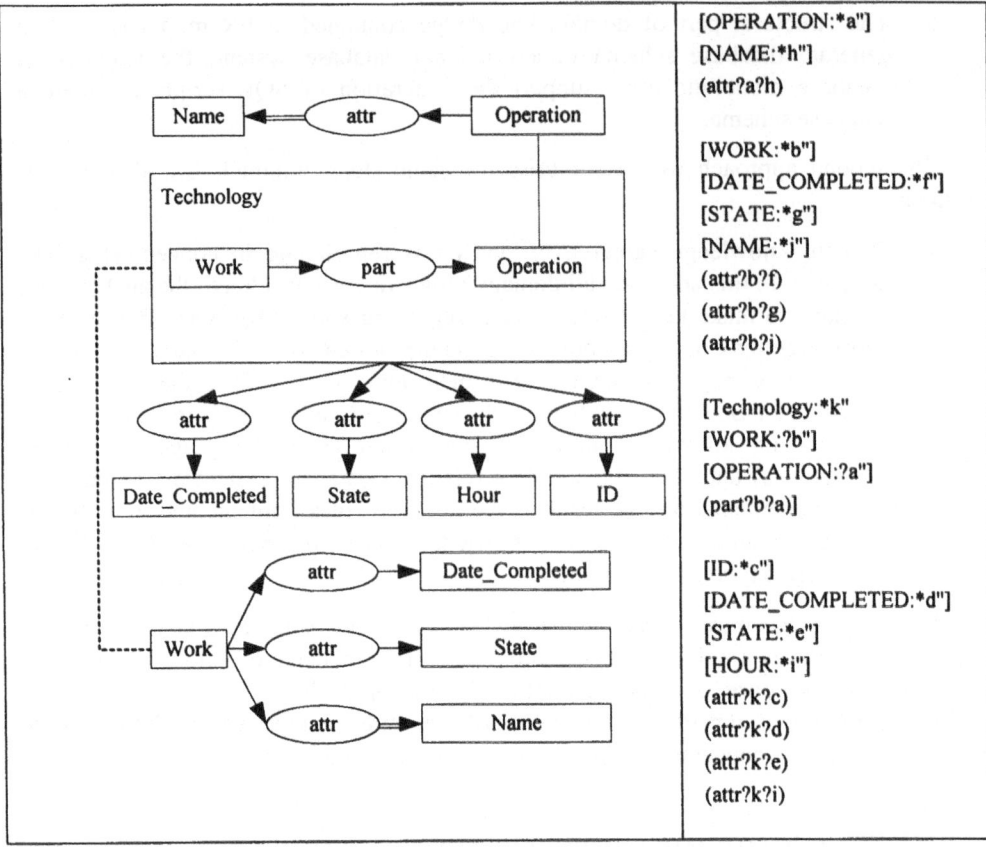

The following CGIF appears to the right of the figure:

[OPERATION:*a"]
[NAME:*h"]
(attr?a?h)

[WORK:*b"]
[DATE_COMPLETED:*f"]
[STATE:*g"]
[NAME:*j"]
(attr?b?f)
(attr?b?g)
(attr?b?j)

[Technology:*k"
[WORK:?b"]
[OPERATION:?a"]
(part?b?a)]

[ID:*c"]
[DATE_COMPLETED:*d"]
[STATE:*e"]
[HOUR:*i"]
(attr?k?c)
(attr?k?d)
(attr?k?e)
(attr?k?i)

Figure 2. The domain structure knowledge model: display form and CGIF

3.2. Business Rules Of The Domain

Two business rules may be applied for the domain:

1. If technology is completed, then completion date must be set to today value.
2. If all technologies of some work are completed, then the state of the work must be set to "completed" and the work completion date must be set to today.

The formulated sample business rules will be modelled using the representation of demons as described Delugach (1991). Demons are the extension to conceptual graphs that captures the idea of a process or transformation. A demon is similar to an actor (actors are part of the original conceptual graphs theory by Sowa). Demon is represented as a double diamond in a graphical form and as <<d>> in a linear form. In this paper a demon representation as <<d>> is used also as a demon representation in CGIF similar to Sowa's original actors notation because CGIF as presented in conceptual graphs standard [Sowa00] does not offer any syntax of demons. Demons represent the statements of the form: if input concepts are ever true, then the output concepts will be true at some future time. If there is more than one input concept, no demon action occurs until all of its input concepts have been asserted. If input concept must be "re-asserted" after a demon has been enabled, the double arrow (pointing also to a concept) is used.

In Figure 3 the first business rule in the display form and CGIF is shown .

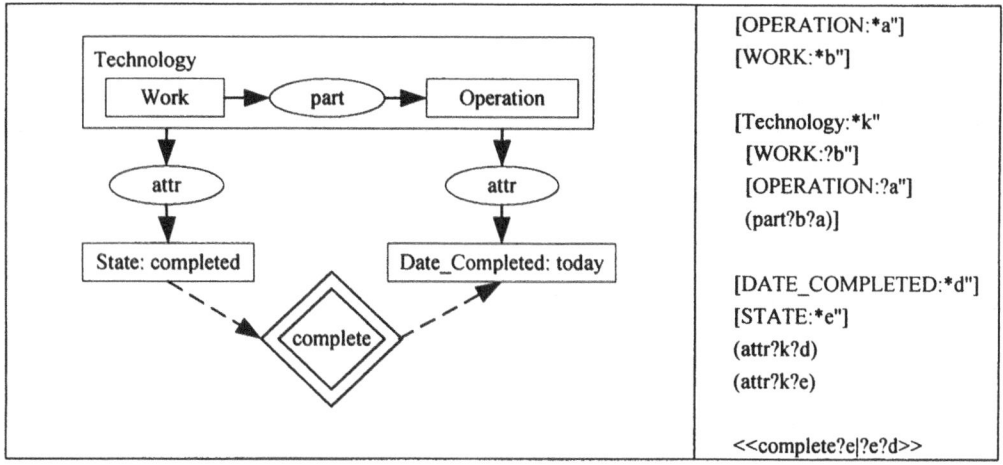

Figure 3. The representation of the first business rule: display form and CGIF

In Figure 4 the second business rule is presented in the display form and CGIF.

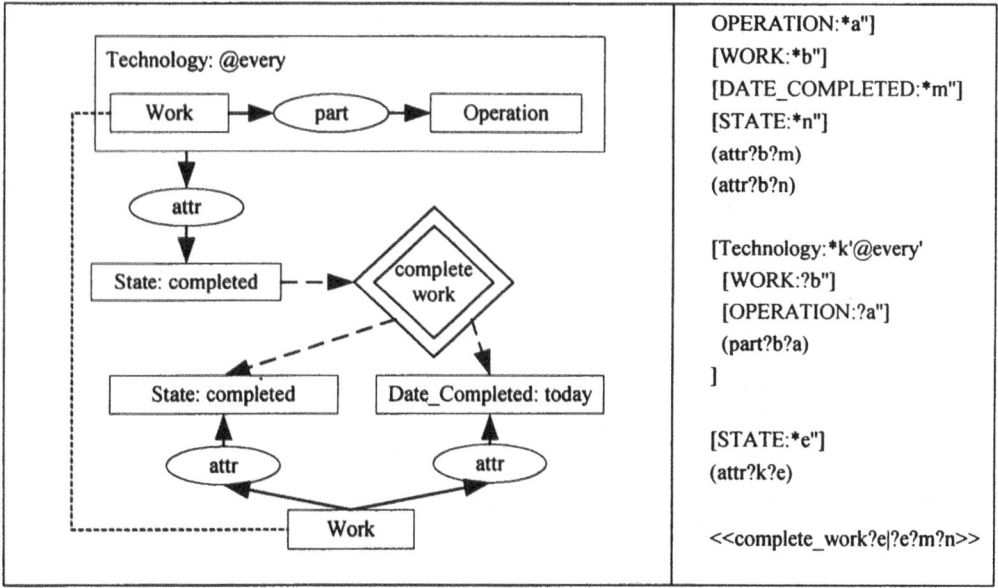

Figure 4. The representation of the second business rule: display form and CGIF

3.3. Translation Of Dynamic Knowledge To Database Triggers

The relational database triggers construction algorithm may be defined by the set of rules listed in Table 1.

Table 1. Database trigger construction rules

Number	Rule
Rule 1	trigger_name = demon_name

Rule 2	for all concepts
	   if concept does not have any relationships "attr"
	   then table_name = concept_name
	   endif
	endfor

Rule 3	for all demon input_concepts trigger_condition = trigger_condition + input_concept endfor
Rule 4	for all demon output_concepts trigger_action = trigger_action + output_concept endfor

Rule 5	if input concept does not have any relationships "attr"
	then
	   if it is assertion of new fact
	   then dml_event_clause = INSERT
	   endif
	   if it is retraction of the fact
	   then dml_event_clause = DELETE
	   endif
	else
	   dml_event_clause = UPDATE ON concept_name
	endif

The set of rules is by no means complete – the simple insert or update or delete action may be performed in the trigger body. The complete set of rules and generic algorithm is yet to be developed. It is feasible because demons and triggers both share the same semantics – they may be interpreted as the ECA (event, condition, action) rules. The demon in Figure 3 may be interpreted as ECA rule as outlined in Table 2.

Table 2. Demons and triggers viewed as ECA rules

Part of a rule	Explanation	Example in Figure 3	Part of trigger
EVENT	assertion of the fact that is represented by input concept	update of technology.state	trigger clause
CONDITION	input concepts	technology.state = "completed"	trigger condition
ACTION	output concepts	technology.date_completed = today	trigger body

4. CONCLUSIONS AND FUTURE RESEARCH

It is shown in the recent publications on the conceptual modelling research that conceptual graphs are suitable for conceptual modelling of application domain knowledge because of several reasons. First, conceptual graphs offer easily understandable, yet powerful constructs to capture all aspects of domain knowledge – structure, behaviour and constraints. Moreover, conceptual graphs allow modelling of all knowledge in the integrated manner thus producing a uniform domain knowledge model. Second, conceptual graphs, being graphical interface to a system of logic, allow to view and use knowledge model presented in conceptual graphs in both formal and informal ways depending on the purpose of potential use. Informal knowledge model in a graphical notation may serve as a basis for communication and involvement of domain experts into information systems development process. Formal knowledge model allows easy description of a domain into executable specifications.

However, there are certain disadvantages with using conceptual graphs in a business practice mainly related to the lack of the effective techniques and commercially available knowledge system as part of the information system development tools. Although many prototypes exist, they serve more for the purpose of scientific research.

For this reason the methodology is presented that outlines the basic steps of information system development process and shows how conceptual graphs may be used in each phase. Although conceptual graphs serve as a basic modelling language, ER language may also be used because it is well-established formalism to model structural domain knowledge. In the proposed methodology, several steps may be implemented as processes to minimise development efforts: automatic translation of relational database schema, generation of database triggers from the conceptual graphs model. The mapping from ER model to relational database schema exist for years, the similar mapping may be introduced from conceptual graphs to relational database schema. Certain questions will arise while developing mapping rules. E.g. conceptual graphs relationships are by default many-to-many while ER language allows specification of relationships cardinality. Conceptual graphs relationships must be augmented with cardinality constraints or cardinality must be modelled as constraints. Database triggers may be generated from dynamic knowledge expressed in a conceptual graphs language. Such algorithm is feasible because of the semantic similarity of demons and triggers – demons may be viewed as ECA rules, event, condition and action parts may be extracted from the conceptual graph representing dynamic process.

The experiment using the example from real business domain was carried out in terms of proposed phases of methodology. It was shown that conceptual graphs with demon nodes are capable of capturing dynamic domain knowledge and constraints. Further, the methods and algorithms for implementation of the proposed modelling and development ideas should be proposed so the presented methodology might be used in real business domains.

5. REFERENCES

Angele, J., Fensel, D., Landes, D., Studer, R., 1998, Developing knowledge-based systems with MIKE, *Journal of Automated Software Engineering 5(4)*, pp. 389-418.

Benn, D.J., Corbett, D., 2001. pCG: an implementation of the processes mechanism and an extensible CG programming language, 2001, Stanford, http://www.cs.nmsu.edu/~hdp/CGTools/proceedings/.

Booch, G., Rumbaugh, J., Jacobson, I., 2000, *The Unified Modeling Language User Guide*. Addison-Wesley, 2000.

Creasy, P., Ellis, G., 1993, A conceptual graphs approach to conceptual schema integration, in: *Conceptual Graphs for Knowledge Representation, ICCS '93, Quebec City, Canada, August 4-7, 1993, Proceedings*, Mineau, G.W., Moulin, B., Sowa, J.F. eds., Springer, Berlin, pp. 126-141.

Cyre, W.R., 1998, Executing conceptual graphs, in: *Conceptual Structures: Theory, Tools and Applications, 6th International Conference on Conceptual Structures, ICCS '98, Montpellier, France, August 10-12, 1998, Proceedings*, Mugnier, M.L., Chein, M., eds., Springer, Berlin, pp. 51-64.

Delugach, H.S., 1991, Dynamic assertion and retraction of conceptual graphs, in: *Proceedings of Sixth Annual Workshop on Conceptual Graphs*, Way, E.C., SUNY Binghamton, Binghamton, New York, pp. 15-26.

Haemmerle, O., Carbonneil, B., 1996, Interfacing a relational database using conceptual graphs, in: *Seventh International Workshop on Database and Expert Systems Applications, DEXA '96*, Wagner, R., Thoma, H., eds., IEEE-CS Press, pp. 499-505.

Kabbaj, A., Frasson, C., Kaltenbach, M., Djamen, J.Y., 1994, A conceptual and contextual object-oriented logic programming: PROLOG++ language, in: *Conceptual Structures: Current Practices, Second International Conference on Conceptual Structures, ICCS '94, College Park, Maryland, USA, August 16-20, 1994, Proceedings*, Tepfenhart, W., Dick, J.P., Sowa, J.F., eds., Springer, Berlin, pp. 251-274.

Kabbaj, A., Lanta-Polczynski, M., 2000, From PROLOG++ to PROLOG+CG: a CG object-oriented logic programming language, in: *Conceptual Structures: Logical, Linguistic, and Computational Issues, 8th International Conference on Conceptual Structures, ICCS 2000, Darmstadt, Germany, August 14-18, 2000, Proceedings*, Ganter, B., Mineau, G.W., eds., Springer, Berlin, pp. 540-554.

Krogstie, J., Ivberg, A., 2000, *Information Systems Engineering : Conceptual Modeling in a Quality Perspective*, Draft of Book, Information Systems Groups, NTNU, Trondheim, Norway.

Lukose, D., 1995, Using executable conceptual structures for modeling expertise, in: *Proceedings of the 9th Banff Knowledge Acquisition for Knowledge-Based Systems Workshop*, Gaines, B., Musen, M., eds., Banff Conference Centre, Banff.

Lukose, D., Mineau, G.W., 1998, A comparative study of dynamic conceptual graphs, in: *Proceedings of the 11th Banff Knowledge Acquisition for Knowledge-Based Systems Workshop (KAW-98)*, Gaines, B., Musen, M., eds., University of Calgary, Calgary, Section VKM-7, 20 pages.

Mineau, G.W., 1998, From actors to processes: the representation of dynamic knowledge using conceptual graphs, in: *Conceptual Structures: Theory, Tools and Applications, 6th International Conference on Conceptual Structures, ICCS '98, Montpellier, France, August 10-12, 1998, Proceedings*, Mugnier, M.-L., Chein M., eds., Springer, Berlin, pp. 65-79.

Mineau, G.W., Missaoui, R., Godin, R., 2000, Conceptual modeling using conceptual graphs, in: *Proceedings of the 7th International Workshop on Knowledge Representation meets Databases (KRDB 2000), Berlin, Germany, August 21, 2000*, Bouzeghoub, M., Klusch, M., Nutt, W., Sattler, U., eds., http://sunsite.informatik.rwth-aachen.de/Publications/CEUR-WS/Vol-29/, pp. 73-86.

Moeller, J.U., 1995, Operationalisation of KADS models by using conceptual graphs modules, in: *Proceedings of the 9th Banff Knowledge Acquisition for Knowledge-Based Systems Workshop, Banff, Canada*, Gaines, B., Musen, M., eds., Banff Conference Centre, Banff.

Rational Software Corp, 2002; http://www.rational.com/media/uml/resources/media/ad970808_UML11_OCL.pdf

Schreiber, G., Akkermans, H., Anjewierden, A., Hoog, R., Shadbolt, N., Van de Velde, W., Wielinga, B., 1999, *Knowledge Engineering and Management: The CommonKADS Methodology*, MIT Press, 1999.

Sowa, J.F., 1984, *Conceptual Structures: Information Processing in Mind and Machine*. Addison-Wesley, 1984.

Sowa, J.F. et al., 2000, Conceptual graph standard, American National Standard NCITS, T2/98-003, 2000, http://www.jfsowa.com/cg/cgstand.htm.

Valatkaite, I., Vasilecas, O., 2002, Knowledge transformation with conceptual graphs, in: *Proceedings of the Conference IT'2002, Kaunas University of Technology*, Technologija, pp. 63-70, (in Lithuanian).

ON MODELLING EMERGING BEHAVIOUR OF MULTIFUNCTIONAL NON-PROFIT ORGANISATIONS

Raul Savimaa [*]

1. INTRODUCTION

Expected behaviour for an organisation is usually described as a set of work processes that has to be carried out. Goal is reached (or task will be fulfilled) when certain pre-described work routines have been completed. For many application domains, for example, in chemical batch processes, automation, control systems, etc., processes are looked at from the technological viewpoint only. Specified relevant work processes do not depend on employees' decisions and therefore their behaviour is not modelled, as a rule.

This paper focuses on human organisations that operate in a dynamically changing environment and that have to react, on operational level, to specified time criteria in minutes or seconds. Also not all possible relevant decisions can objectively be prescribed. Application domain covers military organisations, organisations ensuring public safety (e.g. the police, rescue service), and several other public service authorities or organisations. Organisations have multiple goals and limited resources that often do not correspond to the needs or to tasks.

Relevant behaviour in such organisations is often standardised and described. Suitability of each and every action depends upon the particular situation. Decision-making process is deeply human and subjective. Techniques of process control analysis are not applicable directly and therefore not used, as a rule.

Working in a dynamically changing environment means that work processes and goals are often to be modified to meet the changing requirements and operating conditions. For example, efficiency of organisation's performance may become unsatisfactory or the overall functioning goal may change due to the changes of the requirements of the society. Modelling possible results does not usually precede modifying work procedures. Adjustments are often done intuitively, mistakes are not found quickly and it is difficult

[*] Department of Computer Control, Tallinn Technical University, Ehitajate tee 5, Tallinn, Estonia, 19086.

Information Systems Development: Advances in Methodologies, Components, and Management
Edited by Kirikova *et al.*, Kluwer Academic/Plenum Publishers, 2002

203

to determine their causes. There exist no common solutions on how to model organisational performance and employees' behaviour.

A model that incorporates the organisational goals, structure, components, behaviour (both operational and strategic), and also the grounds of the employees' behaviour can be useful in solving problems. It allows us to model various operational situations and to analyse results of such decisions. It may also assist in strategic decision-making.

A tool enabling to prognosticate the effects of the modifications would reduce potential harmful side effects of the organisational modifications, and would enable to compare objectively alternative reorganisation strategies. Such a tool would be even more important since restructuring is carried out "on-line", i.e. without suspending everyday activities. Quite often modifications are also carried out without increasing day-to-day operating costs, limited additional resources can be used only for exceptional cases. Modelling and simulation using some techniques of process analysis and combined with some theoretical analysis on a sufficiently powerful and regularly updated model of the organisation could form a basis for such a tool.

There are multiple, sometimes very powerful methods assisting in organisational modelling. This paper, however, introduces a new approach that tries to combine classical approaches to modelling work processes with modelling employees as agents. The novelty of this paper lies in the agent technologies that are suggested the basis for modelling the organisational behaviour and structure, while work processes and timing constraints are by modelled using Q-model.

This paper is structured in the following way. Goals for modelling organisations are given and analysing power of models required is examined in the next section. The section gives also a brief survey of suitable methods and tools existing. Suitability of the Q-model for modelling work processes is analysed. Agents are briefly introduced, and the organisation as a multi-agent system is discussed.

The third section of this paper introduces a methodology for modelling above-described organisations and suggests composing for possible modifications. Suitable preliminary model of the organisation is given and issues related to modelling of work processes and behaviour are specified. Aspects related to elaborating modifications are not handled in detail. In the fourth section an example for an illustration is given. Finally, conclusions are drawn.

2.   GOALS FOR MODELLING ORGANISATIONS

2.1. Basic goals for modelling

Main reasons for modelling an organisation and its work processes is, as we know, creating information systems that support the execution of processes and assist in information flow, but often they assist and provide also limits in decision-making. Modelling is also important due to the fact that testing the model is much cheaper than possible errors in the real world.

Models of work processes (together with their starting, interaction, information exchange, and finishing conditions) are mostly used on the level of specialists. These models may be used effectively for creating information systems that support operational performance of the organisation and also give the opportunity for managers to supervise general behaviour of the organisation.

The management uses the information collected for strategic decisions. They have to foresee the possible allocation of resources to allow maximum operational benefit in the framework of overall organisational goals. Whereas decisions about implementation details often modify work processes, as well as the goals and structure of organisation, managers need more sophisticated models that describe the organisation and its more completely.

Comprehending such models by managers becomes essential to enable support for decision-making. As a final result, a suitable software product has to be made for managers as well.

2.2. Required analysing power of models

Models of organisations that support ongoing decision-making and give reasonable realistic information have to meet the following requirements:

1. The ability to model the most important work processes.
2. The ability to deal with hierarchy of business processes (Carley and Prietula, 1998).
3. The ability to observe and check all the essential aspects that form the organisational behaviour, including a generic model for analysing employees' behaviour.
4. Formal communication, related to the goals of the organisation and its official work processes, and informal communication, related to employees' opinions, intentions, interests, etc., has to be described (Ferber, 1999).
5. Straightforward interpretation of modelling results into day-to-day activities and continuous adjustment of the model, in accordance with the changes in the organisation.

Related to the aspects of current research, three critical aspects have been pointed out that have not found sufficient appreciation in modelling organisations.

1. Organisational goals and capacities usually do not match. When giving tasks to organisation, the details of work processes are not specified. Also, the analysis on whether the goals, capacities, and constraints contradict one another or not, is very seldom made – i.e. whether the tasks are achievable. Quite often time constraints are not considered in models, whereas the tasks leading to achieving goals are time-sensitive. Therefore, a model should allow complex hierarchical process and resource analysis as well as of handling sufficiently sophisticated time category.
2. Many decisions at lower levels during the performance of a task are left to structural units of the organisation. Each structural unit in an organisation has its own goal and the employees have personalities of they own, with their own wishes, opinions, and goals. Behaviour is caused by various motivations, not necessarily related to work or organisation. Models should allow analysis of interaction on the levels of structural units as well as on the level of an individual employee or groups of employees.
3. Organisations have to function continuously. Restructuring and structural adjustments are to be carried out "on-line", without suspending everyday activities and quite often the modifications have to be done without increasing day-to-day

operating costs. To guarantee permanent support from the models, they are also to be regularly updated.

The work processes that the organisation carries out during its functioning are not exactly the same as the planned ones. This is caused by the capacities of the organisation as well as the environmental conditions. It is important to model the both work processes – that were planned and that were actually carried out.

For analysing the organisational behaviour, the model (marked as M) of the organisation (O) is needed. Ongoing modelling and model adjustment methodology can be described in this case as function F_{OM}. The function F_{OM} establishes the correspondence between the organisation and the model with minimal efforts. The function F_{OM} can therefore described as

$$F_{OM}: O \rightarrow M$$

The proposed methodology (F_{OM}) is introduced and the model of the organisation (M) described in the third section.

2.3. Existing methods and suitable approaches

There exist several methods for modelling enterprises, organisations, or teams. One powerful approach is the Enterprise Modelling (Bubenko, Persson, and Stirna, 2001) which models business processes, goals, requirements, etc., and gives the framework for incorporating stakeholders' knowledge. The Capability Maturity Model (SEI, 2002) together with its later developments gives solutions for more general (and stabile) cases. However, very general and powerful tool for software development, the Unified Process (Arlow and Neustadt, 2002) does not cover all necessary aspects for current modelling situation. Therefore, there is still a gap for research issues, e.g. finding similar conditions and alternative approaches or suitable solutions from neighbouring application domains.

Modelling cannot be done using conventional algorithmic approach since the behaviour of employees, as well as that of the whole organisation, is often changing, determined by interactions, goals, and knowledge. This influences also the ability to reach given goals.

Models of interactive computing, and agent technology, in particular, seem to be the most suitable approach for this application. Interaction-based models of computing exceed algorithmic models in formal power (Wegner and Goldin, 2000). Interactive computing has the phenomenon of emergent behaviour that is most similar to one existing in real human organisations. Interactive computations possess some other important features that are not present in algorithmic computations (according to Meriste and Motus, 2002) as stream input-output and history-dependent behaviour of the computing agents.

There does not exist a widely accepted model for interactive computations for the time being yet. Several concepts and models have been suggested instead. It is accepted that a primitive component for building a system is not an algorithm, but rather a set of interacting, repeatedly activated algorithms (Motus and Meriste, 2001). This approach is also used in the current research.

2.4. Modelling of work processes using the Q-model

When modelling work processes, the Q-model (Motus and Rodd, 1994) can be used. Q-model approach offers sufficient support for modelling normal procedures and planned behaviour in the organisation. The Q-model supports strongly the analysis of timing questions in models. Carley and Prietula (1998) emphasise also that the task for an agent may be decomposed into simpler tasks. This approach is easily usable in the Q-model.

The Q-model, when used as the only modelling tool, is not sufficient for multi-functional organisations. Model of real organisation in given circumstances may become too complicated very fast to be handled and understood by people. This reduces the efficiency of modelling the actual behaviour of the organisation. As described above, one has to find possibilities for modelling the employees' behaviour as well, that may give many variations in ideal work processes stated initially.

2.5. Representing the organisation as a multi-agent system

Term "agent" usually means a specific piece of intelligent software or robotics (Kido, 2001). In the framework of the organisation modelling, an agent must possess:

- the ability to decide autonomously what actions in the current situation are to be taken to maximize progress towards its (time-varying) goals (Maes, 1997),
- the ability to communicate with other agents (Ferber, 1999), and
- the behaviour that tends to satisfy the agent's objectives (Ferber, 1999).

Whereas there does not exist a common model how to model organisations as agents, then some existing approaches are taken as a basis in the current research. Those are introduced below.

Often an agent has to respond to dynamic changes in the environment in time and therefore to modify their behaviour to achieve the goal. For this, the time is to be modelled for both the agents and the environment in the system. The examples of teamwork between the agents used most come from the military field. Organisations in this field have to deal with rapidly emerging events with high uncertainty and potentially catastrophic impacts (Kang, Waisel, and Wallace, 1998).

Structural units in the organisation may be described as multi-agent systems. To enable modelling the behaviour of a unit, a unit itself as well, may be considered as an agent. According to Ferber (1999), the organisation is thus described as a multi-agent system, which is composed of the structural units and the employees as agents. Barber et al. (2001) state that agent-based system architectures offer modular distribution of decision-making responsibilities. Analysis objective determines what structural units or aspects (views on the behaviour or performance) of the organisation are to be modelled.

Organisational structure characterises the class of the organisation on the abstract level (Ferber, 1999). The concrete organisation is one possible instantiation of this structure. The same organisational structure can act as a basis for defining the multitude of the concrete organisations; this is similar to the generalization and specialization of a role presented by MIT SAG (2001).

3. DESCRIPTION OF METHODOLOGY

The life cycle of the model development for the current approach is based on the Unified Software Development Process, or shortly, the Unified Process (Jacobson, Booch, and Rumbaugh, 1999; Arlow and Neustadt, 2002) with some specialisations, changes, and additions.

Various persons within the organisation (or from outside likewise) may need organisation models. Each of those persons may have different standpoint, knowledge, and ideas, even if they all have the same task to support – the modification of organisational performance. Therefore, the approach, maximising the possible input derived from different information or knowledge sources, is needed. The process of creating a model given below can be regarded as an ontology design. A collaborative approach (Holsapple and Joshi, 2002) is to be used for developing models.

3.1. Model of the organisation

The given formula of modelling methodology introduces a model of the organisation M as a semiformal description of organisation:

$$M = \{G; D; W; Q; A; L; B\}$$

where

G is the organisational goal (given as a prioritised set of descriptions),
D is the semiformal structural description of the organisation,
W is the table of work processes,
Q is the description of work processes in the Q-model,
A is the description of the employees and structural units of the organisation, using agent technology,
L is the catalogue of models for work processes in Q in a simulation software, used for analysis and simulation of existing interaction mechanisms, and
B is the catalogue of models for the organisation as agents in a multi-agent simulation and modelling software, used for analysing and simulating new interactions when the knowledge or goals have been changed.

The goal G is used for generating the analysis and modelling the task and also for distinguishing between the primary and supportive processes in the organisation. G is also used during simulation and composition of modification suggestions.

3.2. Modelling stages

The model of an organisation is developed in four stages. The first stage comprises of the description of the tasks and the existing inner structure of the organisation. This stage corresponds to the inception and elaboration phases of the Unified Process. Since it is important to get sufficiently good and detailed description, the Unified Modelling Language UML (Booch, Rumbaugh, and Jacobson, 1999) is suggested for describing the structure of the organisation and its tasks. First three components of the model M will be completed during this stage: G, D (which also includes models in the UML), and W.

Since the presented methodology stresses the importance of modelling work processes, the component W is described here in more detail. Each row in W describes the work process w which has the following properties:

$$w = \{q; n; d; p; a; \mu; t; \varphi; v; \tau; r\}$$

where

q is the process (sequence) number,
n is the name of the action (string),
d is the description of the action (text),
p is the purpose of actions (free text),
a is the executor, the actor,
μ is the description of the input (free text or set of conditions),
t is the starting time, frequency or period (time),
φ is the duration of process (time),
v is the description of the output (free text or set of conditions),
τ is the estimation about the validity time (permanency) of the output values, and
r is the additional remarks (free text).

The second stage in modelling completes the model of the organisation. It transforms the UML description into the Q-model and multi-agent presentation. The Q-model is used for modelling the work processes and the agents are used for modelling the structural units and some individual employees. Special attention is to be paid to the analysis of the bottlenecks of timing and potential conflicts of interests within the organisation. Here the LIMITS CASE tool is used for the analytical study and simulation of the Q-model. For the study of agents' behaviour, the JADE (JADE, 2002) tool is planned to use in the test-bed for a multi-agent system (presented by Motus et al., 2002).

The third stage of the methodology uses the model M and provides support for the formulation of recommendations and for the development of modification plans. Simulation process itself and results are recorded, and a list of modification recommendations is composed of those. Stored states of M give also estimations about possible behaviour of the organisation during the modification process.

The second and the third stage correspond to the construction phase in the Unified Process. The fourth stage is the implementation of organisational modifications, monitoring and the evaluation of results. This stage corresponds to the transition phase in the Unified Process.

3.3. Modelling, using Q-model and agents

Components D and W, completed during the first stage are used as starting points for the second stage. Normally, the table of work processes W is developed from top to bottom: existing work processes are detailed stepwise until the level of elementary work processes is reached. Modelling in the Q-model starts from elementary processes. As all real actions can be successful or unsuccessful, it is important to model so-called "wrong" but realistic outcomes as well. In the Q-model this may be implemented using selector processes.

The following step is to model larger components of the work processes, such as combinations of the elementary process modules, according to the table of work processes.

Component processes (as well as their models) may also be composed of other component processes i.e. be hierarchical. The whole work process is finally completed as a combination of process modules and/or the modules of elementary work processes. Relation of component processes to the organisation structure is useful to describe.

Table 1. Relation between agents and Q-model

Level	Category	Agent representation	Q-model representation
1	The organisation	Multi-agent system	- (not used)
2	The organisational goals	- (not used)	- (not used)
3	List of actions / processes, requirements	- (not used)	Work process (highest level in hierarchical representation)
4	The structural unit	Agent	- (not used)
5	Plans, work process descriptions, instructions	Agent methods	Work process (component process)
6	An employee	(Elementary) agent	- (not used)
7	Employee's job description (tasks for the role)	Agent methods	Component process
8	Employee's daily routines	Agent methods	Elementary work process

During modelling, each elementary process and component process gets a unique name and their models are collected into one, the indexed repository (components Q and L in model M). The contents of this repository are kept up-to-date with the contents of the table of work processes W.

Another parallel process in modelling is describing employees and/or structural units as agents. All real persons as well as their planned (ideal) roles are described as agents (the components A and B of M). Also artificial agents (robots, software agents, etc.) may be added to this repository. The internal description of the agent may be given in the Q-model and stored in Q and L. If the collective knowledge, goal, motivation, or behaviour of the structural unit as a whole is to be modelled, then the unit is also modelled as one agent.

The agent (the employee, the unit, or the organisation) have methods that may be regarded as job or work processes for that agent (or multiple agents). Those methods are described by the Q-model. Thus the set of the Q-models in the organisation and the set of descriptions using agents are very closely related in the model as seen in Table 1. Each row in the table may also be presented by multiple levels.

4.   EXAMPLES

An example is given for illustration. This example is derived from law enforcement activities where the teamwork and correct timing of actions play an important role, but where also exists a set of given work processes. At first, modelling of work processes and information flow using the Q-model is presented and then agents for modelling roles and employees are described, using the UML.

The example concentrates on the case of a stolen car. A car theft has taken place. The owner becomes usually aware of this from about some minutes to several hours later and then he or she informs the police (by telephone or otherwise). The police begin to search for it immediately, after having obtained the necessary information; but normally the time between the theft of the car and the detection of this fact by the owner causes the main delay in processing and exchanging the information with the Vehicle Register Office, Border Guard and all necessary databases.

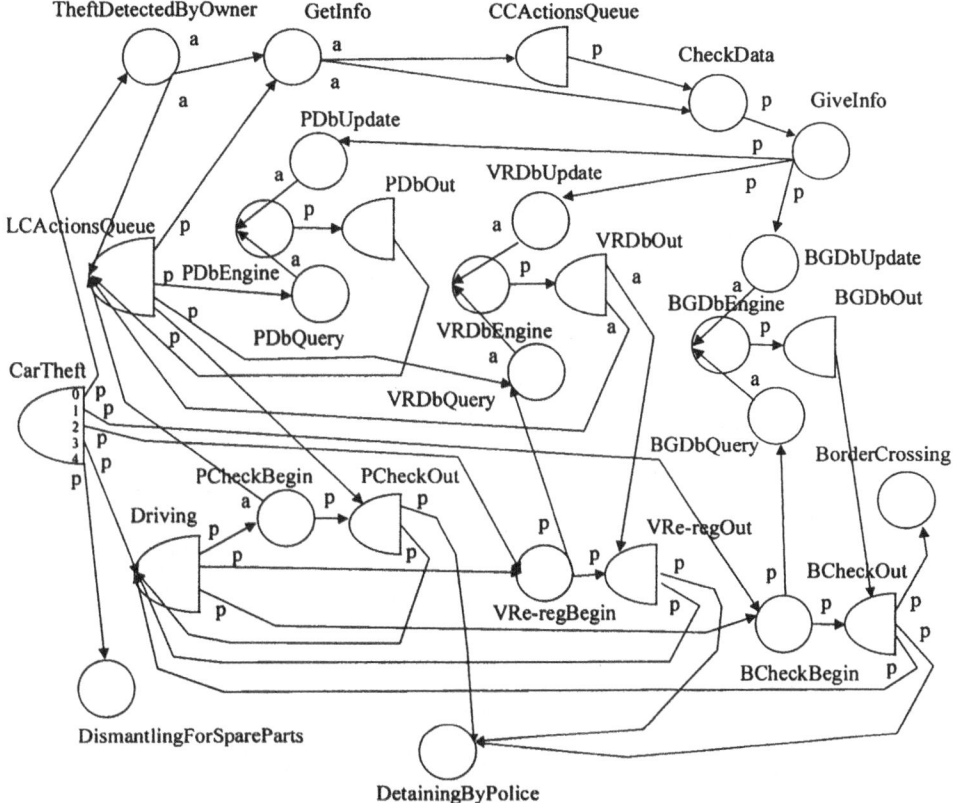

Figure 1. An example on events related to car theft

When a car is stolen, it may have four future possibilities:

- the car is attempted to re-register in the vehicle register for (later) selling,
- the car is taken abroad (either directly or after re-registration),
- the car is dismantled for spare parts, or
- the car will be used within the country for driving for a shorter or longer time, afterwards it will be abandoned or one of the previous actions will be taken.

To prevent those illegal actions and find the car, the police must inform all related authorities as quickly as possible. At first the local (regional) command and control centre (LC) informs the central (national) command and control centre (CC) and then the

CC informs all other local police units, Vehicle Registration Office, and the Border Guard about this to update their databases. Each database engine has own activation time set. During the police check on the street a police patrol officer (PO) asks information about the car from the LC and then the LC, beside other duties, performs the search in the corresponding database(s).

Figure 1 illustrates the above-described example using the Q-model notation. Both human work processes as well as software processes are modelled there. Time parameters of processes and channels are not given, for better clarity and readability.

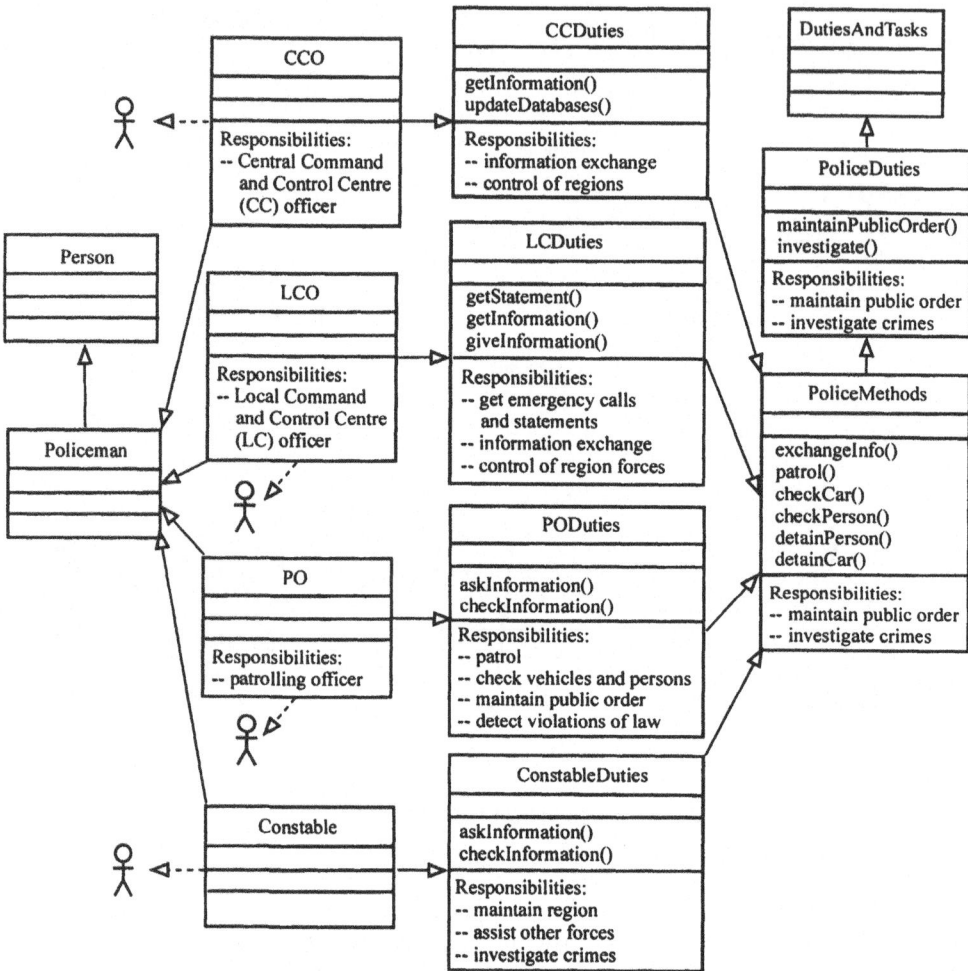

Figure 2. Classes of agents as used in the example

To find the car, the following organisational structure, roles and methods are described for co-operation of a police unit, using classes and agents as shown in Figure 2. Most important classes, attributes, and methods only are specified. Four roles of policemen are described: the LC operation officer (LCO), the CC operation officer

(CCO), the police patrol officer (PO), and the police constable (partially). Some duties of the constable are the same as the ones of the PO. The dual inheritance of those classes is introduced. The role description and common methods as duty descriptions are given in the classes describing duties. Some common police duties are derived from class PoliceMethods and its parent classes. Common human characteristics for policemen are derived from parent class Policeman and its parent class Person.

5. CONCLUSIONS

Some criteria and aspects on how to model the behaviour of a multifunctional non-profit organisation, operating in the dynamic rapidly changing environment, were introduced in this paper. Novelty lies in handling the organisation from the following positions concurrently: on one side, an organisation is modelled as a hierarchical multi-agent system, whereas at the same time the work processes in the organisation are modelled using Q-model.

The proposed methodology stresses the importance of a detailed model for the organisation. The model is to be regularly updated and tested before the real changes in the organisation are effected. When this approach is once taken to the organisation and the model of the organisation is kept updated, then the components of the model completed already can be used for any new analysis. This reduces modelling costs and helps to guarantee stabile quality level of modelling.

The presented methodology could also be suitable for organisations belonging into profit-making organisations. For those organisations, many goals are related to economic and profit issues, therefore the role of algorithmic methods increases and conventional optimisation methods may be more efficient than agent-based methods.

Suggestions made in this paper are not final and further work is needed to elaborate this approach and involve more sophisticated aspects and problems that may emerge.

ACKNOWLEDGEMENT

This research is partially financed by Estonian Science Foundation, grant ETF 4860.

REFERENCES

Arlow, J., and Neustadt, I., 2002, *UML and the Unified Process*, Pearson Education, Boston.

Barber, K. S., Goel, A., Han, D., Liu, T. H., Martin, C. E., and McKay, R., 2001, Sensible Agents Capable of Dynamic Adaptive Autonomy, International Conference on Advances in Infrastructure for Electronic Business, Science, and Education on the Internet, L'Aquila, August 06-12, 2001 (March 25, 2002); http://www.ssgrr.it/en/ssgrr2001/index.htm.

Booch, G., Rumbaugh, J., and Jacobson, I., 1999, *The Unified Modelling Language. User Guide*, Addison Wesley Longman, Inc., Reading, Massachusetts.

Bubenko, J. A. jr., Persson, A., and Stirna, J., 2001, D3: User guide of the Knowledge Management approach using Enterprise Knowledge Patterns - D3 Appendix B: EKD User Guide, Royal Institute of Technology and Stockholm Unicersity, Stockholm (Mai 31, 2002); http://www.dsv.su.se/~js/ekd_user_guide.html.

Carley, K. M., and Prietula, M. J., 1998, Webbots, trust, and organizational science, in: *Simulating Organizations: Computational Models of Institutions and Groups*, M. Prietula, K. Carley, and L. Gasser, ed., AAAI Press / The MIT Press, Menlo Park, California, pp. 3-22.

Ferber, J., 1999, *Multi-Agent Systems: An Introduction to Distributed Artificial Intelligence,* Addison-Wesley, Harlow, London.

Holsapple, C. W., and Joshi K. D., 2002, A collaborative approach to ontology design", *Communications of the ACM,* 45:2, pp. 42-47.

Jacobson, I., Booch, G., and Rumbaugh, J., 1999, *Unified Software Development Process,* Addison-Wesley.

JADE, 2002, Torino, Italy (March 26, 2002); http://sharon.cselt.it/projects/jade/.

Kang, M., Waisel, L. B., and Wallace, W. A., 1998, Team Soar: a model for team decision making, in: *Simulating Organizations: computational models of institutions and groups,* by M. Prietula, K. Carley, and L. Gasser, eds., AAAI Press / The MIT Press, Menlo Park, California, pp 23-45.

Kido, T., 2001, Internet agent for community organization, International Conference on Advances in Infrastructure for Electronic Business, Science, and Education on the Internet, L'Aquila, August 06-12, 2001, 10 pp. (March 26, 2002); http://www.ssgrr.it/en/ssgrr2001/index.htm.

Maes, P., 1997, General Tutorial on Software Agents, Massachusetts (March 25, 2002); http://pattie.www.media.mit.edu/people/pattie/CHI97/.

Meriste, M., and Motus, L., 2002, On models for time-sensitive interactive computing, *Lecture Notes in Computer Science,* no. 2329, Springer Verlag, pp. 156-165.

MIT SAG, 2001, MIT software Agent Group, Massachusetts Institute of Technology, Massachusetts, USA (March 27, 2002); http://agents.media.mit.edu/.

Motus, L., and Meriste, M., 2001, Towards self-organising time-sensitive control systems software, In: *Proc., IFAC Conference on New Technologies in Computer Control,* November 2001, Hong Kong, pp 236-241.

Motus, L., Meriste, M., Kelder, T., and Helekivi, J., 2002, An architecture for a multi-agent system test-bed, IFAC Congress, Barcelona, 2002 (to be published).

Motus, L., and Rodd, M. G., 1994, *Timing Analysis of Real-Time Software,* Elsevier Science Ltd, Pergamon, Oxford.

SEI, 2002, Capability Maturity Models, Software Engineering Institute, USA (June 9, 2002); http://www.sei.cmu.edu/cmm/cmms/cmms.html.

Wegner, P., and Goldin, D., 2000, Coinductive models of finite computing agents, *Electronic Notes in Theoretical Computer Science,* vol.19 (June 10, 2002); www.elsevier.nl/locate/entcs.

HOW TO COMPREHEND LARGE AND COMPLICATED SYSTEMS

Janis Barzdins, Audris Kalnins[*]

1. INTRODUCTION

The basic problem at early analysis stage of the development life cycle is how to quickly comprehend a large and complicated system. One of the ways to comprehend such a system is to build an object model, as it was suggested by the pioneers of object modelling approach such as J.Rumbaugh[1] and J.Martin[2]. In up-to-date terminology it means building a UML class diagram. The authors have got convinced in their everyday practice on extreme efficiency of this type of modelling, though at the same time a significant experience for this job is also required. To make this job easier, a modelling methodology must be developed. The goal of this paper is, on the one hand, to give some methodological recommendations in the conceptual modelling by means of class diagrams, and on the other hand, to discuss requirements for tools which support this type of modelling.

As it is widely known, class diagrams may be used for various purposes. The most popular usage is for object-oriented software design. A lot of books has been devoted to this area,[3-10] and this type of usage is fully supported by such well-known tools as Rational Rose and TogetherJ. The area of building class diagrams for comprehension of complicated systems, started by J.Rumbaugh[1] and J.Martin[2] in the beginning of nineties, in recent years has got significantly less publicity. No widely known methodology here is available, even the books[6-8] covering complete UML-based development life-cycle pay little attention to this area. Most of popular tools support that type of modelling to a significantly lesser degree. One of the tools which is mainly oriented towards conceptual modelling by means of class diagrams is GRADE,[11] in the development of which the authors of this paper have taken part. The GRADE tool has a well developed stereotype mechanism and a number of other facilities which are essential for conceptual modelling support.

[*] Institute of Mathematics and Computer Science, University of Latvia, Raina bulv. 29, LV-1459, Riga, Latvia

Information Systems Development: Advances in Methodologies, Components, and Management
Edited by Kirikova *et al.*, Kluwer Academic/Plenum Publishers, 2002

215

To avoid any misunderstanding, in this paper by term "conceptual modelling" we understand the above mentioned type of modelling, which is required for understanding a system during the early analysis phase.

2. BRIEF NOTES ON SEMANTICS OF CLASS DIAGRAMS

The semantics of class diagrams when applied to software design is quite clear and well documented in UML approach.[2] But the semantics for conceptual modelling is much more vague and intuitive. At the same time this semantics must be understandable for a very broad class of users, because conceptual models frequently must be read unambiguously by non-IT specialists. In this paper we follow the Object Role Modelling (ORM)[12] approach and use natural language sentences as the basis for defining the semantics of class diagrams. We will consider only such class diagrams where the associations present in them express unambiguously readable sentences in a natural language. Consequently, the semantics of a class diagram is defined as a set of natural language sentences which are derivable from the associations of this diagram. We call this semantics the **linguistic semantics**. Let us explain the approach by an example (Fig. 1).

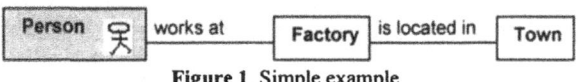

Figure 1. Simple example

This class diagram corresponds to two simple sentences:

Person works at factory.   Factory is located in town.

To enable an unambiguous derivation of sentences from a class diagram, we use one agreement proposed by J.Martin.[2] Namely, the class corresponding to the subject of the sentence we place on the left-hand side, the association name (placed **above** the association line) is used as the predicate, and the object class is placed on the right-hand side. When an association line is rotated in a class diagram, it should be mentally rotated back to the "normal position", so that the name is again above the line. Let us remind that in UML standard the association name direction is marked by a black triangle symbol, but the most popular UML tool Rational Rose doesn't support this mark-up.

3. HOW TO DESCRIBE ACTIVITIES BY CLASS DIAGRAMS

Traditionally, in conceptual modelling class diagrams are used to describe the static structure of a system. Namely this aspect has been thoroughly discussed in early works on object oriented modelling by J.Rumbough[1] and J.Martin,[2] as well as in some recent UML-based publications.[3, 5-7, 10] However, our experience shows that at the conceptual modelling stage class diagrams should cover much broader range of aspects. One of these important aspects is a generalised description of system activities.[13] Now let us cover this aspect in more detail.

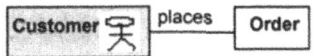

Figure 2. Simple activity

From the linguistic point of view an **activity** is a sentence describing an action, e.g., *Customer places order* (see Fig.2). This example of an activity is extremely simple. In real life activities are much more complicated and frequently can not be described by binary associations in an adequate manner.

Let us consider one more example:

Person submits application to office.

This activity evidently corresponds to a ternary association (Fig. 3).

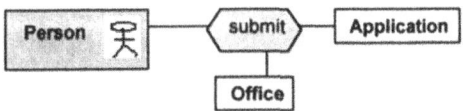

Figure 3. Example of ternary association

In order to have a consistent linguistic semantics definition, a standard representation scheme must be defined for such a ternary association so that the sentence can be unambiguously restored from the association. ORM [12] proposes one way how to do this (see Fig.4).

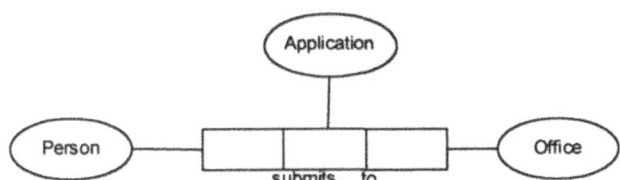

Figure 4. ORM representation of a ternary association

By rephrasing this ORM notation into UML class diagram notation we obtain the form visible in Fig.5.

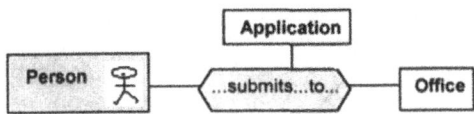

Figure 5. ORM-style representation in UML

The ellipses ... in the sentence template show holes where the corresponding class names must be inserted to obtain a valid sentence in the natural language. This approach works perfectly for small class diagrams where classes can be positioned in accordance with the sequence of ellipses. Therefore the repositioned diagram visible in Fig.6 cannot be considered to be easy readable, though formally the corresponding natural language sentence can be restored easily. However in this particular case the problem can be amended by changing the sentence to *Application is submitted to office by person.*

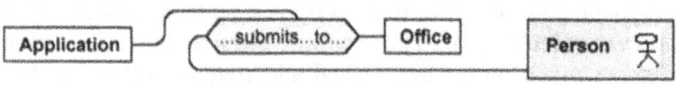

Figure 6. Repositioned representation

Another way how to deal with n-ary associations is by means of role names. However the selection of appropriate role names is not easy, there exists a lot of research in the area of ontologies where this issue is discussed.

A role name classification appropriate for our goals is given by J.Sowa in his book on knowledge representation.[14] This classification is based on Aristotle's four causes or *aitia*, as described in the *Metaphysics*: *Initiator, Resource, Goal, Essence* .

J.Sowa [14] proposes to refine further these four basic causes (roles), depending on the activity type (assumed to be one of *action, process, transfer, spatial, temporal* or *ambient*). Thus the *Initiator* role can be refined to roles *Agent, Effector, Origin* or *Start.* Similarly, *Resource* role can be refined to *Instrument, Matter, Medium, Path* or *Duration.* *Goal* can be refined to *Result, Recipient, Experiencer, Destination* or *Completion.* Finally, the *Essence* role can be refined to *Patient, Theme, Location* or *Point In Time.*

 Most of these new standard roles are self-explanatory, some comments are necessary on what distinguishes *Agent* from *Effector* and *Theme* from *Patient*. The difference is that *Agent* is a voluntary initiator and *Effector* is an involuntary initiator of an action. *Theme* is an essential participant that may be moved, said, or experienced, but which is not structurally changed, but *Patient* is an essential participant that undergoes some structural change as a result of the activity.

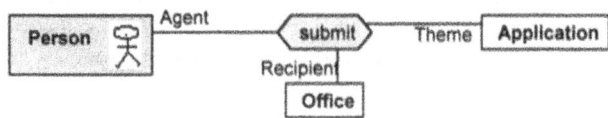

Figure 7. Representation using standard role names

Using these roles, the previous example *Person submits application to office* can be represented as in Fig.7. The usage of roles has one more advantage. By means of roles it is possible to have a simple description of complicated situations which are hard to describe in natural language (Fig.8).

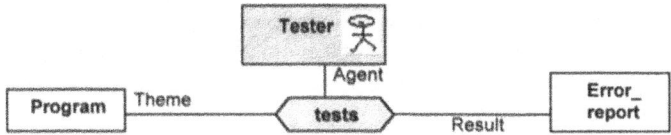

Figure 8. Example of a complicated action

Sometimes it is convenient to represent associations describing activities by separate classes having the stereotype *Activity*. In this case the usage of standardised role names is even more important.

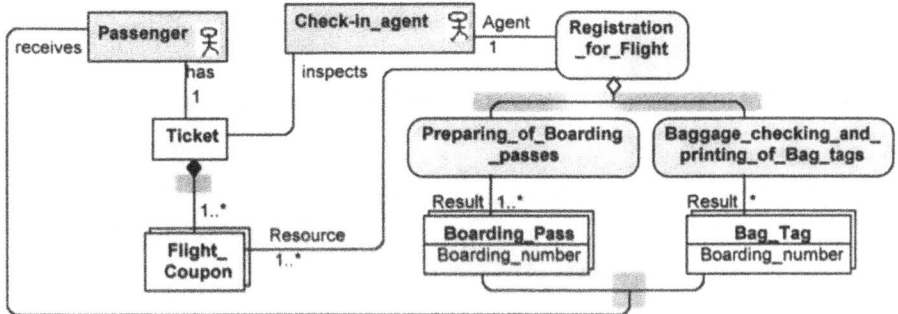

Figure 9. Activity decomposition

One more possibility is demonstrated in the example in Fig. 9, namely, the **activity decomposition**. The activity *Registration for Flight* is shown to be consisting of smaller activities *Preparing of Boarding passes* and *Baggage checking and printing of Bag tags*. Certainly, the activity decomposition can be represented in any typical business-modelling formalism such as ARIS eEPC,[15] but components of an activity there require a separate diagram. This is convenient when the decomposition is very complicated, but not so convenient for simple cases, such as the one in Fig. 9.

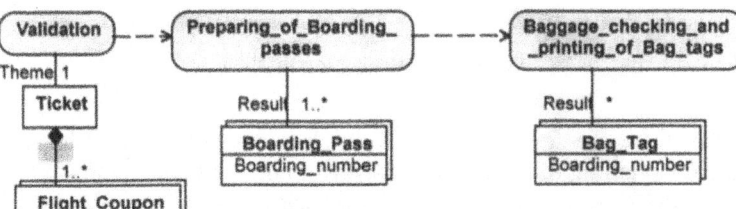

Figure 10. Control flows

If a diagram contains several activities, then frequently also the sequence of actions is of great importance. We propose to represent the sequence by an association with the stereotype <<Control flow>>, graphically depicted as a dashed arrow. Fig. 10 shows an example of control flows.

Frequently the flow of activities is branching, depending on various conditions. This can be represented by attaching guard conditions to control flows. These conditions may be informal texts or formal expressions, both enclosed in square brackets (the standard UML notation for conditions), but anyway they must be mutually exclusive. Fig. 11 shows an example of a branching flow.

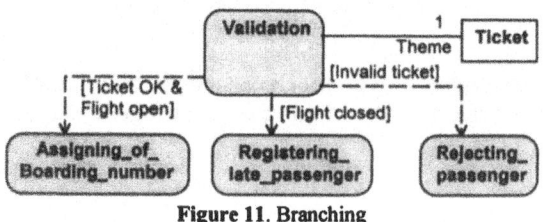

Figure 11. Branching

Guard conditions on associations are useful in other cases too - see Fig. 12.

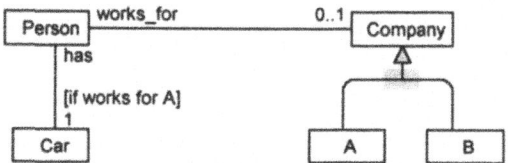

Figure 12. Other case for branching

Class diagram permits also to show the generalisation relation between activities (see Fig. 13) which is useful for understanding, though not present in any business modelling language.

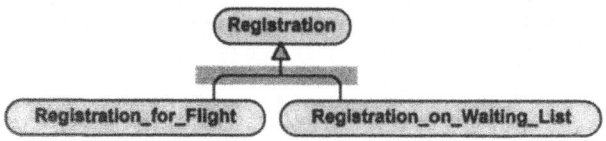

Figure 13. Activity generalization

To sum up, there are several reasons why class diagrams should be used for high level conceptual modelling of activities. Firstly, from the first glance it is not always clear which classes really are activities. Secondly, only class diagram permits to depict also the complete environment of activities – what objects are used, on what they depend, who performs the activity, etc. And only class diagram permits to show vital associations between these environment objects. In addition, class diagram permits to represent far more relations between activities themselves, such as containment and generalisation.

So, our conclusion is that high level view on business activities should be represented by class diagrams, this representation makes them more readable. Specialised business process diagrams can be used after that step, when the main concepts of a system are already understood and categorised.

4. NEW PREDEFINED CONSTRAINTS FOR ASSOCIATIONS

UML contains some predefined constraints for associations, such as {*ordered*}, {*xor*}, {*subset*}. From the formal point of view more predefined constraints are not necessary since all imaginable meaningful constraints can be expressed in OCL.[16] However our practical experience shows that there are more frequently used constraints for which simple graphical shorthands would be of high value. This is confirmed also by the rich set of predefined constraints in ORM, frequently stated as one of preferences of ORM over UML. The goal of this chapter is to investigate whether these constraints can be naturally transferred to UML without "spoiling" the class diagram notation.

Apparently, without any problems in addition to the existing {*xor*} constraint new constraints {*or*} and {*x*} can be added. The latter one expresses the fact that an instance of **A** cannot simultaneously have associations of both types **p** and **r** (see Fig.14).

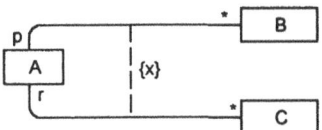

Figure 14. X constraint

This proposal has been already mentioned by Halpin,[13] where this issue is discussed in details. The situation with possible generalizations of the {*subset*} constraint (which exist in the ORM context) is more complicated. In the result of this analysis we propose the following constraint (see Fig.15).

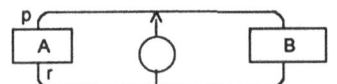

Figure 15. Generalized subset constraint

The meaning of it in the situation of Fig.15 is exactly the same as that of UML {*subset*}. But this new notation can be easily generalized to more complicated situations (see Fig.16). This last notation means that if instances **a** and **b** of **A** and **B** respectively have a connecting path via **C** (i.e., there is an appropriate instance of **C**), then there exists a link **p** directly connecting **a** and **b**. An equivalent OCL statement expressing the same fact would look quite clumsy:

A
```
self.B->includesAll (self.C.B)
```
The same notation can be used to express relationships between longer paths. The necessity for such constraints is frequent in real examples.

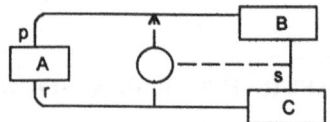

Figure 16. A more complicated case of generalized subset

5. EXISTING MODEL STRUCTURING FEATURES AND THEIR USAGE

The main technique for comprehending complicated diagrams (descriptions, models) is their structuring, i.e., their splitting into more or less independent parts and then refining these parts by subdiagrams. However the practical experience shows that conceptual models (class diagrams) for real domains (such as Internet, Web architecture etc) as a rule cannot be split into independent parts, the whole model is one large "cobweb". Apparently this is the reason why the structuring of class diagrams in its classical sense has not been elaborated. For class diagrams a different approach is used, which in a sense compensates the lack of classical structuring. Firstly, there are mechanisms of generalisation and aggregation, which form the basis for comprehending complicated class diagrams. Another mechanism for structuring - in a sense a completely new approach to structuring - is the concept of stereotype (though actually the role of stereotype is much wider in UML). By defining appropriate stereotypes and assigning easy distinguishable styles (colours, shapes, icons) to them, we can group semantically close classes in a very readable way according to their stereotypes.

It is a function of a good support tool to offer specific symbol **styles** for each of the class stereotypes. Actually, by a symbol **style** here we mean all its graphical style attributes: shape, icon, background colour, border line style, font styles etc. Similarly, stereotypes for associations and other lines must support all line style elements: line colour, line style, end shapes (arrows etc). It should be noted that the official UML recommendations for graphical stereotype notations (a graphic icon, texture or color) which are typically implemented in tools are far too limited for good conceptual modeling.

The GRADE tool has such extended stereotype support. For example, it has a predefined stereotype for the activity (represented by a blue rounded rectangle), and there is an easy facility for defining new stereotypes and assigning styles to them. It should be noted that the formal facility for structuring of UML models is the package mechanism, but it is, in a sense, cutting a large model into pieces by scissors, without any care for semantic independence for fragments (to ascertain, look at the official UML metamodel). The packages are a perfect tool for structuring software design diagrams which must be structured by the very nature of design, but are unacceptable for conceptual modelling.

6. NEW STRUCTURING FEATURE – FRAME

As it was noted in the previous section, traditional structuring mechanisms cannot be used in a proper way for conceptual class diagrams. In this section we offer a principally new structuring feature which we call **frame**. By frame we understand a rectangle, which can be positioned onto a semantically related class diagram fragment and given a readable name. The figure 17 displays (part of) a conceptual model of Web architecture (built by J.Rogovs) and frames *User*, *URL*, *Server*, *Server software* are typical examples of frames. A class diagram enhanced by properly selected frames becomes much more readable and comprehendible.

It is a bit strange that this concept has not been officially included in UML, because such frames are used in everyday practice when we want to present a readable drawing in any area. For the goals of conceptual modelling, which is not for formal processing by computer, but for human understanding, such a little bit fuzzy concept is quite appropriate, if it encourages understanding.

The frame notation becomes especially readable if each frame is assigned a separate colour. The GRADE modelling tool supports the frame feature (it is called there free comment symbol).

7. SOME METHODOLOGICAL ADVICES

Rather comprehensive methodological advices on building class diagrams have already been given by J.Rumbough [1] and we will not repeat them. We will add just some important items, gained from our practical experience.

1. According to J.Rumbough,[1] the building of conceptual class diagram starts with finding the classes. But a new essential criterion for this is offered - only these concepts can be selected for classes, where it is absolutely clear, what are their instances, or in other words, their identity is defined.[17] Not always this can be so easy decided. Apparently, *water* can not be used as a class, but *ocean* can be. In addition, it should be taken into account, that classes may be physical objects (*car*), abstract concepts (*car model*, *flight*) and also activities (*testing*).

2. When classes are chosen, the saturating of diagram by associations can be started. Again the basic principle must be adhered to, that only these associations whose semantics is unambiguous should be added.

3. No anomalies should be directly represented in the class diagram, they can be documented by means of notes (which are official parts of the class diagram).

4. A tool must be used which supports good automatic layouts of class diagrams, since always during the diagram building new classes and associations must be inserted inside the fragments already built. This insertion should never "spoil" the used semantic class positioning principles. If the diagram has 100 or more classes (and real domains are such), it is impossible to draw this diagram "by hand" or by a tool where each class must manually positioned.

8. REQUIREMENTS TO TOOLS FOR CONCEPTUAL MODELLING

Several requirements to tools for adequate support of conceptual modelling were outlined already in previous sections. In this section we will summarise them and give some more requirements.

As it was already mentioned, the most important specific requirement for conceptual modelling is a good stereotype support. First, there should be a set of predefined class and association stereotypes corresponding to widely used modelling concepts, with the most accepted symbol shapes assigned to these stereotypes. For example, activity could be represented by a blue rounded rectangle. Typical association shapes could be dashed arrows for an activity sequence, solid arrows for linguistic links etc. On the other hand, there should be a very convenient facility for introduction of new stereotypes (and their corresponding shapes) by the tool user. These stereotypes should be available model-wide, in order to give easily recognisable graphical notations for system-specific concept groups. For example, the stereotypes could be defined in a model-wide stereotype table. To our mind, an additional, stereotype feature would also be desirable, namely, there should be a possibility to attach to a class stereotype a set of predefined attributes. For example, the stereotype *position* could have predefined attributes *competencies*, *working hours*, *cost per hour*, *number of instances*. The stereotype definition for *activity* should have attributes *duration* and *cost*. The expected tool support for the feature is such that, when a new class with the given stereotype is created, the predefined attributes are prompted in the attribute definition window and the user can select them and define the relevant attribute values. The current UML version 1.4 offers the mechanism of tagged values attached to a stereotype for this purpose, but none of well-known tools implements this mechanism in a usable way, a direct association of default attributes would be much more convenient for end-user.

There are also two purely tool-technical requirements for conceptual modelling, but experience of GRADE usage has shown their great importance in practice.

First, there must be an easy way to maintain the **readability** of diagrams, because conceptual models are built for reading by other humans. There can be several solutions to the diagram readability problem. GRADE solves this problem by its powerful controlled **automatic diagram layout** mechanism, based on sophisticated graph drawing algorithms.[18] This autolayout mechanism permits the user to insert a new class symbol where it is most desired. The existing symbols and lines are automatically moved, to give the necessary space and avoid any overlapping of the new symbol by existing symbols or lines. The movement is "delicate", it doesn't destroy the existing relative placement of symbols. Thus the main graphical aspect of readability – appropriate positioning and clustering of class symbols is supported. Association lines are automatically positioned so that unnecessary line crossings are avoided, thus good traceability of lines is obtained.

Another requirement is a support for **large** class diagrams. Conceptual models of complicated systems tend to be large because human understanding frequently requires to see the whole "big picture". GRADE supports maintenance of extra large class diagrams, firstly, by its autolayout mechanism, which works efficiently also for diagrams with hundreds of classes. In addition, a special diagram zooming feature is provided, similar to that typically available in camcorders. But for extra large diagrams even this may be insufficient. Therefore GRADE supports **views** for class diagrams. The user can maintain a large diagram via views corresponding to subsystems, then any updates will be automatically transferred to other relevant views and to the main diagram.

One novel idea is to add simple multimedia facilities to the class diagram, more precisely, to add **speech**. In this way each class could "explain" itself in a spoken text when selected in the tool. The most important feature in this context is the presentation facility where classes and associations are automatically highlighted (by a moving cursor) in the desired order and spoken comments are given accordingly. This feature employs the inherent human capability to see and listen simultaneously. The automatic highlighting is very useful for understanding large models, since it is the viewing of parts of the diagram in the order conceived by the author that reveals the content in the best way.

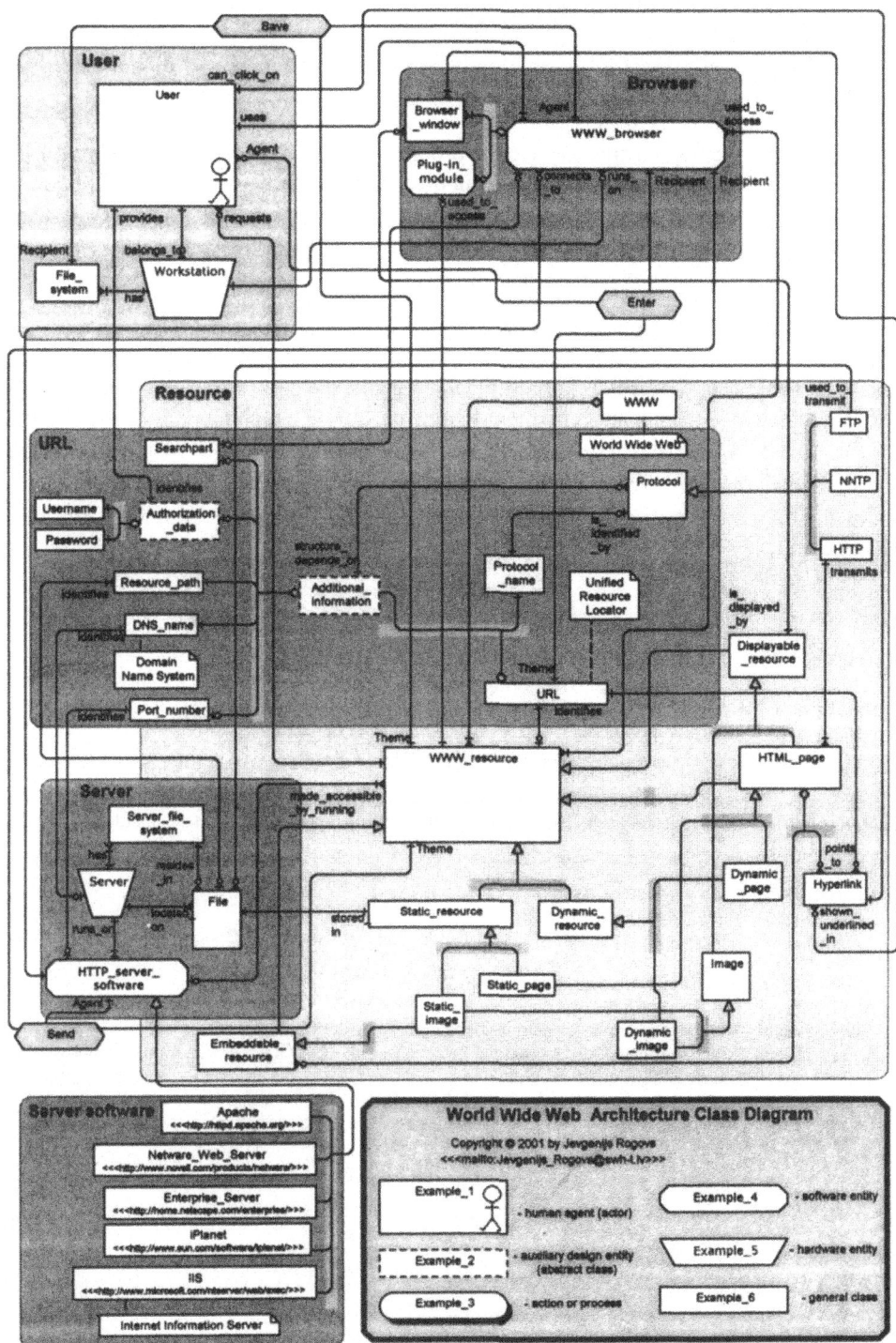

Figure 17. Conceptual model of Web architecture (part)

The presentation feature exists in the GRADE tool since the version 4.0.9 in 1999.[11] The described feature has its highest value in cases when a conceptual model has to be understood by a reader without the presence of the author, e.g. when a model is downloaded via Internet.

It should be mentioned that in fact the problem here is more general – how to link formal modelling methods with multimedia in order to ease the perception of a model.

9. CONCLUSIONS

Evidently, the conceptual modelling is the main facility for comprehending complicated systems (banks, insurance companies, airports etc). In addition to application of conceptual modelling for such important practical goals, we have obtained a good experience in education-related conceptual modelling of Internet architecture, GSM system principles, DCOM component architecture etc. This model building is a very stimulating job for students to understand in details various sophisticated computer-based systems.

REFERENCES

1. J.Rumbaugh, M.Blaha, W.Premerlani, F.Eddy, W.Lorensen, Object-oriented Modeling and Design, Prentice-Hall (1991)
2. J.Martin, Principles of object-oriented analysis and design, Prentice Hall (1993)
3. G.Booch, I.Jackobson, J.Rumbaugh, The Unified Modeling Language User Guide, Addison-Wesley (1999)
4. H.-E.Eriksson, M.Penker, UML Toolkit, Wiley Computer Publishing (1998)
5. P.Harmon, M.Watson, Understanding UML, Morgan Kaufmann Publishers (1998)
6. C.Larman, Applying UML and Patterns, Prentice-Hall,2nd ed. (2002)
7. P.-A.Muller, Instant UML, Wrox Press Ltd. (1997)
8. I.Jackobson, G.Booch, J.Rumbaugh, The Unified Software Development Process, Addison-Wesley (1999)
9. G.Booch, I.Jackobson, J.Rumbaugh, The Unified Modeling Language Reference Manual, Addison-Wesley (1999)
10. M.Fowler, UML distilled , Addison-Wesley (1997)
11. GRADE tools; http://www.gradetools.com/
12. Object Role Modeling ; htpp:// www.orm.net
13. T.Halpin, UML data models from an ORM perspective (part 8), Journal of Conceptual Modeling (April 1999)
14. J.F.Sowa, Knowledge Representation, Brooks/Cole (2000)
15. A.-W. Scheer, ARIS Business Process Modeling, Springer (2000)
16. J.Warmer and A.Kleppe, The Object Constraint Language, Addison-Wesley (1999)
17. N.Guarino and C.Welty, Evaluating Ontological Decisions with ONTOCLEAN, Communications of the ACM (February, 2002) vol. 45, N.2, p.61-65
18. P.Kikusts, P.Rucevskis, Layout algorithms of graph-like diagrams for GRADE, Windows graphical editors,
19. LNCS (1996), v.1027, p.361-364

SOFTWARE ENGINEERING AND IS IMPLEMENTATION RESEARCH: AN ANALYTICAL ASSESSMENT OF CURRENT SE FRAMEWORKS AS IMPLEMENTATION STRATEGIES

Bendik Bygstad and Bjørn Erik Munkvold[*]

1. INTRODUCTION

In the 1990s several new software engineering frameworks were introduced, among them Rational Unified Process (RUP) (Jacobson et al. 1999), OPEN (Henderson-Sellers and Unhelkar 2000), Microsoft Solutions Framework (MSF) (Microsoft 2001), and Catalysis (D'Souza and Wills 2002). A significant feature of these is that the software product is developed incrementally, through a series of iterations. This structure not only mitigates technical risk, but also challenges some of our traditional conceptions of the relationship between software engineering (SE) and information systems (IS) implementation as 'separate worlds', because the SE frameworks also include activities to secure a successful implementation in a complex organisational setting (Kruchten 2000).

Empirical studies have shown that iterative and incremental development of Internet software has significantly increased development speed (Cusomano and Yoffie 1999). Also, in a two year study of 29 projects it was found that evolutionary development and early releases to customers was strongly associated with both product and implementation success (MacCormack 2001), stressing the importance of rapid feedback on design choices.

The important question addressed in this paper is whether the SE frameworks really can serve as implementation strategies, an area traditionally based on IS implementation research. If we accept that three decades of IS implementation research has provided important findings, and we find that these findings are reflected in the implementation mechanisms of the new SE frameworks - then it could be argued that the SE frameworks may lead to better implementation projects.

[*] Bendik Bygstad, Norwegian School of IT, Oslo, Norway.
Bjørn Erik Munkvold, Agder University College, Kristiansand, Norway

Information Systems Development: Advances in Methodologies, Components, and Management
Edited by Kirikova *et al.*, Kluwer Academic/Plenum Publishers, 2002

227

The structure of the paper is as follows. First, the mainstream view of software engineering is briefly presented, and how this relates to IS research. We present the central attributes of the SE frameworks, concentrating on RUP and MSF. This is followed by a brief review of the lessons learned from IS implementation research. For each of these lessons, implications for software engineering are suggested. Lastly, we map these implications to mechanisms in the SE frameworks. Findings are discussed and in the concluding section it is shown that most of the lessons learned are, explicitly or implicitly, integrated into the SE frameworks.

2. SOFTWARE ENGINEERING AND IS IMPLEMENTATION RESEARCH

Software Engineering is defined as:

> *"an engineering discipline which is concerned with all aspects of software production from the early stages of system specification through to maintaining the system after it has gone into use" (Sommerville 2001).*

Software Engineering was first termed in 1968 as a response to the "software crises" (Sommerville 2001). As the systems grew larger during the sixties, it became evident that the informal ways of programming were inadequate to control the increasing complexity. The response to the crisis was a formalisation of the development process, and the introduction of techniques to use in the different steps of the process.

The steps may vary in different process frameworks, but the overall structure is generic, referred to as the 'Waterfall model':

- Requirements analysis and definition: Defining user needs.
- System and software design: A formal specification and design of the system.
- Implementation and unit testing: Write and test program modules.
- Integration and system testing: The programs are integrated into systems, and tested.
- Operation and maintenance: The system is installed and put into use. Maintenance includes improvements as user needs evolve over time.

In the 1980s the waterfall model came under attack for being too rigid (Boehm 1988), but it is important to remember that this rigidity was seen as the central mechanism to control system quality. Most systems in use today are built in line with these principles.

While Software Engineering prescribes methods for the specification and construction of high quality software, IS implementation research focuses on what happens after the system is technically complete, hence often referred to as *organizational implementation* (Walsham 1993). In the rest of the paper we use the term "implementation" as defined in the IS literature:

> *"Implementation is an organisational effort to diffuse an appropriate information technology within a user community"* (Kwon and Zmud 1987).

Though there exist several different perspectives on IS implementation (Marble 2000), the main body of IS implementation research focuses on different aspects of the organisational effort related to diffusing the IT system within the organization. Examples here include the alignment of business and technology (Applegate et al 1999), user acceptance (Agarwal 2000), and user resistance (Markus 1983). These and other related themes are described in more detail later in the paper.

Figure 1 illustrates the traditional view on the relationship between SE and IS implementation as sequential. That is, software engineering builds the product, and when the product is finished it may be taken into use in a user community.

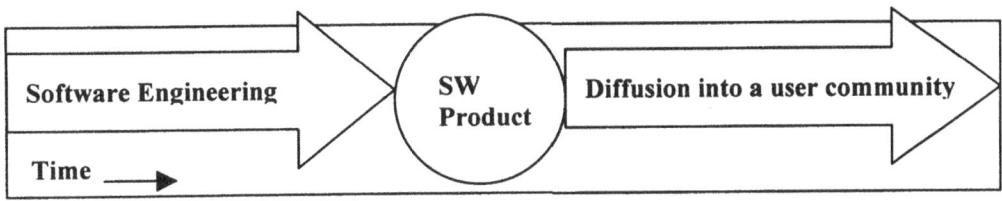

Figure 1. The basic (traditional) relation between software engineering and IS implementation.

The assumption made in this paper is that new concepts and frameworks in software engineering make this sequential perspective less valid.

3. KEY PRINCIPLES IN SOFTWARE ENGINEERING

Unfortunately, the software crisis was not solved by the software engineering methods. Analyses from the mid-1990s documented that the success rate for large software development projects still was very low (Jones 1996). By the mid-1990s there was a strong movement in the software engineering field, campaigning three principles: object oriented development, iterative development and stakeholder focus.

3.1. Object Oriented Development

The principle of *object oriented development* was invented in Norway in the 1960s (Birtwistle et al 1973). It is a way of organising both natural and complex systems, through the concepts of encapsulation and inheritance. To control the complexity of large systems, it is important to modularise it into subsystems and parts. The basic idea - the analogy being the living cell - is that the building blocks are objects, which control their internal processes and thus reduce the need for central coordination. Through this principle of encapsulation, the basic unit becomes much more stable than a procedure, and less vulnerable to change (Booch 1991).

3.2. Iterative and Incremental Development

The principle of *iterative development* was described in a seminal paper by Boehm (1988) and systematically developed during the 1990s. Boehm's point of departure was software economics, where he showed that the symptoms of the software crisis - low software quality, overrun schedules and discontented users - all were consequences of poor risk management. Boehm's solution was a spiral model with four repeating steps:

- Determine objectives and alternatives
- Evaluate alternatives, and identify/resolve risks

- Develop and test prototype
- Plan the next phase

This concept was developed further into different process structures, and iterations are fundamental to the new SE processes in several ways: they decompose a large project into small, manageable parts or versions, they reduce economic and schedule risks, and each iteration provides a clear, short term focus (Kruchten 2000).

Equally important is the feature that iterations solve some important problems connected to the requirements specification, where the waterfall model demanded all requirements at the start of the project. Iterative models "acknowledge a reality often ignored - that user needs and the corresponding requirements cannot be fully developed in front. They are typically refined in successive iterations" (Jacobson et al 1999). This process of mutual learning through the project ideally includes all stakeholders in an iterative project.

The leading iterative object oriented process models are Rational Unified Process (RUP), the Object-Oriented Software Process (OOSP) and the OPEN process (Ambler 2001). This paper will examine RUP and a widely used iterative, but not exclusively object oriented framework, the Microsoft Solutions Framework (MSF). Both RUP and MSF are used as examples, and share several similarities with other current SE frameworks, as noted earlier.

3.3. Stakeholder Focus

The stakeholder focus is the third key feature to understanding the new frameworks. The underlying premise is that modern organisations are no longer best understood as goal oriented, decision-making machines. Rather, they could be viewed as a network of stakeholder relationships, where the formal borders of the organisation are also less precisely defined (Mitroff and Linstone 1993; Kling and Jewett 1994).

Developing information systems in such environments is clearly different from doing this in traditional business hierarchies. Stakeholders need to be identified and involved, the whole problem domain gets more complex, compromises might be negotiated, and most importantly, the implementation process becomes more challenging.

4. CURRENT SE FRAMEWORKS: MSF AND RUP

4.1. RUP

RUP was developed in Rational Corporation in the mid-1990s, building mainly on Jacobson's work at Ericsson in 1987-95 (Jacobson et al 1999). Integrating RUP with the standardisation of UML in 1997, the process was published in 1998. By the end of 1999 more than a thousand companies were using RUP (Kruchten 2000) and it is taught at a number of universities. Table 1 describes the main principles in RUP (Royce 1998; Jacobson et al 1999; Evans 2001).

RUP is structured in four phases: inception, elaboration, construction and transition. Within each phase there is one or several iterations consisting of workflows, starting with business modelling and ending with the physical deployment of software components. Each iteration resembles a small waterfall project, and produces an increment, a release

(Jacobson et al 1999). The release is the key communicating mechanism with the stakeholders, because:

Table 1. RUP principles

Principles	Description
Use Case Driven	a) Each use case represents a need or view from a user group (an actor) and describes how the system will create value for this specific group. b) Use cases drive the rest of the development process: All artefacts through analysis, design, coding and testing are traceable back to the use cases. Use case *realisations* are used at the end to verify that the classes can perform the use cases.
Architecture- Centric	Architecture is described in different views (use case, logical, process, implementation and deployment views): - The structural organisation of subsystems and interfaces - Description of important behaviour - Composition of subsystems into the whole system
Iterative	The problem and solutions are organised into small steps, with each step going through a small, but full development cycle including business and user requirements, analysis and design, coding, test run and user verification
Incremental	The result of an iteration is an increment. The system "grows" from one iteration to the next. The objectives are: - To mitigate risks - Handling changing requirements and allowing for changes - Achieving continuous integration - Attaining learning early
Controlled	The process description requires that all activities are planned: What artefacts are to be produced, how do they relate to other artefacts, who is responsible, and which techniques should be used.

"Users can comprehend a system that operates, even if it does not yet operate perfectly, more easily than a system that exists only as hundreds of pages of documents. (..) Therefore, from the standpoint of users and stakeholders, it is more productive to evolve the product through a series of executable releases..." (ibid)

RUP has been criticised for rigidity, both from the Extreme Programming community (McCormick 2001) and from the OPEN process perspective (Henderson-Sellers et al 2000). More importantly in the implementation context, it has been criticised for concentrating on IT issues and lacking a business perspective. It assumes that all the important business decisions already have been made, and it also has weak support for project management (Henderson-Sellers et al 2000; Ambler 2001).

4.2. MSF

MSF was developed in 1994 by Microsoft Consultant Services, building on best practices from a number of Microsoft projects. MSF is today widely used around the world (Microsoft 2001; Rada 2001). The point of departure is the risk of software projects: "No matter how fast the project team advances, the market, or the technology, or the competition, or the customer's business will advance faster" (Microsoft 1998).

MSF is a rather loose framework, built on a few principles (see table 2). The most important is that a large system is never specified and then built, but rather divided into releases or versions with short development cycles. As in RUP, these iterations represent a small waterfall project (Microsoft 2001).

Compared to RUP, MSF is rather flexible in its implementation and easy to learn. It is mostly used in small and medium sized projects. Though MSF includes several interesting concepts, one might argue that its strength is not really its features but the market position of Microsoft (Rada 2001).

Table 2. MSF principles (Microsoft 2001)

Principle	Description
Risk Manage-ment	a) Risk is inherent in all software projects and must be assessed and managed throughout the lifecycle. Risks may be reduced, transferred or avoided. b) Risk is both a threat and an opportunity
Process Model	a) The system is built iteratively and incrementally, through versioned releases. The most critical parts of the system are built first. b) Each iteration consists of four distinctive phases: Vision, plan, developing, stabilising c) Each iteration is time-boxed: Balancing features against resources and schedule. d) The next iteration (for the next release) usually starts when the current one is completed.
Team Model	The non-hierarchical model focuses on a shared project vision and distinct responsibilities: *Role* *Responsibility* Product management : Satisfied customer Process management : Delivered within restraints Development : Delivery to specifications Testing : Release after addressing all relevant issues User responsibility : Enhance user performance Logistics management : Smooth deployment
Architec-ture	Architecture is seen in four perspectives, showing how applications and data support the business: Business, Applications, Information and Technology

4.3. Summing Up RUP and MSF

There are many similarities between the two frameworks (Microsoft 1997). In the context of implementation the most important are:

- They are both designed to handle risk in a rapidly changing environment.
- The basic mechanism is the short iteration that produces a small release that can be assessed by the user organisation. "Iterations give the stakeholders in the project a learning space: This approach is one of continuous discovery, invention and implementation" Booch in (Kruchten 2000).
- Both processes have very detailed guidelines of the stakeholder roles and responsibilities.
- Both assume that the business analysis has been done prior to the project.

The question addressed in this paper is whether these frameworks may work as strategies for organisational implementation of the systems? Or, to put it with Lehman (1989): Is there, in engineering terms, a link between SE processes and business outcome?

5. SOME LESSONS LEARNED FROM IS IMPLEMENTATION RESEARCH

IS implementation research has over the last 30 years produced a large volume of empirical evidence, but has always been hard to generalise and to classify (Kwon and Zmud 1987; Markus and Robey 1988; Munkvold 1998; Fichman 2000). At the core of IS implementation research is the relationship between technology and organisations, illustrated by the still valid point of departure for this research - Leavitt's diamond

(Leavitt 1965) - describing these relationships and predicting that technology may dramatically change the work organisations.

Different research streams can be identified within IS implementation research, of which the factor and process based research streams have been most dominant (Kwon and Zmud 1987; Prescott and Conger 1995).

5.1. Factor Research

While most researchers agree that the interaction between technology and organisation is "complex", there is a large spectre of perspectives and methodologies in use for studying this interaction. The factor research is based on a positivistic epistemology, using quantitative methods for identifying factors associated with implementation success or failure.

A prominent example of factor research is Swanson's' "Implementation Puzzle", as reviewed in Marble (2000). The model covers the steps from design to business use, and includes nine critical factors to implementation success or failure, as summarized in table 3. An interesting feature is that none of these factors are exclusively under the implementer's control.

Table 3: Swanson's implementation success factors (adapted from Marble, 2000)

IS implementation factor	Lessons learned
1. User involvement	User satisfaction is the most accepted criterion of successful implementation. Users should be involved in all development and implementation issues.
2. Management commitment	Visible management support is critical for resource allocation and the power shift that new systems often imply.
3. Value basis	The user organisation must be confident that the system really contributes to the creation of value.
4. Mutual understanding	The relationship between IS staff and users is critical for implementation success.
5. Design quality	Good design is important for perceived ease of use. Flexible design is important for the system's ability to adapt to changing needs.
6. Performance level	Reliability and responsiveness greatly influence user satisfaction.
7. Project management	The large complexity of implementation calls for structured and controlled project management.
8. Resource adequacy	The technical skills of IT staff are critical in all phases of an implementation project.
9. Situational stability	Implementation will often change work practices, and implementers should be sensitive to users' concerns in respect to these changes.

5.2. The Process Perspective

Factor research has been criticised for being fragmented and for lacking a process perspective (Kwon and Zmud 1987; Pare and Elam 1997). The process oriented research stream focuses more on how the different factors interact during the various stages of the implementation, and on how behavioural and political issues may influence this process. Table 4 lists some key implementation issues discussed in the process based research.

Table 4. Process perspective: Lessons learned

IS implementation issue	Lessons learned
1. The emergent nature of IS implementation	1. Predicting the organisational impact of a new technology is difficult, due to the emergent and situated change processes involved (Suchman 1987; Markus and Robey 1988; Bijker and Law 1992).
2. Organisational innovation	2. The real innovation is the mutual adaptation between the information system and the organisation (Leonard-Barton and Kraus 1988; Davenport 1993).
3. Diffusion of innovation	3. IS implementation is often a diffusion process, involving several steps towards organisational integration (Kwon and Zmud 1987).
4. User acceptance	4. Organisational adoption does not necessarily imply user acceptance. User acceptance is tightly associated with the user's attitudes and beliefs (Davis 1989).
5. Stakeholders	5. Stakeholder interests strongly influence the implementation process (Markus 1983; Kling and Jewett 1994).
6. Organisational mechanisms	6. Structural arrangements ("organisational mechanisms") facilitate users' ability for knowledge creation (Nambisan et al 1999).
7. Actor-networks	7. Successful implementation is dependent on the stabilisation of interests between different actants (Hanseth and Monteiro 1996; Walsham 1999).
8. Context	8. IS implementation is inherently context sensitive (Fichman 2000).

5.3. An Eclectic View on IS Implementation

It may seem difficult to compare, let alone integrate, the lessons learned from factor research and process research. Lessons learned from factor research (table 3) seems to use more precise constructs (although most of them are rather composite constructs, like "user involvement"), while findings from the process based research (table 4) have a stronger organisational and interpretive focus.

The existence of parallel and competing theories is well known in the social sciences (Bernstein 1976) and many researchers take the pragmatic view of using what looks sound from each school. This also seems to be the general attitude of the IS research community (Fichman 2000; Robey and Boudreau 2000). From the perspective of *expert knowledge* Marble (2000) contends that the body of knowledge of IS implementation "should be considered from the point of view of knowledge acquisition and representation", and that we should accept that the findings are complementary and even contradictory:

> "*As long as natural language is involved, an uncompromising expectation of absolute precision in IS modelling is doomed from the start. (..) It is the exercise of judgement, intuition and approximate reasoning that characterises the most valuable (if elusive) trait of expert knowledge. This is what allows an expert to generalise from experience and this is what we ultimately seek for implementation theory*" (ibid.).

Accepting this implies that IS implementation research is inherently context dependent: Findings from this research should be used carefully, by experienced practitioners with thorough knowledge of the context, and capable of selecting the right tools and practices. It is worth noting that this contrasts somewhat with the software engineering principles, which assert (at least implicitly) that the basic techniques can be used by any competent software engineer without domain knowledge.

6. MAPPING THE SE FRAMEWORKS AND THE IS LESSONS LEARNED

We adopt the eclectic view that each stream of IS implementation research has important findings. By integrating table 3 and 4, we suggest the following list of relatively uncontested findings (table 5), which will be called the lessons learned. From these lessons we suggest some implications for software engineering, and then map these to supporting mechanisms in the SE frameworks.

Table 5. The lessons learned from IS implementation research mapped to supporting mechanisms in the SE frameworks.

IS lessons learned (Ref. Tables 3 and 4)	Implications for Software Engineering	Support in MSF /RUP?	Mechanisms in RUP and MSF supporting implications
1. User satisfaction is the most accepted criteria of successful implementation. Organisational adoption does not necessarily imply user acceptance. (3.1, 4.4)	Users should be involved in all development and implementation issues. Practical benefits should be demonstrated early. User attitudes and beliefs should be addressed.	Yes	Users are involved in each iteration. The extensive use of prototypes and the versioning approach together support early learning and adaptation.
2. Visible management support is critical for resource allocation and the power shift that new systems often imply. (3.2, 4.5)	The project needs visible and consistent management support	Partly	A focus on defined stakeholder interests is a central feature in both RUP and MSF.
3. The relationship between IS staff and users is critical for implementation success. (3.4)	The project organisation should facilitate strong cooperation on equal terms between users and IS staff.	Partly	Not explicitly, but prototyping and early versions support this. Both RUP and MSF emphasize mutual learning.
4. Good design, reliability and responsiveness greatly influence user satisfaction. (3.5, 3.6)	The system should be designed, tested and scaled appropriately.	Yes	Both MSF and RUP have extensive design and testing guidelines. RUP covers scalability. Object technology gives higher reliability.
5. The large complexity of implementation calls for structured and controlled project management. (3.7)	The project should be planned and controlled in all phases.	Yes	Both frameworks have extensive process structures. RUP also has a very detailed activity set.
6. Predicting the organisational impact of a new technology is difficult. (4.1, 4.2, 4.3)	The assumptions that careful analysis of user needs ("requirements") predicts implementation success, is doubtful.	Yes	Incremental development acknowledges that the system must be built gradually, as it is being used in a real setting.
7. Structural arrangements ("organisational mechanisms") facilitate users' ability for knowledge creation. (4.6)	Developers need knowledge of the mechanisms that facilitate user creativity.	Yes	Iterative development with short development cycles facilitates several occasions for knowledge creation.
8. Successful implementation is dependent on the stabilisation of interests between different actants. (4.7)	Stabilising an actor-network is a process of negotiation and relation forming, and cannot be planned and controlled in a standard engineering process.	Partly	Not explicitly supported, and the planned and controlled nature of software engineering makes this less probable. In practice, though, the incremental and iterative development makes this possible over time.

IS lessons learned (Ref. Tables 3 and 4)	Implications for Software Engineering	Support in MSF /RUP?	Mechanisms in RUP and MSF supporting implications
9. IS implementation is inherently context sensitive. (3.9, 4.8)	Systems development and implementation need to be sensitive to the political and technological context.	Partly	Not explicitly supported. But iterative development and a stakeholder focus makes this possible over time.

7. DISCUSSION

Though the picture is not uniform, table 5 shows a reasonable "fit" between the lessons learned from IS implementation research and the implementation mechanisms included in the SE frameworks. Most of the implications are explicitly supported, while the rest are found to be at least implicitly or partially supported.

The main supporting mechanism is the iteration approach, which emphasises that software is not a fixed physical product, like a car or a computer, but a conceptual and social construct that changes over time, in much the same way as organisations (and maybe even users). The iterations facilitate a process where there is a potential for mutual adaptation between the software product and the organisation.

However, the SE frameworks obviously have limitations as implementation strategies. Criticism may be formulated from two perspectives, using Dahlbom and Mathiassen's (1993) classification of systems development methods into three paradigms:

- *Construction:* Solving a complex problem through a rational and analytical strategy (ref. the waterfall model).
- *Evolution:* Reducing the risk through an experimental strategy of problem solving.
- *Intervention:* Systems development is seen as an integral part of organizational change, with problem definition as part of the project.

The SE frameworks discussed here are clearly within the evolutionary paradigm. The systems developers "have to interact with the environment, accept the openness of the problem and the system to be developed, take into account the preferences and beliefs of problem owners and users, deal with the economical and political climate of the project, and keep in step with the changes in the kind of technologies on which the project is dependent" (Dahlbom and Mathiassen 1993).

From the construction paradigm it may be challenged that the SE frameworks are not really intended as frameworks for supporting organisational implementation. They represent methods to build software systems, not to conduct organisation development. If we develop software in the "emerging" way made possible by the iterative and incremental approach, we will meet several traditional engineering problems:

- How do we estimate projects which may go 'anywhere' and grow uncontrollably in size?
- How do we handle contracts when the final product is not yet specified?
- How do we integrate these systems into larger architectures?

These problems were actually addressed by Boehm (1988), contending that the iterative approach is not applicable for all kinds of projects and not in all kinds of settings. It is expected to work best in user-driven, in-house projects, where contract terms are not too strongly defined, and where there is a large degree of flexibility in the user organisation.

The RUP community does not agree with this, and documents a series of contract projects that have been using RUP over the years (Royce 1998; Rational 2002). One might suspect that these contract software projects are not 'flexible' in the organisational context that Dahlbom & Mathiassen and Boehm refer to, but only empirical research could really answer this question.

While the construction paradigm thus challenges the engineering ability of the iterative frameworks, the intervention paradigm criticises them for being politically naive and methodologically too rigid. The iterative approach is still *engineering* in the sense that it belongs to the realm of rational problem solving, though perhaps more in line with Simon's "bounded rationality" than pure "rational choice" (Simon 1970). This implies that the iterative frameworks resemble the construction paradigm, in that "the problem is given, and the criterion of success is basically the same. But other kinds of qualities are included. The context of the system is included together with the individual interpretations of different users" (Dahlbom and Mathiassen 1993).

The criterion of success within engineering is that the system conforms to its specification and that it meets the expectation of the customer (Sommerville 2001). In the political IS implementation research this point is challenged: If problem definition is the point of departure, it will be negotiable. Whether implementation is a success or not depends on stakeholders' interests, not on objective technical or diffusion criteria (Markus 1983).

Iterative engineering may also be challenged on a methodological level: If the problem is not clearly defined, the systems development professional can no longer merely be an expert in solving problems. Instead he should be a change agent, on a level similar to other actors in the organisation, where responsibilities and roles are negotiated (Dahlbom and Mathiassen 1993; Markus and Benjamin 1997). The SE frameworks fail to address this issue, focusing more on what should be done, rather than what actually happens.

Concerning the point of political naivism we think the criticism is acknowledged in SE, although in a rather narrow way. Both Sommerville (2001) and Kruchten (2000) admit that organisational expertise is crucial, but that software engineering as a discipline lacks the concepts and tools. To "satisfy user needs" implies that the engineering process must address a wide number of human and business issues, which are critical to success or failure for the overall objectives of the system. As stated by Sommerville (2001), "In reality this is impossible (in the context of SE)."

Regarding the methodological rigidity, we think the criticism is less justified. It is important to bear in mind that the business context of systems development is one of strong competition, tight schedules and customer demands (Royce 1998; Munkvold 2001). SE frameworks are balancing the tightrope between engineering and exploring, and we think the crucial point is not the rhetoric of this balance, but how the underlying principles are applied in real projects.

8. CONCLUSION

This paper has examined central attributes of SE frameworks exemplified by RUP and MSF, focusing on iterative and incremental development, object technology and stakeholder focus. These were then mapped to important lessons learned from IS implementation research. Most of these lessons were found to be incorporated in the mechanisms of the SE frameworks, either explicitly or implicitly. This implies that the current SE frameworks have the potential of leading to better, more successful implementations.

It is argued that our analytical findings are most relevant for in-house projects with a certain degree of organisational flexibility. This does not necessarily exclude contract projects or standardisation packages (like ERP or CRM systems), but the support is weaker, and traditional engineering problems are more likely to occur.

Our assessment of the SE frameworks is purely analytical. Studies of development projects applying these frameworks can provide empirical evidence on these questions. Further research may take several directions. The most important would be to research the practical use of the SE frameworks in projects, preferably over time.

Secondly, the origins of the SE frameworks may be researched. Is the successful mapping of IS implementation research and the SE frameworks a 'coincidence', based on separate assimilation of industry practice? If not, what are the mechanisms through which the lessons learned from IS implementation research filter into software engineering? Or on a larger scale - how do innovations in the IS research field diffuse into the field of applied software engineering?

9. REFERENCES

Agarwal, R. (2000). Individual Acceptance of Information Technologies. *Framing the Domains of IT Management*. R. Zmud. Cincinnati, Pinnaflex: 85-104.

Ambler, S. W. (2001). Completing the Unified Process with Process Patterns. www.ambysoft.com.

Applegate, L., McFarlan, F.W. and J.McKenney (1999). *Corporate Information Systems Management*. Boston, Irwin McGraw-Hill.

Bernstein, R. (1976). *The Restructuring of Social and Political Theory*. London, Methuen & Co.

Bijker, W. E., and Law, J. (1992). *Shaping Technology/Building Society*. Cambridge, Massachusetts, The MIT Press.

Birtwistle, G. M., Dahl, O.-J., Myhrhaug, B., and Nygaard, K. (1973). *SIMULA begin*. Philadelphia, Studentlitteratur, Lund and Auerbach Publ. Inc.

Boehm, B. W. (1988). "A Spiral Model of Software Development and Enhancement." *IEEE Computer* (May): 61-72.

Booch, G. (1991). *Object Oriented Design*. Redwood City, Benjamin Cummings Publishing.

Cusomano, M. A., and Yoffie,D.B (1999). "Software Development on Internet Time." *IEEE Computer* 32 p.60-69.

Dahlbom, B., and Mathiassen,L. (1993). *Computers in context : The philosophy and practice of systems design*. Cambridge, Mass, NCC Blackwell.

Davenport, T. H. (1993). *Process Innovation*. Boston, Ernst & Young.

Davis, F. (1989). "Perveived Usefulness, Perceived Ease of Use, and User Acceptance of Information Technology." *MIS Quarterly* 13:3: 319-340.

D'Souza, D., and Wills, A. (2002). *The Catalysis Approach*. www.catalysis.org.

Evans, G. (2001). *Lightening up a heavyweight*, Rational Corp. 2001. www.rational.com.

Fichman, R. (2000). The Diffusion and Assimilation of Information Technology Innovations. *Framing the Domains of IT Management*. R. Zmud. Cincinnati, Pinnaflex.

Hanseth, O., and Monteiro, E (1996). "Inscribing Behaviour in Information Infrastructure Standards." *Accounting, Management and Information Systems* 7:4: 183-211.

Henderson-Sellers, B., Due, R., Collins, G., and Graham, I. (2000). "Third Generation OO Processes: A Critique of RUP and OPEN from A Project Management Perspective." *IEEE*: 428-435.

Henderson-Sellers, B., and Unhelkar,B (2000). *OPEN Modelling with UML*. Harlow, Addison-Wesley Longman.

Jacobson, I., Booch, G., and Rumbaugh, R (1999). *The Unified Software Development Process*. Reading, Addison Wesley.

Jones, C. (1996). *Patterns of Software Systems Failure and Success*. Boston, International Thomsen Computer Press.

Kling, R., and Jewett, T. (1994). *The Social Design of Worklife with Computers and Networks: An Open Natural Systems Perspective.* San Diego, Academic Press.

Kruchten, P. (2000). *The Rational Unified Process.* Reading, Addison Wesley Longman.

Kwon, T. H., and Zmud, R.W. (1987). Unifying the Fragmented Models of Information Systems Implementation. *Critical Issues in Information Systems Research.* In R. J. Boland and R. A. Hirscheim, ed., Chichester, Wiley: 227-251.

Leavitt, H. J. (1965). Applied Organizational Change in Industry: Structural, Technological and Humanistic Approaches. *Handbook of Organizations.* J. G. March. Chicago, Rand McNally.

Lehman, M. M. (1989). "Uncertainty in Computer Application and its Control Through the Engineering of Software." *Journal of Software Maintenance* 1(1).

Leonard-Barton, D., and Kraus, W. A. (1988). " Implementation as Mutual Adaptation of Technology and Organization." *Research Policy* 17:5: 251-267.

MacCormack, A. (2001). "Product-Development Practices That Work: How Internet Companies Build Software." *Sloan Management Review,* Winter: 75-84.

Marble, R. P. (2000). "Operationalising the implementation puzzle: an argument for eclecticism in research and practice." *European Journal of Information Systems* 9: 132-147.

Markus, M. L. (1983). "Power, Politics and MIS Implementation." *Communications of the ACM* 26:6: 430-444.

Markus, M. L., and Benjamin, R. (1997). "The Magic Bullet of IT-Enabled Transformation." *Sloan Mangement Review.* Winter: 55-68.

Markus, M. L., and Robey, D. (1988). "Information Technology and Organizational Change: Causal Structure in Theory and Research." *Management Science* 34:5: 583-598.

McCormick, M. (2001). "Programming Extremism." *Communications of the ACM* 44(6): 109-111.

Microsoft (1997). *Microsoft Solutions Framework and The Rational Process,* Microsoft, 2001. www.microsoft.com.

Microsoft (1998). *MSF Process White Paper,* Microsoft. 2001. www.microsoft.com/msf

Mitroff, I., and Linstone, H. (1993). *The Unbounded Mind. Breaking the Chains of Traditional Business Thinking.* New York, Oxford University Press.

Munkvold, B. (1998). Implementation of information technology for supporting collaboration in distributed organizations. *Dr.ing. thesis 1998:40,* NTNU, Trondheim, Norway.

Munkvold, B. E. (2001). Perspectives on IT and Organisational Change: Some Implications for ISD, in: *New Perspectives on Information Systems Development: Theory, Methods and Practice,* G. Harindranath et al. (eds.), Kluwer Academic, New York, USA,

Nambisan, S., R.Agarwal, M.Tanniru (1999). "Organizational Mechanisms for Enhancing User Innovation in Information Technology." *MIS Quarterly,* 23:3, 365-394.

OMG (2001). Object Management Group. 2001. www.omg.org.

Pare, G., Elam, J.J. (1997). *Using Case Study Research to Build Theories of IT Implementation.* IFIP TC8 WG 8.2, Philadelphia, Chapman & Hall, 542-568.

Prescott, M. B., Conger, S.A. (1995). "Information technology innovations: a classification by IT locus of impact and research approach." *Data Base for Advances in Information Systems,* 26(2-3): 20-41.

Rada, R. (2001). *Standardizing Management of Software Engineering Projects.* The 34th Hawaii International Conference on System Sciences, IEEE.

Rational (2002). Rational Success Stories. 2002. www.rational.com.

Robey, D.,M.C.Boudreau, M.C. (2000). Organizational Consequences of Information Technology: Dealing with Diversity in Empirical Research. *Framing the Domains of IT Management.* R. Zmud. Cincinnati, Pinnaflex: 51-64.

Royce, W. (1998). *Software Project Management.* Reading, Mass., Addison-Wesley Longman.

Simon, H. (1970). *The New Science of Management Decision.* Englewood Cliffs, Prentice Hall.

Sommerville, I. (2001). *Software Engineering.* Harlow, Pearson Education.

Suchman, L. A. (1987). *Plans and situated actions. The problem of human machine communication.* Cambridge, Cambridge University Press.

Walsham, G. (1993). *Interpreting Information Systems in Organizations.* Chichester., Wiley.

Walsham, G., Sahay, S. (1999). "GIS for District-Level Administration in India: Problems and Opportunities." *MIS Quarterly* 23(1): 39-66.

SOFTWARE DEVELOPMENT RISK MANAGEMENT SURVEY

Baiba Apine[*]

1. INTRODUCTION

Software development is rather complex process consisting of different activities. It is dependent on skills' level of different specialists as well as on usage of different technologies. One of the activities supporting software development process is risk management. Risk management requires knowledge and experience from people involved. This paper addresses software development risk management. The survey among software development experts was performed to find the risks, which are live to software developers in Latvia. This paper summarizes the results of the survey.

2. PROBLEM

There are two activities within risk management process requiring special expertise and experience: risk identification and finding activities for risk mitigation. The task was to identify software development process risks for software developers in Latvia and activities for risk mitigation.

3. BACKGROUND

3.1 Definition of Risk

The definition of risk is very simple: the possibility of loss, injury, disadvantage or destruction, as it is defined in Webster's dictionary[1].

---

[*] Baiba Apine, Riga Information Technology Institute, Kuldigas iela 45, LV-1083 Riga, Latvia

Information Systems Development: Advances in Methodologies, Components, and Management
Edited by Kirikova *et al.*, Kluwer Academic/Plenum Publishers, 2002

241

Risk related to information systems is defined in[2] as the potential that a given threat will exploit vulnerabilities of an asset or group of assets to cause loss of/or damage to the assets.

In the context of software engineering and development, risk can be defined as the possibility of suffering a diminished level of success within a software-dependent development program[3].

Risk is usually measured by a combination of impact and probability of occurrence [2]. Consider a sample, where a potential threat to software development project is leaving of the leading programmer at the peak of the development process. The probability of realizing of this threat is 25%. Let's try to calculate the impact of the threat. We have to arrange new leading programmer from the development team, it will cause the lagging behind the project schedule and paying penalties for $2000, hire new programmer and train him/her for $1200. The total impact of the threat is $3200. The risk is calculated as follow: 25% from $3200 gives $800. This is a quantitative risk assessment. Sometimes it is not possible to assess the impact of risk quantitatively. Then qualitative risk assessment is used, where probability and impact are assessed using terms like low, medium, high etc.

There are a lot of risk management methods, which define when and how to identify threats, how to assess and prioritize them, etc.

3.2 Risk Management Methods

Risk management is a practice with processes, methods and tools for managing risks in a project. It provides a disciplined environment for proactive decision making to assess continuously[4]:

1. What could go wrong (risks).
2. Determine which risks are important to deal with (prioritise risks).
3. Implement strategies to deal with those risks.

Risk management is used in the software development process if it is necessary[5]. Risk management process is iterative process, consisting of three basic activities:

1. Planing of the risk management[6, 5], when potential threats are identified, risk assessment method chosen, responsibility of risk assessment and monitoring assigned, frequency of reassessing risks defined etc.

The identification of potential threats is an activity, which requires expertise and experience. It could be said, that this is the state of the art.

2. Risk analysis, when the probability and impact of each threat is assessed, risks are prioritized according the results of the assessment and preventative, detective and corrective actions planned[6] (in [5] this activity is part of planning).

The most complex part of the risk analysis is identification of preventative, detective and corrective actions. It could be also said, that this is the state of the art requiring knowledge of management and experience.

3. Risk mitigation and monitoring, when developers and managers follow the activities for risk mitigation and responsible managers monitor continuously a situation with the potential threat[6, 5].

There are a lot of risk management methods available, based on standards, but concentrating more on some risk management aspects. For instance, RiskIt method, based on CMM, concentrating on clear and structured definition of risk[7], or method concentrating on maximum involvement of customer in planing of risk management process[8].

It is possible to choose any convenient method, but identification of threat and activities to prevent, detect or correct the risk (or mitigate) will require expertise and experience anyway.

The survey was organized to find out risks, which are actual to software developers in Latvia, and activities for risk mitigation.

4. METHOD USED FOR IDENTIFICATION OF RISKS

Expert polling method was used to find risks, which are important to software developers in Latvia. The "Delphi" method[9, 10] was chosen. "Delphi" is an iterative decision making method. The method has three steps:

1. Forming the group of respondents – the expert group in the terms of the method.
2. Forming of the questionnaire and spreading among the experts.
3. Analysis of the results.

The second and the third steps are performed iteratively, while group of experts has agreed on the topic.

4.1 Group of Experts

Group of experts should consist of 10 to 20 experts having the same level of expertise[10]. Thirteen experts were chosen. Experts where chosen from software development companies (87% of respondents) and from software development units serving other business units within the same company (13% of respondents). Each expert suited the following requirements to ensure the same level of expertise for all respondents within the group of experts.

1. Expert is currently working as software development project manager.
2. Expert has the position of project manager for at least 2 and no more than 8 years.
3. Expert has managed at least 2 software development projects.

4.2 Questionnaire

According to the method used for survey, the first step is to prepare the list of properties experts must agree on. Experts were asked to provide lists of risks having some impact on the software development process. After analysis of the lists of risks, the 12 major risks were highlighted (see **Table 1**).

Table 1. List of risks

No.	Risk
	Customer risks
1.	Lack of hardware on the customer side
2.	Difficult communication with customer
	Requirements risks
3.	Low quality of software requirements
4.	Unstable software requirements
	Project management risks
5.	Unrealistic schedules and budgets
6.	Weak project management
	Developers risks
7.	Software development environment bugs
8.	Lack of developers motivation
9.	Lack of hardware on developers side
10.	Change of qualified personal
11.	Lack of knowledge in software development technologies and environment
12.	Difficult communication among developers

Table 2. Risk relevance matrix

Frequency

Expert viewpoint	Rarely 1	2	3	4	5	6	7	8	9	10	11	Often 12
Low impact 1	1	1	1	1	2	2	2	2	3	3	3	3
2	1	1	1	1	2	2	2	2	3	3	3	3
3	1	1	1	1	2	2	2	2	3	3	3	3
4	1	1	1	1	2	2	2	2	3	3	3	3
5	4	4	4	4	5	5	5	5	6	6	6	6
6	4	4	4	4	5	5	5	5	6	6	6	6
7	4	4	4	4	5	5	5	5	6	6	6	6
8	4	4	4	4	5	5	5	5	6	6	6	6
9	7	7	7	7	8	8	8	8	9	9	9	9
10	7	7	7	7	8	8	8	8	9	9	9	9
11	7	7	7	7	8	8	8	8	9	9	9	9
Heavy impact 12	7	7	7	7	8	8	8	8	9	9	9	9

The next step was to put all the risks from the list in order of decreasing frequency (see Table 4 in "8 Appendix") and decreasing impact (see Table 5 in "8 Appendix").

Risk frequency and impact given by each expert was consolidated using risk frequency and impact matrix (see Table 2). This matrix is prepared using qualitative risk assessment, according to the project of cabinet law[11]. Consolidated frequency and impact forms the risk relevance. The final risk relevance is given in the Table 3.

Table 3. Risk relevance

Risks	Experts													Average relevance
	1	2	3	4	5	6	7	8	9	10	11	12	13	
Difficult communication among developers	2	1	8	2	5	1	5	8	1	5	5	4	5	4
Difficult communication with customer	8	9	9	9	9	9	9	3	1	2	6	9	5	7
Change of qualified personal	7	5	1	3	4	5	1	9	1	8	5	5	5	5
Lack of developers motivation	4	2	1	4	1	1	5	9	1	1	1	2	5	3
Lack of knowledge in software development technologies and environment	3	5	6	4	5	7	5	5	2	6	5	5	5	5
Lack of hardware on developers side	2	1	1	1	1	3	1	9	1	2	1	1	2	2
Software development environment bugs	4	4	5	6	5	2	1	9	7	5	5	5	5	5
Unrealistic schedules and budgets	9	6	9	8	9	9	9	6	6	9	9	8	9	8
Low quality of software requirements	6	8	9	7	9	5	9	4	8	9	8	9	8	8
Lack of hardware on the customer side	2	3	4	2	1	4	1	9	5	2	2	5	5	3
Unstable software requirements	6	9	5	9	9	9	9	4	8	6	9	9	8	8
Weak project management	7	7	2	5	2	5	5	1	5	9	5	5	5	5

Finally the concordance is calculated on the risk relevance according to the "Delphi" method. Concordance rate is between 0 and 1. If the rate is close to 1, it means that experts agree about the topic. If the concordance is closer to 0, then the survey is corrected, updated, the reasons for disagreement among experts are discussed and survey is spread among experts once more. This survey was spread twice and finally has concordance rate 0.91. It means that experts have agreed on risks actual for software developers.

If comparing software development risks identified by our experts and those found in different sources, we can see that mostly risks are the same: unstable or low quality software requirements, unrealistic schedules and budgets, lack of knowledge in software development technologies and environment, change of qualified personal[12, 13, 14, 5, 8].

Risks, which are listed in the expert list, but not often found in sources about software development risks, are:

1. Lack of appropriate hardware on the customer side as well as on the developer's side.
2. Difficult communication with customer as well as among developers.

5. RISK MITIGATION ACTIVITES

Experts were asked to provide activities for each risk mitigation.

5.1 Unstable software requirements

Unstable software requirements might be the king of the risks for software development as it is named in many sources (see above). Nevertheless some creep of requirements is normal during software development process. T. C. Jones comments, that monthly rate of change after the requirements are first identified runs from 1% to more than 3% per month during the subsequent design and coding stages is considered as normal[13].

To prevent creeping requirements, the effective action is contracting of a sliding scale making the implementation of changes financial disadvantageous later in the software development life cycle[13]. Only one expert approved this as a preventative action.

The most popular preventative action is the establishment of project change request board consisting of customers and developers. All experts mentioned this in the questionnaire. Change request board conduct meetings on regular basis. Changes approved by project change request board are implemented only.

Using the iterative software development life cycle with some administratively limited time period of "freezing requirements" is another preventative activity popular among the experts. Dealing with unstable software requirements is the main advantage of the iterative software development life cycle introduced by B. Boehm[15, 16].

5.2 Unrealistic schedules and budgets

The common problem in the software industry is that of intense but artificial schedule pressure applied to the programmers by their managers and customers[13, 17]. This is the significant risk mentioned by the experts as well.

There are four preventative actions proposed by the experts:

1. Use of formal methods for software development cost estimation before the development starts. There are a lot of formal software development cost and schedule estimation models available and the supporting tools. The most popular is COCOMO[18]. This as a preventative action is proposed by T. C. Jones[13] as well.
2. Use the advantages of technology (automated tools for configuration management, project management etc.).
3. Plan the software architecture so that it is possible to use previously developed and tested components.
4. Review of the software development plans, whether all is correct.

The corrective actions proposed were:

1. Negotiate the schedule with customer or executives in order to set the priorities for deliverables or extend the schedule.
2. Increase the workload for experienced team members, which are able to generate original solutions. This gives the result rather quickly, but is not a solution for long term[17].
3. Change inexperienced team members with the experienced ones working in the similar problem area. This is the alternative to adding the extra staff, giving no expected results[19].

5.3 Difficult communication with customer

Preventative actions:

1. Regular meetings on the project management level. It would be better to conduct these meetings on the customer site. If it isn't possible to meet, customer has to be informed about project development by phone or via e-mail. All the experts agreed, that this is the most effective preventative action.
2. More than half of the experts assume that cause of the communication problems is customers' lack of knowledge about software development life cycle. In this case the only action must be taken is education of the customer.

The only corrective action provided was to change of the contact person on the customer side to the person having more procuration in the customer's company and knowing the business area.

5.4 Low quality of software requirements

Preventative actions:

1. Don't cut time for software requirements specification. This job must end with mutually agreed (signed) software requirements specification.
2. Build prototype of the system under development.

The only corrective action provided was to find out more about requirements informally. It could be done by finding informal requirements pioneers on the customer side as well as on the developers' side.

5.5 Other Risks and Mitigation Activities

Lack of hardware on the customer side. Lack of hardware on customer side is one of the risks, which is specific for software developers in Latvia. This risk must be considered carefully during planning, hardware specification must be provided to customer as early as possible and these aspects must be negotiated carefully. All the experts agreed, that this is the most effective preventative action.

If corrective action is necessary, there are two possibilities:

1. Bye or rent hardware specified. This is the most effective corrective action giving the results immediately.

2. Optimization of the software. Half of the experts agreed that this would help. Another half said that this would never help and this was rather risky way, because new bugs would be introduced during optimization.

Software development environment bugs. Preventative actions:

1. Don't use software development environments developers are not familiar with.
2. Don't use new environment versions entering the market before the benchmarking information is available.
3. Establish company wide benchmarking bulletin and motivate developers post there information about problems highlighted during the software development process.
4. Building and using unified components where it is possible.

There are two corrective actions recommended by experts:

1. Find roundabouts on Internet or contact vendors.
2. Change the software development environment and train the developers in using new environment.

Weak project management. The first activity coming in mind is change of the project manager. Experts are rather cautious about change, saying that it may give the expected result as well as aggravate the situation in project. It is the last thing should be done. The other corrective actions proposed are:

Provide the experienced assistant to the project manager covering the areas in the project management field, where project manager is not so successful.

Encourage and assist to project manager in deeper analysis of the situation in the project helping to find out the most painful areas in development process and concentrate on them.

Lack of developers motivation. All the experts agree, that material benefits are important, but it is not enough. It is important, that developers see the result of their job.

Lack of hardware on developers side. This is an issue of planning. The only way is to purchase, rent or use customers' equipment.

Change of qualified personal. Preventative action proposed by experts is:

1. Assign responsibility about development of software component, module, function etc. to two developers always.
2. Document everything during software development process, even if customer doesn't request it.

These two as preventative actions of change of the personal is also recommended in [2].

Lack of knowledge in software development technologies and environment. The preventative action is developers' training. All experts mentioned this. There were two groups within experts' group regarding corrective actions:

1. More than half of experts suggested involving of consultants – software developers, who are able to communicate with the rest of the group and assist during the development process.
2. Another group didn't advise involvement of external experts. They suggested finding a developer or group of developers within software development team, who are able to self-education.

There were three experts cautioning of developers, who claimed of being pioneers.

Difficult communication among developers. This risk has very high probability in the case, when development of some software components is outsourced to the third party. The preventative actions suggested by experts are as follows:

1. Regular meetings of the developers, weekly or twice a week, discussing problems during software development process.
2. Organise small development teams.
3. Define responsibilities.
4. Organize off-hour meetings, sports etc.

6. CONCLUSIONS

Software development practitioners have agreed, that software development risk management is important activity. Top twelve risks are identified and mitigation activities are highlighted. Software development managers could use the identified risks as a checklist for initial risk analysis. The preventative and corrective actions proposed by experts could be used as guidelines for planning software development risk mitigation activities.

There are two very important steps in software development risk management, which could be considered as a state of art:

1. Identification of risk.
2. Finding the appropriate preventative or corrective action.

Communication among software developers as well as communication between customer and developer is very important for successful software development. The further research is needed to provide effective methods for accumulation and appliance of software managers' experience in risk identification and mitigation.

7. REFERENCES

1. G.,P. Babcock, Editor. Webster's Third New International Dictionary: Unabridged (MA: Merrian-Webster, Springfield , 1981).
2. Information Systems Audit and Control Association (2002 CISA Review Manual, 2002)
3. Software Engineering Institute. *"The SEI Approach to Managing Software Technical Risks."* Bridge (October 1992), p.19-21
4. Carnegie Mellon Software Engineering Institute, Software Engineering Risk Management FAQ. (21st of March, 2001); http://www.sei.cmu.edu/publications.
5. C.Mark, B.Curtis, M.B.Chrissis, and C.V. Weber, *Capability Maturity Model for Software, Version 1.1,* (Software Engineering Institute, CMU/SEI-93-TR-24, February (1993)).

6.   IEEE P1540/D11.0, "*Draft Standard for Software Life Cycle Processes – Risk Management*", (IEEE Standards Department, 2000).
7.   R.Basili, J.Kontio, "Riskit: Increasing Confidence in Risk Management" (21st of April 2001); http://satc.gsfc.nasa.gov/support.
8.   B.W. Boehm, "*Software Risk Management: Principles and Practices*" (IEEE Software, Jan 1991), pp. 32-41.
9.   Л.В.Ницецкий, Л.П.Новицкий "*Применение методов экспертного опроса для оценки качества диалоговых обучающих систем*". Методы и средства кибернетики в управлении учебным процессом высшей школы. Сборник научных трудов, (Рига РПИ, 1986).
10. "*Теория прогнозирования и принятия решений*". Под ред. С.А. Саркисяна. (М.: Высшая школа, 1977).
11. Informācijas sistēmu riska analīzes metodika (23rd of February 2002); http://www.lddk.lv,.
12. K.Lockyer, J.Gordon, *Project Management and Project Network Techniques* (Bell and Bain Ltd, 1996) pp. 49-51.
13. T. Capers Jones. Estimating Software Costs._McGraw-Hill, USA, 1998.
14. B.Hetzel, Making Software Measurement Work._John Wiley & Sons, Inc., 1993, 290 p.
15. B. Boehm "*A Spiral Model of Software Development and Enhancement*" (IEEE Computer, vol.21, #5, May 1988), pp 61-72.
16. G. Holt, "*Software Risk Management – the Practical Approach*" (Mei Technology Corporation, 2000, #2)
17. E.Yourdon, *DEATH MARCH. The complete Software Developer's Guide to Surviving 'Mission Impossible' Projects (*Prentice Hall, 1997) 218 p.
18. C.Abts, B.Boehm, B.Clark, S.Devnani-Chulani. *COCOMO II Model Definition Manual_*(University of Southern California) 68 p.
19. R. S. Pressmann, *Software engineering, a practicioner's approach* (McGraw-Hill, 1992).

8.   APPENDIX

Table 4. Risks ordered by frequency (12 - the most frequently, 1 - the least frequently)

Risk	Expert													Average
	1	2	3	4	5	6	7	8	9	10	11	12	13	
Difficult communication among developers	8	3	5	6	5	1	7	8	3	6	5	4	5	5
Difficult communication with customer	6	12	10	10	10	9	10	11	2	6	9	10	8	9
Change of qualified personal	4	5	4	9	4	6	4	12	1	8	6	6	6	6
Lack of developers motivation	3	6	3	2	3	2	5	12	3	4	4	5	5	4
Lack of knowledge in software development technologies and environment	9	8	9	4	7	3	8	8	5	9	8	7	7	7
Lack of hardware on developers side	5	1	1	1	1	11	1	12	3	5	4	4	6	4
Software development environment bugs	1	4	7	11	8	5	3	12	1	7	6	7	6	6
Unrealistic schedules and budgets	10	11	11	8	9	12	9	11	9	10	10	7	9	10
Low quality of software requirements	12	7	12	3	12	7	12	3	6	9	8	11	7	8
Lack of hardware on the customer side	7	10	2	7	2	4	2	12	7	5	6	6	6	6
Unstable software requirements	11	9	8	12	11	10	11	3	6	9	9	9	8	9
Weak project management	2	2	6	5	6	8	6	3	6	10	5	5	7	5

Table 5. Risks ordered by impact (12 - heavy impact, 1 - very little impact)

Risks	Expert													Average
	1	2	3	4	5	6	7	8	9	10	11	12	13	
Difficult communication among developers	1	2	9	4	8	2	7	10	3	5	5	6	6	5
Difficult communication with customer	10	12	12	9	10	12	10	3	1	4	8	10	8	8
Change of qualified personal	9	6	3	3	7	5	4	12	3	10	6	6	6	6
Lack of developers motivation	5	3	1	6	1	3	5	12	3	1	4	3	5	4
Lack of knowledge in software development technologies and environment	4	5	7	8	6	9	8	8	3	7	7	7	7	7
Lack of hardware on developers side	3	4	2	1	2	4	1	12	3	3	4	4	4	4
Software development environment bugs	7	8	5	7	5	1	3	12	11	6	7	7	6	7
Unrealistic schedules and budgets	12	7	11	10	9	10	9	8	5	12	9	9	10	9
Low quality of software requirements	8	11	10	12	12	7	12	8	10	9	10	11	10	10
Lack of hardware on the customer side	2	1	8	2	3	8	2	11	5	2	4	5	5	4
Unstable software requirements	6	10	6	11	11	11	11	8	10	8	9	10	9	9
Weak project management	11	9	4	5	4	6	6	3	7	11	7	6	6	7

RESEARCH NOTES ON DEVELOPING A FORMAL ORGANIZATIONAL LANGUAGE

Panagiotis Kanellis, Dimitris Stamoulis, Panagiotis Makrigiannis and Drakoulis Martakos[*]

1. INTRODUCTION

The emergence of the Information Systems (IS) field and the delineation of its epistemological boundaries are tied to the strong social element that involves the study of Information Technology (IT) in organizations (Land, 1983; Friedman, 1989; Angell, 1991). With the needs of the so-called information-age organization moving further away from those that characterized the industrial organization of the 20th century, research is focusing on where and how new forms of information handling are conceived, planned and implemented (Hammer, 1990; Venkatraman, 1991). The magnitude of the effects that the process of information handling has on the structure and behaviour of organizations is evidenced through its influence on the development of organizational theories (Drucker, 1988; Handy, 1995) and practice (Porter, 1985; Hammer, 1993).

The traditional understanding of the universe of discourse of an IS as mechanical, and consequently suitable for an IS resembling a programmed automaton following a defined set of instructions (Walsham, 1991), does not answer for social interaction within the organization. Nor can the "information systems organization" as perceived in Gallivan (1994), be a closed subset of the organization. In fact, as it is illustrated in Khalil (1994) there are strong links with other aspects of organizational life and structure. Emery (1960), Pugh (1997) and Scott Morton (1991), point out that technology in general has structural effects on organizations, while Cash (1994) identifies a number of changes enabled by IT that fundamentally alter organizational purpose, shape and practice.

The ramification of the above, is that the polymorphic nature of the IT-organization relationship gives rise to a complex web of social phenomena that *Object-Oriented Analysis* (Coad, 1990; Rubin, 1992), *Design* (Booch, 1991), and *Modeling* (Peckham, 1988; Junglaus, 1996) do not cater for. Similarly, *Task* or *Task and Dynamic Modeling* approaches (Mannarino, 1997; Schreiber, 1993) are not entirely focused on the organization.

[*] P. Kanellis, D. Stamoulis, P. Makrigiannis and D. Martakos, Department of Informatics & Telecommunications, National & Kapodistrian University of Athens, Panepistimioupolis, Athens 157 84, Greece.

Information Systems Development: Advances in Methodologies, Components, and Management
Edited by Kirikova *et al.*, Kluwer Academic/Plenum Publishers, 2002

253

An IS cannot develop oblivious to the structural changes that the technologies which will incorporate, cause to the parental organization. Even more so, since change in the parental organization affecting purpose, shape and practice, alter the requirements of the IS itself. On the other hand, the IS requirements do not always take into consideration the business context, i.e. the organization, causing failures to technically impeccable IS (Luff, 1993; Goguen, 1993). A missing link has been identified between the IS requirements methodologies and those for organizational modelling (Stamoulis, 1999), which is partly due to the lack of formal representations regarding the organization. To address this need, conceptual modelling in information systems development has been pursued with a view to create "an enterprise model for the purpose of designing the information system" (Wand, 1997). A concise understanding of the organization is, thus, essential; because a social process of communication, learning and negotiation is involved in IS design and development, as Walsham (1993) explains.

Even before the requirements elicitation phase, organizational "shake-up" projects aim at preparing the ground for the introduction of IS, along the lines of the "don't automate, obliterate" principle (Hammer, 1990). Therefore, frameworks for modelling and analysing organizations have been developed to support business process re-engineering (Yu, 1996).

Finally, a third motivation for studying the business context of an IS, i.e. the organization, comes from the pure IT side. "Legacy systems that support enterprise functions were created independently, consequently do not share the same enterprise models. We call this the Correspondence Problem (Fox, 1997)".

All these research perspectives attempt to shed some light on the same core issue, which is the quest for appropriate organization models. Enterprise modelling has emerged as a new vein of scientific thought addressing the aforementioned particular need. An overview of the most known enterprise models is discussed by Fox (1998). The most modern approach to enterprise modelling is based on an ontology, which is "a formal description of entities and their properties, a shared terminology for the objects of interest in the domain, along with definitions for the meaning of each of these terms (Fox, 1997)". "The development of ontologies for Enterprise Models is more recent. There are few projects whose scope of modelling is rather broad" (ibid). The ontological understanding of an organization can be elicited through the formation and application of a formal organizational language. The first step towards the construction of such a language is the conception of a representation model for the organization.

As a prerequisite, this representation model should be able to model and follow structural change during the development and the application of an IS. Jones (1992) and Jones (1993) have highlighted the benefits of working with organizations in a formal way. Identifying this as a need, this paper presents some research notes on the development of a formal organizational language for public and non-profit organizations. It is believed that through its use a better understanding of the territory, i.e. the organization, can result and a more efficient communication channel between the enabling personnel in any IS development project can be achieved.

The paper begins with a presentation of the basic assumptions and concepts behind the logical formulation of the representation model. Those are applied to a number of possible instances of operation of an organization, and are then generalized so that the components of the model can be defined in such a way that abstraction is made possible. In addition, the model is described as a normative system in terms of its components, which are formally defined. In the third section the social context is modelled through the introduction of a second level of logic so that action and structure duality is taken into consideration. This is deemed necessary for the representation to include the social complexity that characterizes organizational relationships. A way of validating the ontological foundations of our model is presented in section four. In the final section, some concluding remarks and the following steps in this ongoing research are presented.

2. ASSUMPTIONS AND GENERAL CONCEPTS

On an epistemological note, our point of view adheres to *pluralism* using a classification proposed by Mowshowitcz (1981), and our conception of social life within an organization would fall under *segmented institutionalism* according to Kling's (1980) distinction. As stated before, the proposed model described herein aims at the elicitation of a representation of a chosen public and non-profit organization, in order to enhance our contextual understanding in relation to IS analysis and design.

For the purpose of this paper, an "organization" is considered to be an institution, agency or any form of alliance of people engaged in the provision of services, legally bound by law and/or statutes and designated to serve the general public as opposed to specific groups of interest. Any such public organization (O) with a purpose of offering a range of services to the society in general might have a distributed structure overseen by a coordinating committee and facing legal or operational problems. It may include a variety of agencies and/or sub-organizations. In addition it may have established a set of performance indicators in place, and make use of a number of IS. Any attempt for the definition of a formal representation language must cater for this multiplicity of parameters and issues. Furthermore, it must enable an organizational analysis to a level of abstraction that permits a layered development of the representation model. In general the model should be coherent as it is meant to be a reasoning tool and should consider IS as social systems that are technically supported and on occasion implemented (Crap, 1995; Land, 1983).

Since the goals and ways of operation of a public organization is the provision of a service, the basic operation is seen as a response (R) to a request for the rendering of a service. If we see the client (C) as the source of the request (r), then the operation is described in Figure 1a.

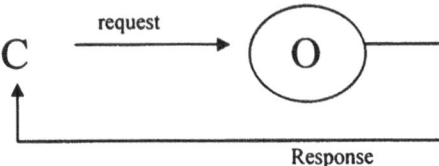

Figure 1a. Requesting the Provision of a Service

It is assumed that an immediate response to the request is the rule rather than the exemption. In most cases there is a pre-defined list of actions to be taken (out of a set of allowed actions) in order for the response to be issued. If the organization can be described as a couple consisting of structure (**S**) and a set of actions (**SoA**) through which it provides the response, then this is depicted in Figure 1b. **C** can be any citizen, a group of citizens, or another organizational entity that by law, or by **O**' s statutes has or is granted the right to request for a service.

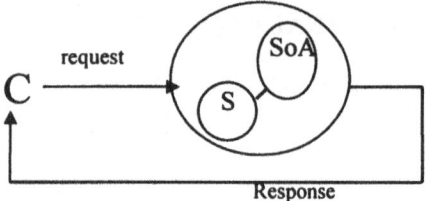

Figure 1b. The Provision of the Service

3. EXTENDING THE REPRESENTATION MODEL

Any organization can be seen as consisting of a number (n) of sub-organizations or agencies A_i, i = 1,2, ..., n with n ∈ **N*** that consist themselves of couples of a structure and a set of allowed actions as depicted in Figure 2. Simply put

$$O = A_1 + A_2 + ... + A_n \qquad (1),$$

with Eq. (1) taken to mean that **S** is constructed from S_i , i = 1, ..., n when a relation "rights" that is described in the next section of the paper and connects the levels of the representation model is applied to the set $\Sigma = \{S_i , i = 1, ..., n\}$.

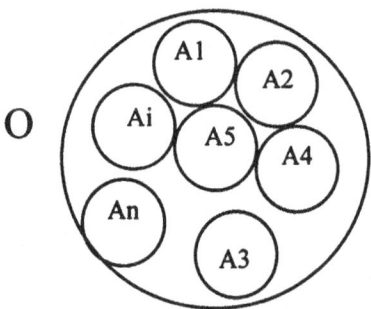

Figure 2. Representing the Organization

Hence, $S_o A \subseteq U \{ S_o A_i , i = 1, ... , n\}.$ \qquad (2)

As depicted in the figure 3, each of the agencies A_i operates in exactly the same manner as **O**.

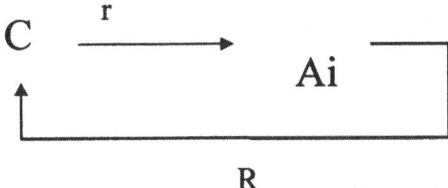

Figure 3. An Agency's Mode of Operation

If C_O the possible set of C's for the organization and C_A the possible set of C's for the agency, then we have:

$$C \in C_A \equiv \{(UAj) \cup C_O , j \neq i, j = 1,...,n\} \qquad (3)$$

Equation (3) means that Customer for Agency Ai can be any of the rest of the Agencies or any of the customers of the organization **O**. It also follows for the request r and the response R respectively that:

$$r \in I'=I_i \cup I_{ij} ,$$

where $I_i \subset I, I=\{\cup r_i, i = 1, ... , n\}$

and $I_{ij}=\{$requests internally accepted by A_i as valid ones, made by Agencies $A_j\}$,

  $R \in B' = B_i \cup B_{ij}$ (similarly with I),

meaning that the requests and responses may be those of the Agency A_i or those of the Agencies A_j that may act as internal customers to Agency A_i.

Thus the process can be described as in figure 4.

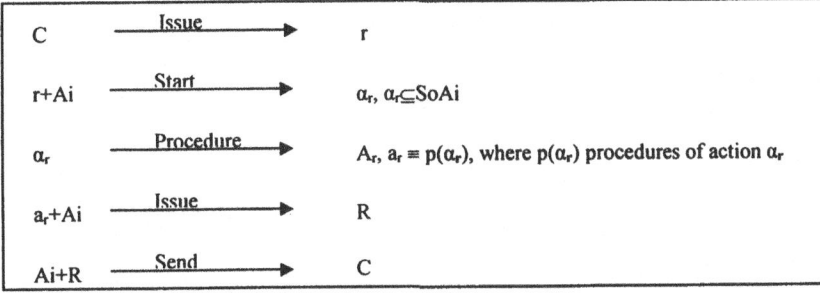

Figure 4. A Layered Representation of the Provision of a Service

We assume that A_i cannot perform an action that is outside the actions that **O** is allowed to perform. Practically, *Start Action* (**Start** →), *Procedure Data/Message* (**Procedure** →), and *Issue Response* (**Issue** →) are special actions and fall under the responsibilities of the S_oA. The explicit description of the intermediate steps between **Issue** and **Send** facilitates an understanding of the organization's environment by the representation of the client base (both internal and external), together with their operations and communications sets. For practical purposes, the process retains the original form as described before and is depicted in the following figure 5.

X is supposed to be a structure, which has taken any actions that is allowed, and regards as necessary for the production of **R** in order to provide a response to C_O's request. Here C is understood to belong to the set of possible clients at X's level, irrespective of whether this is C_O or C_A, or indeed any other set of organization clients. In the same way we can descend down from this level to any number of smaller administrative units, and from there to specific employee posts, where – like an agency – the employees occupying this position have their own methods of working within or even outside any existing formal rules and regulations. Representation at this level is deemed necessary considering the commonly observed phenomenon in public organizations, where due to the scarcity of funds, a single employee may "wear various hats". The possibility where a number of different roles are synthesized into a single one, which is then assigned to an employee, is also a common occurrence.

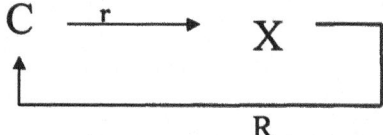

Figure 5. A Structure's Mode of Operation

It follows from the above that the last level in our model should be the **role (L)**, which in this case would consist of two sets, those of methods and of possible actions. The basic representation scheme is applied and is depicted in Figure 6 with **r**, **R** and the other generalizations made previously remaining the same.

In an analogy to Eq. (1),

$$A_i = L_1 + L_2 + \ldots + L_m, \qquad y = 1,\ldots,m, \ m \in N^* \tag{4}$$

Similarly to Eq. (3),

$$C \in C_L \equiv \{(UL_k) \ U \ (UA_j) \ U \ C_O \ , j \neq i, \ k \neq y, \ y = 1,\ldots,m\}. \tag{5}$$

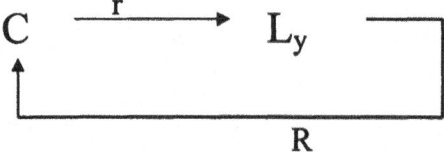

Figure 6. Representation of the role (L)

Up to this point we have not addressed the following issues:

(a) the nature of the «rights» relationship that is mentioned before and connects the levels of our representation model
(b) the issues of work coordination and the management of resources
(c) the issues of data acquisition, storage and maintenance, related to the internal processes of the organization
(d) the issues arising from a possible distributed nature of a particular organization (Dettmer, 1995; Von Simson, 1990).

By addressing the above points, we start to build upon the representation model and move towards a formal language.

If we are to ignore (d) for the moment, assuming that those distribution issues will be addressed by the formal representation of L, our attention will focus on (a), (b) and (c). Considering the latter, any information handling operation can be included in the set of actions that comprise a specific role L, and similarly in every other representation layer. Here, application software and hardware resources that are used by the employee at present (or as future requirements), can be included in the representation. In a similar manner, issues concerning the coordination of work and the management of resources, which should be the responsibility of designated persons, can be included into the role's (L) representation as illustrated in the next section of the paper.

Finally let us consider the "rights" relationship among roles (e.g. from L_1 to L_2) that exists in the set of all the roles in an organization and expresses the rightful action of giving an order or the wish to delegate a task, and define the interrelationships between roles. Although we have represented the structure of an organization, its functional aspects, i.e. the ways through which its goals are attained and its mission is carried out, have not been addressed. Let us consider the three last steps of the "O" analysis collectively and let us name "job" the actual response of the organization to the original request. To describe a job at the role level, we define it as one consisting of a set of activities which are a series of actions from the set linked to a role with a purpose to monitor and control the execution of the activity (and thus for the actions). This removes the need for separate action considerations and could result in the simplification of the model, but we will return to this later in order to serve other purposes first.

We call a *resource* the normal generalization of a role that can include activities, application software and objects, hardware, etc. In reality, only roles have rights, though these rights can be targeted on resources in general. So roles are now perceived as a subset of structure and rights are now relationships from the set of roles in the organization to the set of resources [see (Wieringa, 1991; Crap, 1995)]. Consequently, a role is structured as follows with {} designating a list of optional entries:

Role:
- *name*
- *resources / right: {(resource, right)}*
- *activities: {(activity, ^role, action)}*
- *policies: {(goals), (regulations)}*
- *role's goals: {(activity, ^role, measurements)}*
- *performance: {(role's goals), (policies)}*

Regulations can be any subset of the regulation set adopted by the organization and derived from law, statutes or permanent internal directives championed by people with roles that allow them to do so. *Goals* are projections of the organization goals, and perhaps reactions to internal imperative requests.

The question rising here is how the *goals* and *regulations* define the way an activity has to be carried out. Since a role, in our model, is not yet considered as a subject of action, the *how* is located a step further. The nature of the problem seems to reside within the nature of *regulations*. Walsham (1993) insists that total control is impossible. That doesn't only apply to strategy, but also in the legislation and regulations themselves. Jones (1992), point out the difference between the ideal and the actual. If we consider regulations to be mandatory then we have developed a model to formally describe the ideal of an organization based on its intended nature. Experience however, dictates that actual organizations do not behave like that. Factors like lack of funds, difficulties in acquiring the necessary data, seasonal strenuous workloads, problematic communication channels or clerk malpractice intervene and permeate its formal structure and everyday operation. Still, what we have presented up to this point is a system of norms prescribing how agents ought to behave, specifying how they are permitted to behave, and what their rights are. In short, considering agents to be human individuals (or sets of human individuals) attached to roles and/or other resources, we have a *normative system* of a hybrid nature. In fact, a rational consideration of the following observations could further support our point of view. Namely, the regulations that govern an organization are:

- formulations of norms, designed to regulate the behaviour of individuals (and institutions)
- "open texture", meaning that these regulations cannot be said to be fully explained in terms
- only a part of a "seamless web" of regulations governing the public sector.

In addition to the above, organizations exist in a wider context of laws, principles and other organizations; particular cases may arise, not explicitly mentioned in the statutes, in which impromptu decisions will have to be made by the administrators of the organizations; and there exist provisions for change in the regulations, e.g. decision on difficult cases, hear appeals against allegedly unfair treatment (parliament, administrational associate, court judges, boards of directors). That of course applies for both constitutional (set by law or statutes), and operational (set by internal regulating structures and superiors as far as a role is concerned) rules and regulations. An important point to note is that a descriptive model as ours actually produces a piece of regulation by itself. This is directly derived from legislation using a given number of assumptions.

4. POSSIBLE BENEFITS FROM MOVING UP TO A TWO-LEVEL LOGIC

Up to this point, our model is not addressing the issues that emerge from the duality of action and structure. In order for an actual organization to be fully described, the following are necessary:

$\Rightarrow$ To include in the set of regulations, (up to present approached as mandatory), a number of additional regulations describing how things can be done (permissive rules). It is not uncommon within an organization for a certain agent to be allowed to operate in a given way without the circumstances under which he should do so be clarified in advance (Sergot, 1996). Moreover, in a given set of circumstances an agent may be permitted to act or not in a variety of ways.

⇒ To add to our model a series of practices for the agents that depict the organizational culture on one hand (the set of assumptions and views that are shared from a number of people in the organization), and the way things are done (habits). An example for habit is in the case of multiple regulations applying in a given set of circumstances, when an agent is usually choosing the same each time.

So in this perspective, an agent can be simply defined as follows:

Agent
o   *name*:
o   *role*:
    ...
o   *practices: {(assumptions and views), (habits)}.*

In fact at this point we could simplify the role structure by including pairs of the (resource, right) type in the permissive subset of regulations. In this case we could let the *resources/right* entry to be changed to *available resources,* which would then be a set of resources. Of course, this might not be necessary.

What is important is that our model so far depicts *Content* (from *organization* to *role* and *resource*), and *Social Process* (agent level). If we widen the agent structure with a relations entry of the form

Agent
o   *name:*
o   *role:*
    ...
o   *practices: {(assumptions and views), (habits)}*
o   *relations { (activity, ^agent) }*

that would depict the informal relations among persons that are not included in their formal relationship. Thus we have modelled *Social Context* too, and politics of interpersonal relationships that are important in social surroundings can be accommodated into the *relations* entry. Action and structure duality is also taken into consideration and interpretative schemes are included.

Jones (1992) notes that the difference between the ideal and the actual is suggested to be the realm of deontic logic. We can reason that a deontic logic able to handle both a definitional and a normative component of a model would suffice for the picturing of our model of the organization, its agencies and its roles, by assigning subsets of the deontic logic to each one of them. Extending that to habits (especially the normative component), we can use another logic, of action this time, to describe the organization culture and subculture. This is not an unusual technique. For example, the TROLL specification language uses sub-languages to describe the static properties, the behaviour, and the evolution over time of objects (Wieringa, 1993; Gogolla, 1993; Jungclaus, 1996).

5. VALIDATION BY MEANS OF ONTOLOGICAL COMPETENCY

Among the criteria that have been suggested for evaluating enterprise models, the most important is competency (Gruninger, 1994). Simply put, competency assesses the capability of an ontological representation to answer a set queries, "that are in the form of questions that an ontology must be able to answer" (Gruninger, 1996). Fox (1996) listed a set of competency questions for the organization ontology. This list of competency questions has been copied below, and the answers explain the relevant aspects of our proposed representation model, which, in essence, describes he constructs of a formal language. Answering those competency questions does not mean that our organizational ontology is validated. Instead, the answers provided below simply guarantee that our organizational ontology has all the necessary provisions to cater for these competency requirements. When operationalized, the formal organizational language will be unquestionably validated.

Structure

* Structure of organization. How is decomposed into units?
  The Organization consists of sub-organizations or Agencies, see Eq. (1).
* What are the members of a particular unit of the organization?
  Each Agency has a Structure and a Set of Actions.
* What positions exits in the unit?
  Structural decomposition of an Agency leads to roles.
* What position does person X occupy?
  A Role is played by an Agent.
* Who must person X communicate with?
  Within a Set of Actions SoA$_i$, at any level of the organizational representation, Role's entry: {(activity, ^role, action)} in conjunction with Agent's Role specify who communicates with whom.
* What kinds of information does person X communicate?
  Since information is regarded as a resource, the Role's entry: {(resource, right)} can describe the information which a person has the power to manipulate, e.g. to communicate.
* Who does X report to?
  The "rights" relationship among the roles at a particular layer of the representation model defines the line of reporting.

Behaviour

* What are the goals of: the unit? the position? person X?
  The (ideal) goals and regulations are modelled at the Role construct. The (ideal) goals of an Agency is the union of all the goals of the roles comprising the Agency.
* What activities must: person X performs? a particular position perform?
  Depending on action, an Agent must perform some action according to his/her Role specified activities: {(activity, ^role, action)}.
* Is it possible for an agent to perform an activity in some situation?
  If and only if that activity handles resources over which an Agent is entitled to by means of the rights given to his/her Role: resources/right: {(resource, right)}

Authority, Empowerment, Commitment

- What resources does the person have authority to assign?

  If an Agent has the appropriate rights over a resource and the "rights" relationship can be applied between his/her Role and the assignee, then resources can be assigned.

- What activities may a person execute without explicit permission?

  Those falling into the category of practices, modelled as: Agent's practices: {(assumptions and views), (habits)}. S/he could also make use of interpersonal relationships: relations {(activity, ^agent)}.

- Is an agent allowed to perform an activity in some situation?

  Some answer applies to previous question, for activities without explicit permission. Explicitly permitted situations are described by {(resources, right)} and resource belonging to a situation within permitted Set Of Actions.

- What permission is needed to perform a particular activity?

  Those applying on resources being handled by the Activity, as restricted by Role's policies: {(goals), (regulations)}

- What goals is person X committed to achieving?

  Those modelled by Role's performance: {(role's goals), (policies)}, which fall within the organization's goals as diffused and deployed to the various Roles through policies.

- Is a goal achievable by an agent given its current commitments and the commitments of other agents?

  A Role's Goals are linked to activities that are measured. The activities handle resources, including time. Therefore, Agent's resources can be assessed for adequacy.

- If a goal is unachievable for a given set of agents, how can they be empowered to be capable of performing the activities to achieve the goal? That is, how can the constraints defining empowerment for the agents be modified so as to be able to achieve the goal?

  Through the "rights" relationship among Roles, delegation of power and responsibility can take place.

- What authority constraints are necessary among a set of agents in order to achieve a goal?

  Authority constraints among Agents are the responsibility of the "rights" relationship among Roles.

Goal Achievement

- What goals are solitarily unachievable for a given agent?

  Those not described by a Role's Goals and not fall into the category of an Agent's practices.

- If a goal is solitarily unachievable for a given agent, what agents are required to assist the agent in achieving the goal?

  The need for assistance means that a goal is jointly achieved; therefore, parts of the activities leading to the achievement of the goal must lie with the other Agents' Role activity: {(activity, ^role, action)}. Those are the required Agents.

- What goals are achievable by an agent given the effects of activities that other

agents are capable of performing?
A rationale analogous to the previous one applies here, too.

6. CONCLUSIONS

IS are in essence social systems that cannot be developed independently of their context, i.e. the organization. Recognizing this, we presented a model that its use aims to achieve a better understanding of the organization these systems are intending to serve. We began by presenting the basic assumptions and concepts behind the logical formulation of the representation model. Those were then extended and generalized so that abstraction was made possible. After the description of the model as a normative system in terms of its components, a way was described for representing the social context. Using a two level logic, we catered for the possibility of any issues arising from the duality of action and structure. The validity of the ontology underlying the proposed representation model of an organization is cross-checked with an ontological competency check-list. The use of the model in practice could result in the formulation and provision of coordination and control monitoring facilities in an organization.

We believe that the representation model as described herein has the following characteristics:

> - is general and simple enough to be of use to the practitioner
> - it supports Walsham's (1993) synthesized analytical framework for organizational change
> - is modal enough to map change at any organization level and any extent
> - is formal enough to be handled to test reorganization efforts
> - it enables the interpretation of existing organizational problems and issues through the acquisition of supporting data
> - pictures organizational distribution in a clear and concise way (Galbraith, 1973)
> - includes IT resources at an early stage of the representation, considering them as integrated components of an organization's structure
> - can be used as a tool to aid the conduct of a research effort that falls under the *constructive* and *idiographic* research categories (Iivari,1991)

The computation representation is the ultimate goal of almost all enterprise modelling approaches. Our representation model lies the foundations of a formal organizational language, whose key programming constructs will be the main concepts of the model, such as Role, Agent, Rights relationship etc. that have conceptually been described in this paper. That formal organizational language is readily computational. Therefore, the research notes presented in this paper do not deal with the computational issues. The computability of the ontologies used for enterprise modelling is most usually based on first order predicate calculus, which can be directly translated into Prolog, or other expert systems / artificial intelligence programming languages. Although it is premature to discuss about how the proposed organizational language can be programmed, it seems that object oriented programming languages based on description logics will be the way forward of this research. "Description logics or terminological representation systems are a class of languages that provide an object-oriented format for structured descriptions with associated declarative semantics. These languages are able to support reasoning with definitional and descriptive knowledge" (Fox, 1996), which is the case of our research. Next steps of our research will deal with the translation to a programming language.

Our ongoing research continues to refine the basic model explained in this paper. Relevant feed for thought in the way of expressing the *"rights"* relationships has been found in the ideas of Makinson (1986). Currently we are working on the "rights" relationship, which has only partially and indirectly been described. Moreover, as this model caters for change through its levels in a natural way, it can be further simplified if a logic of action can be embedded so that it substitutes the *(activity, ^role)* system, and if *resources/rights* be considered as a set of local axioms or propositions. Addressing these additional requirements is the immediate next goal of our research.

7. REFERENCES

Angel, I., and Smithson, S., 1991, *Information Systems Management: Opportunities and Risks,* Macmillan Press, London, UK.

Booch, G., 1991, *Object Oriented Design with Applications,* Benjamin/Cummings, Redwood City, CA.

Cash, J.I., Eccles, R.G., Nohira, N., and Nolan, R.L., 1994, *Building the Information-Age Organization: Structure, Control and Information Technology,* Irwin, Homewood, Ill.

Coad, P., and Yourdon, E., 1990, *Object-Oriented Analysis,* Yourdon Press, Englewood Cliffs, NJ.

Crap, H., and Haas, J., 1995, Organizational modelling in distributed corporations, in: *Proceedings of the 3rd European Conference on Information Systems ECIS '95,* G. J. Doukidis et. al., ed., Athens, Greece.

Dettmer, R., 1995, Anyhow, anywhere-the rise of open distributed processing, *IEEE Review,* January.

Drucker, P., 1988, The coming of the new organization, *Harvard Business Review,* **66**:1.

Emery, F.E. and Trist, E.L., 1960, Socio-technical systems, in: *Management Science Models and Techniques,* Vol.2, C.W. Churchman & M. Verhulst, ed., Pergamon Press, Oxford, pp. 83 - 97.

Fox, M.S., Barbuceanu, M., and Gruninger, M., 1996, An organisation ontology for enterprise modelling: preliminary concepts for linking structure and behaviour, *Computers in Industry.* **29**, pp.123-134.

Fox, M.S., and Gruninger, M., 1997, On ontologies and enterprise modelling, *International Conference on Enterprise Integration Modelling Technology 97,* Springer-Verlag. .

Fox, M.S., and Gruninger, M., 1998, Enterprise modelling, *AI Magazine,* AAAI Press, Fall 1998, pp. 109-121.

Friedman, C., and Cornford, D., 1989, *Computer Systems Development: History, Organization and Implementation,* Wiley, Chichester, UK.

Galbraith, J., 1973, *Designing Complex Organization,* Addison Wesley, Reading, MA.

Gallivan, M. J., 1994, Changes in the management of the information systems organization: an exploratory study, in: *Proceedings of the 1994 Computer Personnel Research Conference on Reinventing IS: Managing Information Technology in Changing Organizations,* pp. 65 – 77.

Gogolla, M., Conrad, S., and Herzig, R., 1993, Sketching concepts and computational Model of TROLL light, in: Proceedings of the 3rd International Conference on Design and Implementation of Symbolic Computation Systems, A. Miola, ed., *Lecture Notes in Computer Science,* **722**, Springer Berlin, pp 17-32.

Goguen, J.A., and Linde, C., 1993, Techniques for requirements elicitation, *IEEE International Symposium on Requirements Engineering,* Santiago, California, 4-6 January.

Gruninger, M., and Fox, M.S., 1994, The role of competency questions in enterprise engineering, in: *Proceedings of the IFIP WG5.7 Workshop on Benchmarking - Theory and Practice,* Trondheim, Norway

Gruninger, M., and Fox, M.S., 1996, The Logic of Enterprise Modelling, in: Modelling and methodologies for enterprise integration, P. Bernus & L. Nemes, ed., Cornwall, UK, Chapman and Hall.

Hammer, M., 1990, Reengineering Work: Don't Automate, Obliterate, *Harvard Business Review,* July-August, pp. 104 - 114.

Hammer, M., and Champy, J., 1993, *Reengineering the Corporation: A Manifesto for Business Revolution,* Harper Business Press, New York, NY.

Handy, C., 1997, Trust and the virtual organization, in: *Organization Theory,* 4th ed.,Pugh, D.S.,ed., pp.40-50.

Iivari, J., 1991, A paradigmatic analysis of contemporary schools of IS development, *European Journal of Information Systems,* 1:4, pp. 249 - 272.

Jones, A., and Sergot, M., 1992, Formal Specification of Security Requirements: Using the Theory of Normative Positions, in: *Computer Security-ESORICS 92, Lecture Notes on Computer Science*, Y.Deswarte, G.Eizenberg & J.-J. Quisquater, ed., **648**, Springer-Verlag, Berlin-Heidelberg, pp.103-121.

Jones, A., 1993, Towards a Formal Theory of Defeasible Deontic Conditionals, *Annals of Mathematics and Artificial Intelligence*, Baltzer Science Publishers, The Netherlands.

Jones, A., and Sergot, M., 1996, A formal characterization of institutional power, *Journal of the IGPL*, **4**(3), pp.429-445.

Jungclaus, R., Saake, G., Hartmann, T., and Sernadas, C., 1996, TROLL-A Language for Object-Oriented Specification of Information Systems. *ACM Transactions on Information Systems.* **14**(2), pp.175 - 211.

Khalil, O.E.M., 1994, Information Systems and Total Quality Management: Establishing the Link, in: the *Proceedings of the 1994 Computer Personnel Research Conference on Reinventing IS: Managing Information Technology in Changing Organizations*, pp. 173.

Kling, R., 1980, Social analyses of computing: theoretical perspectives in recent empirical research, *ACM Computing Surveys*, **12**(1), pp.61-110.

Land, F.F., and Hirschheim, R., 1983, Participative systems design: rationale, tools and techniques, *Journal of Applied Systems Analysis*, **10**, 91 - 107.

Leavitt, H.J., 1965, Applied Organizational Change in Industry, Chapter 27, *Handbook of Organizations*, Rand-McNally, Chicago

Luff, P., Jirotka, M., Health, C., and Greatbatch, M., 1993, Tasks and social interaction: the relevance of naturalistic analysis of conduct for requirements engineering, in: the *Proceedings of the 1st IEEE International Symposium on Requirements Engineering.* San Diego, USA. 4-6 January, pp. 187-190.

Makinson, D., 1986, On the formal representation of rights relationships, *Journal of Philosophical Logic*, **15**, 403-425.

Mannarino, G.M., Henning, G.P., and Leone, H.P., 1997, Metamodels for Information System Modeling, in: *Production Environments*, IFIP, J.B.M. Goosenaerts, F. Kimura, & H.Wortman, ed., Chapman & Hall.

Mowshowitcz, A., 1981, On approaches to the study of social issues in computing, *Communications of the ACM*, 24:2, pp. 146 - 155.

Peckham, J, and Maryanski, F., 1988, Semantic data models, *ACM Computing.Surveys*, **20**:3, pp.153 - 189.

Porter, M.E., and Millar, V.E., 1985, How Information Gives You Competitive Advantage, *Harvard Business Review*, **63**:4, pp.149 - 160.

Pugh, D.S., 1997, The measurement of organization structures: does context determine form?, in: *Organization Theory*, 4th ed., Pugh, D.S., ed. Penguin, London, UK.

Ramaprasad, A., and Rai, A., 1996, Envisioning management of information, *Omega*, **24**:2, pp. 79 - 193.

Rubin, K.S., and Goldberg, A., 1992, Object behavior analysis, *Communications of the ACM*, **35**:9, pp.48-62.

Schreiber, G., Wielinga, B., and Breuker, J., 1993l, K*ADS: A Principled Approach to Knowledge-Based System Development*, Academic Press Inc.

Scott Morton, M.S., 1991, *Corporation of the 1990s: Information Technology and Organizational Transformation*, Oxford University Press, New York.

Sergot, M., and Prakken, H., 1996, Contrary-to-duty obligations, *Studia Logica*, **57**:112, pp.91-115.

Stamoulis, D.S., and Martakos, D.I., 1999, Searching for the missing link between business modelling and I.S. development methodologies, in: *Proceedings of the 25th International Conference on Computers & Industrial Engineering, March 29-31, 1999*, M. I. Dessouky, ed., New Orleans, Louisiana

Venkatraman, N., 1991, IT-Induced Business Reconfiguration, in: *The Corporation of the 90's: Information Technology and Organizational Transformation*, Scott Morton M.S., ed., N.Y Oxford University Press.

Von Simson, E.M., 1990, The centrally decentralized IS organization, *Harvard Business Review*, July-August, pp. 158-160.

Walsham, G., 1993, *Interpreting Information Systems in Organizations*, Wiley, Chichester, UK.

Walsham, G., 1991, Organizational metaphors and information systems research, *European Journal of Information Systems*, 1:2, pp.83-94.

Wand, Y., Monarchi, D. E., Parsons, J., and Woo,C.C., 1997, Theoretical foundations for conceptual modelling in information systems development, *Decision Support Systems* 15:4, pp.285-304, Elsevier SBV

Wieringa, R.J., 1991, A conceptual model specification language, *Technical Report IR-248*, Vrije Universiteit, Amsterdam, NL.

Wieringa, R.J., Jungclaus, R., Hartel, P., Hartmann, T., and Saake, G., 1993, OMTROLL-Object Modeling in TROLL, in: *Proceedings of the International Workshop on Information Systems Correctness and Reusability IS-CORE '93*, U.W.Lipeck and G.Koschorreck, ed., pp. 267-283.

Yu, E. S. K., Mylopoulos, J., and Lesperance, Y., 1996, AI models for business process reengineering, *IEEE Expert*, August 1996, pp. 16-23.

APPLYING SYSTEM DEVELOPMENT METHODS IN PRACTICE
The RUP example

Sabine Madsen and Karlheinz Kautz[1]

1. INTRODUCTION

System development methods have already long been controversially discussed, but there is still a lack of knowledge and understanding based on empirical studies about how systems development is actually conducted in practice, how system development methodologies and methods are used and to what degree they are used as proposed in the literature (Floyd, 1986; Nandhakumar & Avison, 1999). The purpose of this paper is to contribute to this understanding. It reports how and to what degree Rational's Unified Process (RUP) was used in two commercial development projects.

RUP is considered a state-of-the-art, object-oriented methodology with a focus on iterative and incremental development features and has been promoted as a solution to problematic issues in systems development such as unfinished projects, budget and time overruns, erroneous systems and systems with lacking functionality (Boehm, 1988; Jacobsen et al., 1999).

This paper presents an empirical case study in a consultancy firm. The case study is based on interviews with experienced project managers and systems developers, who participated in the two projects. The paper is structured as follows: Section 2 introduces the background and related work of the study. Section 3 introduces the conceptual framework, which is used to analyze the empirical findings from the case study. In section 4 RUP is explained and section 5 describes the research approach, which has been used for data collection and analysis. Section 6 presents the case study and section 7 contains a discussion of the empirical findings. The last section summarizes the main conclusions from our investigation.

---

[1] Sabine Madsen and Karlheinz Kautz, Copenhagen Business School, Department of Informatics, Howitzvej 60, DK-2000 Frederiksberg, Denmark.

Information Systems Development: Advances in Methodologies, Components, and Management
Edited by Kirikova *et al.*, Kluwer Academic/Plenum Publishers, 2002

267

2. BACKGROUND

The major part of systems development literature concerns methodological development. Numerous books provide a vast number of different methods specifying prescriptive guidelines on how to develop systems. These methods are by and large based on the assumption that system development is a rational, goal-driven and managed process and it is taken for granted that there is a need for a method to facilitate this process (Truex et al., 2000). The theoretical proposition is that the use of methods reduces production time and complexity and improves the development process as well as the quality of the final system.

Another stream of literature concerns amethodical systems development. The term amethodical as coined by Truex et al. (2000) does not entail an anti-methodological or method-less approach to systems development. It rather presents a critique of the methodological literature. As the authors put it, it implies management and orchestration of systems development without strictly predefined structure, sequence control, rationality or claims for universality, but it does not mean chaos or anarchy. The main argument is that the assumptions behind the prescriptive methods do not resemble how systems development is conducted in practice. According to the amethodical view systems development is a unique, negotiated and opportunistic process driven by accident (Truex et al., 2000). Such a position has earlier been formulated by Floyd et al. (1989) and Kautz (1993) using the concept of evolutionary systems development and have, in addition, recently been reframed as adaptive (Highsmith III, 2000) or agile development (Cockburn, 2002). This point of view is also supported by empirical studies, which suggest that systems development can be characterized as a somewhat amethodical activity, where formalized methods are not used at all or where only certain tools or techniques from a particular method are used (Bansler & Bødker, 1993; Fitzgerald, 1997, 1998).

Based on their study Bansler & Bødker (1993) conclude that there is a wide gap between how the Structured Analysis Method is proposed in the literature and how it is used in practice. Stolterman (1994) and Fitzgerald (1997) further report that experienced developers adapt and apply methods in a pragmatic way. Systems developers tailor the method to the project at hand by omitting method aspects, which are too time-consuming, cumbersome or irrelevant for the particular situation. Also indicating a problematic relationship between prescriptive methods and practice, Wastell (1996) reports from a case study in which the Structured Systems Analysis and Design Method (SSADM) was used rigorously. However, instead of improving the development process, the method inhibited creative thinking and caused the developers to focus on details instead of on the overall aim of the project.

Recently the discussion regarding system development methods has gained renewed interest in the context of web development. One stream of literature argues that traditional development methods are applicable for web development (Chen et al., 1999; Murugesan & Deshpande, 2001), while another stream argues that development of web-based systems is fundamentally different and therefore entirely new methods and approaches are required (Braa et al., 2000; Greenbaum & Stuedahl, 2000; Baskerville & Pries-Heje, 2001; Carstensen & Vogelsang, 2001). Between these extremes it has been suggested that front-end oriented web development (of the user interface) requires new methods and approaches, but back-end oriented and technically complex web development (of the functionality) requires traditional methods (Pressman, 1998). This is supported by Eriksen (2000). Based on an empirical study he concludes that traditional development methods are useful for development of back-end functionality, but provide little guidance with regard to the web-based front-end.

The two projects in this case study were large-scale back-end oriented web projects. However, due to their scale and complexity we do not perceive them to be any different from traditional systems development projects and below we will refer to them as such.

3. RESEARCH FRAMEWORK

Numerous definitions of the concepts methodology and method exist (Avison & Fitzgerald, 1995). To analyze the findings from our case study we draw upon Mathiassen et al.'s (1990) definition of a method[2]. They define a method as a disciplined, structured approach to solve a problem, which is characterized by (1) its area of application, (2) the underlying perspective and (3) guidelines for performing the process with the help of a) techniques, b) tools and c) principles of organization.

A method has an area of application for which it is suitable depending on type of information system, on project size, on team size etc. Furthermore, a method is based on an underlying perspective, i.e. on a set of assumptions. These assumptions determine the type of questions, which are asked to analyze the problem at hand, the type of solutions, which are proposed etc. At the more practical level a method consists of a number of guidelines regarding techniques, tools and principles of organization. A technique indicates how an activity should be undertaken; a tool is linked to a technique and is used to ensure that an activity is undertaken in the most effective way. Principles of organization indicate how people and groups should work together and how limited resources should be allocated. Examples of principles of organization are: division of the project into phases and guidelines about user involvement.

4. RATIONAL'S UNIFIED PROCESS

In the early 1990's a large number of different object-oriented methods had emerged. Jacobsen, Booch and Rumbaugh had each authored their own method, but in the mid 1990's they joined forces to unify their different notations into one consistent modeling language, now known as the Unified Modeling Language (UML). They continued collaborating and further unified the prevailing ideas about systems development into Rational's Unified Process, a full-fledged process model that claims to support the entire systems development life cycle (Jacobsen et al., 1999).

RUP was originally developed for traditional and large-scale systems development, but it is not only a single process. RUP provides a generic process framework, which can be customized to fit many different projects, different types of organizations, different

---

[2] The terms methodology and methods are much debated in the literature, but a discussion of their distinguishing characteristics is not part of this paper. For the purpose of this paper we will use the terms interchangeably.

levels of competence and different project sizes (Jacobsen et al., 1999). Thus, RUP's application area is claimed to be very broad.

RUP is characterized as a *use case driven, architecture centered, iterative* and *incremen-tal* process model. Use cases define the functionality of the system and each use case describes the step by step actions, which the system performs to provide the user with a result. Use cases are used for requirement specification and for splitting a project into suitable and manageable increments. For each increment one of the most important activities is to find, test and evaluate the architecture, i.e. the technical systems design consisting of the infrastructure, components and interfaces, which make up the system.

The RUP terminology contains 4 phases, called the inception, elaboration, construction, and transition phase. The inception and elaboration phases are also labeled the engineering stage and during these phases the analysis, design and planning activities are undertaken. During the construction and transition phases, also called the production stage, the coding, testing and deployment activities are performed. A project is divided into a number of iterations and for each iteration the project goes through all four phases. Furthermore, RUP is based on incremental coding, which means that for each iteration an increment, i.e. a part, of the over-all system has to go through all four phases. This allows the project team to incorporate the lessons learnt, when the next iteration is initiated and the idea is to help the project team discover major obstacles in time. Thus, RUP provides a framework for tailoring an iterative and incremental process and for selecting tools and techniques to fit a given project.

RUP has a strong focus on documents and the activities in the inception and elaboration phases mainly concern the creation of diagrams and writing of textual descriptions. The UML notation and the software program, Rational Rose, support this work.

5.   RESEARCH METHOD

The research presented in this paper is based on empirical data from a case study in a consultancy firm in Norway, which used RUP in two projects. The case study consisted of seven interviews. Each interview lasted 45-90 minutes and the participants were the Director of Process and Technology (responsible for system development methods at the company), a Method Consultant, two Project Managers, one Chief Programmer and two Systems Developers. The participants covered a wide range of roles and activities in systems development and had between 4-10 years of experience with systems development projects. Below we refer to the participants with the more general term of consultants.

Data collection was carried out using semi-structured interviews and each interview was taped. The interviews were structured around an interview guide, which focused on their experiences with RUP as a development perspective as well as its techniques, tools and principles of organization.

After the interviews had been conducted, the main topics were identified and a detailed, descriptive account of each interview was written up. Each participant received a copy of this account for correction and approval. Furthermore, a management summary outlining the main topics and conclusions from the case study was written. This, too, was sent to the participants for correction and approval.

6. CASE STUDY

The case study was performed in a large consultancy firm with more than 500 employees and considerable experience with systems development. RUP was 'officially' introduced in the company in January 2001. Top management had decided that the initiation of RUP should take place via a number of pilot projects and at the company level RUP had to be adapted to the company's approach to system development. Thus, a number of guidelines on when and how to use RUP had to be developed. However, when this study was conducted in October 2001 the management decisions regarding the introduction of RUP had not been clearly communicated to the organization and the development of guidelines and a formal adaptation of RUP had not yet taken place.

The case study concerns the experiences, which the interviewed consultants had gained while using RUP on two large-scale projects, project A and B. Project A lasted 12 months and involved 18 people. At the time of the interviews the project was ready for the final delivery to the customer. Nine of the team members were from the consultancy firm, including the project manager. The other nine team members were from a supplier of a large ERP system, which constituted the basis software for the project. The consultancy firm had the overall responsibility for the project. Furthermore, they had the responsibility for the presentation layer, i.e. the design and coding of the front-end, and the integration of the presentation layer and the data from the back-end. The supplier had the responsibility for the back-end and for supplying data to the presentation layer.

Project B was initiated in May 2001 and the first phase of the project had just finished, when the interviews for this case study were conducted in October.12 people had been involved in the project, 6 of these working full time. The second phase of project B is expected to last another 15 months and will involve around 10 people full time.

With regard to project A RUP was chosen as the development methodology due to an internal wish to try this method, where as for project B RUP was chosen due to a request from the customer. Below the experiences, which the two project teams have had with RUP, will be presented. The description will be structured according to the following RUP features: the development case document, iterations, use cases, architecture, documents and as a connecting link RUP's relation to formal development contracts.

6.1. Development Case

The purpose of the development case document is to help the project team tailor the process to fit the project. The development case document describes the kind and number of tools, techniques and documents to be used in a particular project. One consultant stated that RUP recommends that the team dedicates time - RUP proposes 2 weeks - in the beginning of the project to perform this activity and that the development case document is updated throughout the entire project. Thus, the methodology itself indicates that it takes a lot of time to plan for using RUP. The team has to plan how many iterations they need, which documents they need and later in the process they have to plan for incremental coding.

In both projects a development case document was developed and at the beginning of each phase an attempt was made to use it, but in both projects it resulted in too many documents. In project A some of this documentation was actually not used during coding and implementation and one consultant expressed that it takes a lot of skill and experience with RUP to select the right documents and to find the right level of detail. The B team did not find the development case template particularly helpful for the purpose of tailoring an iterative process to fit the project. They felt that the focus was far too much on documents and less on iterations.

6.2. Iterations

The purpose of iterative development is to ensure a learning process, where experiences from one iteration can be incorporated in the next. The aim is to learn about project risks, changing and emerging requirements as well as technical obstacles as early in the process as possible.

In both projects the development process was divided into two main phases. Using the company's own rhetoric these two phases were a specification phase covering the inception and elaboration phase according to RUP terminology, i.e. the engineering stage and an implementation phase covering the construction and transition phase, i.e. the production stage. In both projects the contract was re-negotiated between the two phases.

In project A RUP was used on request of the project team, but when the contract was re-negotiated after the specification phase, the customer was not interested in iterative and incremental development. Instead the customer wanted a traditional development process with a design phase, an implementation phase, test etc. and this development process had to be a part of the contract. Even though the contract was based on a traditional waterfall model during the second phase of the project, the templates from RUP were still used for design. However, the team found it difficult to plan for and use iterative and incremental coding. They were running on a very tight schedule and did not have any experience with RUP from previous projects. They therefore ended up following the process, which they were familiar with. So due to the customer and the contractual circumstances as well as lack of experience with RUP, iterations were not used during the last part of project A.

In project B there were three iterations during the specification phase, where the customer received use cases and other documents for approval after each iteration, but after the specification phase the project team experienced difficulties in getting the customer to accept the final set of deliverables. The project team realized that they had not paid sufficiently attention to getting the customers' full accept after each iteration. Furthermore, they became aware that the customer was expecting the deliverables to be very detailed, while the B project team had tried to work at a broader level as recommended by RUP. The customer expected the outcome of the specification phase to be a detailed documentation of what was to be the final system, as if the system was developed according to the traditional waterfall model. In contrast, the B project team expected the outcome of the specification phase to be use cases and architecture documents, which had to be further refined during the following phases. In other words, there was a mismatch between expectations due to different development perspectives and even though the customer had requested that the system should be developed according to RUP, the interviewed B team members felt that the customer had not really understood the RUP process. The customer lacked an understanding of the process as a learning process, where decisions are very abstract and broad in the beginning, but get continuously more and more detailed as the project team and the customer get a better understanding of the system.

6.3. The Contract

The purpose of the contract is to formally establish the economic and legal context in which development of a given project can take place. However, both teams experienced a mismatch between the assumptions, which are underpinning the legal contracts and traditional ways of doing business and the assumptions, which RUP is based on. The consultancy firm has normally followed the traditional business rules for commercial systems development, where the contract and the payment is tied together with and based on a requirements specification. According to these business rules it is assumed that the outcome of the first phase is a complete and final description of the system, which will be the basis for the rest of the project and a fixed price contract. RUP, however, assumes that the project team does not really know what they are going to develop before much later in the process. Therefore, they have to start at an abstract, general level and work their way to a more and more detailed understanding of the system via iterations, a focus on architecture and early coding on core parts of the system. Thus, when developing according to RUP, the project team learns about the project throughout the process and therefore it is not possible to have a complete and final requirement specification early on in the project.

The interviewed consultants experienced this as a huge challenge with regard to the customer. The customers in project A and B wanted a fixed price contract. They wanted to be sure that they would get, what they were paying for. The consultants explained that the problem with a fixed price contract is that the price is estimated based on a number of assumptions about project scope, scale and functionality, but with RUP's iterative development perspective it is explicitly recognized that these assumptions will change during the process. However, when the project is based on a fixed price contract, the project team has to be critical towards changing requirements, which will increase the project costs, and consultants from both teams stated that a fixed price contract inhibits a true learning process.

RUP's focus on systems development as a learning process was not only a challenge with regard to the customer. It was also a challenge for the systems developers, because they too felt uneasy and out of control when working at the more abstract levels. They were used to working according to the traditional waterfall model, so they also assumed that they should have a complete description and understanding of the final outcome early on in the project.

6.4. Use Cases

The purpose of use cases is to help the project team with requirement specification and the division of the project into suitable increments. Starting by identifying the different types of users in order to develop an apprehension of the system's purpose, use cases are refined to describe the system's functionality at a more detailed level, understandable and appropriate for both users and developers.

However, with regard to requirement specification it was a challenge for both teams to identify and describe the use cases. One consultant stated that the examples in the books from Rational are very easy and obvious, but in practice most requirements do not come as neat and simple use cases. Instead the requirements had to be split into several use cases and for both teams it was a time-consuming and challenging task to identify the relevant ones. Furthermore, one of the interviewed B team members stressed that it is important to determine the purpose of the use cases. The experience from project B – although intended differently by the methodology - was that the use cases were too focused on functionality and less on who the user is and what the user's goal is. The consultants themselves thought that the use cases in the beginning should have focused on the user in order to help the developers understand who and what the system was projected for. This would have been a better starting point for a subsequent refinement with a focus on functionality.

In both projects use cases were used for requirements specification, but in project A they were not used as a tool for planning an incremental development process. With regard to project B it is the intention to plan for and use iterative and incremental development based on use cases, when the project enters the implementation phase.

6.5. Architecture

The purpose of RUP's strong emphasis on architecture is to guide the project team in establishing a stable architecture, which the project can be based upon and to ensure that the team discovers technical difficulties as early as possible. Therefore, RUP prescribes that the architecture is tested and evaluated continuously throughout the development process and both teams dedicated time to perform this activity.

However, in project A the customer abandoned the architecture, which was recommended by the project team. One of the main components in the recommended architecture was an ERP system from a particular company, but the customer had already established a business relationship and a preference for another supplier. The chosen architecture was not tested and evaluated anew, and the project team did, therefore, not discover the major technical obstacles before late in the implementation phase. Furthermore, the recommended architecture was based on an object-oriented technology, but the chosen ERP system was not object-based. This meant that the design documents, which had been created in the specification phase, were useless for actual coding and implementation.

In project B the architecture was also tested and evaluated in the specification phase, but the project had not entered the implementation phase, when this study was conducted, and therefore we are unable to report further from their experiences.

6.6. Documents

The purpose of RUP's analysis and design documents is to create a documentation set, which is useful and valuable throughout the entire development process.

Both teams used Rational Rose for drawing use case diagrams and in both projects it was not difficult for the customer to understand the use cases. But the interviewed consultants experienced that due to the amount of pages with use cases it was difficult for the customer to evaluate them all and to give useful feedback to the project teams.

Furthermore, both teams used architecture documents, including class diagrams and domain models, as RUP suggests, and they did not experience difficulties in the outset when using the templates and drawing diagrams in Rational Rose. But the architecture documents did pose a challenge for both teams later. The A team experienced difficulties with the usage and maintenance of the class diagrams, because the chosen architecture was not object-based and in the end the class diagrams were discarded altogether. For the B team the architecture documents turned out to be a challenge for the customer. The team had made a number of architecture documents from different perspectives, as recommended by RUP. However, it was difficult for the customer to understand these different perspectives, and especially to see them as part of one coherent description, and therefore they were reluctant to accept them.

7.   DISCUSSION

Even though the interviewed consultants experienced difficulties with the usage of RUP, they all stated that they had a positive impression of the method and would like to use it again. However, when comparing their experiences with Mathiassen et al.'s (1990) definition of what characterizes a method, the project teams only used 2 out of 5 characteristics. In both project A and B the techniques and templates from RUP had been used, but the project teams – although they had planned and attempted to - did not succeed in using RUP for structuring the development process and they did not adopt the underlying development perspective. RUP was not used for tailoring and managing an iterative and incremental process. Instead the two projects under investigation ended up following a traditional development process, i.e. a waterfall model, supplemented with tools and techniques from RUP. RUP was primarily used as toolbox, from which tools and techniques were selected and applied in a pragmatic way. This is in keeping with Fitzgerald (1998), who concludes that in practice the most evident contribution of a system development method is as a toolbox, and not as a process framework. It also supports Truex et al. (2000), who argue that systems development is an opportunistic, negotiated and compromised activity, which does not follow the rationalistic ideal of a predefined process. In our study opportunism, compromise and negotiations became obvious and visible through the developers' and customers' lack of experience with the methodology, the contractual circumstances and the general behavior of the customers.

Whether a project team with more experience with RUP could have avoided the described situation, has however to be questioned. The project members were after all very experienced IT professionals. Thus, the role of the development contracts and that of the customers has to be revisited.

Our case study indicates that iterative development and the explicit focus on systems development as a learning process cause difficulties, when systems development is performed according to a fixed price contract. This type of contract necessitates strict cost control, and thereby inhibits a true learning and development process based on intermediate, not fully documented specifications. Mathiassen & Bjerknes (2000) present a similar argument and discuss the balance between trust and control with regard to contracts and client-contractor relationships. While trust promotes creativity and mutual learning, a contract promotes decisions and monitoring of progress according to the agreement. Mathiassen & Bjerknes (2000) therefore suggest that there is a need for a well-adjusted relation between trust and control in order to create an environment for learning. They conclude that it is impossible to improve systems development practices without changing the current form of contracts. This case study supports that argument.

Mathiassen & Bjerknes (2000) reason more about the customer-supplier liaison. In our case the customers had a tremendous influence on the course of the projects and the utilization of the methodology. They had a different understanding of the methodology; they were only partly interested in incremental development and more in favor of detailed written specifications as results of distinct phases. In one project they even made a technical, architectural decision, which was in conflict with the methodology's development approach. In such an environment the development organization and the developers could not apply the methodology as described in the method guidelines and as intended by themselves. They had to make concessions, go through negotiations and find a pragmatic and practical way to deliver the demanded product. As our case demonstrates, systems development has to reconsider the customer-supplier relationship to enhance practice.

As an object-oriented and iterative method RUP has been marketed as a solution of the problems in systems development. But it appears that even though RUP is claimed to be a modern method with a seemingly broad area of application, it pays little attention to the context in which commercial systems development takes place. However, whether it is feasible to incorporate all activities performed in systems development in one methodology has to be doubted. In line with Cockburn (2002) our study shows that systems development, as actually performed, is such a complex process that it cannot be accurately described. And if it could, no one would be able to read such complicated description and learn from it, how to perform systems development. The challenge is to find an equilibrium between the methodical and the amethodical elements of systems development.

8. CONCLUSION

This case study has described how and to what degree RUP was used in two large-scale development projects in a consultancy firm. In summary we put forward three main conclusions.

In both projects tools and techniques from RUP were used, but the project teams did not succeed in using RUP as a framework for structuring and managing the process and they did not adopt the underlying development perspective. Thus, RUP was used as a toolbox, not as a process framework. These findings are in accordance with other empirical case studies and lend support to the proposition that in practice system development is a somewhat amethodical activity, where methods are used for selecting and applying tools and techniques in a pragmatic way. In the literature method is primarily related to the concept of process (Truex et al., 2000), but in these two projects the prescriptive process seemed to play a secondary, if not insignificant, role.

Furthermore, we have reported that the project teams experienced difficulties with RUP's iterative development features when developing according to a fixed price contract, because a fixed price contract requires strict cost control, thereby inhibiting the learning process, which is the aim of iterative development. RUP is claimed to have a broad area of application, but this case study indicates that it pays little attention to the contractual and economic issues of system development.

Finally, we discussed the customer-supplier relationship in systems development. RUP promotes the active involvement of clients and future users. This had severe consequences for the cooperation of the different stakeholder groups and the application of a methodology like RUP. To enhance systems development there is clearly a need to rethink the customer-supplier relationship, both in the context of methodologies and beyond, and future research is needed to further explore the advantages and disadvantages of iterative development in a commercial setting.

REFERENCES

Avison, D., & Fitzgerald, G., 1995, "Information Systems Development: Methodologies, Techniques and Tools", McGraw-Hill, London, UK.

Bansler J. & Bødker K., 1993, "A Reappraisal of Structured Analysis: Design in an Organizational Context", ACM Transactions on Information Systems, 11(2), pp. 165-193.

Baskerville R. & Pries-Heje J., 2001, "Racing the E-Bomb: how the Internet is redefining Information Systems Development", IFIP TC8/WG8.2 Working Conference, July, Idaho, USA.

Boehm BW, May 1988, "A spiral model of software-development and enhancement", IEEE Computer, pp. 61-72.

Braa K., Sørensen C. & Dahlbom B., 2000, "Changes - From Big Calculator to Global Network", In: Planet Internet, Studenterlitteratur, pp. 13-39.

Carstensen P. & Vogelsang L., 2001, "Design of Web-Based Information Systems - New Challenges for Systems Development?", European Conference on Information Systems (ECIS).

Chen L., Sherrell L.B. & Hsu C., 1999, "A Development Methodology for Corporate Web Sites", First ICSE Workshop on Web Engineering (WebE-99), Los Angeles, USA.

Cockburn A., 2002, "Agile Software Development", Addison-Wesley, Boston ,USA.

Eriksen L.B., 2000, "Limitations and Opportunities of System Development Methods in Web Information System Design", IFIP TC8/WG 8.2 Working Conference, Boston, USA, pp. 473-486.

Fitzgerald B., 1997, "The use of Systems Development Methodologies in Practice: A Field Study", Information Systems Journal, 7(3), pp. 201-212.

Fitzgerald B., 1998, "An Empirical Investigation into the Adoption of Systems Development Methodologies", Information & Management, vol. 34, pp. 317-328.

Floyd C., 1986, "A Comparative Evaluation of Systems Development Methods", In: Information Systems Design Methodologies: Improving the Practice, (eds.): Olle et al., North-Holland, pp. 19-37.

Floyd C., Reisin F.-M., Schmidt G., 1989, "STEPS (Software Technique for Evolutionary, Participative System Development) to Software Development with Users", In: Ghezzi C. & McDermid J. A., European Software Engineering Conference (ESEC) '89, pp. 48-64, Springer-Verlag, Germany.

Greenbaum J. & Stuedahl D., 2000, "Deadlines and work practices in New Media Development", Proceedings of IRIS 23, University of Trollhättan Uddevalla, pp. 537-546.

Highsmith III J. A., 2000, "Adaptive Software Development: A Collaborative Approach to Managing Complex Systems", Dorset House Publishing, New York, USA.

Jacobsen I., Booch G. & Rumbaugh J., 1999, "The Unified Software Development Process", Addison-Wesley.

Kautz K., 1993, "Evolutionary System Development - Supporting the Process", Research Report 178, Dr. Philos. Thesis, Department of Informatics, University of Oslo, Norway.

Mathiassen L. & Bjerknes G., 2000, "Improving the Customer-Supplier Relation in IT Development", The 33rd Hawaii International Conference on Systems Sciences (HICSS).

Mathiassen L., Kensing F., Lunding J., Munk-Madsen A., Rasbech M. & Sørgaard P., 1990, "Professional Systems Development: Experience, Ideas and Action", Prentice Hall.

Murugesan S. & Deshpande Y., 2001, "Web Engineering: A new Discipline for Development of web-based systems", In: Web Engineering - Managing Diversity and Complexity of Web Application Development, Springer-Verlag.

Nandhakumar J. & Avison D., 1999, "The fiction of methodological development: a field study of information systems development", Information, Technology & People, 12(2), pp. 176-191.

Pressman R.S., 1998, "Can Internet-Based Applications be Engineered?", IEEE Software, pp. 104-110, September/October.

Stolterman E., 1994, "The 'transfer of rationality', acceptability, adaptability and transparency of methods", Proceedings of the 2nd European Conference on Information Systems (ECIS), Nijehrode University Press, Breukeln, pp. 533-540.

Truex D., Baskerville R. & Travis J., 2000, "Amethodical systems development: the deferred meaning of systems development methods", Accounting Management & Information Technologies, 10(1), pp. 53-79.

Wastell D., 1996, "The Fetish of Technique: Methodology as a Social Defense", Information System Journal, 6(1), pp. 25-40.

SCALABLE SYSTEM DESIGN WITH

THE BCEMD LAYERING

Leszek A. Maciaszek, Bruc Lee Liong*

1. INTRODUCTION

Modern software production is *incremental and iterative*. Systems are developed in successive *iterations* delivering *incremental releases* of the product. Iterative and incremental development can only succeed if the scalability is built into the system architecture in the first iteration and carefully managed in successive iterations.

Incremental releases imply the steady growth in system complexity as the *extra requirements* are added. With each extra feature, there comes a need to re-evaluate how the feature will interact with the other modules. All these extra functionality requires a system wide evaluation to determine the impact on the system as a whole, which accompanies a minor change to the product's requirements.

The greater the number of classes within a system, the more interaction that occurs that has to be accounted. For example, for two classes, there are two interactions that occur (i.e. A to B and B to A), for three classes, there are six (i.e., A to B, B to A, A to C, C to A, B to C and C to B), and so on. Since each new feature normally results in at least one new class that must interact with the rest of the system, the complexity of the project can increase rapidly. Obviously this is a worst case equation and in reality it would be less but it does highlight the potential to quickly turn a small scale system into a medium sized one by the addition of a few more features. The objective of the *BCEMD (Boundary – Control – Entity – Mediator – DBInterface)* architecture is to restrict the growth of complexity to a manageable size.

The paper is structured as follows. In the next Section we explain the *BCEMD* principles. Then we go on to describe related research in Section 3. Next in Section 4 we introduce the metric called *cumulative class dependency* (CCD). We use this metric to demonstrate the advantages of the *BCEMD* framework. In Sections 5, 6 and 7 we show how the CCD can be minimized when adding new inner classes, interfaces and composition relationships. In Sections 8 and 9 we explain how to avoid an increase in CCD due to indiscriminate use of delegation and acquaintance in the run-time program structures. In

---

* Macquarie University, Sydney, NSW 2109, Australia

Information Systems Development: Advances in Methodologies, Components, and Management
Edited by Kirikova *et al.*, Kluwer Academic/Plenum Publishers, 2002

279

Section 10 we present a global *BCEMD* model for the examples used in the paper. The model provides a case-study evidence that the *BCEMD* approach scales up.

2. *BCEMD* PRINCIPLES

The *BCEMD* layering is an extension of the *BCED* approach introduced in Maciaszek (2001). However, the *BCED* approach in Maciaszek (2001) is only presented as a variation and elaboration of the MVC (Model/View/Controller) paradigm introduced (and enforced) in Smalltalk-80 (Krasner and Pope, 1988). Although advantages of the *BCED* layering in reducing complexity of object-oriented designs are documented throughout Maciaszek's book, no attempt was made there to treat the topic in its own right.

BCEMD focuses on dividing software into *layers of class packages*. Each layer implements a well-defined common functionality. The structure provides direct *peer-to-peer communication* between neighboring layers through a pre-defined chain of command. The peer-to-peer communication between non-neighboring layers is indirect (in order to lower the cumulative class dependency).

Internally each layer (package) is likely to have a *dominant class* that implements the interface that the package realizes. Channeling message passing through a dominant class allows a significant level of information hiding which is the main technique for managing complexity. The vertical *supervisory-subordinate communication* between classes within a package uses message passing along inheritance and composition hierarchies as well as association links. In initial iterations, the vertical *BCEMD* structure can be simplified by using *inner classes*.

The stratified hierarchies of classes within packages are from highest to lowest level of abstraction or from coarse to fine level of modular functionality. As we descend from the apex of the hierarchy (dominant class), the action is formalized into more and more detail by each successive lower class (i.e. information hiding occurs between vertical levels of classes in the package).

The *BCEMD* *layers* represent dimensional differences that can occur when scaling systems. Horizontal layers establish a *fixed structure* between packages – invariant for all iterations of the incremental development process. Vertically, inside each package, we permit inheritance, composition and association hierarchies (and networks to lesser extent) between classes.

While horizontal structure defines the rules of the game, vertical structures inside packages define the course of the game. A horizontal structure defines the permitted moves; vertical structures leave room for relatively flexible strategies how to decide on the choice of the actual move.

The *BCEMD* approach enforces that each class in a system is assigned to one of the five packages. To make this assignment a thoughtful process and to easily recognize the package to which a class belongs, each class name is prefixed with the first letter of the package name (e.g. E_Invoice is a class name in the Entity package).

The Boundary package contains classes that define GUI objects. In Microsoft Windows environment, many boundary classes would be subclassed from the MFC (Microsoft Foundation Classes) library.

The Control package consists of classes responsible for the program's logic, algorithmic solutions, main computations, and processing user's interactions. A class containing the program's main function is also housed in the Control package.

The `Entity` package contains classes representing "business objects". They store (in program's memory) objects retrieved from the database or created in order to be stored in the database. Many entity classes are container classes.

The `Mediator` package establishes a channel of communication that mediates between entity and dbinterface classes (ref. Mediator pattern in Gamma *et al.*, 1995). Mediation serves two main purposes. Firstly, to isolate the two packages so that changes in any one of them can be introduced independently. Secondly, to eliminate a need for control classes to directly communicate with dbinterface classes whenever new entity objects need to be retrieved from the database (such requests from control classes are then channeled via mediator classes).

The `DBInterface` package is responsible for all communications with the database. This is where the connection to the database is established, all ODBC/JDBC/SQLJ queries and database calls are constructed, and the database transactions are instigated.

3. RELATED WORK

Related research is in system design, in object-oriented design patterns and in software metrics. There exists a commonly accepted set of principles and patterns that should be satisfied by a system design to be understandable and scalable. The original book by Gamma *et al.* (1995) documented many of these principles and patterns. Arguably, the main contribution of Gamma *et al.* (1995) was a recommendation that good designs should "favor *object composition* over *class inheritance*". This observation was enhanced by another principle – "program to an *interface*, not an implementation".

Gamma's *et al.* design patterns increased awareness of proper application of common techniques for reusing functionality, for information hiding and for working with abstraction in object-oriented systems. This awareness has been evident in the work of others, e.g. Lakos (1996), Maciaszek (2001), Page-Jones (2000), Szyperski (1997).

Layers simplify the architectural design by establishing a *Subscribe/Notify* protocol for peer-to-peer communication. They decouple packages so that they can evolve (scale up) independently. Layers hide the internal complexity in a *dominant class* (Rumbaugh *et al.*, 1999).

The *Subscribe/Notify* protocol underpins the peer-to-peer communication between BCEMD packages. Neighboring packages subscribe to each other services and implement message and object passing via direct association links. If necessary, changes to a package state can be notified to a neighboring package. Communication between non-neighboring packages is possible via a chain of *Subscribe/Notify* commands.

The complexity is hidden in a *dominant class* of a BCEMD package. The dominant class implements main interfaces and abstract classes of a package. In words of Rumbaugh *et al.* (1999, p.219) – "A dominant class subsumes the interface of the component".

The scalability is achieved by a combination of a fixed hierarchical structure of packages and by encapsulated organization of vertical class hierarchies inside packages. The next section formalizes our discussion of designing for scalability. It introduces the metric called *cumulative class dependency (CCD)*. The definition of CCD is based on the metric labelled *cumulative component dependency* in Lakos (1996). The idea of both metrics is similar but Lakos used his metric as a measure for the link-time cost of incremental regression testing.

4.   CUMULATIVE CLASS DEPENDENCY IN *BCEMD*

The *cumulative class dependency* (*CCD*) provides a numerical value that characterizes the relative complexity associated with iterative and incremental development and evolution of object-oriented systems.

> DEFINITION: *Cumulative class dependency (CCD)* is the sum over all classes C_i in a system of the number of classes C_j to be potentially changed – according to statically-defined relationships between classes in the immediate neighborhood – in order to modify each class C_i. The immediate neighborhood condition eliminates double calculation.

Consider a system design with only five classes in which no BCEMD framework was used and in which each class in the system depends on all remaining classes. Assuming that each *BCEMD* package contains one class only, Figure 1 illustrates the dependencies.

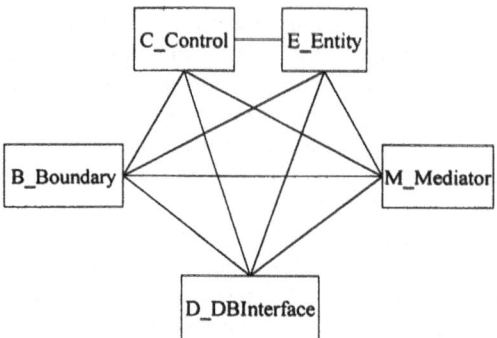

Figure 1. Graph with CCD = 25

In the presence of cyclic dependencies as in Figure 1, it may be necessary to modify each class in order to introduce a change to any one of them. There are five classes in the design. The cost of changing any one of these classes is five. Therefore, CCD on a cyclically dependent graph in Figure 1 is 25 (i.e. the total number of classes (5) multiplied by the cost of changing a class (5)).

Admittedly, CCD=25 is the worst case scenario but the mere fact that the scenario is possible inhibits scalability. A change to any class can potentially impact on all remaining classes and therefore the change impact analysis must consider all classes.

Consider now a layered design as in Figure 2. Although the design is layered, it does not fully conform to the BCEMD approach, as will be explained later. The cost of changing B_Boundary and D_DBInterface is 2. The cost of changing each remaining class is 3. Therefore, CCD on a graph in Figure 2 is 13 ((2 * 2) + (3 * 3)).

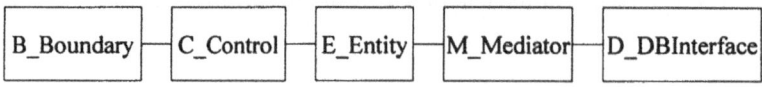

Figure 2 Graph with CCD = 13

The *BCEMD* approach uses slightly more complex design. Figure 3 illustrates *BCEMD* dependencies between five *dominant classes* in the *BCEMD* packages. In general, the same dependencies apply between packages (hence, for example, no class in the Boundary package can directly communicate with a class in the Entity package). The cost of changing B_Boundary and D_DBInterface is 2. The cost of changing E_Entity is 3. The cost of changing C_Control and M_Mediator is 4. Therefore, CCD on a graph in Figure 2 is 15 ((2 * 2) + (1 * 3) + (2 * 4)).

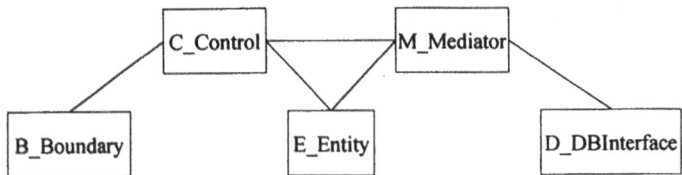

Figure 3. BCEMD Graph with CCD = 15

The elevation of Control and Mediator packages in *BCEMD* requires explanation. There is more than one reason for this but the main one has to do with program's initialization. In a typical scenario, the program has to connect to a database before entity classes can be instantiated. The database connection is initiated from a control class containing the main function. We call such a class C_Init. The C_Init object creates C_Control and M_Mediator objects. M_Mediator instantiates a D_DBInterface object before D_DBInterface can establish a database connection and retrieve data to populate E_Entity objects.

Figure 4 demonstrates the CCD increase in the presence of C_Init. The new CCD equals 20. Note that the CCD with C_Init for the network structure as in Figure 1 would be 36 (almost twice as complex).

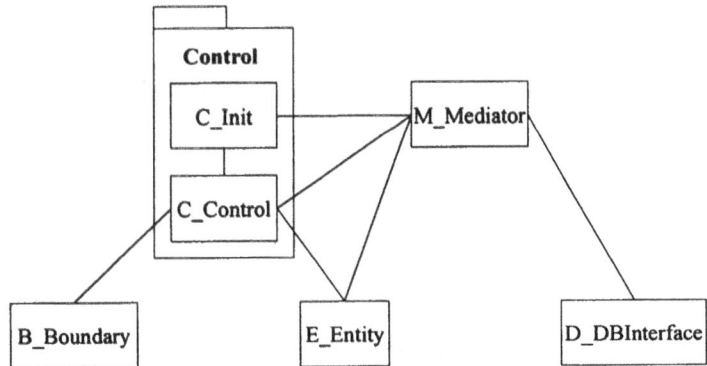

Figure 4. BCEMD Graph with two classes in Control package (CCD = 20)

5. MINIMIZING CUMULATIVE CLASS DEPENDENCY WITH INNER CLASSES

Clearly, the *BCEMD* design in Figure 4 minimizes CCD by restricting communication paths between packages (represented in Figure 4 by dominant classes). The `Control` package includes – apart from a dominant class - a program's initialization class `C_Init`. The run-time structure is further simplified because all these classes are likely to be *singleton* classes (i.e. only one object can be instantiated from them; ref. Singleton pattern in Gamma *et al.*, 1995).

As the incremental software development demands increasingly complex solutions in successive iterations, new classes will have to be added to each package. The `Entity` package presents the most immediate difficulty. This package is an in-memory representation of business objects retrieved from various database tables. Each table is likely to map to a separate entity class. Clearly, the dominant `E_Entity` class cannot properly represent all these objects.

The *BCEMD* recommended solution is to control the inevitable increase of CCD by placing class definitions corresponding to database tables within the definition of `E_Entity`. This means, in Java parlance (Eckel, 2000), that `E_Entity` becomes an *outer class*. The *inner classes* corresponding to database tables can be made private to hide them from classes in other packages. In practice, however, most inner classes will be public so that `C_Control` and `M_Mediator` can communicate directly with inner objects.

Regardless, only the outer `E_Entity` class can construct inner objects according to the *Producer/Product* principle (SourcePro, 2001) employed in *BCEMD*. The paradigm uses objects of one class (*Producer*) to instantiate objects of another class (*Product*). Rather than invoking public constructors, the producers use private constructors. In effect, instantiation relationships implement run-time *acquaintance links* between producers and products.

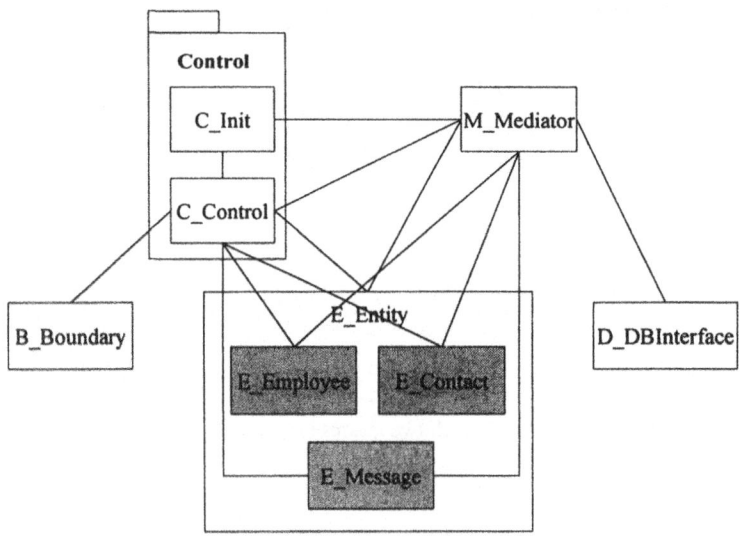

Figure 5. BCEMD Graph with public inner classes in Entity package (CCD = 35)

Figure 5 shows the *BCEMD* design in which three inner classes were introduced in the `Entity` package. The application domain in Figure 5 is about messages stored in a database and scheduled for emailing to contacts (customers) by designated employees (the application users). The database contains three tables: `Employee, Contact, Message`. The corresponding inner classes are `E_Employee, E_Contact` and `E_Message`.

CCD in Figure 5 is 35 (CCD for the network structure as in Figure 1 would be 81). Note that inner classes are distinctly different from aggregation/composition (Eckel, 2000). Aggregation/composition is a superset/subset relationship between objects of different classes (e.g. `Book` object contains `Chapter` objects). Outer/inner classes apply to class definitions. Typically, an outer class contains a reference to an inner class (admittedly, not considered in Figure 5 in the CCD calculation). An inner class has an automatic access to all the elements in the outer class. The Java syntax (`E_Entity.this`) directly supports that access (as an aside, note the difference with a name-hiding mechanism in C++ called *nested classes*).

6.   MINIMIZING CUMULATIVE CLASS DEPENDENCY WITH INTERFACES

Even though the early iterations of the project may conform to the designs in Figures 4 or 5, the successive iterations will introduce new classes within the five packages. So, our next consideration is to propose principles for class structures within packages such that CCD is minimized and at the same time the overall design demonstrates a proper balance of class cohesion and coupling (Maciaszek, 2001; Page-Jones, 2000).

A fundamental technique of simplifying in-package design to enhance future scalability is to "program to an interface, not an implementation" (Gamma *et al.*, 1995). Interface is a "pure" abstract class with no implementation at all. An interface can be implemented in many classes and – to support "multiple inheritance" – many interfaces can be implemented in a single class.

In *BCEMD* an interface can be an almost perfect substitution for a dominant class in its capacity of reducing CCD. Consider the model in Figure 6 where the interface `B_Boundary` replaced the class `B_Boundary`. Depending on the iteration of the project, `B_Boundary` is implemented in `B_Console` or `B_Window` class (the latter is used when the project moves into the phase when the console window is replaced by a full-blown GUI window implementation).

An instance variable in `C_Control` (of `B_Boundary` type) implements the association in support of peer-to-peer communication between `Control` and `Boundary` packages. This variable will be referencing a `B_Console` object in early project iterations and a `B_Window` in later phases. In fact, this can be an object of any subclass of `B_Boundary` such as `B_MessageBrowser`. The simplifying outcome of using interface is that `C_Control` is insulated from changes to the implementation of `B_Boundary`.

An interface can also be used to factor out all program constant values such as to name a driver to connect to a database, a database url, mail host name, or any hard-coded program values, such as user's authorisations or fixed information displayed in a window. An interface works well in that capacity because all fields in it are implicitly static, final and public.

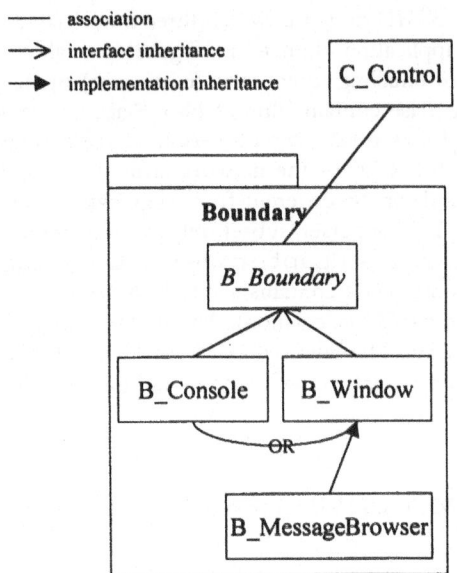

Figure 6. Using interface as a replacement for a dominant class

Figure 7 demonstrates *C_Constants* interface implemented in concrete classes C_Init and C_Control. This ensures that the concrete classes inherit the same constant values. Any changes to a database url etc. will be automatically inherited by the concrete classes, thus facilitating scalability and portability of the application.

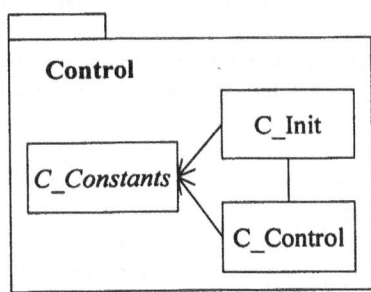

Figure 7 Using interface to factor out program constants

7. MINIMIZING CUMULATIVE CLASS DEPENDENCY WITH COMPOSITION

After an initial overindulgence of OO community in the benefits of inheritance, much has been written in recent years about the disadvantages of inheritance and the need to "favor object composition over class inheritance" (Gamma *et al.*, 1995). Composition can minimize CCD by allowing an independent growth of in-package class structures without increasing class dependencies between packages.

The vertical in-package structure of classes may be "rooted" on a superset (aggregate) class that effectively takes the role of a dominant class. Figure 8 shows an incremental development of the DBInterface package. The package contains now three additional classes responsible for connecting to the database (D_Connection), for retrieving data from the database (D_Selector) and for updating the database (D_Updater). The dominant class D_DBInterface "has" these three subclasses and it serves as the sole entry point for external classes (M_Mediator).

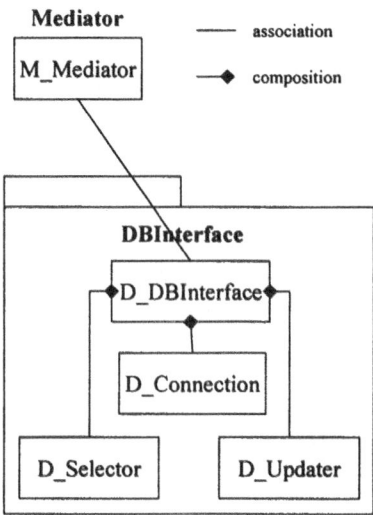

Figure 8. Superset class in the role of the dominant class

For example, because D_DBInterface keeps a D_Selector instance variable, a request from M_Mediator to D_DBInterface to retrieve Employee information from a database will result in a D_DBInterface object *delegating* retrieval task to a D_Selector object.

In effect, the superset D_DBInterface class in Figure 8 does not only take the role of the dominant class but it also encapsulates the subset classes. The CCD of the overall model increases only by the number of composition relationships in the composition. Changes to subset classes are not affecting their superset class (except for changes to operation signatures).

8. COMPOSITION CAN INCREASE CUMULATIVE CLASS DEPENDENCY

In Maciaszek *et al.* (1996a) and Maciaszek *et al.* (1996b) we demonstrated the benefits of composition over inheritance. Whenever inheritance is "considered harmful" we can replace it by composition. However, like inheritance, composition must not be overused. Both techniques have their own rightful place in modeling.

The main advantage of composition over inheritance is that it allows composing behaviors at run-time. Consider a variation of Figure 8 in which D_Selector is specialized into D_JDBCSelector (when JDBC is used to select from a database) and D_SQLJSelector (when SQLJ is used to select from a database) (Figure 9). Because JDBC and SQLJ statements can be intermixed in a program, there is not exclusivity between the subclasses (as it was between B_Console and B_Window in Figure 6).

In Figure 9, our D_DBInterface object can be retrieving from a database via JDBC calls or SQLJ statements by simply replacing its D_Selector instance with D_JDBCSelector or D_SQLSelector. Solution in Figure 9 enhances software flexibility and therefore also scalability, without impacting unduly on CCD.

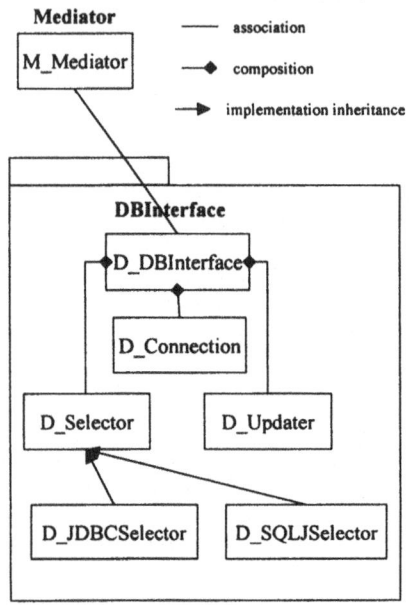

Figure 9. Composing run-time behaviors via delegation

However, composition and delegation can also lead to rapid and undesirable increase in CCD. Consider a modified version of Figure 6 in which implementation inheritance is replaced by composition (Figure 10). B_MessageBrowser reuses the behavior of B_Window by keeping a B_Window instance variable and delegating generic B_Window behaviors to it (such as openWindow() or closeWindow()).

Composition in Figure 10 complicates the model. The price for the possibility of composing behavior at run-time is a significant increase in CCD. In order to act on B_MessageBrowser, C_Control must have a B_MessageBrowser instance variable. A B_Boundary instance variable is not sufficient any more because B_MessageBrowser is not a kind of B_Boundary.

While the reason for the difficulty in Figure 10 is intuitively obvious, we do not endeavor in this paper to give rules when to use composition in lieu of inheritance. Such rules are very hard to establish (ref. Gamma *et al.*, 1995; Maciaszek, 2001). However, when there is no contention between inheritance and composition, like in Figure 8, composition should be used to minimize CCD.

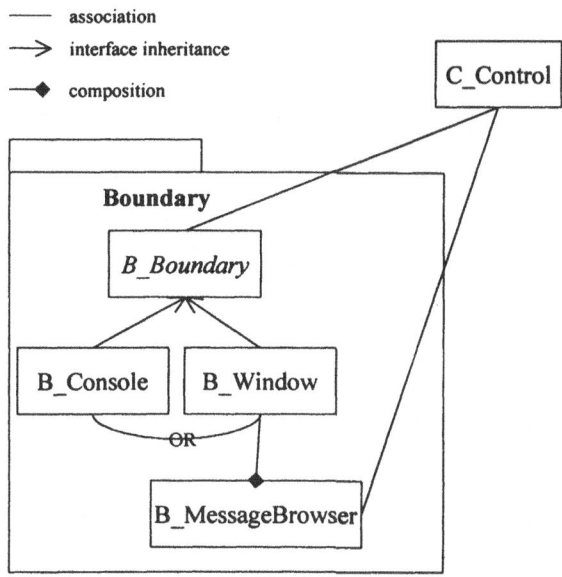

Figure 10. Composition that increases CCD

9.   INTERPLAY OF COMPILE-TIME AND RUN-TIME *BCEMD* STRUCTURES

So far we have concentrated on program's structures frozen at compile-time. The whole idea of *BCEMD* is to force a pre-defined class organization that minimizes CCD and enhances scalability for iterative and incremental development. However, behaviors can be composed at run-time. In the previous two sections, we explained how delegation could compose behaviors at run-time, hopefully without increasing CCD in a significant way.

A compile-time structure of *BCEMD* is realized in three kinds of relationships: association, inheritance and composition. But a program's run-time structure may or may not use these relationships for object intercommunication. Moreover, as objects are passed between operations, objects tend to *know* about other objects whether or not they are connected by compile-time relationships. Gamma *et al.* (1995) calls such run-time knowledge an object *acquaintance*.

Figure 11 shows acquaintance between three entity classes in the Entity package. In reality, E_Employee, E_Contact and E_Message will be container classes. The object content of these containers will be quite dynamic; therefore we may opt against designing associations between these classes. So, if the program has just retrieved messages from a database to be emailed to a particular contact by a particular employee, the run-time knowledge of who is connected to whom may be obtained via acquaintances.

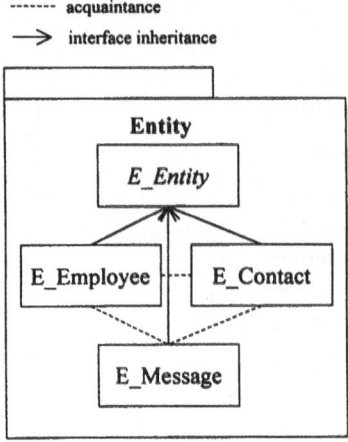

Figure 11. Acquaintance inside a package

The danger of acquaintance is that acquainted objects may invoke operations on each other to the point that the analysis of compile-time code structure will not reveal merely enough about how the system works.

Consider Figure 12 where the programmer acquainted B_MessageBrowser and E_Message. Although this may seem at first to be a good idea, the acquaintance breaks the *BCEMD* framework and it rapidly increases the CCD.

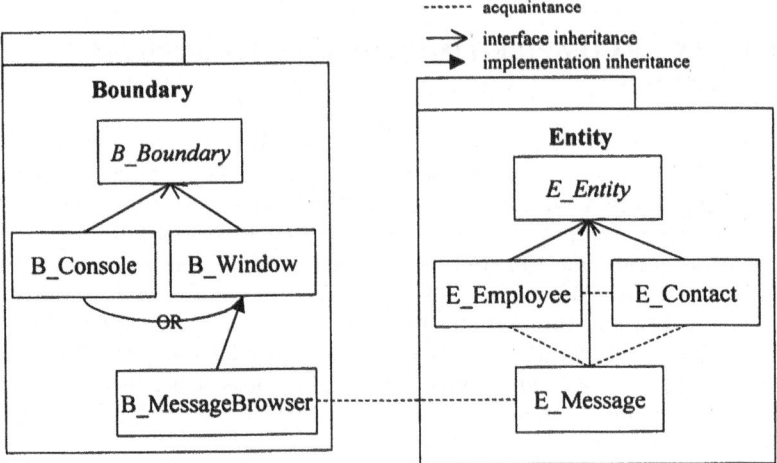

Figure 12. Acquaintance between non-neighboring packages

Acquaintance is a bad choice for two reasons. Firstly, the *BCEMD* principle that only neighboring packages can directly communicate is broken. Secondly, the run-time program structure cannot be understood any more from the compile-time structure. Note that if association were used instead of acquaintance, only the first principle would be at stake. The compile-time structure would still help understand the run-time structure.

10. THE *BCEMD* EXAMPLE RECAPPED AND CONCLUSION

This paper presented the *BCEMD* framework in support of a scalable system design necessary in iterative and incremental development. We demonstrated the advantages of using *BCEMD* in terms of minimizing the Cumulative Class Dependency (CCD) in the system. We illustrated our approach with an example from a single development project.

Figure 13 shows a complete model for our example. Despite expanding in-package structures, the CCD equals 37. This compares very favorably with simplistic structures in Figures 4 and 5.

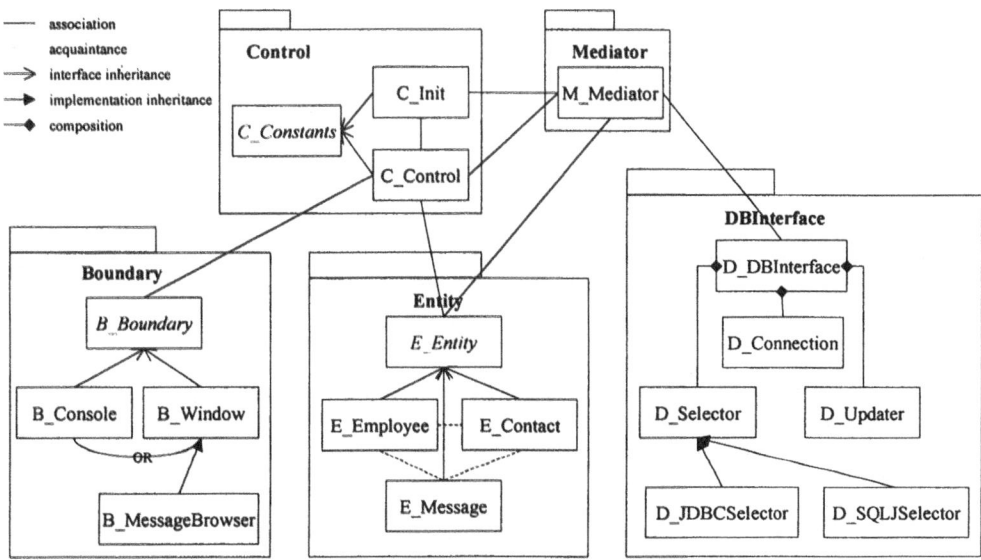

Figure 13. BCEMD graph with desirable internal package structures (CCD = 37)

When compared with Figure 4 (with CCD = 20), the CCD increase in Figure 13 results from:

- increase of 3 in Boundary due to:
  - o implementation inheritance (a subclass depends on its superclass but not vice versa) – CCD for B_Window is 2,
  - o interface inheritance (changes to B_Boundary need to be reflected in an interface implementation in B_Console or in B_Window)

- no increase in Control as the interface C_Constants represents constant values that are automatically reflected in concrete classes (changing in C_Constants has no impact on other classes),

- increase of 9 in Entity due to acquaintances,

- no increase in Mediator,

- increase of 5 in DBInterface due to:
  - ○ compositions (CCD for D_DBInterface is 5; it is 0 for the subset classes),
  - ○ implementation inheritance – CCD for D_Selector is 2 (it is 0 for subclasses).

Although not illustrated in this paper for space reasons, we expanded significantly the model in Figure 13 in successive iterations of the emailing project. Despite introducing many new classes in each package, the *BCEMD* framework evolved gracefully. The *BCEMD* framework ensures scalable system design demanded by iterative and incremental development.

REFERENCES

B. Eckel, (2000): Thinking in Java, 2nd ed., *Prentice-Hall*, 1128p.

E. Gamma, R. Helm, R. Johnson, and J. Vlissides, (1995*):* Design Patterns. Elements of Reusable Object-Oriented Software, *Addison-Wesley*, 395p.

G.E. Krasner, and S.T. Pope, (1988): A Cookbook for Using the Model View Controller User Interface Paradigm in Smalltalk-80, *J. Object-Oriented Prog.*, Aug-Sept, pp.26-49.

J. Lakos, (1996): Large-Scale C++ Software Design, *Addison-Wesley*, 846p.

L.A. Maciaszek, (2001): Requirements Analysis and System Design. Developing Information Systems with UML, *Addison-Wesley*.

L.A. Maciaszek, O.M.F De Troyer, J.R. Getta and J Bosdriesz, (1996a): Generalization versus Aggregation in Object Application Development - the "AD-HOC" Approach, *Proc. 7th Australasian Conf. on Information Systems ACIS'96*, Vol. 2, Hobart, Tasmania, Australia, pp.431-442.

L.A. Maciaszek, J.R. Getta, and J. Bosdriesz, (1996b): Restraining Complexity in Object System Development - the "AD-HOC" Approach, *Proc. 5th Int. Conf. on Information Systems Development ISD'96*, Gdansk, Poland, pp.425-435.

M. Page-Jones, (2000): Fundamentals of Object-Oriented Design in UML, *Addison-Wesley*, 458p.

J. Rumbaugh, I. Jacobson, and G. Booch, (1999): The Unified Modeling Language Reference Manual, *Addison-Wesley*, 550p.

SourceProDB (2001): SourcePro DB. Offering Power and Productivity for C++ Database Applications, White Paper, Rogue Wave Software (accessed from www.roguewave.com on 20-May-2001)

C. Szyperski, (1997): Component Software. Beyond Object-Oriented Programming, *Addison-Wesley*, 411p.

REFINING OEM TO IMPROVE FEATURES OF QUERY LANGUAGES FOR SEMISTRUCTURED DATA

Pavel Hlousek and Jaroslav Pokorny[1]

1. INTRODUCTION

Semistructured data can be explained as "schemaless" or "self-describing", indicating that there is no separate description of the type or structure of the data. This is in contrast with the structured approaches, such, e.g. relational databases, where the data structure is usually designed first and described as a database schema. Semistructured data is data whose structure is irregular, is heterogeneous, is partial, has not a fixed format, and evolves quickly. These characteristics are typical for data available in the Web (HTML pages, e-mail message bases, bookmarks collections etc). The research of semistructured data aimed at extending the database management techniques to semistructured data in the late 90's (Suciu, 1998).

An independent but strongly relevant development of XML (W3C, 1998), the eXtensible Markup Language, resulted in an agreement that XML became the *de facto* representation for semistructured data. Consequently, the data base research focused on XML as a source of new database challenges (Pokorny, 2001). Particularly, new XML data models and XML query languages have appeared and so called "native XML databases" have come into common usage among companies (Bourret, 2001).

In the research community, initial work on the semistructured databases was based on simple graph-based data models such as the Object Exchange Model (OEM) (Papakonstantinou, et al, 1995). Though XML and OEM are similar, there are some differences, and one of the most significant of them concerns data ordering. OEM and other original semistructured data models are set-based: an object has a set of subobjects. However, since XML is a textual representation, any XML document specifies the order inherently: an element has a list of subelements. Of course, some applications may treat the order as an irrelevant artefact of the serialization "forced" by an XML representation. In other words, we can use OEM model as a simpler alternative for management of some semistructured data collections.

OEM model can be still simplified. In OEM, cycles in the data graph are allowed. Since query languages for this data need to count with a possibility of cycles, they

---

[1] Dept. of Software Engineering, Faculty of Mathematics and Physics, Charles University, Malostranske nam. 25, Praha 1, Czech Republic, email: {hlousek|pokorny}@ksi.ms.mff.cuni.cz

Information Systems Development: Advances in Methodologies, Components, and Management
Edited by Kirikova *et al.*, Kluwer Academic/Plenum Publishers, 2002

become really complicated. Also the size of an answer to a query may contain data that are useless for the user, because with each node there is its whole subgraph returned. This is because the data in OEM model is held by nodes without outgoing edges.

On the other hand, data is often of a tree structure in terms of a part-of relationship.[2] If the data is modeled by a graph, then cycles can appear to represent some added information.[3]

Therefore, in Section 2, we refine the OEM model not to lose the notion of part-of relationships in the cyclic graph. This affects answers to queries, which with each node no longer return its whole subgraph, i.e. all accessible nodes, but they rather return only nodes that represent just the node's subparts. Because part-of relationships form a tree, we do not have to bother with cycles any more, and can make our query language evaluation much simpler. This can be very suitable for query languages with emphasis on the semantics of the data represented in XML, where the part-of relationship corresponds to the element-subelement relationship and cycles are realized through IDREF(S) attributes.

The fact that we can work with data as if it were a tree, creates many possibilities for query languages for semistructured data. One of them – default structuring – is presented in Section 4. Where other similar languages, like Lorel (Abiteboul, et al, 1996), XML-QL (Deutsch, et al, 1998), UnQL (Buneman, et al, 1996), and recently XQuery (W3C, 2001), force the user to use some "construct" clause to explicitly specify the result structure (or the default structure of these languages is a set), our proposed language provides default structuring of the result nodes by keeping the minimal structural context which these nodes had in the source data graph.

In the following text, we use an email message base as an example of semistructured data, and associated query language MailQL, which was developed as the author's master thesis (Hlousek, 2000). In the email message base, part-of relationships are e.g. folder-subfolder, folder-message, message-fields, and cycles are caused by edges representing e.g. message threads (described later). There are many other areas in the semistructured data where part-of relationships can be found: image and its subregions, XML data with elements containing elements, etc.

1.1 Goals

The goal of this paper stated in a single word is "simplicity" and is divided into two subgoals.

One is to reduce the complexity of evaluation of query languages for semistructured data by letting the evaluation work with data in a tree instead of a cyclic graph (though the cycles are not lost). This is achieved by the OEM refinement.

The other is to utilize the refinement to find a simpler syntax of a query language for semistructured data. To be more precise, the current query languages usually provide some construct clause to specify the structure of the result. They also provide some default structuring of the result when the construct clause is omitted, which means the

---

[2] By part-of relationship we mean the situation in part-subpart relationship, where parts may not appear in multiple aggregation. We can take a directory structure with files without any kind of links as a good example. Here a subdirectory (or file) is in a part-of relationship with its parent directory.

[3] Providing symbolic links can result in cycles in our graph that represents the directory structure with files. However, the tree structure behind the cyclic graph is not lost.

result is a set of nodes. Our goal is to find more structurally expressive default structuring of the result by reflecting the structure of the nodes in the source.

1.1 Structure

In Section 2 we present the OEM model and its proposed refinement. Section 3 briefly explains the syntax of MailQL that is used in examples. Section 4 introduces our proposed default structuring of the result based on keeping the minimal structural context, which the result nodes had in the source data base. Finally, in Section 5 we give some conclusions.

2. DATA MODEL

In this section, we define the OEM model and its refinement.

2.1 OEM Model

The Object Exchange Model (OEM), first appearing in the TSIMMIS project (Papakonstantinou, et al, 1995), is the de facto standard in the modeling of semistructured data. The following definition of OEM is borrowed from (Abiteboul and Suciu, 2000).

Definition 1. An *OEM object* is a quadruple (label, oid, type, value), where label is a character string, oid is the object's unique identifier, and type is either complex or some identifier denoting an atomic type (like integer, string, etc.). When type is complex, then the object is called a *complex object*, and value is a set of oids. Otherwise, the object is an *atomic object*, and value is an atomic value of that type.

Thus the data represented by the OEM model is held by the OEM objects of atomic type that are referred to by the OEM objects of complex type.

The OEM data is usually understood as an oriented graph with labeled nodes, where the OEM objects correspond to nodes, and for each node n representing the complex OEM object o = (label, oid, *complex*, value) there are edges leading from n to nodes that represent the OEM objects in o's value field. Figure 1 shows how this model can describe a part of our tree-like message base. It is also apparent from the picture that we use very common modification of the OEM model - labels are attached to edges, rather than to nodes. Thus we take our message base as an oriented edge-labeled data graph.

Definition 2. We say that m is an *l-subobject* of n if there exists an edge labeled l leading from n to m.

Henceforth, the OEM objects are simply called *objects*. We should also note that in the following text we mix the terms node and object, while according to the OEM definition a node represents an object, and vice versa.

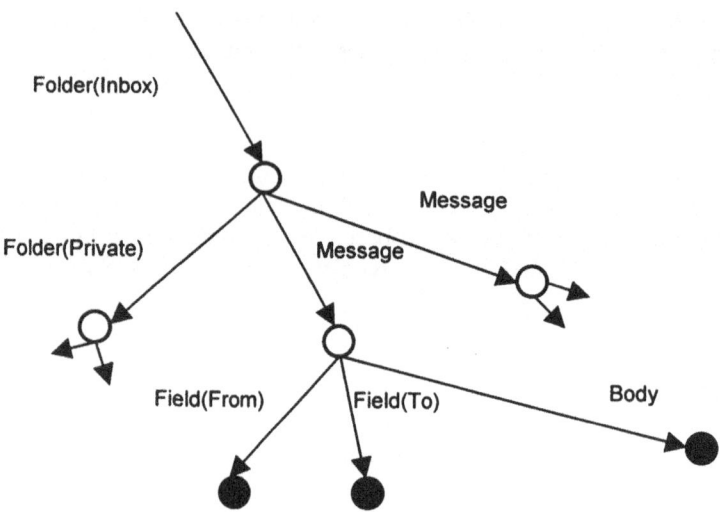

Figure 1. Part of a message base modeled by OEM. We can see folder Inbox which contains two messages, and subfolder Private. One message contains fields From and To, and a node representing its body. (Other children are omitted.) The complex nodes are distinguished from the atomic ones by empty circles. Since it is not essential to distinguish these two types of nodes in the following figures, we draw all nodes using filled circles.

2.2 Refining OEM Model

Now let us turn our attention to our example of an email message base. Let us enrich the message base model (which so far consists only of part-of relationships like a folder-message), with edges providing us with some added information. Let us add to the model message threads, which help us organize email messages by dialogs in which these messages appeared. A bit more formally, a *message thread* of message m is a set of messages T_m that is generated by a reflexive, symmetric, and transitive closure of a binary relation *is-reply-to* between messages from the message base.

In the message base model the message thread of m is expressed by edges labeled thread, leading to all nodes that represent a message in the message's thread T_m.[4]

We mentioned earlier that with each node there is its subgraph returned as an answer to a query in other OEM-oriented languages. So with each message, all messages from its thread must always be returned. However, this OEM specific behavior does not satisfy us, because we might be interested in the messages only, without caring about their threads, which can be in some cases very large, and so they could wastefully enlarge the size of the result.

Therefore we refine the OEM model to distinguish between two types of edges. This refinement is described by Definition 3.

---

[4] In figures throughout this paper, we omit edges that represent identity.

Definition 3. (i) *Core edges* are edges describing the tree structure of data, given by part-of relationships. (ii) *Secondary edges* are edges describing added information. *Core (secondary)* paths are oriented paths consisting only of core (secondary) edges.

Definition 3 says that core edges are edges describing part-of relationships (e.g. edges from the message to its fields), while secondary edges are edges describing other than part-of relationships (e.g. edges from message to its message thread).

Definition 4. By *core data tree* of data graph $G = (V, E)$ we denote its subgraph $G' = (V, E')$, where $E' \subseteq E$ consists of all core edges.

We should note that the data graph and the core data tree share the same nodes. The only difference is that the core data tree is composed exclusively of core edges, and therefore it is always a tree, because the core edges represent the part-of relationships, while the complete data graph can contain cycles which are caused by the presence of secondary edges.[5]

We should also note that the refined OEM model can hold the information about the types of edges, but the information about an edge type must come from the outside world.

Now back to our example. It is clear that the thread edges will be marked as secondary. By enriching the treelike message base model by these edges (that provide some additional information to the hierarchical structure of data), the data graph is no longer a tree: it is a cyclic graph. But using Definition 3 and Definition 4, we do not lose the notion of the previous tree in the data graph (due to the core edges). Figure 2 illustrates this situation on our message base model enriched by thread-edges.

The introduced refinement was done to achieve this feature: let the query language evaluator work with the core data tree instead of working with the (possibly cyclic) data graph. Specifically, with each result node there will be its subgraph from the core data tree in the result. So asking for a message, we will not put the messages from its thread in the result, as is the case in OEM. Furthermore, the result, just before being shipped to the user, will be provided with those secondary edges where both nodes of each secondary edge were preserved. Thus the messages from some message's thread that got into the result, will remain connected with the thread-edges as they were in the source data base.

2.3 Path Expressions

Now that our semistructured data model has been refined, we need a way to navigate through the data graph. Nowadays, the most suitable tool seems to be *path expressions*. Their presence in languages for the semistructured data is almost a rule because of their navigational syntax.

If o is object and l is label, then by expression $o.l$ we denote the set of l-subobjects of o. We should notice that $o.l$ always denotes a set of objects. This semantics of path expressions is typical for semistructured data.

---

[5] Here we have made a simplification talking about a data tree. Instead of tree, we might talk more generally about a rooted acyclic graph, but there is no major consequence for this paper in distinguishing these two but the number of roots. Henceforth, by the root of a data tree we will understand any root in the rooted acyclic graph.

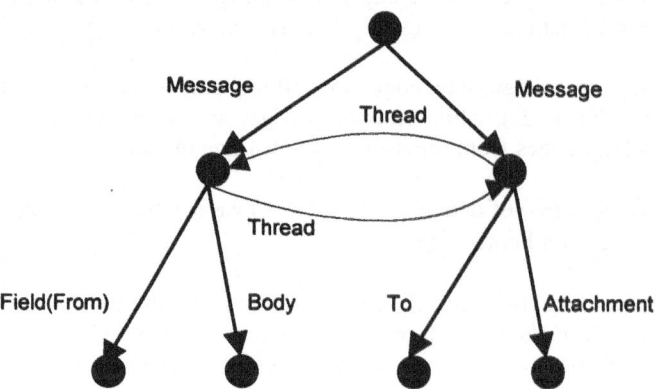

Figure 2. Part of a message base with core edges (drawn with thicker lines) and secondary edges (drawn with thinner lines). Here one message is a reply to the other. The message-nodes have their outgoing edges heavily reduced in order to simplify the figure; normally, there would be many more of them.

Simple path expression is expression $r.l_1 ... l_n$, where r is root node and $l_1 ... l_n$ are edge labels. *Data path* is a sequence of $o_0, l_1, o_1, ..., l_n, o_n$, where o_i are objects, and for each i there is an edge between o_{i-1} and o_i labeled l_i. According to these definitions we can see that there can be more than one data path that satisfies some simple path expression.

Semantics of simple path expressions is very intuitive. We will explain it on the example of root.A.B. Expression root denotes the starting object. Expression root.A denotes set X of all objects for which there exists an edge leading to them from root and labeled A. Expression root.A.B denotes then set Y of all objects for which there exists an edge leading to them from any object in X and labeled B.

Thus each simple path expression denotes a set of objects, even if there is no data path satisfying it. In such a case, the path expression denotes an empty set.

General path expressions enhance the power of simple path expressions by enabling use of wild cards and regular expressions in path expressions.

With a wild card we can substitute either an edge label (using %) or a sequence of edge labels (using *). Thus expression root.%.B means root.*any_label*.B, and expression root.*.Z means root.*any_path*.Z.

Regular expressions enable use of path expressions such as root(.A|.C).B which matches two simple path expressions root.A.B and root.C.B and so it results in the union of two sets of objects.

Usually, there are many more constructs that are typical for regular expressions, and the usage of wild cards could be widened much more. But talking about this is not the goal of our paper.

Using path expressions, we often refer to a common prefix of two or more path expressions. First we define predicate *IsPrefix*, a definition of a common prefix follows. We should note that by a common prefix we always mean the longest common prefix.

Definition 5. Let pe_1 and pe_2 be path expressions. The $IsPrefix(pe_1, pe_2)$ predicate is true if $pe_1 = X.l_1...l_n$ and $pe_2 = X.l_1...l_{n+m}$, where $n \geq 0$ and $m \geq 0$.

Definition 6. Let pe_1 and pe_2 be path expressions. By *common prefix* we denote path expression $pe = X.l_1...l_k$, where both $IsPrefix(pe, pe_1)$ and $IsPrefix(pe, pe_2)$ are true, and there is no such path expression $pe' = pe.l_{k+1}$ for which both $IsPrefix(pe', pe_1)$ and $IsPrefix(pe', pe_2)$ would be true.

3. EXAMPLE SYNTAX

As introduced earlier, MailQL is a query language for an email message base which we use here as an example of the semistructured data modeled by the refined OEM model. MailQL queries borrow their syntax from OQL (Cattel et al, 2000) and its semistructured-data-oriented successor Lorel (Abiteboul et al, 1996).

```
SELECT list_of_path_expressions
FROM list_of_aliases_for_path_expressions
WHERE boolean expression
```

The following example shows a simple query in MailQL, which returns fields From and Date of all messages that contain string 'MailQL' in its Subject field.

```
SELECT m.From, m.Date
FROM Inbox.Message: m
WHERE m.Subject CONTAINS 'MailQL'
```

Note that FROM clause in a MailQL query plays a different role than in other syntactically similar languages: here it only defines aliases (m) for path expressions (Inbox.Message). Before the query is executed, all occurrences of aliases are substituted by appropriate path expressions in the SELECT and WHERE clause, therefore the FROM clause is no longer needed after that.

4. AUTOMATIC CONSTRUCTION OF THE RESULT STRUCTURE

In this section, we introduce an interesting use of the OEM refinement. So far, we considered what to return with nodes specified in the SELECT clause. It means we were inspecting the part of the core data tree on the path from these nodes to leaves. Now we switch our attention to the other part of the core data tree: the path from the root to nodes specified in the SELECT clause. The question is: In what structural relationships should the nodes, specified in the SELECT clause of a query, be?

Current languages for semistructured data usually provide some "construct" clause, which explicitly defines the structure of the result. Some of them provide default structuring, which means returning a set of tuples. But having the core data tree, we can improve the default structuring by keeping the minimal structural context which the specified nodes had in the source data tree. Simply said, the path expressions from the SELECT clause specify nodes in the data graph. All these nodes will be returned (with their subgraphs, as described in Section 2). And we also want all these nodes to stay in the same structural relationships to each other as they did in the source data tree. This could be surely realized even in such a way that we would preserve the whole paths leading to these nodes from the root of the core data tree. But our solution vertically reduces these paths and keeps from them only the "interesting" nodes, i.e. nodes, in which the path forks to reach the specified nodes.

Compared with other languages for semistructured data, we might miss the strong result restructuring features, but we think that in many cases the user needs to see the minimal structural context of the data, which is just what our language provides. Furthermore, all introduced features of our language could be incorporated into any query language with strong formatting options and thus provide default structuring of the result.

4.1 Minimal Structural Context

Let us first formalize vertical reduction.

Definition 7. Let $T = (V, E)$ be a tree where V is a set of nodes, and E is a set of edges. We say that tree $T' = (V', E')$ is a *vertically reduced tree* of T if both the following conditions are true:

i) $V' \subseteq V$
ii)   for each $e' = (v'_1, v'_2) \in E'$ it is true that either (a) $e' \in E$ (an edge preserved from T); or (b) there exists sequence $e_1,...,e_n$ of edges *from* $E \setminus E'$ and sequence $v_{k2},..., v_{kn}$ of nodes from $V \setminus V'$, such that $v'_1, e_1, v_{k2},..., v_{kn}, e_n, v'_2$ is a path in T.

Definition 7 says that T' was obtained from T by leaving out some nodes in such a way that if there was a path between two nodes in T and if both nodes were preserved in T', then there must exist a path between them in T', as well. (T' is sometimes called a *minor* of T in the graph theory.)

Now let us turn our attention to the "interesting" nodes. As declared at the beginning of this section, the only nodes that are preserved from paths leading to the queried data are those, where these paths fork. Given two path expressions, we can determine the path expression of nodes where the paths fork by finding out the common prefix of these two path expressions. So by the "interesting" nodes we understand those nodes that are represented by the common prefixes of path expressions from the SELECT clause.

To describe the minimal structural context of the result nodes, we use a data structure called a *result structure tree* (*RST*), which helps us specify the (vertically reduced) structure of the result. Formally, a result structure tree is a tree RST = (V, E) with mapping PEI (path expression infix) defined for all its nodes. Each node is mapped by PEI to a part of path expression in such a way that the common prefix is empty for each pair of siblings in RST. PEI of the root node always maps to an empty path expression.[6]

To get the structure of the result for a certain query, we take all path expressions from the SELECT clause of the query and construct an appropriate RST of them. The construction is directed by the common prefixes of path expressions. We start with the

---

[6] If the data tree forms a rooted acyclic graph, then the *RST* forms a rooted acyclic graph as well, but it has always just one root, which represents an empty path expression. So if there are more roots getting to result from the source data graph, then they are represented as children of the root in *RST*.

root node which is always present and is always assigned an empty path expression. The first path expression will be represented by a child node of root, where `PEI` of that node will be the path expression itself. When adding every other path expression to the RST, we find its common prefix with each of the path expressions that are already represented by RST. If there is a non-empty common prefix, then some splitting must take place in order not to violate the definition of RST. Figure 3 illustrates in three steps how RST is created for the set of path expressions {Inbox.Private.Message.To, Inbox.Private.Message.From, Inbox.Private.Name}.

Now, the inner nodes of RST, which represent common prefixes of a path expression in the SELECT clause of a query, represent the path expressions of the "interesting" nodes that we spoke about above – they are the ones which express the minimal structural context of the returned data – while the leaf nodes represent the data which are to be returned.

Henceforth, we will denote the structure of the result using XML-like syntax (according to Buneman at al, 1996). So the structure of the final tree in Fig. 3 will be written down like this:

```
{<Inbox.Private>
    {<Name>...</>}
    {<Message>
            {<To>...</>}
            {<From>...</>}
    </Message>}
</Inbox.Private>}
```

4.2 Structure-Forcing Operator

Sometimes, it may be convenient to keep more of the structure in the result. For example, for the query

```
SELECT Inbox.Message.To
```

the result structure will be a set of recipients {<Inbox.Message.To>...</>}. But we may want to distinguish the recipients according to the messages they come from. Therefore we introduce a *structure-forcing operator* <...>, which forces the "interestingness" of a path expression. Thus the desired query can be formulated

```
SELECT <Inbox.Message>.To
```

4.3 From RST To Result

Once we have the RST representing the result structure, we can easily generate its result tree. Each node from RST may correspond to several nodes of the core data tree (e.g. Inbox.Message denotes a set of messages). Thus the result for RST in Fig. 3 can look in XML syntax like this:

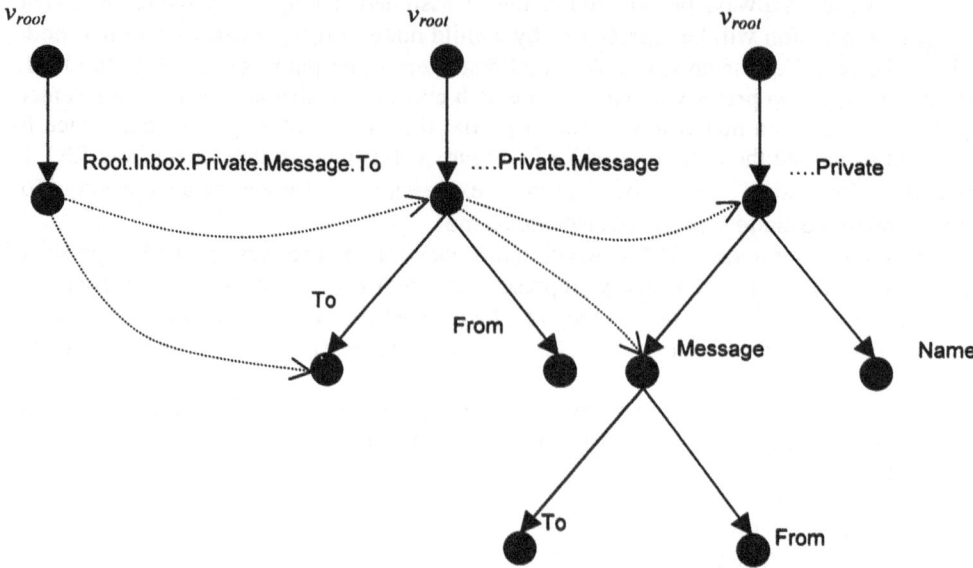

Figure 3. Example of RST construction. The dotted lines indicate node splitting, the text attached to the node is the value of the node's PEI. The three dots in the second and the third RST are an abbreviation for Root.Inbox.

```
<Inbox.Private>
    <Name>Private</>
    <Message>
            <To>julius.satinsky@theatre.sk</>
            <From>fan@home.cz</>
    </Message>
    <Message>
            <To>julius.satinsky@theatre.sk</>
            <To>milan.lasica@theatre.sk</>
            <From>fan@home.cz</>
    </Message>
</Inbox.Private>
```

4.4 The Added Value

At the beginning of this section, we stated that the MailQL query language is simpler than other languages for the semistructured data, because it provides a feature of keeping the minimal structural context of the queried data. To demonstrate this fact, we compare here the syntax of two queries, which we write in MailQL and Lorel, to get the result structure of the final tree in Fig. 3.

Using MailQL, we have two basic options to express the query. One that uses the FROM clause

```
SELECT P.name, M.to, M.from
FROM inbox.private: P, P.message: M
```

or the other without the FROM clause

```
SELECT  inbox.private.name,
        inbox.private.message.to,
        inbox.private.message.from
```

Both of them have the same output of the desired structure.

To get the same result in Lorel, we have to write a query that is much more complex, because (i) we have to specify the structure of the result in the SELECT clause of the query and (ii) in the FROM clause we have to specify all the bounding variables to keep the structure of the queried data.

```
SELECT inbox.private: {name: N, message: {to: T, from: F}}
FROM inbox.private: P
P.name: N,
P.message: M,
M.to: T
M.from: F
```

It is obvious that our query language wins in all situations where the result structure has to reflect the structure of the queried data. And on the other hand, MailQL looses in all situations where some major restructuring of the queried data is needed. Therefore we propose the mechanism of the automatic creation of the result structure to be used as part of some query language with stronger restructuring capabilities.

5. CONCLUSIONS

Leaning on the fact that most semistructured data (e.g. XML data) is of tree structure in terms of part-of relationships, we refined the OEM model not to lose the notion of this tree in a complex data graph with cycles. This refinement allows us, and generally all languages for the semistructured data, to work with the data as if it were a tree, and so it allows us not to bother with the complexity caused by cycles in the data graph. As other use of our OEM refinement, we introduced bases of a query language that provides default structuring of a query result based on keeping the minimal structural context as a different approach to the default result structuring of languages for the semistructured data.

It also has to be said that our refinement is useful in cases, in which there is a rule helping us decide, whether an edge is a core one or a secondary one. Here, XML can serve as a good example with the rule: let us represent the element-subelement relationships by core edges, and let us represent the cycles-causing IDREF and IDREFS attributes by secondary edges.

The query language presented here (originally MailQL) was designed and implemented for an email message base as part of a master thesis and is further under development.

6. ACKNOWLEDGEMENT

This work was supported in part by the GACR grant No. 201/00/1031.

7. REFERENCES

Abiteboul, S., Quass, D., McHugh, J., Widom, J., and Wiener, J., 1996, The Lorel query language for semistructured data, *International Journal on Digital Libraries.* 1(1), pp. 68-88.

Abiteboul, S., and Suciu, D., 2000, *Data on the Web: From Relations to Semistructured Data and XML*, Data Management Systems, 1st edition, Morgan Kaufmann.

Bourret, R., 2001, XML and Databases; http://www.rpbourret.com/xml/ /XMLAndDatabases.htm.

Buneman, P., Davidson, S., Hillebrand, G., and Suciu, D., 1996, A query language and optimization techniques for unstructured data, (JAGADISH, H.V. and MUMICK, I.S. Eds.), SIGMOD, pp. 505-516. ACM Press.

Bray, T., Paoli, J., and Sperberg-McQueen, C. M., 1998, Extensible Markup Language (XML) 1.0, February 1998; http://www.w3.org/TR/1998/REC-xml-19980210.

Cattell, R.G.G. et al., 2000, *The Object Database Standard: ODMG 3.0*, Morgan Kaufmann Publishers, Inc.

Deutsch, A., Fernandez, M., Florescu, F., Levy, A., and Suciu, D., 1998, XML-QL: A query language for XML; http://www.w3.org/TR/1998/NOTE-xml-ql-19980819.html.

Hlousek, P., 2000, MailQL, query language for an email message base, Master's thesis, Charles University, Prague. In Czech.

Papakonstantinou, Y., Garcia-Molina, H., and Widom, J., 1995, Object exchange across heterogeneous information sources, in *Proceedings of the Eleventh International Conference on Data Engineering*, Yu, Ph. S. and Chen, A.L.P. eds., pp. 251-260, IEEE Comp. Soc.

Pokorny, J., 2001, XML: a challenge for databases? Chap. 13 in: *Contemporary Trends in Systems Development*, Sein, M. et al, eds., Kluwer Academic Publishers, Boston, pp. 147-164.

Suciu, D., 1998, An Overview of Semistructured Data, SIGACTN: SIGACT News (ACM Special Interest Group on Automata and Computability Theory), 29.

W3C, 1998, Extensible Markup Language (XML) 1.0; http://www.w3.org/TR/REC-xml

W3C, 2001, XQuery 1.0: An XML Query Language W3C Working Draft 07; http://www.w3.org/TR/xquery/.

DERIVING TRIGGERS FROM UML/OCL SPECIFICATION*

Mohammad Badawy and Karel Richta[†]

1. INTRODUCTION

The term integrity is used to refer to the accuracy or correctness of the data in a database. In other words, integrity involves ensuring that the data stored in the database are in any time correct. The database management system needs to be aware of certain rules that users must not violate. Those rules are to be specified in some suitable language, and have to be maintained in the data catalogue[1].

Integrity enforcement can be divided into two categories - static and dynamic. Static integrity enforcement is the task of ensuring that the data in a database are in a legal state. Dynamic integrity enforcement is the task of ensuring that a user transaction applied to a legal database state leads to a new state, which is also legal.

The common rationale of research in this area is to centralize the management of data integrity. One possible solution of this problem is to extract the data integrity management from application programs and bringing it into an ad-hoc component, which may be incorporated into the *active database management system*[2].

The specification of what data are semantically correct constitutes one of the most important tasks in the database design process[3, 4]. In this process, data correctness requirements are gathered from users, business rules, and applications developers, and are translated into integrity constraint specification[5].

An active database management system continually monitors the database state and reacts spontaneously when predefined events occur. Functionally, an active database management system monitors *conditions* triggered by *events* representing database events

* This work has been partially supported by the research program no. MSM 212300014 "Research in the Area of Information Technologies and Communications" of the Czech Technical University in Prague (sponsored by the Ministry of Education, Youth and Sports of the Czech Republic.

† Mohammad Badawy, Dept. of Computer Science & Engineering, Faculty of Electrical Engineering, CTU in Prague, Karlovo ndm. 13, 121 35 Prague 2, Czech Republic, badawm1@cslab.felk.cvut.cz_._Karel Richta, Dept. of Computer Science & Engineering, Faculty of Electrical Engineering, CTU in Prague, Karlovo ndm. 13, 121 35 Prague 2, Czech Republic, also Dept. of Software Engineering, Faculty of Mathematics & Physics, Charles University, Prague 1, Malostranské ndm. 25, 118 00, Czech Republic, richta@fel.cvut.cz

Information Systems Development: Advances in Methodologies, Components, and Management
Edited by Kirikova *et al.*, Kluwer Academic/Plenum Publishers, 2002

305

or non-database events (e.g. hardware failure); and if the condition is satisfied (evaluates to true), the action is executed[6].

Active databases are taking a prominent role in commercial database applications[7, 8, 9, 10]. With client/server solutions, applications are being developed by small, autonomous groups of developers with narrow views of the overall enterprise; the enterprise information system is very vulnerable to integrity violations because it lacks strict enforcement of the enterprise business rules[2]. An active database management system should support constraints as well as event-driven application logic.

Triggers that run automatically when certain events occur in the database are used to enforce database rules and to actively change data[11]. Triggers have a very low impact on the performance of server, and are often used to enhance applications that have to do a lot of cascading operations on other objects[12]. Triggers provide a very powerful and flexible means to realize effective constraint enforcing mechanisms[5]. On the other hand the trigger approach is not yet well based. Triggers have limitations and pitfalls you should be aware of. Otherwise trigger systems can do what they want and not what the user wants.

The goal of this paper is to give the rules for implementing triggers from UML/OCL integrity constraint specifications. These rules are independent of any particular commercial database. A comparison of advantages of declarative constraints and triggers is also discussed. The text is structured as follows: Section 2 discusses the relative advantages of constraints and triggers. In Section 3, we give the trigger syntax. Section 4 gives some rules used to derive triggers from integrity constraint specifications, which are discussed in Section 5. An applicable example is given in Section 6, and finally, Section 7 gives the conclusions.

2. INTEGRITY CONSTRAINTS AND TRIGGERS

The database engine has to provide two ways of enforcing data integrity - declarative constraints and procedural constraints (triggers). Declarative constraints should be used in lieu of triggers whenever possible. Several triggers may be required to enforce one declarative constraint; even then, the system has no way of guaranteeing the validity of the constraint in all cases[2]. Consider the database load utilities in which the database checks the declarative constraints against the loaded data before it can be accessed. There is no way to determine which triggers should be checked since triggers are also used for transitional constraints and for event-driven application logic. This behavior also applies when constraints and triggers are added to a database with pre-existing data.

The declarative constraints provided by most relational systems are defined in SQL-92 (the same constraints are defined in SQL: 1999) to support only a small, although useful, set of static constraints that define the acceptable states of the value in the database. They do support only a limited subset of transitional constraints that restrict the way in which the database value can change from one state to the next. They do not support event-driven invocation of application and business logic. Hence, triggers are required to enhance the declarative constraint constructs and to capture application specific business rules. Triggers provide a procedural means for defining implicit activity during database modifications. They are used to support event-driven invocation of application logic, which can be tightly integrated with a modification and executed in the database engine by specifying a trigger on the base table of the modification. Triggers should not be used as a replacement for declarative constraints. However, they extend the constraint logic with transitional constraints, data conditioning capabilities, exception handling, and user defined repairing actions[2].

In summary, there are advantages to using both declarative constraints and procedural triggers, and both types of constructs are available in many commercial systems. It is not feasible to expect applications providers to either migrate their existing applications to use only triggers or partition the tables in their database according to the type of constraints and triggers that are required. It is therefore imperative to define and understand the interaction of declarative constraints and triggers[2].

3. TRIGGER SYNTAX

SQL: 1999[13] provides the concept of triggers. A trigger is a procedure that is automatically invoked by the DBMS in response to specified database events. Triggers can be viewed as event-condition-action (ECA) rules that allow users to implement application logic within the DBMS. Triggers can be used to monitor modifications of the database, to automatically propagate database modifications, to support alerts, or to enforce integrity constraints[5].

A trigger in SQL: 1999 has the following components:

- A unique *name*, which identifies the trigger within the database,
- A triggering *event*, which are for our purposes INSERT, DELETE, or UPDATE on a database table,
- An *activation time*, which is BEFORE or AFTER executing the triggering events (it specifies when the trigger should be fired),
- A trigger *granularity*, which is FOR EACH ROW or FOR EACH STATEMENT,
- A trigger *condition*, which can be any valid SQL condition involving complex queries (The WHEN clause contains the violating condition),
- A triggered *action*, which can be any valid sequence of statements.

A trigger is implicitly activated whenever the specified event occurs. Thus, the database can react to changes made by applications or ad-hoc users. Several triggers may refer to the same event.

4. INTEGRITY CONSTRAINTS SPECIFICATION

Typically, an integrity constraint can be formulated in such a way that all qualified rows from a table or a combination of tables have to satisfy a condition. We will use Object Constraint Language (OCL)[14, 15, 16] for integrity constraints specification. Let us suppose that we have objects of a type T. The constraint specification in OCL has the following form:

```
context <T> inv <name> : <condition>
```

where name is the constraint identification, and condition is an OCL expression over elements of T. The condition may include OCL functions, and it can include quantifiers, typically used to express general conditions over variables. From the logical point of view we can look at the OCL invariant specification as to the first-order logical formulae:

$$(\text{<name>}) \quad \forall x \in T : \text{<condition>}$$

Let us suppose a sample application - we have to manage information about employees and departments. Employees are uniquely identified by id. We assume that every employee has a first name and a last name. Every employee has a supervisor, except the top manager who supervises himself. We require remembering job, salary, bonus, and department for any employee. The declaration of employee (EMP) is as follows:

```
EMP(id,lastname,firstname,job,supervisor,salary,bonus,dept)
```

Whereas, each department has a unique id, a unique name, and a location. The declaration of the department (DEPT) looks as follows:

```
DEPT(id,name,location)
```

Example 1. Let C1 be an integrity constraint stating that *"the salary of a manager must be greater than the salary of an employee"*. This constraint can be expressed in OCL as follows:

```
context EMP inv C1 : forAll(e1,e2 |
e1.id = e2.supervisor implies e1.salary > e2.salary)
```

The equivalent logic formula can be:

```
(C1) ∀e1,e2 ∈ EMP: e1.id = e2.supervisor ⇒
              e1.salary > e2.salary
```

Example 2. Suppose there is another integrity constraint C2 defining *"Every department has at least one project managed by the department"*, in OCL:

```
context DEPT inv C2 : exist( p: PROJ | d.id = p.managed_by)
```

Example 3. Finally, assume there is an integrity constraint C3 stating, *"The total salary of the employees working in a department must not exceed the department's budget"*, in OCL:

```
context d:DEPT inv C3 : d.budget >=
iterate(e:EMP sum = 0 | e.dept = d.id implies e.salary +
                         sum )
```

5. DERIVING TRIGGERS FROM CONSTRAINT SPECIFICATIONS

Before going into the details how triggers can be derived from constraint specifications, it is worth mentioning that this part assumes only integrity constraints that cannot be implemented using any declarative specifications.

The SQL standard used in this paper is SQL: 1999[13]. The reference DBMSes used are Oracle8i Server (Release 8.1.6),[17] IBM DB2 Universal Database (Version 7),[18] Informix Dynamic Server (Version 9.1),[19] Microsoft SQL Server (Version 7.0),[20] Sybase Adaptive Server (Version 11.5),[21] and Ingress II (Release 2.0)[22].

5.1. The Deriving Rules

Given an integrity constraint C, the procedure to derive triggers has to determine the following:

- Trigger event,
- Trigger granularity,
- Trigger condition, and
- Trigger activation time.

5.1.1. Determining a Trigger Event

The first task is to determine critical operations; that is, database modifications that can lead to a violation of C. Such operations are insert, update, and delete statements on tables. These operations eventually determine triggering events on base tables as part of any trigger specification. If one misses a critical operation, then the integrity constraint can be violated without the system reacting to the violation. By analogy, if an operation has been identified that actually can never violate the integrity constraint, implementing a respective trigger would be meaningless and only leads to a decrease in the system performance. So how can such critical operations be determined?

Let us suppose the integrity constraint C1 *"The salary of a manager must be greater than the salary of an employee" expressed* in OCL:

```
context EMP inv C1 : forAll(e1,e2 |
e1.id = e2.supervisor implies e1.salary > e2.salary)
```

Obviously, a deletion of an employee or manager from table EMP will not violate this integrity constraint. Only insertions into the table EMP and updates of the column salary can lead to a constraint violation. Updates on the column salary can even be refined further. Only salary increases for employees and salary decreases for managers can lead to a constraint violation.

Suppose the integrity constraint C2 *"Every department has at least one project managed by the department"*, in OCL:

```
context DEPT inv C2 : exist( p: PROJ | d.id = p.managed_by)
```

Again, for universally quantified variables, only insertions (and updates, here on the column id of table DEPT) on the associated table can lead to a constraint violation. In contrast, for tables covered by existentially quantified variables, deletions (and updates, here on the column managed_by of the table PROJ) from the associated table can lead to a constraint violation. But if the tables covered by negation of the existentially quantified variables, insertion (and updates, here on the column managed_by of the table PROJ) from the associated table can lead to a constraint violation.

Finally, assume the integrity constraint C3 *"The total salary of the employees working in a department must not exceed the department's budget"*, in OCL:

```
context d:DEPT inv C3 : d.budget >=
iterate(e:EMP sum = 0 | e.dept = d.id implies e.salary +
                     sum )
```

Of course, decreasing a department's budget can violate this integrity constraint. Also, insertions into the table EMP and updates of the column salary of table EMP can lead to a violation of this integrity constraint.

As a general rule for deriving critical operations from integrity constraint specifications we thus have:

- For tables that are covered by universally quantified variables (and by negation of the existentially quantified variables), insert operations are critical.
- For tables that are covered by existentially quantified variables, delete operations are critical.
- In both cases update operations on columns used in comparisons are critical.

Note that critical update operations can be refined further based on the type of comparison columns are involved in. Such refined updates prove to be useful for specifying triggering conditions. Critical operations are then based on the underlying base tables. Determining a critical set of operations for a set of integrity constraints eventually results in a set of ⟨table, operation⟩ pairs that specify critical operations on tables.

5.1.2. Determining Trigger Granularity

The second task is to determine for each such pair whether the check of the underlying integrity constraint can be performed for each individual row from the table affected by the operation, or only for all rows affected by that operation. In the former case, a *row-level trigger* should be used, and in the latter case a *statement-level trigger*. Again, some general rules can be given that are applicable to all reference systems[5]:

- Almost all types of integrity constraint verifications based on ⟨table, operation⟩ pairs can be accommodated in *statement-level triggers*. For performance reasons, however, *row-level triggers* are preferable because they allow tailoring of verifying conditions to modified rows only.
- Integrity constraints that include aggregate functions always require at least *one statement- level trigger*, defined for the table and rows over which the aggregation is performed.
- In all reference systems, only *row-level triggers* allow a **WHEN** clause verifying properties of rows to be checked. This holds in particular for rows involved in the verification of state transition constraints. Typically, state transition constraints and their critical operations can only be verified using *row-level triggers*.

5.1.3. Determining a Trigger Condition

Once triggering *events* and trigger *granularities* have been determined, the next step in implementing constraint enforcing triggers is to formulate SQL statements that verify whether the integrity constraint from which a ⟨table, operation⟩ pair has been derived is violated. This is a crucial design task since one cannot use the original constraint but has to use its negation. That is, one has to specify a condition that checks whether there exists a row (or a combination of rows) for which the integrity constraint is violated. If such a row (or combination of rows) exists, the triggering action specifies how to react to the constraint violation, which is typically a rollback of the transaction. For instance, the violating condition for the integrity constraint C1 in Example 1, is formulated in SQL as follows:

```
CHECK (NOT EXISTS
        (SELECT * (FROM EMP e1, EMP e2
                   WHERE e1.id = e2.supervisor
                     AND e1.salary <= e2.salary)))
```

Note that in some commercial database systems (e.g., DB2), the trigger condition specified in the **WHEN** clause can contain such a verifying condition. In other systems (e.g., Oracle), the condition must be evaluated in the trigger body.

5.1.4. Determining Trigger Activation Time

Associating the "right" activation time with a trigger is highly system dependent. In almost all cases, **AFTER** triggers are sufficient. **BEFORE** triggers associated with a table are typically evaluated before declarative constraints are evaluated for the table. Note that there are several subtle differences among the implementation of the SQL: 1999 activation time feature in current systems. For example, in DB2 the granularity **FOR EACH STATEMENT** is not supported for **BEFORE** triggers and rows cannot be modified in **BEFORE** triggers, whereas in Oracle modifications of affected rows are only allowed in **BEFORE/FOR EACH STATEMENT** triggers.

Here, we would like to point out one important aspect of grouping verifying conditions into single triggers. Different integrity constraints, of course, can have the same critical operations. Thus, it seems reasonable to specify respective verifying conditions in one trigger associated with a ⟨table, operation⟩ pair, assuming they can be enforced using the same trigger granularity. For Sybase and Informix, this is the only option since only one trigger per event is allowed. Having all verifying conditions in one trigger also is beneficial regarding the performance of the system. However, practice often shows that integrity constraints are modified, added, or become obsolete. If the verifying conditions for a ⟨table, operation⟩ pair has been specified in one trigger; the whole trigger needs to be modified. Also, if an integrity constraint has to be (temporary) disabled, a trigger can only be disabled as a whole. In certain situations, it is desirable to disable integrity constraints of a table temporarily for performance reasons, e.g. when loading large amounts of data into the table using a loader, when performing batch operations that make massive changes to the table, or when exporting or importing a table.

5.1.5. Deriving Summary

Finally, we have a set $C = \{C_1, C_2, ..., C_n\}$ of integrity constraints that can only be enforced using triggers. Then, the following steps are applied to C to derive the respective constraint enforcing triggers (we assume that per event several triggers can be specified):

1. For each integrity constraint C, determine the set of critical operations in a set of ⟨table, operation⟩ pairs. Refine critical update operations according to the type of condition(s) specified in C.
2. For each critical operation, determine whether a *row-level trigger* is sufficient or a *statement-level trigger* is necessary.
3. Determine whether a **BEFORE** or **AFTER** trigger is necessary (system dependent).
4. For a *row-level trigger*, define a triggering condition in the trigger *condition* (**WHEN** clause).
5. Determine the violating *condition* from the original integrity constraint C, and formulate the condition either in the **WHEN** clause (if allowed/provided by the systems) or in the trigger body.

6. AN EXAMPLE

In this section we give an example of how to derive triggers from constraint specifications according to the above rules. Given the integrity constraint C1:

"The salary of a manager must be greater than the salary of an employee "

Firstly, the specification of this constraint in the OCL is as follows:

```
context EMP inv C1 : forAll(e1,e2 |
e1.id = e2.supervisor implies e1.salary > e2.salary)
```

Secondly, we convert the OCL formulation into triggers (we use Oracle as a reference system) according to the deriving steps given above, as follows:

1. By applying the general rules for deriving critical operations pairs, we have the following pairs: ⟨EMP, insert⟩ and ⟨EMP, update (salary)⟩.
2. For the two critical operations above **row-level triggers** are sufficient but we also use in this case a **statement-level trigger** to overcome the mutating table problem.
3. In our reference system both **BEFORE** and **AFTER** are available.
4. It is not possible to define the triggering *condition* in trigger condition (**WHEN** clause) because it has a subquery, and subqueries are not allowed in the **WHEN** clause of the reference system.
5. The violating *condition* is inserting an employee with salary more than the manager's salary or updating the salary column so that the salary of any employee becomes more than the manager's salary. The violating *condition* has to be formulated in the trigger body.

From the above five steps, we can derive from the constraint specifications the triggers used to implement the above constraint C1. In this case we will use three triggers: one **row-level trigger** and two **statement-level triggers,** as follows:

- The first trigger *man_emp_sal1* is a **BEFORE/STATEMENT** trigger. It handles the temporary table, which we used to avoid the mutating/constraining table problem. We create a temporary table as follows:

```
CREATE GLOBAL TEMPORARY TABLE TEMP_EMP(
supervisor    number(38),
id            number(38),
salary        number(7,2))
ON COMMIT PRESERVE ROWS;
```

- The second trigger *man_emp_sal2* is an **AFTER/ROW** trigger. It is used to test the integrity violation and if so, it stores the old state of the table in the temporary table (TEMP_EMP) to retrieve it later.
- The last one *man_emp_sal3*, which has an effect only when an integrity violation occurs, is an **AFTER/STATEMENT** trigger. It is used to retrieve the old state of the table from the temporary table to repair the violation. The three triggers are as follows:

```
CREATE OR REPLACE TRIGGER man_emp_sal1
BEFORE INSERT OR UPDATE OF salary ON EMP
DECLARE
    c1    INTEGER;
BEGIN
    SELECT COUNT(*) INTO C1
    FROM TEMP_EMP;
    IF (c1=0)
    THEN - initiate TEMP_EMP
        INSERT INTO TEMP_EMP
        SELECT supervisor,id,salary
        FROM EMP;
    END IF;
END;

CREATE OR REPLACE TRIGGER man_emp_sal2
AFTER INSERT OR UPDATE OF salary ON EMP
FOR EACH ROW
DECLARE
    employeecnt INTEGER;
    managercnt INTEGER;
BEGIN
    SELECT count(*) INTO employeecnt FROM TEMP_EMP tmp
    WHERE tmp.id = supervisor AND salary >= :new.salary;

    SELECT count(*) INTO managercnt FROM TEMP_EMP tmp
    WHERE tmp.supervisor = id AND salary <= :new.salary;
```

```
    IF((employeecnt > 0) OR (managercnt > 0))
    THEN
            IF UPDATING THEN
                    UPDATE TEMP_EMP
                    SET TEMP_EMP.salary = :old.salary
                    WHERE TEMP_EMP.id = :new.id;
            END IF;

        ELSE

            IF UPDATING THEN
                    UPDATE TEMP_EMP
                    SET TEMP_EMP.salary = :new.salary
                    WHERE TEMP_EMP.id = :new.id;
            END IF;

            IF INSERTING THEN
                    INSERT INTO TEMP_EMP
                    VALUES (:new.id, :new.job, :new.salary);
            END IF;
        END IF;
    END;


CREATE OR REPLACE TRIGGER man_emp_sal3
AFTER INSERT OR UPDATE OF salary ON EMP
DECLARE
    c INTEGER;
BEGIN
    IF INSERTING THEN
            DELETE FROM EMP
            WHERE id NOT IN (SELECT id FROM TEMP_EMP);
    END IF;

    IF UPDATING THEN
        SELECT COUNT (*) INTO C
        FROM EMP e, TEMP_EMP t
        WHERE e.id = t.id AND e.salary <> t.salary;
        IF(c>0) THEN
                UPDATE EMP
                SET salary = (SELECT salary
                              FROM TEMP_EMP
                              WHERE TEMP_EMP.id = EMP.id);
        END IF;
    END IF;
END;
```

7. CONCLUSIONS

In this paper we have provided some rules used to derive triggers from constraint specifications. By using these rules, deriving triggers becomes simpler. These rules are also general, so they can be used by any reference system.

When using declarative integrity constraints there are many problems. SQL: 1999 defines an unlimited CHECK clause, but current implementations do not allow it. In our point of view, the best way is to convert the CHECK clause into triggers. We suggest that integrity constraints can be converted into a more formal description (e.g. OCL), decomposed, and then converted into triggers, as we have illustrated in this paper.

REFERENCES

1. C. J. Date, *An Introduction to Database Systems.* Addison-Wesley Publishing Company, 6th edition, 1995.
2. R. Cochrane, H. Pirahesh, and N. Mattos, Integrating Triggers and Declarative Constraints in SQL Database Systems. *In Proc. of the 22nd International Conference Very Large Data Bases (VLDB)*, (Bombay, India, September 1996), pp. 567-578.
3. J. Biskup, *Achievements of Relational Database Schema Design Theory Revisited.* In: Semantics in Databases, Lecture Notes in Computer Science, Vol. 1358, (Springer-Verlag, Berlin, 1998), pp. 29-54.
4. D. Maier, *The Theory of Relational Databases.* Computer Science Press, Rockville, MD, 1983.
5. C. Türker and M. Gertz, Semantic integrity support in SQL: 1999 and commercial (object-) relational database management systems. *The VLDB Journal 10*: 241-269 (2001).
6. R. Elmasri and S. B. Navathe, *Fundamentals of Database Systems*, 2nd edition. Benjamin / Commings Publishing Comp any, 1994.
7. B. Von Halle, Uncovering Business Rules. *Database Programming and Design*, 8(7), (December 1995), pp. 13-18.
8. C. S. Mullins, The Procedural DBA. *Database Programming and Design*, 8(7), (December 1995), pp. 40-45.
9. *Database Programming and Design.* The 1996 Business Rules Summit, February 1996.
10. R. Ross, *The Business Rule Book.* Database Research Group, 1994.
11. R. Jenkins, *Constraints and Oracle Server.* Oracle Corporation, 1998.
12. B. Branchek, P. Hazlehurst, S. Wynkoop, and S. L. Warner, *Using Microsoft SQL Server 6.5, Special edition.* Que Corporation, 1996.
13. International Organization for Standardization (ISO) & American National Standards Institute (ANSI), ANSI/ISO/IEC 9075-2: 99. ISO International Standard: *Database Language SQL - Part 2: Foundation (SQL/Foundation)*, September 1999.
14. OMG Documents: *UML 2.0 OCL*; http://www.omg.org/
15. B. Demuth, H. Hussmann, *Using OCL Constraints for Relational Database Design.* Proc. of UML'99, Fort Collins, Colorado, Springer-Verlag, October 1999.
16. B. Demuth, H. Hussmann, and S. Loecher, OCL as a Specification Language for Business Rules in Data Base Applications. *Proc. of the Fourth International Conference on the Unified Modeling Language 2001*, (Toronto, Canada, October 1-5, 2001).
17. Oracle Corporation. *Oracle8i SQL Reference, Release 8.1.6*, December 1999.
18. IBM Corporation. *IBM DB2 Universal Database.* SQL Reference, Version 6, 1999.
19. Informix Software, Inc., Menlo Park, CA. *INFORMIX-Universal Server: Informix Guide to SQL Syntax, Version 9.1*, March 1997.
20. Microsoft Corporation. *Microsoft SQL Server, Version 7.0*, 1999.
21. Sybase Inc. *Transact-SQL User Guide, Version 11.0*, 1999.
22. Ingress Corporation. *Ingress Database Administrator's Guide, Version II*, 1999.

THE FUTURE OF INFORMATION TECHNOLOGY – HOPES AND CHALLENGES

Jacek Unold[1]

1. INTRODUCTION

Although the information revolution is in full swing, development of information technology may rest on one question: can silicon-based computer technology sustain Moore's law beyond 2020? In 1965 – three years before he co-founded Intel with Bob Noyce – Gordon Moore published an article that turned out to be uncannily prophetic. Moore wrote that the number of circuits on a silicon chip would keep doubling every year. He later revised this to every 18 to 24 months, a forecast that has held remarkably well over several decades (Unold, 2001).

As of today, Moore's law is the engine pulling a trillion-dollar industry. But how will it hold up in the future? A typical wire in a Pentium chip is now 1/500 the width of a human hair, the insulating layer is only 25 atoms thick. The laws of physics suggest that this doubling cannot be sustained forever. Physicists predict that by 2020 transistors will become so tiny that their silicon components will approach the size of molecules. At these incredibly tiny distances, the bizarre rules of quantum mechanics take over, permitting electrons to jump from one place to another without passing through the space between (Heisenberg's uncertainty principle). Like water from a leaky fire hose, electrons will spurt across atom-size wires and insulators, causing fatal short circuits. And this will mark the end of Moore's law.

The article explores the basic concepts of the alternative, non-silicon, computer and the prospects for nanotechnology, the ultimate form of information technology. It also points to the new forms of future communications, when the Internet is expected to become wireless and ubiquitous, to eventually... disappear. Disappear by becoming ubiquitous.

---

[1] Jacek Unold, College of Business and Economics, Boise State University, Boise, Idaho, USA

Information Systems Development: Advances in Methodologies, Components, and Management
Edited by Kirikova *et al.*, Kluwer Academic/Plenum Publishers, 2002

317

2. WHAT WILL REPLACE SILICON?

There are currently no known and proven solutions to the problem of the limits of miniaturization. The search for a successor to silicon has become a kind of crusade. Some of the theoretical options being explored include:

- the optical computer,
- the DNA computer,
- molecular and dot computers,
- the quantum computer (Kaku, 2000).

The optical computer replaces electricity with laser light beams. Unlike wires, light beams can pass through one another, making possible three-dimensional microprocessors. An optical transistor has already been invented; unfortunately, the components are still very large and clumsy. The optical of a desktop computer would be the size of a car.

The DNA computer seems to be one of the most ingenious ideas pursued in the field of information technology. It treats the double-stranded molecule as a kind of biological computer tape. Instead of encoding 0s and 1s in binary, it uses the four nucleic acids, represented by A, T, C, G. This approach holds much promise for crunching big numbers. Hence large banks and institutions may one day use it. However, a DNA computer is an unwieldy contraption, consisting of a jungle of tubes of organic fluid, and is unlikely to replace a laptop in the near future.

Molecular and dot computers. Other exotic designs replace the silicon transistor with a single molecule in the molecular computer and with a single electron in the quantum dot computer. Those molecules and electrons then act as tiny logic gates and switches. These approaches face formidable technical problems, such as mass-producing atomic wires and insulators. No viable prototypes yet exist.

The quantum computer is the darkest horse to emerge in this race. It is sometimes dubbed the ultimate computer. The idea is to direct a laser or radio beam on a carefully arranged collection of atomic nuclei, each of which is spinning like a top. As the beam bounces off the atoms, it flips the spins of some of them. Complex computations can be performed by analyzing how the spins have been flipped.

US intelligence agencies are nervously eying these new designs. Quantum computers, in particular, could be so powerful that they might one day break the most intricate secret codes the CIA can concoct.

These computers seem to be exquisitely sensitive, however. The tiniest disturbance - even a passing cosmic ray - can change the orientation of their computational atoms, spoiling the calculation. At present, quantum computers can perform only trivial calculations on perhaps five atoms. To do any useful work, they would need to calculate on millions of atoms.

Clearly, none of these designs are ready for prime time. Most are still on the drawing board, and even those with working prototypes are too crude to rival the convenience and efficiency of silicon.

3. NANOTECHNOLOGY AS THE ULTIMATE FORM OF IT

The research in the area of miniaturization and new IT has become a priority in most developed countries. In January 2000 President Clinton declared a National Nanotechnology Initiative, promising $500 million for the effort (Lemonick, 2000).

In fact, nanotechnology[2] has an impeccable and longstanding scientific pedigree. It was back in 1959 when Richard Feynman, arguably the most brilliant theoretical physicist since Einstein, gave a talk titled *"There's Plenty of Room at the Bottom"*, in which he suggested that it would one day be possible to build machines so tiny they would consist of just a few thousand atoms.

Within 25 years, nanotechnologists expect to create real, working nanomachines (nanorobots, nanobots), complete with tiny "fingers" that can manipulate molecules and with miniscule electronic brains that tell them how to do it, as well as how to search out the necessary raw materials. The fingers may well be made from carbon nanotubes - hairlike carbon molecules (discovered in 1991), that are 100 times as strong as steel and 50,000 times as thin as a human hair.

Potential applications of nanobots include:

- information technology,
- manufacturing,
- molecular medicine,
- environmental cleanup,
- creating futuristic materials

Information technology. The advantages of smaller computers are well known – more speed, more memory. But building matchbox-size supercomputers is too delicate a job for conventional mass manufacturing. Nanobots could do it easily, laying down circuits, made of nanotubes, molecule by molecule without a single mistake.

Manufacturing. The conventional approach to manufacturing is top-down: with large clumps of steel, wood, plastic, masonry, and shaping them into desired forms. Nanotechnology is, by contrast, bottom-up: stacking individual atoms into useful shapes. We know that the bottom-up approach is possible because that's what biology does, assembling proteins from individual atoms and molecules, putting them together to form cells and layering cells upon cells to form large, complex objects. For example, a nanomachine will be able to take raw carbon and arrange it, atom by atom, into a perfect diamond.

Molecular medicine. Streaming through the body by the billions, nanobots could chip plaque from arteries, gang up on bacteria and viruses, scour toxins from the bloodstream, repair broken blood vessels – and dozens of jobs doctors have not dreamed of yet. Today, nevertheless, it is hard to imagine a device that cruises the human bloodstream, seeks out cholesterol deposits on vessel walls and disassembles them.

Environmental cleanup. Specialized nanobots dumped into an oil spill or a polluted stream could seek out and find dangerous molecules, remove them or change their chemical structure one by one to render them harmless - or even beneficial.

---

[2] The term **nanotechnology** comes from nanometer, or a billionth of a meter; a typical virus is about 100 nanometers across.

Creating futuristic materials. A diamond's extraordinary clarity and strength make it an ideal building material, especially in the field of electronics. But this material is also very hard to work with. Nanobots could make diamonds in any shape. And because the basic feedstock is ordinary carbon, these diamonds would be as cheap as glass.

The greatest challenge in the field of technology, as of today, is the issue of self-replication of those tiny machines. To accomplish any sort of useful work, nanotechnologists will have to unleash huge numbers of nanomachines to do every task – billions in every bloodstream, trillions at every toxic waste side or to put a futuristic computer together. No assembly line could crank out nanobots in such numbers. But nanomachines could do it. Nanotechnologists want to design nanobots that can do two things: carry out their primary tasks, and build perfect replicas of themselves. And the question is, what if the nanobots forget to stop replicating? Without some sort of built-in stop signal, the potential for disaster would be incalculable. A fast-replicating nanobot circulating inside the human body could spread faster than a cancer, crowding out normal tissues. One idea to prevent this danger is to program a nanobot's software to self-destruct after a set number of generations. Still, some critics contend that the potential dangers of nanotechnology outweigh any potential benefits (Lemonick, 2000). Yet those benefits are so potentially enormous that nanotech, even more than today computers, could be the defining technology of the new century.

4. WHAT WILL REPLACE THE INTERNET?

The next question is, what is the future of telecommunications, especially of the Internet? Early versions of the Net have been around since the 1960s and '70s, but only after the mid-1990s did it begin to have a serious public impact. Since 1994, the population of users has grown from about 13 million to more than 300 million around the world. About half are in North America and most still reach the Internet by way of the public telephone network.

According to Vinton Cerf, who co-invented the Internet protocol TCP/IP, the Internet will first become wireless and ubiquitous, crawling even under our skin. Eventually, it will disappear. Disappear by becoming ubiquitous.

Most access will probably be via high-speed, low-power radio links. We already have Internet-enabled mobile telephones and personal digital assistants (PDA). According to V.Cerf (2000), in several years a PDA can serve as an appliance-control remote, a digital wallet, a cell phone, an identity badge, an e-mail station, a digital book, a pager and perhaps even a digital camera. There are already plans to call such a device WIDGET (Wireless Internet Digital Gadget for Electronic Transactions). Nokia has already constructed a prototype of such a device, Nokia Communicator 9110. It offers access to the Internet, telephone, e-mail and the possibility to send high quality, digital photographs.

By 2020 so many appliances, vehicles, buildings will be online that it seems likely there will be more things on the Internet than people. More and more "smart devices" will have access to the GPS (global positioning system), increasing the value of geographically indexed databases.

The advent of programmable, nanoscale machines will extend the Internet to things the size of molecules that can be injected under the skin, leading to Internet-enabled people. Such devices, together with Internet-enabled sensors embedded in clothing, will avoid hospital stay for medical patients who would otherwise be there only for observation. The speech processor used today in cochlear implants for the hearing impaired could easily be connected to the Internet; listening to Internet radio could soon be a direct computer-to-brain experience.

The Internet will undergo substantial alteration as optical technologies allow the transmission of many trillions of bits per second on each strand of the Internet's fiber-optic backbone network. The core of the network will remain optical, and the edges will use a mix of access technologies, ranging from radio and infrared to optical fiber and the old twisted-pair copper telephone lines. By then, the Internet will have been extended, by means of an interplanetary Internet backbone, to operate in outer space (Cerf, 2000).

Table 1 presents the basic stages of the development of information technology.

Table 1. Development of information technology

Decade	Implementation	Technology
1950-	Electronic brain	Programming languages
1960-	Commercial computer	Operation systems
1970-	Computer databases in business	Databases, terminals, computer networks
1980-	Software engineering	CASE
	Personal computers	PC, user-friendly interface, Local Area Networks (LAN)
1990-	Client-server applications	Distributed processing
2020 ?	Collapse of Moore's law The end of the silicon chip era Nanorobots	Nanotechnology Optical computer DNA computer Molecular computer Quantum computer

Source: Own research

The best illustration of the speed of the information revolution is the fact that a musical birthday card contains more processing power than the combined computers of the Allied Forces in World War II (Kaku, 2000).

5. CONCLUSIONS

A natural question which arises from such bright prospects is, are there any downsides to a society suffused with information, which is accessible instantly and from virtually anywhere, and the tools to process it? Computers, biotechnology and nanotech don't work the traditional way – they are self-accelerating. It means that the products of their own processes enable them to develop ever more rapidly. New computer chips are immediately put to use developing the next generation of more powerful ones; this is the inexorable acceleration expressed as Moore's law. The same dynamic drives biotech and nanotech – even more so because all these technologies tend to accelerate one another. Computers are rapidly mapping the DNA in the human genome, and now DNA is being explored as a medium for computation.

Technologies with this property of perpetual self-accelerated development[3] create conditions that are unstable, unpredictable and unreliable. And since these particular autocatalytic technologies drive whole sectors of society, there is a risk that civilization itself may become unstable, unpredictable and unreliable.

There are scenarios, however, in which technology may break itself. In the aging population of the developed world, many people are already tired of trying to keep up with the latest cool new tech. The market for change may one day dry up, and "lock-in" (characteristic for the traditional achievements: automobile, TV, plane etc.) may again become the norm.

Religious and cultural factors are very important as well. Radical new technologies are often seen as moral threats by conservative religious groups or as economic and cultural threats by political groups. In general, change that is too rapid can be deeply divisive. If only an elite can keep up, the rest of the society will grow increasingly mystified about how the world works.

With so many powerful forces in play, IT could hyperaccelerate with stunning rapidity, or it could stall completely. The most likely scenario, according to the golden rule, is that it will do both, with various technologies proceeding at various rates. The new technologies may be self-accelerating, but they are not self-determining. They are the result of ever renegotiated agreement with society. Because they are so potent, their paths may undergo wild oscillations, but the prevailing trend should proceed in the dynamic middle.

6. REFERENCES

Cerf V. , 2000, What will replace the Internet? *Time*, 3 July
Kaku M. , 2000, What will replace silicon? *Time*, 3 July
Lemonick M.D., 2000, Will tiny robots build diamonds one atom at a time? *Time*, 3 July
Michael Journal, 2000, Rougemont, Canada, Nr 7
Plus GSM, 2000, Roaming. *GSM Plus Journal*, June
Unold J. (2001): *Marketing Information Systems* (in Polish). The Wroclaw University of Economics Press,
      Wroclaw, Poland

---

[3] sometimes termed „autocatalysis".

RECOMMENDATIONS FOR THE PRACTICAL USE OF ELLIOTT JAQUES' ORGANIZATIONAL AND SOCIAL THEORIES IN THE INFORMATION TECHNOLOGY FIELD: TEAMS, SOFTWARE, DATABASES, TELECOMMUNICATIONS AND INNOVATIONS

Sergey Ivanov[1]

1. INTRODUCTION AND THE PROBLEM OF UNIVERSALS

The problem of universals is a profound abstract question that quests into the nature of our knowledge, which our civilization has been querying for the past several millennia – this large abstract problem, Artz (2002)[2] writes, has been known to the modern world from the ancient Greek philosopher Plato. Such as, how do we know the true identify of the object and fit the identity into a true classification so that we could understand and attribute to this object? There are multiple problems with identifying a true identity, for example, how do we know what the true identity of the object is? Is there such a thing such as the true identity or are there multiple true identities? Could we really classify the objects even if we knew the true identity(s) of them? And what is classification? Does it exist in the world as a true relationship between identifiable objects or it is just a human way of 'languaging' a common understanding? Artz (2002) elaborates in depth on the problem of universals in information modeling, finding that both essence in the same issue – difficulty of classification, or how do you know a thing is a thing and that it belongs to the class of the thing? How do we know what makes a thing thing; and what properties relate things?

The problem of universals is at the core of information and social sciences today – to make progress in either or both, it is required to understand the basic constructs and logically relate them together to construct a more abstract and logical relation of the constructs to form a theory how things ought to work to allow humanity to achieve its purposes in the specified field(s). The word progress in this context means that society is

[1]   Sergey Ivanov, The George Washington University, Washington, D.C., 20052, U.S.A.
[2]   Artz, John (2002). Information Modeling and the Problem of Universals: an Analysis of Metaphysical Assumptions. The George Washington University: Unpublished Paper.

able to achieve its goals within a certain time, such as building a better information system within one year, or creating a new technology product within three years, and so on. Furthermore, the necessity to understand basic constructs or "smallest universals" does not imply an inductive approach – the author believes that a deductive approach, such as a greater understanding how larger things work together, helps identify smaller units, but this paper avoids discussing deductive or inductive approaches to understanding and establishing theories – this discussion is out of scope of this research.

The major issue of the information and social studies of today is the deduction of the problem of universals in these fields – there is no well-tested, defined and explained understanding how to group and relate things, and furthermore, many phenomena remain unexplained and ambiguous, for example, not well defined concepts of phenomena of organization, conflict, excellence, intelligence, capability and many others that are used in management and information systems studies nowadays. Tackling the problem of universals in the fields of information technology and management could transform our society for the reasons of gaining understanding how things that constitute each field relate, and how it is possible to influence some things to achieve results for the larger society within a reasonable time. Solving the problem of universals is very difficult, though, some significant progress has been achieved, and is presented in this paper; also items remaining to be discovered and clearly identified are briefly discussed in this paper as well.

2.   THEORETICAL BACKGROUND

2.1. Two-dimensional Time

Dr. Elliott Jaques, a Research Professor at The George Washington University and author of many books, is the prime discoverer of key findings that offer science-based theoretical propositions for the social sciences. These discoveries, some of which are not new to the civilization – many have been discussed by St. Augustine and other ancient thinkers, are new to the modern world, and include a new understanding of time[3], biological life, and some ideas about complexity, but this last issue is still in the works.

The first proposition is that our present understanding of time is inaccurate and not sufficient to understand the biological life. The clock time, the one that is most understood by the researchers and the society in general, measures how long it took for the events to occur – Elliott Jaques calls it time of succession[4]. Dr. Jaques mentions that ancient Greeks called this aspect of the time phenomenon Chronos, and he proposes to think of it as a dimension of time. The other (or perhaps future researchers may call it another should they discover more time dimensions) is the time of intention, or as Dr. Jaques writes ancient Greeks called it Kairos, the time of opportunity. This is one of the most crucial findings that allows for a better understanding of social sciences: time is two-dimensional consisting of the dimensions of succession and intention, or how long it took for the event to occur (natural sciences event), and by when someone intends to achieve certain results (social sciences event).

---

[3]   The author understands that the discoveries mentioned postulate and imply a quantum leap in the social
       sciences with wide implications, and asks the reader to evaluate these ideas with an open mind when
       reading and analyzing this paper; all comments, suggestions, and criticisms are respectably welcome.
[4]   Jaques, Elliott (1982). The Form of Time. New York, New York: Crane, Russak & Company.

To elaborate on the idea of two dimensions of time, it is necessary to distinguish that the idea that there exist past, present and future is invalid. Dr. Jaques writes in "The Form of Time" that St. Augustine also recognized this phenomenon – the only thing that exists is present past, present present, and present future; both, the past and future are with us today – they do not exist separately from us. The following chart elaborates and explains the time phenomenon further:

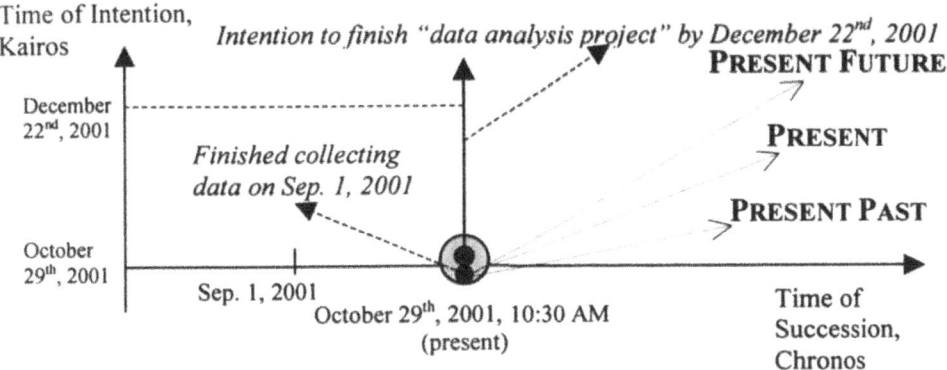

Figure 1. Dimensions of Time.

Let's assume that today is October 29th, 2001, 10:30 AM. Today, on October 29th, 2001 I know I finished collecting data for a project – I keep a record of this event, finishing collecting data. At the same time, today, I am intending to analyze the data collected by December 22nd, 2001 – this is the intended future event that can be measured with a ratio scale measure – 55 days – this is the future that is with us in the present; when December 22nd comes, I will record whether I am done with the task or whether I re-schedule it, and eventually would record an actual date of finishing on the axis of succession. The time of succession may feel more real because as a generation, we have become used to it, but which one is more real, intention or succession? Let's assume that I indeed finished the intended project on December 22nd. The following time chart would help explain the events:

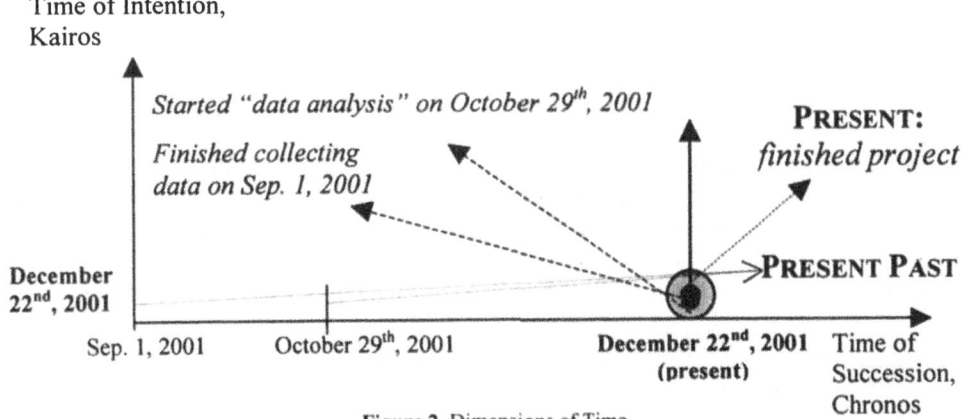

Figure 2. Dimensions of Time.

Today, on December 22nd, 2001, I have the records of the events on September 1st and October 29th, 2001, and I finished the 55-day assignment as I intended on October 29th (to finish by December 22nd) – all these events exist now, in the present past. Additionally, including the time of intention, we can measure goals with precise ratio-scale data – by when! This is one of the most profound premises of the scheduled to be published in June/July 2002 book, "A Theory of Life" by Dr. Elliott Jaques[5] – the premise is that the difference between inanimate physical objects and living organisms is intentions: living organisms intend to do something by a certain deadline, while inanimate objects have no intentions, and thus, exist in a four-dimensional world, rather that five-dimensional of the living biological creatures.

Before we proceed further, it is important to understand the concept of the present or more correctly the constant continuous present or present present – which is a continuous (living) present that includes past, present, and future, as depicted in the following diagram:

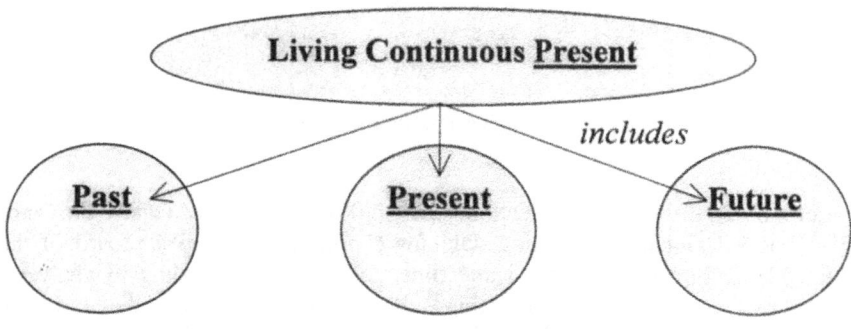

Figure 3. Present.

Analyzing the diagram, it is evident that physical objects exist in a four dimensional world: time of succession, and three space coordinates; or as Dr. Jaques calls it (3 + 1) dimensions.[6] The living organisms, on the other hand, have intentions (goals to achieve), and thus, live in a (3 + 2) dimensional world, having added another time coordinate: the axis of intentions, which are measurable with ratio-scale measures – by when[7]!

2.2. Human Cognitive Abilities

Another major discovery also comes from Dr. Elliot Jaques, and most explicitly and clearly in his most recent book, "The Life and Behavior of Living Organisms: a General Theory," published in 2002. The main proposition and finding are that all humans (and all biological organisms) develop cognitively in precise patterns – this paper concentrates on issues pertaining to human lives and societies, and leaves the discussion of other biological creatures to other researchers and possible future endeavors. The discovery has

---

[5]   Jaques, Elliott. "A Theory of Life: An Essay on the Nature of Living Organisms, Their Intentional Goal-Directed Behavior, and Their Communication and Social Collaboration." 2002.

[6]   Jaques, Elliott (1982). The Form of Time. New York, New York: Crane, Russak & Company.

[7]   The discussion of actual measuring and collecting of data is discussed later in the paper.

found that humans' cognitive abilities develop from birth through old age in predictable patterns of mental processes, as depicted on the chart below:

More Complex Mental Processing

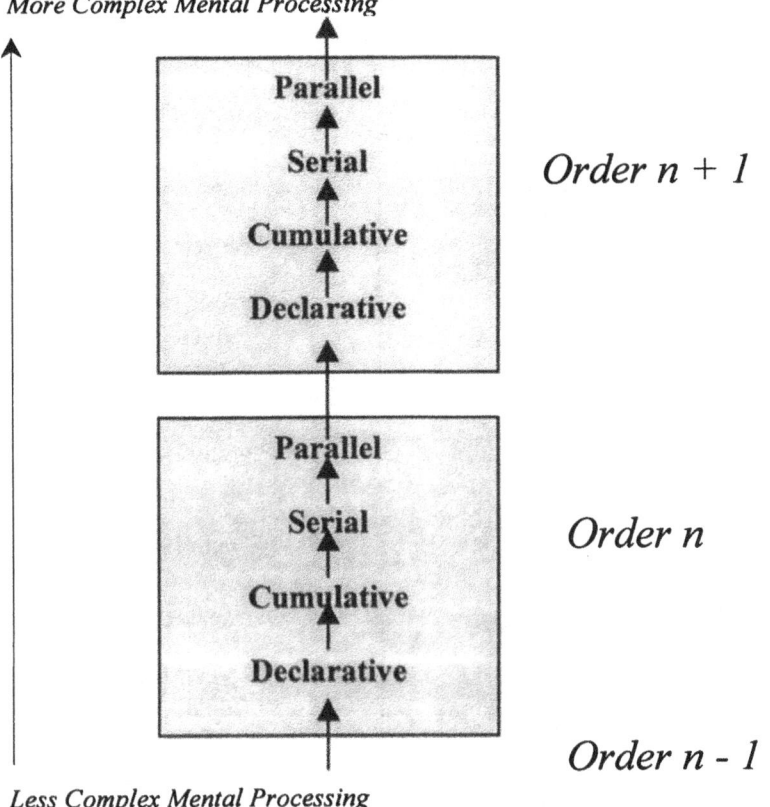

Less Complex Mental Processing

Figure 4. Cognitive Development.

Every human is born to a certain cognitive trajectory that the person's development goes through certain mental stages as depicted above, in the sequential succession, from declarative state, to cumulative, to serial, to parallel, and then to declarative of a different order – please refer to "Human Capability" by Elliot Jaques for an in-depth discussion and research into these processes. The mental trajectory will determine through how many stages the person will develop and at what age.

These ideas of distinct cognitive levels are not new – other researchers have noticed disparate cognitive levels before. For example, Blooms Taxonomy describes six cognitive levels, the description of which is very similar to Jaques' research.[8] Humans have become the only known species able to disengage further in the time of intention than other species known to mankind based on the development of cognitive processes

---

[8]   "Blooms Taxonomy." http://www.arch.gatech.edu/crt/lln/Wordsworth/bloomstaxonomy.htm: Georgia College of Technical Architecture Web Site, 1998.

that support planning (intending) objectives up to a certain time of the present future, as depicted in the chart below:

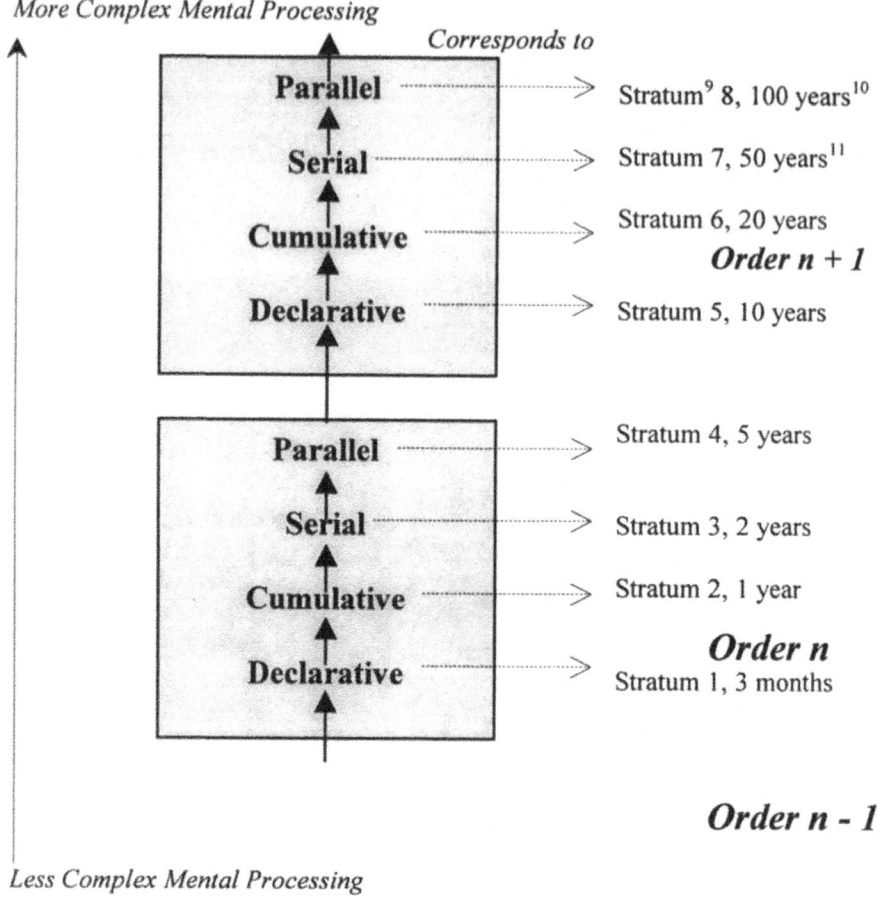

Figure 5. Cognitive Strata.

In his book, "A Theory of Life," Dr. Jaques describes how human babies develop cognitively through the strata and orders of mental processing. Dr. Jaques argues that adults also continue developing through similar patterns long into the old ages, and depending on the acceleration trajectory, some reach extraordinary cognitive capability

---

[9]   Dr. Elliot Jaques called each new level stratum.
[10] This time of intention represents the maximum distance a person is able to project and create goals into the present future, and to plan, execute, and fulfill these intentions.
[11] Jaques, Elliott. "A Theory of Life: An Essay on the Nature of Living Organisms, Their Intentional Goal-Directed Behavior, and Their Communication and Social Collaboration." 2000.

with age. The cognitive potential capability determines how far into the present future[12] the individual can realistically plan for to achieve actual goals, which in other words, is that maximum potential capability of the individual determines the longest distant objective on the axis of time of intention the individual can cope with, which is significant for our analysis, and the theory of social sciences.

One way to determine the maximum potential capability, which Dr. Jaques states is an in-born capacity, the analyst must involve an individual into a "vicious argument" and observe the pattern and structure of that person's presenting ideas spontaneously – the language structure would show whether the constructs are declarative (you are wrong!), cumulative (this is right, and this is right, perhaps this is right too), serial (if this is ok, and this is not, then perhaps the conclusion is this), or parallel (considering this idea, we may come to this conclusion, but on the other hand, this idea leads to a different conclusion). Looking at the speech constructs further, it is possible to determine to which order of information complexity (complexity of mental processing) a person belongs to – see "Human Capability" by Elliott Jaques[13] for more information on precise evaluating of individuals. In this paper, the author is assuming that the measuring instruments of the person's development are correct to proceed further to the analyses reaching beyond to the depth structures of our society.

2.3. Rigorous Refutable Definitions

Dr. Elliott Jaques introduced another concept for acceptance by social scientists, which at the present time indeed may be attributed to a major discovery within the social sciences – this innovation is to create and use univocal, universal, and rigid definitions of concepts in order to be able to compare, refute, and advance theories, case studies and hypotheses within social studies similar to the way it is done in the natural sciences. Presently, no keyword used in most studies have uniform definitions and understandings, thus, making comparing similar-oriented researches impossible; which makes it impossible to refute some studies while accepting and improving good ideas. For example, such notions like organization, manager, team, bureaucracy, employee and many others have unknown meanings, while most of the studies refer to the concepts of organization, manager, team, employee and others freely, which furthermore, has created and made acceptable ideas that it is all right not to define and/or understand key assumptions, and further, created a culture that it is impossible to measure precisely and even understand social processes.

For example, in Economics, everyone understands what a dollar is, and the amount of money can be accurately measured with its monetary amount – for example, $1,000,000 dollars designated and budgeted for a certain public program. But, the statement that a virtual team of cohesive members has been assigned to run this program indeed seems impossible to understand under the present no-definitions-allowed policy.

Dr. Jaques compares the state of the social sciences today with the state of natural sciences in the 17th century, when no measuring tools were available to measure universally observed phenomena, such as speed, temperature, weight, and others – the re-

---

[12]   The author is trying to be as precise as possible defining concepts and words to ensure the reader may come to similar conclusions or disprove the findings through testing the theoretical propositions presented in this paper.

[13]   Jaques, Elliot & Cason, Kathryn (1994). Human Capability. Rockville, MD: Cason Hall.

discovery of the time of intention allows for a precise measurement of work (the definition of work is also not known at the present time), which altogether is believed to be the starting point for the social sciences to launch into the new millennium. Thus, before proceeding further, I am going to define the necessary concepts for evaluating the ideas in this paper by independent reviewers objectively, and with possibilities to test and refute all of the ideas discussed.

The first crucial definition explains the concept of work. Dr. Elliott Jaques defines work as the "exercise of judgment and discretion in making decisions in carrying out goal directed activities."[14] This precise wording is directly related to the time of intention – work is everything we do to achieve our goals set some time into the present future – achieve what by when, and it is no different in the employment-related activates (please see any of Elliott Jaques' works for a complete set of definitions)! Organization is defined as a "system with an identifiable structure of related roles,"[15] which may be divided into bureaucracies and associations. An association is a member-based institution, either voluntary, such as church, or community, or stockholder member, or non-voluntary, such as a country (citizens, elected officials) – no one can be fired or laid off from such an organization. The other type of the organization is bureaucracy, which is organized by an association(s) to work on its behalf (notice that work is clearly defined, such as achieving set objectives!) with a reporting structure – for example a company with hired employees: stockholders constitute an association, which elects board members to organize a corporate bureaucracy to continue and proceed with business activities (the board hires a CEO, etc.). For example, university faculty members without tenure are employees of the university, while the faculty professors with tenure have become members of the institution. Similar analysis applies to law firm partners – they are members of the firm, while the non-partner attorney is an employee.

Having defined all concepts clearly and without ambiguities, it is possible to set on a course of conducting studies and comparing research and theories of similar phenomena to advance the state of the current thought.

2.4. Measuring in Social Sciences

Despite a general understanding of measurements and measuring, it is integral to re-visit the measurements theory and understand measuring in the social sciences. It is crucial to understand and elaborate what a measure is, what types of measures there are, and what the differences among different types of measures exist to ensure reliable, accurate and meaningful depiction of reality measured. Sarle (1995) argues that proper use of various measuring and statistical techniques and methods is necessary for a "responsible real-world data analysis."[16] He distinguishes between measures and actual attributes measured – the idea is that the measures should accurately depict a real-world phenomenon. The example the author provides is measuring lengths of sticks with a ruler – if one stick is 10 cm, and the other is 20 cm, then the second stick must be twice longer than the first – thus, we have drawn an accurate conclusion about the sticks' lengths.

---

[14]   Jaques, Elliott. "A Theory of Life: An Essay on the Nature of Living Organisms, Their Intentional Goal-Directed Behavior, and Their Communication and Social Collaboration." 2000.
[15]   Jaques, Elliott. "Requisite Organization." Arlington, VA: Cason Hall & Co, 1996.
[16]   Sarle, Warren S. (1995). Measurement theory: Frequently asked questions. Disseminations of the International Statistical Applications Institute, 4, 61-66.

Sarle defines measurement as "assigning numbers or other symbols to the things in such a way that relationships of the numbers or symbols reflect relationships of the attribute being measured."

There are various types of measurements that are known – the types vary by their degree of accurate reflection of the real world phenomenon. These types are: nominal, ordinal, interval, log-interval, and ratio numbers. Despite most researchers know and use statistical measures mentioned above, for the purpose of this study and to extenuate the discovery of a new measure in social sciences, it is necessary to define and explain the differences between the measures.

Nominal measures are less useful – they are just an enumeration and have nothing more than symbolic values. Ordinal type is also not very useful[17] – the ordinal measures show whether one property is less or more than the other, and depict the following relationship, that if things X and Y with attributes a(X) and a(Y) are assigned numbers n(X)and n(Y), in such a way that   n(X) > n(Y), then a(X) > a(Y).[18] Interval measures become more useful than ordinal, though even interval measures may still be inadequate for a precise scientific research – the main property of the interval-level variables is that the differences between numbers reflect similar differences between the attributes. Log-interval measures are such that the ratios between numbers reflect ratios between attributes.

Ratio measures are most interesting and in-demand in every scientific field. Ratio scale numbers depict accurately the differences and ratios between the attributes and have a concept of zero, such as zero means nothing. For example, a stick, which length is zero centimeters equals to the length of zero meters, and is nothing – it doesn't exist! This is important to note because in interval-level numbers, zero does not mean that the property does not exist.

The following diagram demonstrates the usefulness (or preciseness) of measures' types:

Less Useful   **More Useful**
Nominal   *Ordinal*   *Interval/Log-Interval*   *Ratio*

Figure 6. Preciseness of Measures.

At the present time, it has become acceptable in social sciences to manipulate and calculate numbers to analyze information using ordinal-level numbers, and various statistical techniques have been developed to make the analysis depicting reality as close and accurate as possible. The main reasons for using the ordinal-level measures have been the lack of measuring instruments to observe ratio-type data, until the recent past. Dr. Elliott Jaques found a scientific way to collect ratio-scale data within the social sciences, which is a phantom leap forward towards social sciences catching up with natural sciences in data analyses and mathematical propositions.

The new instrument to obtain ratio-scale data within organizational science is called time-span, which measures the level of work in a role by identifying the longest task or

---

[17]   It is the author's opinion that ordinal scale measures are not very useful as they are imprecise depicting a real-world relationship.
[18]   Sarle, Warren S., Ibid.

project within the role assigned by the manager to a subordinate, for which the subordinate has discretion and authority to complete the assignment. Dr. Elliott Jaques defines time-span as the "targeted completion time of the longest task or sequence in the role,"[19] and it is quite easy to measure. To measure a role, a researcher has to interview the manager and learn what is the actual longest assignment s/he assigned to the subordinate. Having measured over eighty organizational roles, the author learned it takes about five minutes to interview the manager – please see "Time-Span Handbook"[20] by Elliott Jaques for an exact guide how to go about using the time-span instrument, and its comprehensive description and examples of various types of roles, such as accounting, machinist, technologist, and many others.

Time-span is a ratio-scale measure of the time of intention, with the absolute concept of zero. If the role's time-span is zero, that means that the role does not exist. If role A is measured at 6 months, and role B is measured at 1 year, then $t(A) = \frac{1}{2} t(B)$ (t stands for time-span) – this means that role B is twice bigger than role A. Thus, all roles within a bureaucracy can be measured with time-span, and thus, analyzed in a new light. For example, a Canadian firm, Capelle Associates Inc. has based its management consulting business primarily upon the theory and measures that Dr. Elliott Jaques has developed, and they are quite successful with research papers confirming the findings measuring organizational productivity and performance; their research papers are available at their corporate web site at www.capelleassociates.com.

The time-span instrument is the first one in its kind in social sciences that allows measuring and comparing precisely levels or roles within various types of organizations, industries, and countries – it is universal. A project manager's role in company A, country X measured at 3 years is accurately comparable to the database designer's role in company B of country Y should time-span of this role be found to be 3 years as well.

In another example of divergent roles, it may take a day to prepare a small proposal – thus, the targeted completion time of this task is one day, and should this be the longest task in the role, it is a one-day role. In another role it may take seven years for the following task: expand into the Eastern market, build and create an Eastern-European home for the corporate products, and possibly merge and acquire emerging and competing companies with comparable products and potential – thus, the targeted completion time of this task is seven years, and should this be the longest task in the role, we will have measured the role at a seven years time-span. The following figure depicts the measurement through the target completion time:

[19]   Jaques, Elliott. "A Theory of Life: An Essay on the Nature of Living Organisms, Their Intentional Goal-Directed Behavior, and Their Communication and Social Collaboration." 2000.
[20]   Jaques, Elliott (1964). Time-Span Handbook. Rockville, MD: Cason Hall.

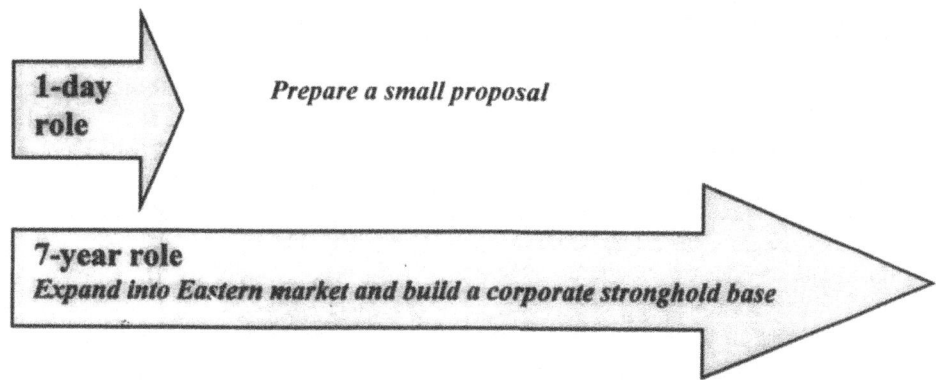

Figure 7. Role Differences.[21]

Furthermore, it is possible to evaluate a person's potential capability via the instrument called time-horizon, which is defined as a "method of quantifying an individual's potential capability, in terms of the longest time-span s/he could handle."[22] Dr. Jaques' book on human capability describes methods of determining an individual's potential[23], though, there is no a discovered instrument to obtain a ratio-scale number yet, though it is possible to evaluate the potential cognitive stratum, and thus, estimate a potential time-horizon, as it will be within the bounds of the stratum. For example, a person at a certain age measured at stratum 3 would have a potential time-horizon between 1 year and 2 years (see chart 5 above).

There are other instruments that are still being discovered, in addition to time-horizon, such as complexity. Presently there is no way to measure complexity with ratio-scale numbers, and instead, there are various methods available to estimate complexity, such as, function-point analysis in information systems. Dr. Elliott Jaques thinks of complexity as number and rate of variables manipulated over time, but there is no ratio-scale instrument to measure the complexity of the task precisely at the present time.

Despite the lack of instruments, the discovery of time-span, and having come closer to measuring time-horizon and eventually task's complexity creates a new paradigm in social sciences that allows a real possibility to collect ratio-scale data for testing theories and hypotheses scientifically to create a new promising and possible future for mankind through a different organizational design, revised social sciences, and within it, many fields, including information systems technology.

---

[21]   For example, in a multiple-task role, the level of work would be defined by the time-span of this role, which would be the longest phase of a project or entire project for which the person may make decisions to lead it to a conclusion – for example, install and configure a corporate firewall within the next four weeks – if four weeks were the longest assignment in this role, then the time-span of this role would be four weeks (the time of intention). At the same time, each person has an in-born capability to work into the present future on the axis of intention, which Jaques measures with the maximum distance the individual can cope with, or time-horizon.

[22]   Jaques, Elliott. "A Theory of Life: An Essay on the Nature of Living Organisms, Their Intentional Goal-Directed Behavior, and Their Communication and Social Collaboration." 2000.

[23]   Jaques, Elliot & Cason, Kathryn (1994). Human Capability. Rockville, MD: Cason Hall.

3. RECOMMENDATIONS

3.1. Team Building

According to the theoretical propositions above, which are described at length and details in the Requisite Organization theory (ROT) in Jaques' writings, the author has concluded that team building is a scientific endeavor, easily resolvable and testable. Such as, a work team within a bureaucracy is created to achieve a certain purpose by a certain deadline within a specified budget. In normal times (normal non-emergency projects, excluding war, extreme hardship, survival of the company and so on), according to the ROT, the project's time-span would determine what type of people should constitute a team, and what type of manager should the team members report to. If the entire team is assembled to carry on the entire project, while the manager is juggling several projects (including this one) to complete a larger project, then mathematically, the above propositions should be described as follows:

Theorem 1:   Project of [t(prj)] of stratum n should consist of team members of $p = n$, and manager of $p = n + 1$, where

Time-span of the Project:	t(prj)
Time-horizon of the Individual/Potential:	p
Stratum:	n

According to the theory, the derived theorem above should help create a team in which every member is capable of adding value to the entire project and has the necessary capability to cope with the complexity of the project. The manager should have the capability one stratum higher than the project's time-span to coordinate projects of complexity of stratum n to complete a larger project of complexity $(n + 1)$, of which the projects of complexity n are part of.

If, on the other hand, the project is handed to a manager, who is authorized to assign the parts of the projects to his/her subordinates and assemble the team of subordinates to delegate parts of the project to, then mathematically, the above propositions should be described as follows:

Theorem 2:   [t(prj)] of stratum n should consist of team members of $p = n - 1$, and manager of $p = n$

Theorem 1 should apply in partnerships, associations, college study groups and other types of organizations which are not bureaucratic managerial hierarchies because each member should have the capability to determine and add value to the entire endeavor, and should be able to work at the complexity required to complete the project successfully. Theorem 2 applies to a managerial bureaucracy, in which a manager is assigned a project by his/her manager (manager once removed), and the manager should delegate pieces of the project to his/her subordinates accordingly, and make decisions for the course and endeavor of the entire assignment.

3.2. Software Development

Developing software may be a complex or not a complex endeavor, requiring skills and knowledge of various tools, programming languages, algorithms, and experience. Under normal circumstances (abnormal circumstances are discussed later in the paper), developing good software applications does not differ from building successful teams – the main idea is to put the right people on the project.

For example, let's assume that project's estimated length is three years. Under normal circumstances, it is likely that the complexity of this endeavor corresponds to the complexity of the task's stratum as follows (derived from theorem 2):

Theorem 3:   A person of $p = n$ should manage the software development project of
[t(prj)] of stratum n, with reporting subordinates (software programmers)
of $p = n - 1$

Thus, the manager who is in-charge of the project should have time-horizon at least matching the complexity of the project, with subordinates with time-horizons of one stratum lower.

The theorem above is incomplete because the concept of complexity (or simplicity) has not been defined, well understood or operationalized so that it could be accurately measured with ratio-scale data. It is assumed that during normal times, a reasonable estimation of a project's length corresponds to its complexity, which will default in abnormal times, like war, extreme pressure, and other not normal circumstances, which are discussed later in the paper.

Function-Point Analysis, an empirical method of estimating software complexity is not a precise way of measuring the complexity of a software development task – it is an estimation to approximate how long the project might take and how much it might cost based on the number of variables participating in the endeavor. Not having a precise measuring instrument to measure complexity, function-point analysis allows a rough estimate, and suggests that a more 'complex' project should take longer to implement (the one which receives more function-points), and thus, provides some support for the propositions above. Pressman (1997) writes "FP has no direct physical meaning – it's just a number."

3.3. Database Development

Database development is no different from software development – database development may also be of various levels of complexity. The key to success is assembling a team of right people for the entire project or part of the project, with the project manager in-charge whose capability matches the complexity stratum of the project. Function-point analysis is also used in database development to estimate roughly the size of the project, and under normal circumstances, the formula developed for the previous paragraph corresponds to database development as well, as database applications are a type of software applications with specialized requirements.

3.4. Designing and Implementing Telecommunications

Building, upgrading, and designing telecommunications infrastructure are the processes that are in essence no different than designing software or databases, just requiring a slightly different set of skills and knowledge, but on the abstract level these endeavors are the same – they require to complete a project within a certain deadline, and manage all technology pieces to put them together to complete the development's goal. Theorem 2 and 3 formulas should apply within managerial bureaucracies – the length of the project should constitute the magnitude of the person who should manage the project.

3.5. Creating Innovations

Creating something new – developing a brand new product for the market, designing and implementing a new technology, is a high-capability endeavor, requiring efforts of people of highest capability. It is arguable that any new project is an innovation because the specific requirements have never existed before – it is unique, requiring the people working on it make unique decisions (Jaques, lectures). On the other hand, similar endeavors have happened in the past – they may have differed in some ways, but in general there is some experience how to proceed. In managerial bureaucracies, theorem 2 depicts who should constitute a team or a person responsible for the task of a certain estimated complexity.

Let's assume that it is indeed possible to measure the complexity (c) of any endeavor similar to the discontinuous strata Jaques has pointed out, and let's assume that when c = 0, the project does not exist. Therefore, to achieve the complexity of level 1 would at the minimum require a person whose time-horizon is at least at level 1. Thus, the complexity of the highest technological innovation would require a person of the potential capability at least comparable with the complexity of the project.

According to the hypothesis above, it is evident that greater innovations may be achieved only by highest-capability people, and complex innovations may take a long time, and likely resources, at the minimum to support the developers. The issue is that there is no discovered instrument to identify various complexity levels at the ratio-scale level.

The author's hypothesis is to identify all steps and their sequences: determine the steps necessary, their relations (or, and, sequence or parallel), and then determine the order of the steps. This is also not a precise method because it requires interpretation of the order of the step (step's stratum) and step by itself is not well defined – something to get done within a specific time as the smallest unit of assignment in the project.

Thus, estimating, it is possible to evaluate the magnitude of the required innovation – the goal is clear and unambiguous – to develop or create something by a certain date, and it is manageable to estimate the order of a specific innovation, thus, needing a person of at least matching capability, interest and commitment to the idea, and allowing the necessary time and resources.

3.6. Exceptions: Abnormal Circumstances

There are exceptions to normal course of life – such as war, extreme competition for survival of the organization, or some other emergency – situations Jaques describes as compressed time (lectures), which mean that the project that normally should take a certain time, gets completed a lot sooner by people of higher capability than required by the project's complexity level, because they can perform a lower-complexity task faster. Thus, in extraordinary circumstances, it is necessary to engage people of the highest capability available to complete a task, as follows:

Theorem 4:   *A person of $p = n + m$ should manage the project of [t(prj)] of stratum n, with reporting subordinates of minimum $p = n$ or $n + m - 1$, where $m > 0$ (stratum).*

For example, in an emergency, a three-year project, normally requiring a person with potential capability at stratum 4, may get completed in two-years by a person working at stratum 5 or higher.

4.   CONCLUSION

Understanding the basic constructs of the problem of universals in management studies, such as potential capability of the person estimated through time-horizon, and the level of working role in a managerial bureaucracy through time-span of the role, and estimates of complexity could help match people with technology projects better. Projects of different complexity would require people of matching capability – understanding both could help fit people better. Such as, a lengthier project (with a higher number of function points) may hint to assign people of a higher and corresponding time-horizon than a smaller project – to which is best to assign people with a lower, and also corresponding time-horizon, and so on. Teams should be assembled with understanding of the capabilities of people to achieve success.

5.   RECOMMENDATIONS FOR FUTURE STUDIES

The theoretical theorems discussed in this paper need to be tested empirically. In addition, there is a need to develop a measuring instrument to measure the complexity of the project accurately, and match complexity levels with the ability of people at different time-horizons to work at the abstract complexity levels to solve the problem of universals within their scope of the assignment, according to their current potential capabilities.

6.   REFERENCES

Capelle Associates Inc.. http://www.capelleassociates.com/text/index.html: Web Site

Blooms Taxonomy, http://www.arch.gatech.edu/crt/lln/Wordsworth/bloomstaxonomy.htm: Georgia College of Technical Architecture Web Site, 1998.

Abdel-Hamid, Tarek K, Sengupta, Kishore, and Swett, Clint, *The impact of goals on software project management: An experimental investigation*, MIS Quarterly 23.4 (1999): 531-555.

Agarwal, Ritu, and Karahanna, Elena., *Time flies when you're having fun: Cognitive absorption and beliefs about information technology usage*, MIS Quarterly 24.4 (2000): 665-694.

Ahituv, Niv, Neumann, Seev, and Zviran, Moshe., *Factors Affecting The Policy For Distributing Computing Results*, MIS Quarterly 13.4 (1989): 389

Ahuja, Manju K., and Carley, Kathleen M., *Network structure in virtual organizations*, Organization Science 10.6 (1999): 741-757.

Alberts, Michael, Agarwal, Ritu, and Tanniru, Mohan., *The Practice of Business Process Reengineering: Radical Planning and Incremental Implementation in an IS Organization*, Communications of the ACM (1994): 87-96.

Aronson, Elliot., *Readings about the Social Animal*, New York, NY: Worth Publishers,   1999.

Aronson, Elliot., *The Social Animal*, New York, NY: Worth Publishers, 1999.

Artz, John., *Information Modeling and the Problem of Universals: an Analysis of Metaphysical Assumptions*, The George Washington University: Unpublished Paper, 2002.

Bharadwaj, Anandhi S., *A resource-based perspective on information technology   capability and firm performance: An empirical investigation*, MIS Quarterly 24.1 (2000): 169-196.

Blanton, J. Ellis, Watson, Hugh J, and Moody, Janette., *Toward a better understanding of information technology organizations*, MIS Quarterly 16.4 (1992): 531

Bleandonu, Gerald., *Wilfred Bion: His Life and Works*, New York, NY: The Guilford Press, 1994.

Bloom., *Major Categories in the Taxonomy of Educational Objectives*, faculty.washington.edu/krumme/guides/bloom.html: University of Washington, 1956.

Brause, Alison., *Summary of an Investigation of Presidential Elections Using the Jaques/Cason Construct of Mental Complexity*, University of Texas, Austin, TX: Unpublished Doctoral Dissertation, 2000.

Brockers, Alfred, and Differding, Christiane., *The Role of Software Process Modeling in Planning Industrial Measurement Programs*, 3rd International Metrics Symposium, Berlin, 25-26 March 1996 IEEE, 1996.

Campbell, Donald T., and Stanley, Julian C., *Experimental and Quasi-Experimental Designs for Research*, Boston: Houghton Mifflin Company, 1963.

Cosier, Richard A., *The Effects of Three Potential Aids for Making Strategic Decisions on Prediction Accuracy*, Organizational Behavior and Human Performance 22 (1978): 295-396.

Daxis, P., *Realizing the Potential of the Family Business*, Organizational Dynamics 12 (1983): 47-53.

Feeny, David F, Edwards, Brian R, and Simpson, Keppel M., *Understanding the CEO/CIO relationship*, MIS Quarterly 16.4 (1992): 435-449.

Fisher, Sven J., Achterberg, Jan, and Vinig, Tsvi G., *Identifying Different Paradigms for Managing Information Technology*, 1993. 37-55.

Gash, D. C., *Negotiating IS: Observations on changes in structure from a negotiated order perspective*, Special Interest Group on Computer Personnel Research Annual   Conference: Proceedings of the ACM SIGCPR conference on Management of information systems personnel (1988): 176-182.

Grant, Rebecca., *E-Commerce Organizational Structure: An Integration of Four Cases*, Communications of the ACM (2000)

Gregson, Ken., *Realizing organizational potential*, Work Study 44.1 (1995): 22-28.

Hall, Alfred Rupert., *Isaac Newton*, http://www.newton.cam.ac.uk/newtlife.html Microsoft Encarta: Microsoft Corporation, 1998.

Hamilton, J Scott, Davis, Gordon B, and DeGross, Janice I., MIS doctoral dissertations: 1997-1998, MIS Quarterly 23.2 (1999): 291-299.

Harvey, Jerry., *How Come Every Time I Get Stabbed in the Back, My Fingerprints Are on the Knife?*, San Francisco, CA: Jossey-Bass, 1999.

Harvey, Jerry., *The Abilene Paradox*, CA: Lexington Books, 1988.

Harvey, Jerry B., *The Abilene Paradox and Other Meditations on Management*, San Francisco, CA: Jossey-Bass Publishers, 1988.

Igbaria, Magid, Parasuraman, Saroj, and Badawy, Michael K., *Work experiences, job involvement, and quality of work life among information systems personnel*, MIS Quarterly 18.2 (1994): 175-202.

Jaques, Elliott., *A Theory Of Life: An Essay on the Nature of Living Organisms, Their Intentional Goal-Directed Choice Behavior, and Their Communication and Social Collaboration*, 2000.

Jaques, Elliott., *Requisite Organization: The CEO's Guide to Creative Structure and Leadership*, Arlington, Virginia: Cason Hall and Co., 1989.

Jaques, Elliott., *Time-Span Measurement Handbook*, Cason Hall, 1964.

Jaques, Elliott., *Requisite Organization: A Total System for Effective Managerial Organization and Managerial Leadership for the 21st Century*, Arlington, Virginia: Cason Hall & Co., 1996.

Jaques, Elliott., *Personal Interview*, 2001.

Jaques, Elliott., *A General Theory of Bureaucracy*, London, UK: Heinemann Educational Books, 1976.

Jaques, Elliott., *The Life and Behavior of Living Organisms: a General Theory*, Westport, CT: Praeger Publishers, 2002.

Jaques, Elliott., *Personal Communication*, 2001.

Jaques, Elliott., *The Form of Time*, New York, New York: Crane, Russak & Company,   1982.

Jaques, Elliott, and Cason, Kathryn., *Human Capability*, Rockville, MD: Cason Hall, 1994.

Judkins, Jennifer ., *The Aesthetics of Musical Silence: Virtual Time, Virtual Space, and the Role of the Performer*, Anri Arbor, Michigan: UMI Dissertation Information   Service, 1987.

Lott, Christopher M., *Technology trends survey: Measurement support in software engineering environments* Int. Journal of Software Engineering and Knowledge Engineering 4.3 (1994)

Mahmood, Mo A.;Becker, Jack D., *Impact of organizational maturity on user satisfaction with information systems,* Special Interest Group on Computer Personnel Research   Annual Conference: Proceedings of the twenty-first annual conference on Computer   personnel research 1985. 134-151.

Mennecke, Brian E., Crossland, Martin D., and Killingsworth, Brenda L., *Is a map more than a picture? The role of SDSS technology, subject characteristics, and problem complexity on map reading and problem solving,* MIS Quarterly 24.4 (2000): 600-629.

Misra, R, and Banerjee, R., *Use of Time-Span Instrument in Job Analysis and Measurement of Responsibility (India),* Journal of the Institution of Engineers XLII.8, Part GE 2 (1962)

Moore, Jo Ellen., *One road to turnover: An examination of work exhaustion in technology professionals,* MIS Quarterly 24.1 (2000): 141-169.

Mort, Joe, and Knapp, John., *Integrating workspace design, web-based tools and organizational behavior,* Research Technology Management 42.2 (1999): 33-41.

Nissen, Mark E., *Redesigning reengineering through measurement-driven inference,* MIS Quarterly 22.4 (1998): 509-534.

Pressman, Roger S., *Software Engineering: A Practitioner's Approach,* New York: McGraw-Hill, 1997.

Reich, Blaize Horner, and Benbasat, Izak., *Measuring the linkage between business and information technology objectives,* MIS Quarterly 20.1 (1996): 55-82.

Richardson, Roy., *Fair Pay and Work,* London: Heinemann, 1971.

Robey, Daniel., *Computer Information Systems and Organizational Structure,* Communications of the ACM 24.10 (1981): 679-687.

Roepke, Robert, Agarwal, Ritu, and Ferratt, Thomas W., *Aligning the IT human resource with business vision: The leadership initiative at 3M,* MIS Quarterly 24.2 (2000): 327-353.

Sarle, Warren S., *Measurement theory: Frequently asked questions,* Disseminations of the International Statistical Applications Institute 4 (1995): 61-66.

Segars, Albert H., and Grover, Varun ., *Strategic information systems planning success: An investigation of the construct and its measurement,* MIS Quarterly 22.2 (1998):   139-163.

Shrednick, Harvey R, Shutt, Richard J, and Weiss, Madeline., *Empowerment: Key to IS world-class quality,* MIS Quarterly 16.4 (1992): 491-506.

Swanson, E Burton, and Dans, Enrique., *System life expectancy and the maintenance effort: Exploring their equilibration,* MIS Quarterly 24.2 (2000): 277-298.

Trochim, William., *The Research Methods Knowledge Base,* Atomic Dog Publishing, 2000.

THE IMPACT OF TECHNOLOGICAL PARADIGM SHIFT ON INFORMATION SYSTEM DESIGN

Gábor Magyar and Gábor Knapp[*]

1. INTRODUCTION

The evolution of ISD methods from water flow model to object-oriented techniques was catalyzed by the need of managing several uncertainties in user requirements, changes in software and hardware technology as we have reported in our previous case study [1]. However it was assumed, that there are only slight changes in the static and dynamic behavior of the completed system during operation.

The technology driven paradigm shift in audiovisual content industry caused dramatic changes in workflow. This predicts several novel ways of future applications that are completely unknown at design and implementation phase. So design philosophy has to be prepared to build a system that can serve new functionality by re-arranging the original, unchanged system components.

The current paper shortly describes the effects of the change in technology of audiovisual content industry, introduces the „value-oriented" model and its consequences in workflow management. The second part outlines the principles, requirements of a suggested component based information systems design philosophy based on the concept of „semantic hook" that can handle the problem of supporting several parallel and partly unknown workflows.

2. THE PARADIGM SHIFT

A paradigm shift takes place in the audiovisual content management industry. This paradigm shift is originated from technology development, and nowadays results dramatic changes in the workflow of the audiovisual content management industry. The

---

[*] Gábor Magyar, Department of Telecommunications and Telematics, BUTE, Budapest, Hungary.
Gábor Knapp, Centre of Information Technology, BUTE, Budapest, Hungary.

Information Systems Development: Advances in Methodologies, Components, and Management
Edited by Kirikova *et al.*, Kluwer Academic/Plenum Publishers, 2002

341

traditional linear workflow is being replaced by a new, central model (often called to "value-oriented" model).

For decades the audiovisual (TV and radio) content management workflow followed a linear, „water flow-like" model:

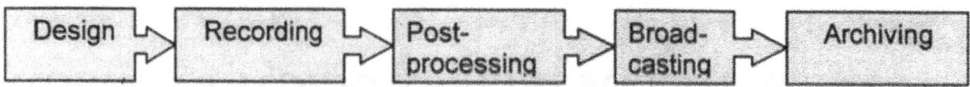

Figure 1. The conventional archiving workflow

This linear model is quite simple and has the following features:

- each phase of the workflow uses the output of the previous phase as input,
- each phase has its own technical criteria (there is, of course a strong interdependence between the different criteria, but the different phases can have different technical priorities; e.g. recording needs high bandwidth for best recording quality, while broadcasting confronts to the limited bandwidth of transmission channels; archiving needs as much metadata as it can be generated, while the previous phase have no special interest in providing these meta data, etc.)
- each phase itself can be defined as an independent workflow, connected by defined interfaces to the neighboring phases.

There were significant achievement in the technology of the audiovisual (TV and radio) content production and broadcasting, but all these could be implemented within the same frame of this linear model.

Two revolutionary changes were taken place during the last 6-8 years.

The first one is the emerge of the digital technology. Digital technology appeared first in devices, later, more and more digital devices could be interconnected, but all these devices were developed simply to replace their analog counterparts within the same, unchanged workflow. The main advantages of developing and using digital devices were:

- simple and cheap copy of the recordings, using standard computer storage systems instead of special digital video formats.
- controlled and stable quality of the copies (a duplicate does not differs from the original; so the original record can be copied without quality loss),
- efficient maintenance, fault tolerance (the storage system and even media controls itself)
- Using digital devices authors created new expressive forms of art.

The second one is the diffusion of the information technology in the realization of different tasks. First IT promised only better management of the (same, unchanged) processes, but it was clear, that the extreme flexibility of IT based workflow leads to a new structure. IT gave a new view of the whole process, so new workflow (and new value chain) could be established. The aims of using IT in audiovisual industry were

- a flexible content production,
- the reuse of the content,

- multiple usage of content elements,
- better content retrieval.

The driving forces behind these changes have both *technical* and *business* nature:

- The proliferation of the recorded audio and video content resulted huge, hard-to-manage archives. (Ten millions of hours of sound and video materials are stored in the archives, but the majority of these values are practically inaccessible, due to the lack of the up to date, detailed catalogues.) New recording formats and technologies comes out year by year, archives have to manage a heterogeneous "iceberg" of audio and video materials: only a small portion of the content is visible (like the upper 10 percent of the real iceberg). And even if you know what you have, and know how to find it, the retrieval time is too long.
- The increasing number of broadcasting channels (including CATV and broadband internet, so better to write communication channels) require more content, but the content production industry lag behind the increase of the distribution capacity in volume. Content distribution (broadcasting, CATV distribution, internet) must reuse old content many times.

3. THE NEED FOR A NEW MODEL

A new concept for planning and designing the information system of audio/video archives is needed. Future archives should be integral parts of computer- and network based broadcast areas instead of isolated islands, they should give direct access to their content using networks. Future audio/video archive would not consist of specific audio/video storage media anymore. Audio/video recordings are not related any longer to the audio/video carrier they were recorded on, but as a pure audio/video information – called "essence" – that must be saved in the best way possible. Audio/video archives should no longer try to find the "eternal audio/video carrier" to preserve the recording in its specific and optimal technical quality. Instead, archives should start to preserve audio/video as data files that should not be changed but can easily be transferred or copied without any modification of its original information, the "eternal audio file".

An archive based on a digital mass storage system, controlled, managed and accessed by an open architecture ("evolutionary") information system automatically satisfies several requirements, but not all.

Aspects of proper operation and availability are important for different specialists (like journalists, music or film editors, etc). They will need the archive for their daily work even as they use on-air systems nowadays. So the archive has to serve the demands of different jobs (accessing and loading text, music, video, data content), including jobs which are not known at the moment of planning, designing and implementing the archive information system (interactive systems, virtual collections etc.).

A new kind of production chain is emerging. The new systems have to be ready to involve new workflows without the change of the massively and continually used existing systems.

As defined by Porter, the value chain is a collection of activities that are performed by the firm to design, produce, market, deliver, and support a product or service. [4] The configuration of a firm's value chain – the decisions relative to the technology, process, and location and whether to 'make or buy' each for each of these activities – is the basis of competitive advantage. The production chain, in turn, is a part of a larger, the real value system that incorporates all value-added activities from raw materials to buyer distribution channels, through component and final assembly. [11]

In other words, the new value chain should be configurable.

4. THE NEW MODEL

Engineers planned and developed new technical model for audiovisual production information systems. The new model is not linear, but centralized. The core of the model is a digital audiovisual archive. [10] This is the audio and video essence management system that consists of media archive software components and different hardware components: database server, media server, conversion servers, online storage and near-line storage. [2,3]

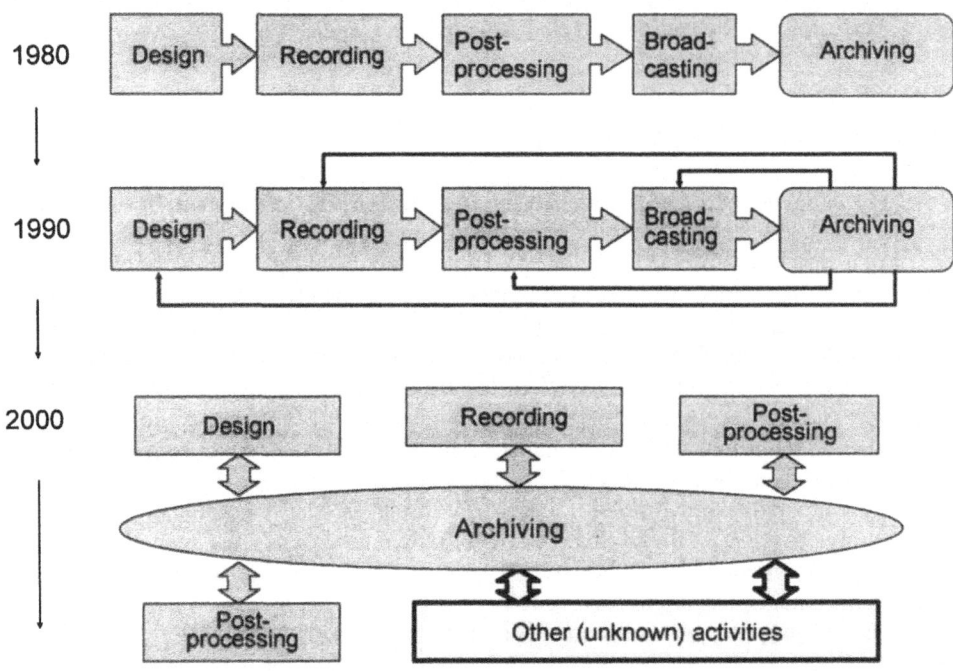

Figure 2. Shift to central model

The new architecture allows easy but controlled access to archive content, to all existing recordings for all users connected. Any content needed can be automatically delivered on demand. Other applications (supporting design, recording, post-processing, play-out) integrated in the system, are requesting services from the content (essence) management system.

The essence management system will optimize access time to the archived files. The applications have several user interfaces for the documents catalogue, the display formats are designed for the special needs of different user groups. The document forms may also contain http links to other applications. E.g. the user can check whether the recording he is looking for is processed by other users or not. The user can start a pre-listening/previewing or rough-cut operation, or initiate the export of a production quality audio/video file to his workstation. [5]

5. THE NEED FOR A NEW DESIGN METHOD

The workflow within the centralized architecture differs significantly from that of the linear model. Parallel and independent actions can be performed in the same system enhancing efficiency, but also enhancing management problems. Furthermore, new applications are supposed due to the widening of the range of potential authors and customers. These new applications have to be assembled without influencing the existing ones, the change of functionality of software components is not possible.

New functionality should be added without changing the existing frame.

The centralized archiving structure was initiated by user demands (efficiency) and was made possible by technology. It is not evident that the change in structure fundamentally changes the workflow also. However when data are concentrated, novel relationships can be recognized and new ways of usage can arise, so new functionality has to be implemented on the same system.

It's hard to give an example for a functionality that is not exist yet, but let's try!

Nowadays radio and television archives are usually not prepared to exploit the possibilities of the Digital Video Broadcast (DVB). Interactive e-learning or on-demand services can appear soon that primarily use the search and broadcast capabilities of the production chain, but the monitoring of user statistics can effect Recording, Design and Production also, not mentioning the need for new modules like internet gateway.

5.1. Static versus dynamic approach data element representation

The data elements in the archive generally have very different properties. The 'essence' can be managed (e.g. searched) easily in the case of text (free text search), the processing of sound is more difficult (e.g. voice to text conversion), and the search of video streams has still several uncompleted problems (visual search languages, pattern recognition). If we use attached metadata to classify essence, efficiency of search is automatically enhanced, because human intelligence is added and structured or semi-structured metadata schemas provide specific search capabilities. The metadata represent the properties of the essence, so they also have to be of different kind. However to provide simultaneous search, at least a subset of metadata have to be similar. For multimedia content the 15 basic elements of Dublin Core Metadata Initiative are suggested. [6] The differences can be denoted by more specific elements (e.g. MARC for books), or by the refinement of the Dublin Core structure by qualifiers. [7]

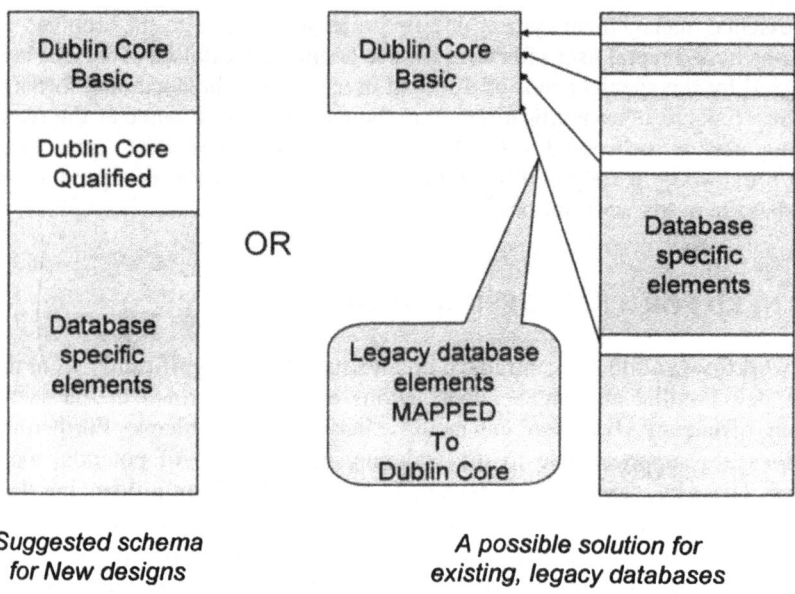

Suggested schema   A possible solution for
for New designs   existing, legacy databases

Figure 3. Static approach of classification

The main problem with metadata structures is, that they are static. If novel properties are to be added, the complete re-design of the database is probably unavoidable. However the intensive use of archives is expected to stimulate a large number of novel ways of application, and the hard-wired classification (indexing) can be a bottleneck. So the databases should be designed to allow additional classifications or indexing without the change of basic structure (stable structure is required, because several independent usages are assumed, and it is not allowed to disturb other workflows). To support ordering according to new aspects or creating new navigation lines (not known at design phase) the database has to provide 'semantic hooks' that makes possible the connection of new indexes with unchanged basic data. [8]

A database can aware for several aspects, but not all. If somebody studies extinct animals he should be allowed to create a new relation, a new navigation path (without any change in the basic database structure) and the value of the database can be enhanced by the publication of this new feature.

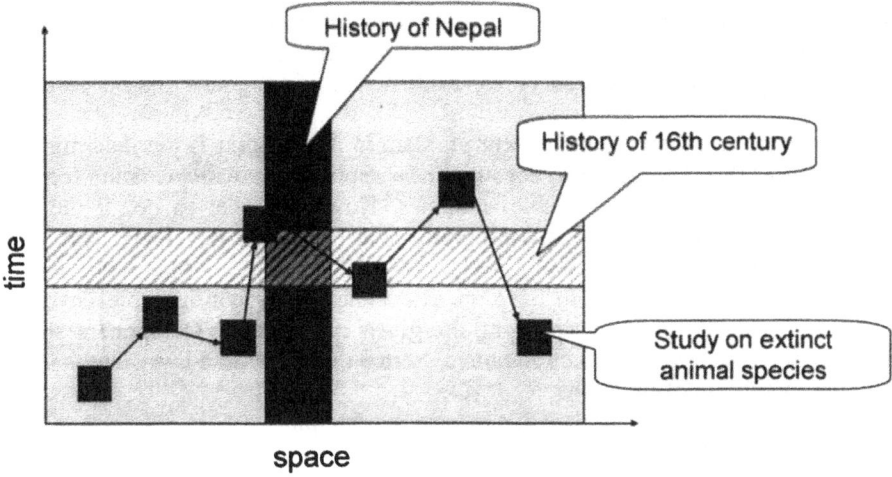

Figure 4. Example for the flexibility in requirement space

The task described above, require a component-oriented approach. Functional components (implemented by different software modules) can be identified in the model, and – in principle – several workflows (including new ones) can be constructed using different sets of system components. New components can be added to the model, extending the variety possible workflows.

5.2. Consequences in functionality

In the centralized archive several different types of 'essence' are stored (text, video, image, sound, or database records) so standard archiving functions (store, modify, search and retrieve) differ significantly for each type at the syntax level, however – to assure simultaneous, type independent search – they have to be equivalent at the semantic level. The essence and related essence-type-specific functions have to be managed together, therefore object oriented philosophy seems to be essential for these types of applications.

Another consequence of difference in program codes is, that the entry points of the functions (methods) managing the content embedded in objects have to be defined at semantic level (so the same methods have to be implemented in each object classes). [9] Workflows are assembled of components, however different, possibly quite new functions need novel order or component hierarchy, and – just like in the case of databases – the function-primitives has to remain unchanged. So, software elements in the archive, has to be designed to allow the building of new applications, without any affect on the existing components.

The flexibility of the information system can be guaranteed by observing the following simple rules

- components should be minimal in implemented function
- unified and simple interfaces should be defined between the components (interface means communication and control primitives)

Problems of system construction, reconstruction are based on system primitives.

Construction of procedures should be based on statements, construction of software modules should be based on procedures, construction of programs should be based on modules.

The information system development problem in this context is the determination of the level of modules. The task is to construct new applications, using existing (or existing and new, but conform, interoperable) modules. However, what to do, if the module interfaces don't fit? In this case, the program must be further developed (evolutionary program development).

In other words, the goal is to have an extensible, easily configurable development environment extension without changing the given components. Component should be designed to have semantic connection points, 'semantic hooks' to a layer that is in charge of the interoperability of functions.

We need a method to generalize this extension principle.

6. THE PRICLIPLES OF A NEW METHOD

First, we should understand, that *no local specification can be given for global functionality*. One is allowed to modify aspects independently from each other. Separation of aspects is a fundamental principle of engineering.

Good and simple example of this principle is the planning of a building. Separate plans are made for different aspects, like structure, water, gas, electricity supply, etc.

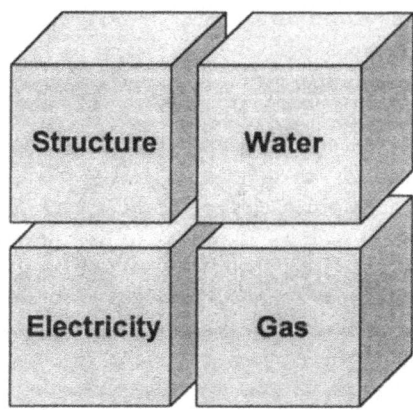

Figure 5. Separate plans are needed for different functionalities

We have to **weave** them together for constructing the building.

We are looking for something similar for continually develop IT based audiovisual information systems.

Important feature of this approach is, that the methodology is useful either for reengineering legacy systems, or the development of new product families. Such a methodology can be used also as a starting point for developing of more abstract and expressive composition languages.

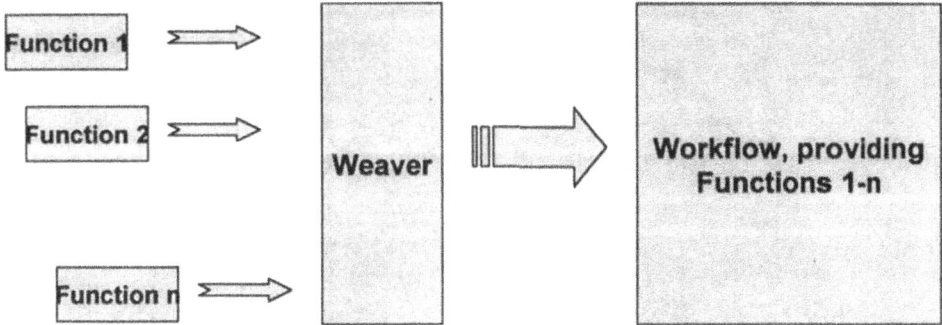

Figure 6. Weaving the functions into a workflow

Weaver connects different functional aspects. Local specifications ease adaptation of components by exchange of aspects. Weaving leads to systems with extensible aspects.

Overall: "Weaver" approach gives a flexible and efficient method for developing applications within a certain architectural IT model.

But a number of practical questions arise in connection with the realization of the approach:

- How does one extend components in a flexible way using minimal number of additional layers for mapping views into the system?
- How does one merge aspects into components without using specially constructed systems?
- By which mechanisms beyond inheritance and delegation can components be adapted?
- Do we need for each aspect a special language, for each combination of aspects a special weaver?

To eliminate, or at least handle these problems, we introduce the concept of "semantic hook". The semantic hook is an element of the abstract syntax of a component. The method completed by hooks is able to adapt and extend components at hooks by system function transformation.

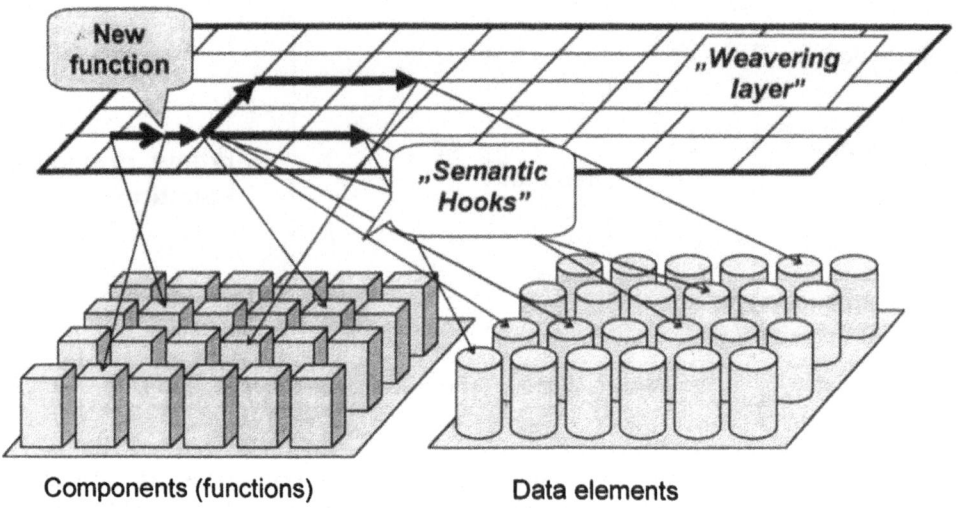

Figure 7. The layer of semantic hooks

(We can find simple examples for hooks in software engineering: one can consider communication calls as semantic hooks.)

Software technical details are out of the scope of this paper, so we mention here without evidence, that we need interfaces for weaving different components that encapsulate the component such that it is replaceable by arbitrary variants. This method is then by principle more flexible as with modules, frameworks, architectural languages in traditional methods.

This approach can be referred as "aspect oriented information system development", since its aim is to provide new or improved functionality of an existing system without any changes in system architecture.

7. CONCLUSION

Moving from low level syntax ("hardware") to high level syntax (high level programming tools, application generators) and towards even higher (semantically smart) level of abstraction is reducing efficiency in operation but sometimes worth: rewarding in skilled human resource savings, on the whole: in the efficiency in planning, design and redesign of systems.

Principles of a new approach were introduced in the paper. System designers and software engineers use formal techniques in design and implementation tasks. They have heuristic, ad-hoc techniques only for the interpretation of user ideas during very beginning phase of the development. There is a need for formal representation of the early ideas, which can be used as a tool for managing the changing ideas resulting flexible, variable system features (in functionality and in database schemas). That kind of tool can be used in new system-designs and systems-redesigns as well. This formalization requires a new layer of abstraction: the waving layer. Semantic hooks are elements of this layer. Existing systems can be enriched with new functions using waving, and novel system designs produce better (more flexible) result.

The high-level modeling tool allows users to define rules, which can be used to control the flow of and monitor functional events. The model applying a "wavering layer" gives a method for flexible description of new (business, functional) rules and makes possible to modify the existing rule-set without the need to stop and reengineering the systems. This form of dynamic rule installation can underpin the ability of an enterprise to retain critical business flexibility.

The implementation of these principles in system design have to be done in standard and open platforms (e.g. XML).

REFERENCES

1. Magyar, G., Szakadát, I., Knapp, G.: "Information Systems in the e-Age", Presented at ISD 2001, Egham, London
2. György, P., Knapp, G., Kovács, A.B., Magyar, G., Rozgonyi, K., Szakadát, I.: „National Audiovisual Archive Pilot Project" Poster at the VI. European Conference on Archives, Florence, 2001Cantor, Murray. Object-Oriented Project Management with UML. New York: John Wiley & Sons, Inc., 1998.
3. Knapp, G., Kovács, A.B., Magyar, G.: National Audiovisual Archive Pilot Project and Feasibility Study. Presented at the Europrix – IST CEE Regional Workshop, Budapest, 2001
4. Porter, M. E. (1985), *Competitive Advantage: Sustaining and Creating Superior Performance* (New York: The Free Press).
5. The Revolution Will Not Be Televised: Personal Digital Media Technology and the Rise of Customer Power. White paper. KPMG Digital Media Institute. April 15. 2000.
6. The Dublin Core Metadata Element Set, NISO Press, Bethesda, Maryland, USA, 2001
7. Dublin Core Qualifiers, DCMI Recommendation, 2000. http://www.dublincore.org
8. Shabajee et al.: Adding Value to Large Multimedia Collections Through Annotation Technologies and Tools, presented at "The Museum and the Web", Boston, USA, April, 2002.
9. Swinton Roof: The Semantic Web and Further Thougths About Where Software Meets the Hardware in Ergonomican of AbraxPsiPhi (A Compendium of Essays), May, 2001 http://home.earthlink.net/~sroof/
10. A look at the Streaming Media Value Chain. YANKEE GROUP, Internet Computing Strategies, REPORT Vol. 6, No. 1—May 2001, by Amy Prehn
11. Robert L. Phillips : The Management Information Value Chain., Perspectives, Issue 3, 2001

KEY ISSUES IN INFORMATION TECHNOLOGY ADOPTION IN SMALL COMPANIES

Jože Zupančič and Borut Werber[*]

1. INTRODUCTION

The influence of small companies in the entire economy is increasing. Small companies employ more people than ever, and many more are starting their own businesses. Small companies sometimes act as incubators for future economic giants. Realizing the importance of new information technology (IT) small companies are increasingly investing in their information systems, encouraged also by declining cost of contemporary IT. They are replacing their existing manual systems or old legacy systems with new, more flexible and reliable systems, to run their daily operations. A general belief exists that this enhances the flexibility of small companies, although some investigations (e.g. Levy and Powell, 1998) indicated that IT increases the effectively and effectiveness of the business and strengthens a creative way of thinking, but does not improve their flexibility.

Percentage of work force employed by small companies varies from country to country: in European Union (EU) from 56,2% in Belgium to 81% in Spain. In Singapore, small companies employ 53% of total workforce and contribute 34% to Singapore's GPD. In Slovenia, small companies employ 34% of the workforce (SWB, 1999), but their number and importance is still increasing. Their ability to adapt their business practices to changes in the environment and customer demand improved the quality and diversity of services and products on the market. Although the structure and business practices of small companies in Slovenia is comparable to EU, some major differences are still evident. For example, in EU only 10% of small companies come for the construction industry, in comparison to 45% in Slovenia (Lesjak and Lynn, 2000).

[*] Jože Zupančič and Borut Werber, University of Maribor, Faculty of Organizational Science, 4000 Kranj, Slovenia, E-mail: joze.zupancic@fov.uni-mb.si

No generally accepted definition of a small firm can be found in the research literature. The most common criterion for a small company is the total number of employees, often combined with some financial indicators, such as gross annual sales and firm's assets. In absence of a precise definition, small organizations are defined in different context in various business cultures. Sometimes, the definition of a small business depends on the industry: for example up to 200 total employees for the service sector, and 500 for the manufacturing sector (e.g. Pollard and Hayne, 1998). In some investigations, companies with less than 15 total employees are classified as small (Ibrahim and Goodwin, 1986), while others treated the size less than 50 employees as small (e. g. Verber and Zupančič, 1993; Lai, 1994; Seyal et.al., 2001). Some authors set the limit for small companies as high as 600 employees (Vijayaraman and Ramakrishna, 1993). Osteryoung et al. (1995) studied characteristics of companies with up to 500 employees and concluded that businesses with less than 50 employees in general differ from larger companies in terms of organization, internal and external communication, management style, and business practice. Therefore, they recommended that such firms should be classified as small companies.

For the purposes of our investigation we used the definition of small company stated in the accounting legislation in Slovenia: a small company is any enterprise that meets at least two of the following three criteria: (1) the company has less than 50 employees, (2) annual gross sales are less than 280 millions SIT (about 1.3 million €) and (3) companies total assets are less than 140 millions SIT (about 0.65 million €).

2. BACKGROUND

Small companies cannot be treated as downscaled versions of large companies, due to differences in organization, management style, business practice and information systems. They are frequently managed by owners who are usually CEOs. The decision-making process is often more intuitive than based on reliable, precise and unambiguous information. Small companies often lag behind larger businesses in in-house adoption of information technology (IT), due to severe constraints on financial resources, lack of in-house expertise, and a short-term management perspective imposed by a volatile competitive environment. On the other hand, they demonstrate a high level of ability to adapt to changes in the environment. Small companies usually don't use a computerized IS for communication among organizational levels and units. They use IT for automation of existing processes, rather then for decision support, or to increase flexibility of the firm and gain competitive advantage.

Many studies indicate that findings related to key success factor in IT and IS implementation and use in large companies cannot be simply generalized to small companies (e.g. Doukidis et al., 1994; Thong et al., 1997; Razi 2001; Hunter 2001). Implementation and operation of IT in small companies was investigated by several authors (e. g. DeLone 1988; Yap 1992; Verber and Zupančič, 1993; Winston and Dologite 1999; Seyal et al. 2001, Hunter 2001). These studies focused on issues related to IT, and considered success factors such as number of PCs, number of programs and/or software tools used by the company, number of users, history of use of IT in the company, security and safety, and computing knowledge and skills of managers and employees. Other studies focused on issues related to use of IS: ease of use and maintenance, ability to create and shape additional information, IS implementation, user training and support (Doukidis et al., 1994; Chau 1994, Cragg and Zinatelli and Cavaye, 1995; Razi 2001), or considered external factors: competitive environment, impact of government policy, market situation, impact of global factors (Thong 1999, Papazafeiropoulou and Pouloudi, 2000). Several models of key factors in IS implementations were also proposed (e. g. Palvia et al., 1994; Thong 1999; Winston and Dologite 1999).

In this paper, results of an empirical study among 122 small companies in Slovenia, focusing on key success factors in implementation and use of computerized IS in small companies are presented.

3. RESEARCH APPROACH AND CHARACTERISTICS OF THE COMPANIES PARTICIPATING IN THE STUDY

Data for the investigation was collected via structured interviews with owners or top managers of small companies. Several other studies (e. g. Verber and Zupančič, 1993; Doukidis et al., 1994; Seyal et al., 2001) showed that this group plays a dominant role in management and decision-making in small companies. In the interview, mostly closed-response questions were asked. Except for demographic data, respondents either rated quoted statements on scale 1 to 5, or responded to multi choice questions. To avoid bias in the responses, businesses focusing on software development and/or IS implementation, and education and training in IS area, were eliminated from the randomly selected sample. In total, 136 interviews were conducted, and 122 of them met the criteria for inclusion in the investigation. The rest was eliminated because the interviewee was not owner, a close relative of the owner, or top manager of the company, or because responses to some key questions were missing.

Companies that had no computers. Out of 122 participating companies, 28 (22,9%) didn't use computers. In most cases their owners/managers indicated two reasons for not using computers: insufficient knowledge ("nobody knows how to use computers") and lack of financial resources. Companies without computers were mostly from the manufacturing sector (64%), and Pearson Chi Square test showed that the formal education of their owners/managers was lower then the education of other managers/owners (p=00.5).

Hardware and software. On average, companies from our sample who used computers had 3.5 computers (one per 1.6 employees) and have been using computers for 4.7 years on average. In 43 (45.3%) out of the 94 companies using computerized IS, computers were networked, while 15 (15.8%) used modems to connect computers within the company. Most companies (70,3%) exchanged some data in electronic format with their business partners and/or government institutions. MS Windows was the prevailing operating system (84%), followed by DOS (14%) and Linux (2%). More than one third (39%) of all programs supporting business functions, such as accounting, bookkeeping, general ledger and sales management, used character oriented user interfaces. In nearly 80% of all companies general purpose tools (spreadsheets, databases, ...) were installed, but only a few of them used them to analyze data from the database or to prepare customized reports. These tools were mostly bought together with the computers, which may explain why they were not much used. Insufficient computing knowledge and skills of the owners/managers and employees may be a possible explanation for the nonuse of software tools. Relatively low self-assessment of computing knowledge and skills, found in our study, supports this assumption.

Use of Internet. Figure 1 presents how Internet was used in the surveyed companies. Not surprisingly, electronic mail and searching for information were the most widely used Internet facilities. Nearly one fourth (22,3%) of the companies started to use Internet for e-business, mainly for electronic banking.

A comparison with results of similar studies from other countries shows that small businesses in Slovenia are not much different from small businesses in developed countries in terms of IT availability. In USA and in Australia computers were used in 83% and 84% respectively (Burgess and Schauder, 2001), compared to 77,1% of Slovenian small businesses. The portion of Slovenian businesses that have Internet access was 70,2% (Figure 1). Slovenian small businesses mostly use the Internet for e-mail (68,1%), followed by search for information (62,8%).

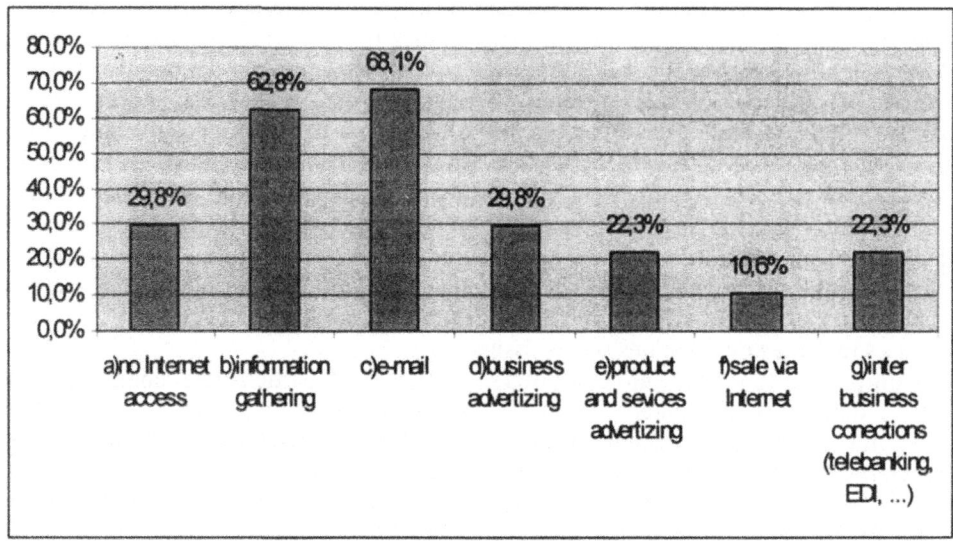

Figure 1. Small business use of the Internet.

A study of Brazilian small businesses in 1997 (Lébre La Rovere, 1998) found that only 22% of small businesses had access to the Internet. An investigation in Singapore (Thompson and Tan, 1998) from 1996 showed that 69,7% of mixed size business (75 % with more than 300 employees) had Internet access. In Singapore, 29,1% of surveyed companies offered their customers the possibility to buy products/services over internet, compared to 10,6% in Slovenia. An investigation among small and medium enterprises (SMEs) in Great Britain (Lawrence and Hughes, 2000) showed that 79% were using Internet. Comparing companies with 1-49 employees and companies with 50-250 employees, the survey found slightly different levels of Internet use in each of the two groups. Table 1 presents a comparison of some of these findings.

Table 1: Comparison of Internet use

	Our study	*(Lawrence et al., 2000)*	*(Kurbel, 1998)*
Number of employees	*Up to 50*	*Up to 49*	*From 1 to 1000+*
Information gathering	56,4%	77%	90,6%
Communication (including e-mail)	68,1%	63%	
Advertising of business	29,8%	35%	83.4%
Advertising of products and services	22,3%	34%	76,3%
Sale via Internet	10,6%		34,8%
Electronic business with banks, ...	22,3%	30%	43,6%

Data collected via interviews was statistically analyzed using factor analysis, and primary component method was used to reduce the number of variables. Details of the analysis are presented in (Werber, 2000).

To measure efficiency and effectiveness of IS in companies under consideration, we used the following indicators, which have been applied also in several other investigations:

1. Number of applications and software tools that a company uses (Henson, 1995)
2. How the owner/manager uses his/her personal computer (Morris et. al, 1989, Magal and Lewis, 1995)
3. Satisfaction with business programs which was measured by the following four indicators:
    *   Cost of computing versus benefit (Henson 1995)
    *   Impact of computers on the business practice (based on six sub-questions
    *   Alignment of business programs with company's business practice
    *   Assessment of the quality of business applications used in the company by owners/managers
4. The available IT in the company: number of computers, servers, modems, UPS, ...

4. SUMMARY OF FINDINGS

Statistical analysis of data, collected in the interviews, revealed the following success factors related to the implementation of IT and IS in small companies:

Knowledge and skills of the owner/manager. Formal education of the owner /manager and his/her self-assessment of computing skills and knowledge are positively related to IS success and IT usage in the company, while the success was negative related to the age of the owner/manager. Similar conclusions were found also by several other studies (e. g. Palvia et. al., 1994; Igbaria et al., 1997; Thong 1999; Winston and Dologite 1999).

Total number of employees. Our study showed that larger companies more likely would use IT successfully. Similar results published in several other studies (e. g. Palvia et.al., 1994; Thong 1999).

IS safety measures. Companies that paid more attention to safety and security used their IS more successfully. This result could not be checked with other studies because we found no comparable investigation in the literature. Issues related to IS safety and security in small companies in Slovenia are discussed in (Werber 2002) in more detail.

Business relations with large partners. A belief exists that companies that strongly depend on one or a few large business partners use there IS more efficiently and effectively, because their strategic partners encourage or press them to adopt more advanced technology and approaches. Like some other studies (e. g. Henson 1995; Igbaria et al., 1997; Winston and Dologite 1999) our investigation supported this assumption.

Use of Internet. Internet provides companies with a platform for electronic commerce and allows them free access a wide range of information. Previous investigations (e. g. Poon and Swatman 1999; Kurbel and Teuteberg 1998; Deschoolmeester and Hee, 2000) mostly indirectly support the hypothesis that companies that use Internet are more successful in using their computerized IS. Finding of our investigation support this hypothesis.

Competitive environment. Competition in general stimulates the development. Results of our study and some previous studies (e. g. Thong 1999) do not support the hypothesis that competitive pressure faced by the company, contributes to successful use of IS. Other investigations (e. g. Igbaria et al., 1997) support the opposite assumption.

Number of customers. Some small companies sell most of their products and/or services to one or a few business partners. Our hypothesis was that such companies lag behind in the implementation and use of there computerized IS. Analysis of data collected in the study could not confirm this.

Duration of use. The longer a company uses it's computerized IS, the higher are the demands for relevant information and comprehensive support of business processes. Analysis of data collected in our investigation confirmed this statement. Several other studies came to the same conclusion (Yap et. al., 1992, Palvia et al. 1994; Winston and Dologite 2000, Riemenschnieder and Mykytin 2000).

Government support. Several countries (Singapore, Hong Kong, EU countries) organized campaigns and provided funds to support small companies in successful computerization of their businesses. Some investigations indicate that such support contributed to successful implementation and use of IS (Yap et al, 1994; Mulhern 1995; Sherif 1996; La Rovere 1998) while others could not confirm this impact (e. g. Yap et al., 1992). Our study indicated that companies that received no support of any kind from the government were more successful in using their IS. Because their number was very small this relation was not statistically significant.

Education and training of employees. Knowledge and skills are considered a key success factor in implementation and use of computerized IS. For example the study (Palvia and Palvia, 1999) highlighted that lack of training as the most important among the 11 investigated factors for dissatisfaction of IS users in small companies. Our investigation showed that companies where owner /manager and/or employees attended training courses on computing and IS used their IS more successfully.

Participation of employees in decision-making and purchase of software. Our study showed that companies where managers encourage the employees to participate in the selection of software and hardware are more successful in use of there IS. Some other authors (Yap et. al, 1992; Igbaria et al., 1997) reported similar findings. This may be related to knowledge and skills of employees, as it was found also by (Buckland 1995; Igbaria et al., 1997; Thong 1999).

IT vendors as knowledge providers. Like most other investigations that considered this issue, data from our study supported the hypothesis that perceived knowledge of the software vendor positively impacts the use and success of computerized IS in small companies. According to Yap (1992) the same applied also to use of consultants. Our study could not confirm this finding because only a few companies used consulting support.

Type of contract with the software supplier. We expected that companies that signed long-term maintenance contracts with their software supplies were more successful in use of ISs. Analysis of data collected in our study showed the opposite: companies that buy the software and pay for each upgrade or enhancement separately, used their IS more successfully. A possible explanation for such a finding are criteria for successful use of IS adopted in our study. Companies with long-term contracts who are paying a monthly fee for maintenance usually buy all the software and hardware from one supplier. Software tools, such as scheduling programs, drawing tools, spreadsheets or word processors, are often sold together with the package. The whole package was considered as one program by the owner/manager of the company who provided data for our study. In our investigation, the number of different programs installed in a company was used as one of the indicators for successful implementation and use of IS. Other investigations, to the knowledge of the authors, didn't address this issue.

Outsourcing of the accounting function. Most companies from our sample (70%) outsourced accounting and bookkeeping to an external partner who specialized in selling accounting services to small companies. Because some of the interviewees quoted them as primary sources of information about computing and IT, we asked participating owners/managers to rate knowledge and skills of their providers of accounting service. We expected that companies that rated their knowledge and skills higher were more successful in using the IS. Data collected in the study didn't confirm this relation; it showed even a (statistically non-significant) negative corelation with most indicators of success in IT implementation. But data from our study indicated that companies who outsourced the accounting function were less successful in using IT and IS. A possible explanation could be that outsourcing partners provide companies with enough information, so that they do not feel any need for higher level of computing and IS activities in the company.

5. CONCLUSION

Small companies in Slovenia are relatively well equipped with information and communication technology, but they are rather far from exploiting its full potential. Our investigation showed that implementation and successful operation of computerized IS is strongly related to characteristics of the owner/manager (formal education, age, IS and IT knowledge and skills, participation in additional training, how he/she is using the computer, ...). Small business owners/managers in general recognize and value IT, but they use it mainly to support daily operations of the business rather than to support decision-making, and are mostly not aware of its strategic and organizational impact.

Most of them also don't use end-user friendly PC based tools, such as spreadsheets or databases, which are readily available to most companies participating in our study. This may also indicate the lack of computing knowledge and skills in small firms.

In order to take full advantage of the existing information and communication technology, current and future owners and managers of small companies should increase the level of their computing skills and knowledge, stay informed about new trends, developments and applications in the IT and communication area, and acquire technical and managerial competences needed to effectively manage the IS. Training in using of PC based end-user tools, and their use for data analyses may also help owners, managers and employees of small companies to gain self-confidence in using IT, and better exploit the available technology.

Equally important for successful application IT in small companies are training and education of employees and their participation in decisions related to implementation and operation of firm's IS. Employees may, through their knowledge and demands for IT support on the workplace, largely contribute to the selection and implementation, or reengineering of the firm's computerized IS. Together with support and knowledge of the IT vendor, this may to some extent compensate for owner's lack of IS and IT knowledge. Government support, for example by organizing and financially supporting IT training courses and briefings for owners/managers and employees (e.g. through local Chambers of Commerce), may enhance the diffusion of relevant technical and organizational knowledge within small companies.

Knowledge and skills of software vendors helps small companies to use the available IT more effectively and efficiently. But, letting the vendor to make decisions related to IT purchases may be an expensive strategy. When the seller of software and hardware designs an information system, it is likely that a computer salesperson will err on the side of recommending a larger and more expensive computer than is actually needed. Our study showed that many small companies purchased software tools, such as spreadsheets, databases and scheduling tools, but were not using them at all. Therefore, a better strategy for the owner/manager would be to separate the design of company's IS from the purchase of a system (Tagliavini et. al., 2001). System specifications should be developed by businesses, to meet their individual needs and to be aligned with the development perspectives. Consultants, local peers, employees, friends and partners to whom the company outsourced the accounting function, may help them to define the needs of the business, which would also strengthen their negotiating position with the vendor.

To establish a firm partnership with their software and hardware vendors, we believe that small companies should opt for long-term maintenance contracts with their software supplies. Because other investigations didn't address this issue we cannot refer to other studies to support our assumption.

Against our expectations, no relation was found between the estimated knowledge of external partners who were selling accounting and bookkeeping services to small companies, and successful use of IS in small companies. These partners could encourage small companies to use IT and company's data resources more effectively and efficiently, for example for decision support, and by sharing their knowledge and ideas with owners of small companies. A possible explanation for this observation is the business practice of accounting and bookkeeping companies. End of eighties and in the beginning of nineties, when the political and economic transition in Slovenia and other East and Central European countries began, some self-confident accounting employees noticed the increasing demand for accounting and bookkeeping services among the fast growing number of small firms. Many of them started their own small accounting businesses that focused on data entry and production of minimum accounting documentation required by the law, using purchased software packages.

Because data needed for decision support and business planning is mostly available in company's databases, shaping this information and offering it to small companies' owners and/or managers may be a new business opportunity for accounting firms who are in most cases also small companies. Selling additional services to their customers, such as shaping of accounting information for decision-making purposes, financial consulting, preparation of development plans, and financial planning, could differentiate them from their competitors and open new development perspectives.

REFERENCES

Buckland B., 1996, *Testing the role of knowledge barriers in the diffusion of information technology within and among small businesses*, Dissertation, UMI Dissertation Services.

Chau P., 1994, Selection of packaged software in small businesses, *European Journal of Information Systems*, **3**:292.

Cragg P. and Zinatelli N., 1995, The evolution of information systems in small firms, *Information & Management*, **29**:1.

DeLone W. H., 1988, Determinants of success for computer usage in small business, *MIS Quarterly*, **12**:12.

Deschoolmeester D. and Hee J., 2000, SMEs and Internet: on the strategic drivers influencing the use of the Internet in SMEs, in: *Proceedings of 13th Proceedings of 13th Bled Electronic Commerce Conference*, edited by Klein S., O'Keefe B., Gričar J., Podlogar M., (Moderna Organizacija, Kranj, Slovenia, 2000), pp. 754 – 769.

Doukidis G., Lybereas P. and Galliers R., 1994, Information System Planing in Small Business, in: *Proceedings of The Fourth International Conference Information Systems Development - ISD'94, Methods & Tools, Theory & Practice*, edited by: J. Zupančič and S. Wrycza, (Moderna organizacija, Kranj, 1994), pp. 3-19.

Gadena D., 1998, Critical success factors for small business: an inter-industrial comparison, *International Small Business Journal*, **17(1)**:36.

Henson J. M, 1995, *Factors determining the level of computerization in a small business: an empirical examination*, Dissertation, (UMI Dissertation services).

Hunter M. G., 2001, Small business adoption of information technology: unique challenges, in: *Managing Information Technology in a Global Economy*, edited by M. Khosrowpour (IDEA Group Publishing, 2001), pp. 126-131.

Igbaria M., Zinatelli N., Cragg P. and Cavaye A., 1997, Personal Computing Acceptance Factors in Small Firms: A Structural Equation Model, *MIS Quarterly*, **21**:279.

Kurbel K. and Teuteberg F., 1998, The current state of business Internet use: results from an empirical survey of German companies, in: *Proceedings of the 6th European Conference on Information Systems*, (Aix-en-Provence, France) edited by Baets, pp. 542-556.

Lai S.V., 1994, A survey of rural small business computer use: Success factors and decision support, *Information & Management* **26**:297.

Lawrence J. and Hughes J., 2000, Internet Usage by SMEs: a UK Perspective, *Proceedings of 13th Bled Electronic Commerce Conference, Electronic Commerce: the End of the Beginning*, (Bled, Slovenia), edited by Klein S., O'Keefe B., Gričar J. and Podlogar M., (Moderna organizacija, Kranj, Slovenia), pp. 738 – 753.

Lébre La Rovere R., 1998, Diffusion of information technologies and changes in the telecommunications sector, *Information Technology & People*, **11**:194.

Lesjak D. and Lynn M., 2000, Small Slovene firms and (strategic) information technology usage, *Proceedings of the 8th European Conference on Information Systems ECIS 2000 – A Cyberspace Odyssey*, Vienna, Austria) edited by Hansen H., Bichler M., Mahrer H., pp. 56 – 63.

Levy M. and Powell P.1998, SME flexibility and the role of information systems, *Small Business Economics*, **11**:183.

Mulhern A., 1995, The SME sector in Europe: a broad perspective, *Journal of Small Business Management*, July 1995, pp. 83 – 87.

Osteryoung J. S., Pace D. and Constand R., 1995, An empirical investigation into the size of small businesses, *The Journal of Small Business Finance*, Vol. **4(1)**:75.

Palvia P. in Palvia C., 1999, An examination of the IT satisfaction of small-business users, *Information & Management*, **35**:127.

Palvia P., Menas D. and Jackson W., 1994, Determinants of computing in very small businesses, *Information & Management*, **27**:161.

Papazafeiropoulou A. and Pouloudi A., 2000, The government's role In improving electronic commerce adoption, *Proceedings of the 8th European Conference on Information Systems ECIS 2000 – A Cyberspace Odyssey* (Vienna, Austria), edited by: Hansen H., Bichler M., Mahrer H., pp. 709 – 716.

Poon S. and Swatman P., 1999, An exploratory study of small business Internet commerce issues, *Information & Management*, **35**:9.

Razi M. A., 2001, ERP: Challenges for small-to-medium (SMC) companies, in: *Managing Information Technology in a Global Economy*, edited by M. Khosrowpour (IDEA Group Publishing, 2001), pp. 937-938.

Riemenschneider C. and Mykytyn P., 2000, What small business executives have learned about managing information technology, *Information & Management*, **37**:257.

SBW, The Chamber of Commerce and Industry of Slovenia - Slovenia Business Week. Nr. 15/1999 April 12th, 1999, http://www.gzs.si/eng/news/sbw/default.htm.

Seyal A. H, Rahim M.M. and Rahman M-N.A., 2001, An empirical investigation of use of information technology among small and medium businesses organizations: a Bruneian Scenario, *The Electronic Journal on Information Systems in Developing Countries*, **2**:1, http://www.ejisdc.org

Tagliavini, M., Ravarini, A., Buonanno, G., 2001, Information system management within SME: an Italian Survey, in: *Managing Information Technology in a Global Economy*, edited by M. Khosrowpour (IDEA Group Publishing, 2001), pp. 578-583.

Thompson T. and Tan M., 1998, An empirical study of adopters and non-adopters of the Internet in Singapore, *Information & Management*, **34**:339.

Thong J., 1999, An integrated model of information systems adoption in small businesses, *Journal of Management Information Systems*, **15**:187.

Verber B. and Zupančič J., 1993, Application of information technology in small business in Slovenia, in: *Proceedings of the United Kingdom System Society Conference on System Science: Addressing Global Issues*, edited by F. Stowell, D: West, J. Howell (Plenum Press, 1993), pp. 493 - 504.

Vijayaraman B. in Ramakrishna H., 1993, Disaster preparedness of small businesses with microcomputer based information systems, *Journal of Systems Management*, June 1993, pp. 28-32.

Werber B., 2002, Computer Security in Small Business – An Example from Slovenia, 2002, in: *Managing Information Technology in Small Business: Challenges & Solutions*, edited by S. Burgess (IDEA Group Publishing, 2002), pp. 156-175.

Werber B, 2000, *Ključni dejavniki uspeha računalniško podprtih informacijskih sistemov v malih podjetjih* (in Slovenian), disertation (University of Maribor, 2000).

Winston E. and Dologite D., Achieving IT Infusion, 1999, A Conceptual Model for Small Businesses, *Information Resources Management Journal*, January - March 1999, pp. 26 – 38.

Yap C., Soh C. and Raman K., 1992, Information systems success factors in small business, *OMEGA*, **20**:597

Yap C., Thong J. in Raman K., 1994, Effect of government incentives on computerization in small business, *European Journal of Information Systems*, **3**:191.

'TEAMWORK': A COMBINED METHODOLOGICAL AND TECHNOLOGICAL SOLUTION FOR E-WORKING ENVIRONMENTS

Jenny Coady, Larry Stapleton, Brian Foley[*]

1. INTRODUCTION

The word "virtual" describes work that spans one or more discontinuities. The term has been applied to work where people are in discontinuous physical work locations, where work is done in discontinuous time frames, where people have discontinuous organizational affiliations (Watson-Manheim, Crowston & Chudoba (2002)).

Community is an important aspect of life for most people. Cooley (1983) states that all normal humans have a natural affinity for community. He suggests that the primary factor inhibiting the formation of communities, no matter what their scale, is that they are difficult to organise. Extending the moral ideas inherent in nearly all individuals to the notion of community requires a system or institutional framework. The development and maintenance of such institutions sap the energy of the members of the would-be community and confuse the moral ideals inherent in the notion of community with the project of the institution itself (Fernback & Thompson (1995)). Given the development of new communications technologies we need to continuously examine the new relations and their potential for new or renewed relationships that arise in this context. Alongside these developments researchers have been advocating the need for a fundamental revision of software technology support in organisations for some time (Wood & Wood-Harper (1993)). However, few have proposed a basis upon which this progress can be made. This paper argues that new technological forms must recognise that organisational members act in fluid ways and, that organisations are moving towards flexible forms in virtual space. It thus attempts to address this gap in the academic literature, as well as in common practice. The paradigm proposed here is based upon a model of IT, which views organisations in terms of roles operating in particular settings or 'scenarios'. The remainder of this paper briefly reviews a new solution based upon this role-scenario paradigm.

[*] Jenny Coady, I.S.O.L. Research Group, Waterford Institute of Technology, Cork Road, Waterford, Rep. Of Ireland. Dr. Larry Stapleton, I.S.O.L. Research Group, Waterford Institute of Technology, Cork Road, Waterford, Rep. Of Ireland. Brian Foley, Tecnet, Glanmire, Cork, Ireland.

Information Systems Development: Advances in Methodologies, Components, and Management
Edited by Kirikova *et al.*, Kluwer Academic/Plenum Publishers, 2002

1.1. The differences between remote and traditional working environments:

Over the years researchers have highlighted the need for an alternative approach to support technology for organisational information processing. Indeed, the relationship between organisations and the technologies they use to process information has been, and continues to be, a major problem for researchers (Stapleton (2002)). In the 1990's some researchers have returned to fundamental questions of IT support and reviewed many of the basic principles that underlie IT provision. For example, Wood & Wood-Harper (1993) argued for a 'radical rethink' of IT support, reflecting in many ways the issues raised by Hedberg & Jonnson (1978) fifteen years earlier, and suggested that software researchers needed to create completely new paradigms for information technology. Since the earliest days of computer-based IT scholars such as McLuhan (1964) have noted that the development of electronic communication technologies has essentially abrogated space and time so that we live in a boundless "global village". Boorstin (1978) argued that communication technology creates ties; bind nations into a new type of community, which he terms the "Republic of Technology". The most recent communication technology development within the post-industrial era is commonly termed Computer-Mediated Communications (CMC). Comprised of different systems such as Electronic mail, bulletin board systems, and real-time chat services, CMC is both an interpersonal, one-to-one medium of communication as a one-to-many or even many-to-many form of mass communication. With an estimated 25 million CMC users worldwide, and this still growing (Calem (1992)) computer mediated communications have the potential to affect the nature of social life in terms of both interpersonal relationships and the character of the community. Virtual communities encompass the economic, political, social and cultural dimensions of community postulated by Van Vliet & Burgers (1987). Communities have been affected by electronic media's undermining of the relationship between location and access to information so that 'physical location now creates only one type of information-system, only one type of shared but special group experience" (Meyrowitz (1985) p 143-4)

Furthermore, it is well understood that electronic group dynamics are not the same as face-to-face and tend to have unpredictable consequences for those attempting to streamline organisational communication (Sproull & Kiesler (1993), (Stapleton (2001)).

2. ROLES & SCENARIOS

Most modern organisations cannot function effectively without information systems support, but they struggle to live with them in a harmonious relationship (Mc Bride (1999)). They experience a series of gaps, or mismatches, associated with the assumptions, which underpin information systems practice as regards organisational structure, dynamics and the various realities with which organisations must grapple. There are also conflicts associated with the system models, the resulting technological artefacts and many other aspects of the IT domain. Indeed, it has been shown on many occasions that our approach to eliciting and understanding technological and system requirements is typically at variance with the social environment the systems are intended to support (Stapleton (2000)).

In systems design, system components identified by an analysis of the functions to be performed by the whole are intended to fit each other so as to work together harmoniously as well as efficiently and effectively (Hamilton (1997)). However, when it comes to the interaction between users and the technology, this is rarely the case (Davenport (1998), Stapleton (2001)). Researchers have argued that the mechanistic approaches which typify information systems methods are based upon a functionally rationalistic paradigm which needs to be reviewed and revised. This paper proposes the adoption of roles & scenarios (i.e. a model of the social/task setting in which roles are played out) as basic modelling constructs. These constructs are then used to determine system functionality in terms of an organisation of documentation and contracts. This concept drives information systems thinking in a new direction and fundamentally plants both the design process and the technology itself in the organisational space, both in terms of process and language. So, whilst in itself role-based methods and scenario modelling (as separate approaches) are not new, the unique combination adopted here, and the configuration of the methodology with the technology, is highly innovative.

Scenarios can be understood in a number of ways. They can be a sequence of activities, or a more or less richly branded structure of such sequences. They can represent parallels or alternatives, or various intermediate options (Alexander (1999)). Branches can represent alternatives or parallels, or various intermediate options. Scenarios can assist in all phases of the systems life cycle. They can clarify systems scope; help to identify stakeholders, situations, and needs. They can also aid in the organisation of requirements, guide design and coding and also provide scripts for testing. But throughout the cycle their main contribution is perhaps simply to improve project communications between all the groups of people involved.

3. MODELLING ROLES & SCENARIOS IN A VIRTUAL ENVIRONMENT

The model is a common language: the users who are providing the information about the system readily understand it. Using working models, users and analyst's work together to reach an identical understanding of the requirements. Once the model is agreed, the system is implemented by building a real-world version of the model (Kerth (2001)). In Systems and Software Engineering, there are perhaps more schools of thought concerning IS notations and method that on any other topic (Alexander (1999)). One key function of scenarios is to help people communicate about the systems they are involved in. The choice of representation would ideally be a pragmatic one - people selecting the best for their situation.

This paper proposes a role-centred approach for primarily pragmatic reasons. Firstly, people identify themselves better with roles than functions and thus feel more able to understand the system and, hopefully, helping them to focus upon to role ownership. Evidence for this appears in (Warboys, et. al. (1999)). Role models design roles and identify explicitly information flows between them. This information flow based architecture is needed for later implementation in an remote working organisation (such as over the intranet or internet), which is the essence of a virtual work setting. Workflow based structures can be easily derived from role models following a defined set of rules which can be pre-determined and are widely used in industry (Warboys, et. al. (1999))

A scenario workshop can represent a sequence of tasks directly by role-play. A task can be described verbally by a player, or acted out in a variety of ways. The execution path can be symbolised by passing a token, such a juggling a ball to the player whose turn it is next. Both singly- and multi-threaded scenarios can be simulated in this way. Stakeholders see, often for the first time, how their tasks relate to the overall process, and many misconceptions and confusions can be banished in a playful workshop (Alexander (1999)). A similar approach was adopted in object oriented BPRE as laid out in Graham (1994), but has received little attention in the context of roles as specified here.

One of the best-known representations is the idea of Use Cases and UML (Jacobson (1992)). This supposes that the world of business is divided into neatly addressable cases where a user interacts with a system. Graphically a stick person connected to an elliptical bubble depicts it. Each bubble depicting a use case, and a number can be arranged into a list to form a diagram. The norm is to arrange them vertically, which does not imply sequence and this can visually be misleading. There has been much debate about whether a use case is simply another synonym for scenario (Alexander (1999)). The question could be answered by obtaining a precise definition of both terms, were it not for the elasticity of both terms. Hence UML notation can be helpful, but currently only goes part of the way towards resolving modelling differences.

4. 'TEAMWORK'

The TeamWork initiative grew out of developments in the automotive industry in the early 1990's (Messnarz et. al. (1999)) Large firms in this sector realised that in supply chains they work together in kinds of virtual organisations and that borders could not be easily drawn between enterprises. This mirrored experiences elsewhere in industry as outlined in Darnton & Giacoletto (1992), Davidow & Malone (1993), Drucker (1988, 1993) and elsewhere. The TeamWork research project involves three main axes, as shown in figure 1. This paper focuses upon the information systems development support aspects of the project, and therefore addresses itself to the Methodology and Technology components of TeamWork. It recognises the importance of new organisational forms and is targeted at communities of workers whose members come together for specific projects, but may not be part of the same organisation or inhabit the same geographical space. The power of such communities, and the risks involved in such organisational forms, are outlined elsewhere (Sproull & Kiesler (1992)). TeamWork delivers a holistic solution based upon the concepts embodied in a role/scenario organisational perspective. This holistic solution comprises BESTREGIT as a methodological component and NQA as an information and coordination technology based upon organisations as comprising roles and scenarios.

The basic advantages of this approach to information systems development and organisational support include the following:

People will be more aware of their responsibilities and clearly understand communications interfaces to other virtual organisation members. Role-based solutions are specifically adopted with this in mind.

New staff can be more easily integrated through clear role assignments, skill acquisition for the roles and clear appreciation for information and communication flows within the community. Sproull & Kiesler (1993) show that this is a specific opportunity presented by CSCW systems which support virtual communities.

Information technologies can be deployed which support virtual community communications, documenting of activities and configuration of results in a highly effective manner. This is primarily achieved by making the communications interfaces highly visible, a central deficiency of many current CSCW and ISD approaches which focus on highly functionally rationalistic views of organisational behaviour (Stapleton (2000), Stapleton & Murphy (2002)).

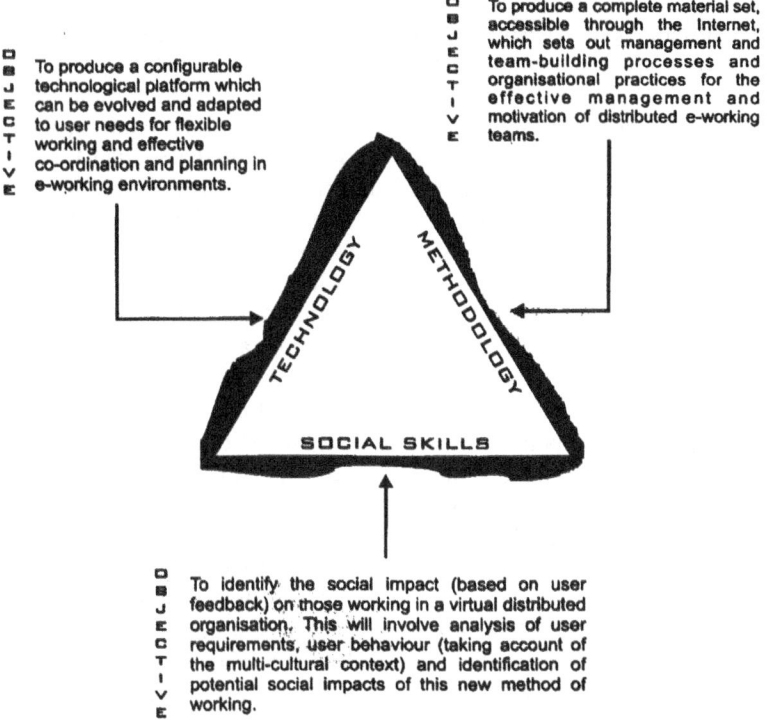

Figure 1. The TeamWork Research Project – Major Axes

Studies examining the effectiveness of this paradigm suggest that significant advantages exist where such an approach is adopted. Indeed preliminary studies of software development projects conducted in Ireland, Austria, Spain and Germany show potential benefits to firms, which adopt technologies based upon this approach (Messnarz & Tully (2000); Messnarz, et. al. (1999)).

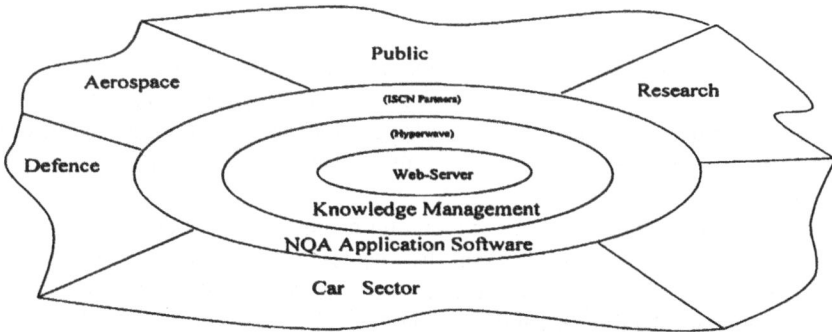

Figure 2. The Basic Technology Platform Architecture

Technologically TeamWork is constructed using Hyperwave's Knowledge Management solution. The key innovation is in the organisational coordination utilities supplied via a combination of NQA technology and the BESTREGIT Methodology.

5. BESTREGIT

The Teamwork research project believed that the Best Regional Innovation Transfer (BESTREGIT) modelling approach could be used as a trigger for the development of an 'Innovative Organisation' through a life long learning philosophy. The research approach adopted in the TeamWork Project (2001) had to do with formally examining the work scenarios behind an organisations business fields. Through this analysis TeamWork came to a greater understanding of the roles people in organisations play, how those roles interact to produce results, and how TeamWork could track the progress of these results. Once a team view, defined to be a commonly understood mission and goals, has been achieved, then the next logical step was to improve the work scenarios so that those goals can be met. In the modern era work goals are as such that a single individual, operating alone, cannot easily perform their work (Nurmi (1996)). Communication and work coordination is needed, and scenarios are designed aiming at productivity. It is apparent that this approach is based upon a view that business and organisational modelling takes precedence over more traditional information models as are used in SSADM or other structured methods (van Reijswoud & Mulder (1999)). Instead, the research adopted a view that role/scenario modelling (as a case of business/organisational modelling) can usefully depict information processing in an organisation, and be used directly to configure a complex IT solution.

A major success factor in the transformation from goal trees into TeamWork-based processes is to set the right priorities for the organisation (Warboys, B., Kawalek, P., Robertson, I., & Greenwood, M. (1999)). In real business cases, unlike the BESTREGIT research, it is often not possible to model all work scenarios, often due to time demands and limited resources. However if a business field with a typical work scenario is selected then there is a *re-use factor* whereby, once defined, they can be re-used in various models.

Each organisation is viewed as a set of work scenarios, typical examples of these include customer handling and service delivery. A work scenario is therefore a description of the best-perceived way in which to conduct a certain business case in the organisation. In BESTREGIT work scenarios are described with two complimentary views. Firstly as *role models*, which depict roles as enacted by individuals. One person can play many roles as well as many people playing just one role. Roles exchange information and work results. This information flow between the roles forms the role model. The second view is the *workflow models*, which consist of a network of work steps. These produce results that can be used by other work steps, each step requiring resources.

BESTREGIT uses an integration of both of these views. Firstly the role models are analysed and designed. Secondly the role models are transformed into workflow views. Then both models are integrated so that a work scenario according to BESTREGIT can be defined as so that people are assigned to roles, roles are assigned to activities, activities are part of a network of work-steps, activities produce results, and roles use resources to perform the activities.

These relationships are then defined for a certain business case of the organisation to have a description of the best way to perform the business case. Once process have been modelled using BESTREGIT it becomes possible to think about optimally organising all their elements for greater effectiveness and efficiency.

6. NQA – A NEW DEPARTURE IN CMC

NQA is a CMC system based upon a role-scenario view of organisational behaviour. The initial concepts, which are reified in NQA, are the basic building blocks upon which

BESTREGIT was constructed. NQA comprises important concepts in the context of information systems development through configuration:
1. Development by Configuration
2. Re-Use Pool - Organisational Function Based Configuration

6.1 Development By Configuration

This paradigm takes data independence objectives as its underlying principle with the added notion that data can be assigned with functionality by the user through specific configuration techniques. Such an approach overcomes problems often associated with component based approaches to software development, particularly those associated with the object oriented paradigm in which configuration of data (attributes) and methods (functions) is inseparable because both data and functions are encapsulated into a single, inviolable object (Graham (1991), Atzeni et. Al. (2000)). The Object Oriented paradigmatic approach has often been mooted for advanced coordination systems and methodologies for CSCW, Business Process Re-Engineering (BPRE) and other approaches associated with organisational transformation (Malone (1992), Shelton (1994)). However, this research present recognises that, whilst superficially the purely object-oriented paradigm may seem attractive, it does not reflect basic organisational information processing as it occurs in virtual communities. This fit between organisational information processing (as well as other organisational properties) has been shown to be important elsewhere, particular for groupware-type systems (Kraut et. al. (1994), Rogers (1994)). NQA must maintain the power of object-orientation by storing document objects etc. but a paradigmatic shift towards data independence assumptions is also required in order for the system to adequately reflect organisational behaviour in virtual spaces. This is an important theoretical shift which is required in order to effectively place NQA as a solution for virtual communities.

In role-based views, the data and function in organisational information processing are fundamentally separated and this must be reflected in the supporting technological architecture. In this approach users can insert and maintain document or result templates and adapt the system to their own specific documentation requirements without any change or customisation of code (just by configuration of data). Whilst documents are stored in an object-based technological solution (ensuring that the power of objects is harnessed), NQA also provides for a data independence rationality, which is usually seen as fundamentally opposed to object orientation. The reconciliation of these two rationalities is fundamental to NQA.

6.2 Re-Use Pool Concept (Organisational Function Based Configuration)

The Re-use pool contains a series of objects that can be reused in order to configure NQAs virtual office and enable the user to tailor the system. At the moment the following basic elements can be configured to which the above functionality is generated.

Tasks - A submission of a document and/or report automatically selects roles (role based configuration) which should receive this information and creates tasks for the users playing these roles. These tasks can then be traced.

Documents - There is a standard user interface for offering documents in the virtual office. Documents fall under the configuration management utility, and they can be linked with reports and other documents, can be downloaded, edited, and uploaded, and submitted to a team.

Reports - Reports usually are forms to be filled in, linked to a document, and submitted to a team.

Linked Reports – these are treated the same as regular reports, plus the link report is automatically linked by its creation backward and forward to a predefined set of reports and documents.

Depending on the user needs the elements are configured within a project administration structure. For example, Feature Requests (FR) can be linked with User Requirements Documents (URD), so that an URD is automatically created by the links to accepted FRs. Further basic elements are under consideration for insertion into the NQA configuration pool in later releases. This is possible because of the relatively easy extensibility provided by the object-oriented structure in NQA.

7. PRELIMINARY STUDY

7.1 An outline of the Research Approach

The preliminary training of users of the NQA configured system took place over 3 days involving 20 people enacting various roles within the system. The research data was gathered by open-ended questions, which were not defined with any particular structure. Rather, the participants were encouraged, in their own words, to describe their experiences with working with the NQA system. No assumptions were made as to the likely experiences of the participants in working with the system. This provided oral, open feedback to the research team. This was felt to be more appropriate than a questionnaire based approach, due to the exploratory nature of the work and follows other similar approaches where a preliminary study was utilised (Stapleton (1999a), Stapleton (2001)). This data gathering approach was informed by hybridised grounded theory as per Miller & Dunn (1998) and Stapleton (2001) and loosely informed by Glaser & Strauss (1967).

7.2 User Responses

The key findings of this preliminary study can be summarised as follows:
1. The general approach used to organise document exchange was useful due to the explicit definition of roles and the scenarios in which the documents were to be exchanged and processed
2. The security aspects of the system were important for virtual office scenarios. Participants felt that the explicit management of security according to the role of users was very suitable for virtual working where perceptions of security were critical
3. The ability to lock documents, ensuring they can be read but not modified, when they are presently being worked on in a shared environment, was also critical to the perceived usefulness of the system and helped organise user interactions in the virtual space, which would otherwise have been difficult to manage coherently
4. Political realities, which were typically hidden to members of rapidly created virtual work groups, were highlighted explicitly by the use of role-based modelling, particularly when combined with the scenarios in which these roles would be enacted.
5. The use of goal-trees to organise the document navigation system was inappropriate and lead to a deal of confusion and difficulty in document processing

6. People were often unhappy with the explicit codification of certain roles. The research assumes that roles will be adopted without question by the participants. Instead, participants often questioned the roles as outlined in the NQA system, even when they participated in the definition of these roles during design and pre-configuration stages

The above findings illustrate the strengths and weaknesses of the research position adopted here. Despite this only being a prototype, trial system it is apparent that, despite some fairly serious caveats, users did find the approach and the technology useful for informing and supporting virtual office activities. Firstly, the importance of security and integrity in the virtual environment increases. This is possibly due to the fact that people are working remotely and lack the tangible contact of face-to-face contexts so that certain sensitivities about information are amplified. Whilst this was not explicitly tested in the preliminary study, it would fit with the findings of other researchers, especially the work of Sproull & Kiesler (1993). Little research has been done on the perceived importance of integrity and security in virtual work environments as they envisaged in this study, and these issues should be scrutinised during a more formal social analysis (see discussion below). This is particularly true of integrity, as the particular dynamics of virtual offices creates an environment in which version management is critical but can create severe problems where there are quite informal processes such as those associated with the creation of a research proposal in virtual space. The importance of integrity for complex, dynamic and distributed objects is well documented elsewhere and reflected in developments in object oriented database technology and other database technologies (Brown (1991), Loomis (1992)).

Users described how the system helped them to make explicit the political realities associated with the processes in which they were involved. This was a direct result of the emphasis upon roles (loci of power and influence) and scenarios (enactments of power and influence). The researchers had not expected this issue to emerge as important as it receives relatively little attention in the literature on ISD. This therefore needs further analysis and will probably require refinements to the methodology. These findings do reflect some work on politics in ISD, in particular early work by Lynn Markus (Markus (1984) and some more recent work suggesting that a greater emphasis upon political realities is required during ISD (Cernetic & Hudovernik (1999), Stapleton (2001)).

The most serious problem which emerged during the training was associated with the use of goal-trees. The literature on virtual organisations suggests that it can be easy for members of virtual, asynchronous groups to become distracted by the needs of their physical settings and that attention needs to be focussed in order to ensure that people devote the necessary energy to the virtual project group (Sproull & Kiesler 1993), March (1999)). In order to ensure that the virtual group remained 'on-track', and in compliance with a role-based approach, goal-trees were adopted as a means by which document processing activities could be organised according to explicitly defined objectives. This goal-orientation was seen to be a critical aspect of the NQA system and was intended to obviate problems associated with typical virtual group problems as documented in the organisational studies literature. It is apparent from the preliminary feedback that this assumption may be inappropriate for virtual work settings. Furthermore, the codification of certain roles proved to be quite problematic. It became difficult to obtain agreement of users as to the roles they had adopted, even when they had been part of the process by which the system was configured according to those roles. Further research will be required to assess the extent to which these issues can be successfully addressed. Indeed, it is possible that the extent to which any technology can successful codify the activities of a virtual group may have been reached in NQA. Researchers have argued for sometime that the functionally rationalistic perspective of organisational information processing, as reified in IT solutions, is not always appropriate, and that alternative perspectives require development, particularly as codification (Moreton & Chester (1999), Probert & Rogers (2000)).

Whilst the idea of a role based system is not new (Warboys, B., Kawalek, P., Robertson, I., & Greenwood, M. (1999)) it is readily apparent that more research is needed in order to understand the information processing dynamics associated with this approach. This is especially true for technologically mediated groups which operate in virtual settings, especially where those technologies are developed according to a combinatory role/scenario model. In the next stage of this research, a formal social analysis will be carried out in order to more fully understand the relationship between the technology and the participants, in order to address the issues identified above.

Finally, there were difficulties associated with the use of the primitive prototype. Some of these difficulties included the delving deep into an un-user friendly interface to access documents and folders. The ability to assign access rights, leave messages and update notes came with the price of having to navigate through several windows before perhaps stumbling upon the correct screen. Whilst these problems are also associated with the use of goal-trees, it was predicted that some difficulties would be experienced with the use of such a prototype. Further research and testing will be required in order to isolate the specific problems in this space.

8. CONCLUSION & FUTURE DEVELOPMENTS

This paper presents a new technology, which has been developed to prototype stage, and is based on an alternative view of organisational information processing. The next stage is to perform a full system test of the technology, and its underlying methodology. To this end, the research team have organised a full test with naïve users. This test will involve a scenario in which several research partners develop a research proposal across several countries. Participants in the test include research agencies as well as researchers in technical universities across Europe. Specifically, the countries involved are Ireland, Austria and Slovenia. This test is organised and will get underway shortly. The success of this test will not only be measured by user feedback, but also by the successful completion of a research project to proposal (and hopefully funding) stage. Thus the test process utilises NQA in a real-life scenario, and involves major stakes for those involved. It is intended that such an approach will yield the best possible usability data and reflects recent thinking in information systems development as regards training and learning (Stapleton (2000), O'Keeffe (2001)).

The Teamwork approach combines role and scenario based approaches into a coherent methodology. This methodology is then reified in a new virtual office technology called NQA. The virtual office environment and, in particular, the social relationships which make such an environment work, are strongly reflected and supported by such an approach. Consequently, NQA is a very promising system with regard to highly dispersed organisations, even though it is only in its preliminary phase. Preliminary studies show that there is empirical evidence for merit in such an approach, although the evidence still requires further corroboration. Consequently it is important for this research to progress to a full test in a relatively complex user environment. This full testing phase is scheduled and will be conducted for a remote work environment where a complex set of social interactions must be supported. With certain caveats, the evidence to date suggests that the basic methodology and technology will be successful although it is likely that modifications will be necessary for the prototype NQA system. This paper presents the results of a research study which has culminated in a working prototype based upon a new IT view of organisational behaviour, a view which is far closer to the reality of virtual work than previous systems or approaches. Further research must be carried out in order to obtain substantial test results across a large community of users. Furthermore, other researchers must build upon role-scenario based approaches in order to confirm, or otherwise, the general applicability of the approach across a variety of user domains. One area of potential development is the inclusion of intelligent agent technologies, which interpret and update user profiles, thus obviating the extensive tree searches, which are required in goal tree-type systems such as adopted in NQA. Whatever the case, it is readily apparent that NQA represents a radical new approach to IT support in remote working environments. This approach is both innovative and potentially opens an array of new possibilities – which are the subject for future research.

9. ACKNOWLEDGEMENTS

The authors gratefully acknowledge the financial support for this project provided by the European Commission as part of the IST programme of FP5 (IST-2000-28162). The authors also gratefully acknowledge the help of the reviewers in revising early drafts of this paper.

10. REFERENCES

Alexander, I. (1999), Migrating Towards Co-operative Requirements Engineering. CCEJ, 1999, 9, (1), *Scenario Plus User Guide and Reference Manual*, http://www.scenarioplus.org.uk. pp. 17-22

Boorstin, D. (1978), *The Republic of Technology: reflections on our future community*, Harper & Row, USA.

Brown, A. (1991). *Object-Oriented Databases: Applications in Software Engineering*, McGraw Hill: NY.

Calem, R. (1992), The network of all networks, *New York Times*, December 6th, p. 12.

Cernetic, J. & Hudovernik, J. (1999). The Challenge of Human Behaviour Modelling in Information Systems Development, in *Evolution and Challenge in Systems Development*, Zupancic, J., Wojtkowski, W., Wojtkowski, W.G., Wrycza, S., (eds.), Kluwer Academic/Plenum Publishers: New York, pp. 557-567.

Cooley, C. (1983), *Social Organization: A Study of the Larger Mind*, Transaction Books, New Brunswick, NJ.

Darnton, G. & Giacoletto, S.(1992), *Information in the Enterprise: It's More than Technology*, Digital: MA.

Davidow, W. & Malone, M. (1993), *The Virtual Corporation: Structuring and Revitalising the Corporation for the 21st Century*, HarperBusiness: NY.

Davenport, T (1998). Putting the Enterprise back into the Enterprise System.' *Harvard Business Review*, August, 76, 4, pp. 121-131.

Drucker, P. (1988). 'The Coming of the New Organisation', *Harvard Business Review*, Jan-Feb pp.45-53.

Drucker, P. (1993). *Post-Capitalist Society*, Butterworth-Heinemann Ltd: Oxford.

Fernback, J. & Thompson, B. (1995), Computer Mediated Communication and the American Collectivity: The Dimensions of Community Within Cyberspace, presented at the *Annual convention of the International Communication Association, Albuquerque*, New Mexico.

Glaser, B. & Strauss, A.(1967).*The Discovery of Grounded Theory: Strategies for Qualitative Research*, Aldine: Ill.

Graham, I. (1994). 'Business Process Re-engineering With SOMA', in *Proceedings of Object Expo Europe Conference*, SIGS: NY, pp. 131-144

Hamilton, A. (1997), *Management by Projects: Achieving Success in a Changing World*, Oak Tree,Dublin.

Hedberg, B. & Jonsson, S. (1978). 'Designing Semi-Confusing Information Systems for Organisations in Changing Environments', *Accounting, Organisations & Society*, 3/1 pp. 47-64.

Jacobson, I (1992), *Object-Oriented Software Engineering: A Use Case Driven Approach*, Addison-Wesley.

Jones, Q. (1997), Virtual Communities, virtual settlements and cyber-archaeology: a theoretical outline, *Journal of Computer Mediated Communication*, 3(3).

Kerth, N. (2001), *Project Retrospectives: A Handbook for Team Reviews*, Dorset House, NY.

King, M. (1996), On Modelling in the Context of I, *JTCI Workshop on Standards for the use of Models that Define the Data & Processes of Information Systems.*

Loomis, M. (1992). Advanced Object Databases, *Object Expo European Conference Proceedings*, SIGS: NY.

March, J.G. (1999). *The Pursuit of Organisational Intelligence*, Blackwell: Oxford.

Markus, L. (1984). Systems in Organisations: Bugs & Features, Pitman.

Markus, L. & Robey, D. (1988). 'Information Technology and Organisational Change: Causal Structure in Theory and Research', *Management Science*, 34/5, May, pp. 583-598.

Mc Bride, N. (1999), *Chaos Theory and Information Systems Track 4: Research Methodology and Philosophical aspects of IS research*, www.cms.dmu.ac.uk/~nkm/CHAOS.htm

McLuhan, M. (1964), *Understanding Media: The Extensions of Man*, McGraw-Hill, NY, USA.

Mehrabian, A. (1971). *Silent messages.* Wadsworth, Belmont, California

Meyrowitz, J. (1985), *No Sense of Place: The Impact of Electronic Media on Social Behavior*, Oxford University Press, NY, USA.

Miller, K. & Dunn, D. (1999). 'Using Past Performance to Improve Future Practise', in Zupancic, J., Wojtkowski, W., Wojtkowski, W.G., Wrycza, S., (eds.), *Evolution and Challenge in Systems Development*, Kluwer Academic/Plenum Publishers: New York, pp. 99-107

Moreton, R. & Chester, M. (1999). 'Reconciling the Human, Organisational and Technical Factors of IS Development', in *Evolution and Challenge in Systems Development*, Zupancic, J., Wojtkowski, W., Wojtkowski, W.G., Wrycza, S., (eds.), Kluwer Academic/Plenum Publishers: New York, pp. 389-404

Nocera, J. (1998), *Cultural Attitudes Towards Communication and Technology*, University of Sydney, 193-195.

Nurmi, R. (1996), Teamwork and Leadership, Team Performance Management: An International Journal, 2, 1.

Probert, S. & Rogers, A. (2000). 'Towards the Shift-Free Integration of Hard & Soft IS Methods', in *Systems Development for Databases, Enterprise Modelling & Workflow Management*, Wojtkowski, W., Wojtkowski, W.G., Zupancic, J., (eds.), Kluwer Academic/Plenum Publishers: NY.

Rheingold, H. (1993), *The Virtual Community: Homesteading on the electronic frontier*, Addison-Wesley, MA.

Sproull, L. & Kiesler, S. (1992), *Connections*, MIT Press: MA.

Stapleton, L. (1999). 'Positivism, Interpretivism & Post-Phenomenology - Views of an Organisation in Flux: An Exploration of Experiences in a Large Manufacturing Firm', in *Proceedings of the European Colloquium of Organisational Studies* (EGOS 1999), University of Warwick: UK.

Stapleton, L. (1999a). 'Information Systems Development as Interlocking Spirals of Sensemaking', in Zupancic, J., Wojtkowski, W., Wojtkowski, W.G., Wrycza, S., (eds.), *Evolution and Challenges in Systems Development*, Kluwer Academic/Plenum Publishers: New York, pp. 389-404

Stapleton, L. (2000). 'From Information Systems in Social Settings to Information Systems as Social Settings', *Proceedings of the 7th IFAC Symposium on Automation Based on Human Skill: Joint Design of Technology and Organisation*, Brandt, D. & Cerenetic, J. (eds), VDI/VDE-Gesellschaft Mess- und Automatisierrungstechnik (GMA): Dusseldorf, pp. 25-29.

Stapleton, L. (2001). *Information Systems Development: An Empirical Study of Irish Manufacturing Firms*, Ph.D. Thesis, National University of Ireland.

Stapleton, L. & Murphy, C. (2002). 'Examining Non-Representation in Engineering Notations: Empirical Evidence For The Ontological Incompleteness of The Functionally-Rational Modelling Paradigm', in *Proceedings of the International Federation of automation and Control World Congress*, Forthcoming.

Teamwork (2001), Periodic Progress report No.1 of the TeamWork Project.

Van Reijswoud, V. & Mulder, H. (1999). 'Bridging the Gap Between Information Modelling & Business Modelling for ISD', in *Evolution and Challenge in Systems Development*, Zupancic, J., Wojtkowski, W., Wojtkowski, W.G., Wrycza, S., (eds.), Kluwer Academic/Plenum Publishers: New York, pp. 317-330.

Van Vliet, W. & Burgers, J. (1987). Communities in transition: From the industrial to the post-industrial era. In I. Altman & A. Wandersman (Eds.), *In Neighborhood and Community Environments.* Plenum: NY..

Warboys, B., Kawalek, P., Robertson, I., & Greenwood, M. (1999), *Business Information Systems: A Process Approach*, McGrath Hill, UK.

Watson-Manheim, M. B., Crowston, K. & Chudoba, K. M. (2002), A new perspective on "virtual": Analysing discontinuities in the work environment. Presented at *HICSS-2002*, Kona, Hawaii, USA.

Wood J.R.G. & Wood-Harper, T. (1993). 'Information Technology in support of individual decision making', *Journal of Information Systems*, 3/2, April, pp. 85-102.

SOURCING AND ALIGNMENT OF COMPETENCIES AND E-BUSINESS SUCCESS IN SMES: AN EMPIRICAL INVESTIGATION OF NORWEGIAN SMES

Tom R. Eikebrokk and Dag H. Olsen[*]

1. INTRODUCTION

The purpose of this work is to investigate the proposition that sourcing and alignment competencies are positively related to e-business success for small and medium sized enterprises (SMEs). We review and operationalize critical competencies in sourcing and alignment related to the use of e-business i SMEs, and we report the exploratory empirical findings from a survey. The results contribute to our understanding of adoption of e-business in SMEs, and will have implications for programs, which aim to stimulate e-business adoption and success in SMEs.

The growth of business conducted over the Internet is one of the most significant developments in today's economy. Online businesses are established all over the globe at a staggering rate. Many companies, both existing ones and new ventures are seeking to reap the benefits of this revolution both in terms of increased effectiveness and efficiency as well as access to new markets. However, there is an uncomfortable fit between SMEs and e-business. Studies show that SMEs are reluctant to embrace the principles of e-business (Bode and Burn, 2001). Other studies have shown that SMEs need help in understanding how e-business can enhance the competitive position (Owens and Beynon-Davies, 2001). A recent study Amit and Zott (2001) argue that amongst IT-investments in general Internet technologies have the highest potential for value-creation through linking companies, its suppliers and customers in new and innovative ways. Amit and Zott further argue that this value potential of e-business is poorly understood both theoretically and in practice. In a European survey among Internet experts and online traders it was found that most companies did not consider e-business as a priority (Forrester Research, 1999).

---

[*] Tom R. Eikebrokk and Dag H. Olsen, Agder University College, Kristiansand, Norway N-4604

Information Systems Development: Advances in Methodologies, Components, and Management
Edited by Kirikova *et al.*, Kluwer Academic/Plenum Publishers, 2002

375

The prevailing business models are based on research done primarily on large (american) companies. Studies that have addressed the size factor find differences that indicate that the business practices promoted in the literature may have limited applicability in SMEs (Rodwell and Shadur, 1997). The literature has not described the differences or adjusted the practices to SMEs.

Research within the field of information systems (IS) economics show that the business value from IT investments in general is positive and highest in companies that not only invest in technology but also in organizational measures including training programs to leverage ITs potential (Brynjolfsson and Hitt, 1998). Most studies of the effects of IT-investments are done in the context of large firms, usually Fortune 500 companies, and there are very few studies of IT-value in smaller companies. In a recent study based on 441 SMEs in Spain, Dans (2001) found that IT-investments also gave a significant and positive business value in SMEs.

Small and medium sized enterprises suffer from resource poverty. In an European survey among SME managers it was found that the main obstacles to increased competence were high costs of courses, problems of internal organization of the attendance to the courses, bad quality of the training supply as well as difficulties in identifying training needs (E/1224, 1997). With the knowledge of key competencies, SMEs may spend their limited resources on factors that contribute the most to business value.

We will investigate the effects of sourcing and alignment of competence on e-business success in SMEs. SMEs are known to have limited resources and therefore a limited ability to build up competencies needed to implement business innovations. It poses both a resource and a competence problem. It will therefore be important for SMEs to identify and utilize external sources of competence. Clearly, this also requires competence. Building competence on sourcing of competence will improve SMEs' ability to exploit e-business possibilities. Furthermore, alignment can be seen as the process of bringing together competencies from different sources and putting them into action. Sourcing and alignment can therefore be classified as meta competencies that make organizations able to identify and combine other competencies in a relevant way. We will therefore investigate the importance of these two meta competencies to e-business success.

2. THEORY

There is general agreement in the literature that the successful use of IT assumes organizational and individual capabilities. There are different approaches to conceptualizing this phenomenon. Many terms are used in IS to describe organizations' capabilities. Some researchers describe this with terms like core capability (Feeny and Willcocks, 1998; Amit and Schoemaker, 1993; Penrose, 1959; Teece et al. 1997). Others use core competence (Sambamurthy and Zmud, 1994), explicit and tacit knowledge (Basselier et al., 2001) or distinguishes between cognitive, skill-based and affective dimensions of competence (Marcolin et al., 2000; Kraiger et al., 1993). Only a few studies have examined the relationship between information technology competence in general and business value. Also, the user competence construct is largely absent from prominent technology acceptance and fit models, poorly conceptualized, and inconsistently measured (Marcolin et al., 2000). Recent works show increased academic interest in this phenomenon. In this study we will define organizational competencies as organizational capabilities and individual know-how and skills.

Many practitioners and researchers have been concerned with the importance of alignment over the years. In several studies, business executives have consistently ranked alignment as one of the most important tasks (e.g. Watson and Brancheau, 1991). In the literature, there are also many recommendations that sub strategies relating to areas like manufacturing and IT should be aligned with business strategies in order to contribute to business performance (e.g. Skinner, 1969; Henderson and Venkatraman, 1993). Only recently have empirical studies been able to document a positive relationship between alignment and business performance. Papke-Shields and Malhotra (2001) report a positive relationship between alignment of manufacturing and business strategy and business performance. In the IS field, empirical results have been mixed and inconclusive. The studies have mostly investigated one company or industry and have not been able to conclude across companies or industries. There is also a clear lack of studies of alignment focusing on small companies and on e-business solutions.

Although there has been an extensive discussion of alignment in the literature, there is a lack of empirical evidence as to what factors best explain how business and IT strategies can be aligned. Research by Luftman et al. (1999) has identified a set of enablers and inhibitors of alignment. Executive support, leadership, IT-business relations and understanding the business are identified as both enablers and inhibitors. In later work, Luftman (2000) argues that the knowledge of how to maximize enablers and minimize inhibitors can be viewed as organizational capabilities that mature over time. If this holds true, we can expect that business performance will increase with competence in alignment. Since it is also likely that both alignment maturity and competence in general increase with size, through increased specialization and resources, we might also expect that business performance would increase with company size.

Rodwell and Shadur (1997) found that larger firms have a stronger tendency to outsource activities than smaller firms. They argue that outsourcing as a policy might be a product of the growth of an organization. As a company grows, important activities that are non-core to the company's business may be outsourced. This will make the company able to grow in terms of revenue while maintaining a clear strategic focus. Rodwell and Shadur further propose the implication that as firms grow, they go through a process of clarifying the focus of the business. On the other hand, in the digital economy, one may expect to see sourcing of activities as well as competence as a way of growing in terms of revenue and profit rather than in terms of number of employees. This factor should therefore be addressed more in future research.

SMEs generally are dependent on a small number of customers to whom they sell a limited number of products. SMEs will thus often be heavily influenced by the e-business strategy of the dominant actor in the value chain, and be forced to conform to their solutions and technology (Ballantine et al., 1998), and thus the competence requirements defined by this.

As e-business evolves we should expect to see more integration of companies in electronic networks and partnerships. Clearly, in such partnerships there will be several decision-makers concerning business strategy and IT investments. How IT-investments are governed in such constellations will vary. In general, we should expect that small businesses would have to give up some control over IT-investment decisions in the presence of dominating partners. When the importance of cooperation in electronic networks increases, both in terms of activities and value, it is likely that small companies will be more influenced by dominating business partners. A substantial customer may want to dictate the IT-investment decisions, and thus make such decisions a matter of compliance rather that a process where the investments are tailored to the long-term needs and the resources of the small company. In some cases, external governance could be a threat to the ability to align IT with business, and thus a threat to e-business performance.

Amit and Zott (2001) have explored the theoretical foundations for value creation in e-business in a recent study. Their review suggests that no single theory available in the entrepreneurship or strategic management literature today can fully explain the value creation potential of e-business. Rather, they argue that an integration of theoretical perspectives on value creation is necessary. Their proposed model is based on the concept of a business model, which describes how one or several companies (e.g. a company and its suppliers and customers) in co-operation can exploit business opportunities through the design, structure and governance of transaction content. With this as the level of analysis, they describe the potential of value creation in e-business in four interrelated dimensions, which are efficiency, complementarities, lock-in, and novelty. Efficiency relates to the potential of reducing transaction costs as a result of utilizing e-commerce technologies. Complementarities describe the value potential of companies that cooperate in an e-business strategy and combine their products and services, technologies and activities. This combination can create value to customers as a result of reduced costs related to finding and ordering products and services, and to business partners as a result of utilizing interconnectivity of markets and possibilities of process integration. The third dimension, lock-in, describes the value potential as a result of switching costs where customers are motivated to repeat their transactions, and where business partners extend their relationships and also improve their current associations. As the last dimension in Amit and Zott's model, novelty describes how e-businesses can create value through innovations in the way business is conducted. Examples of such innovations are new ways of structuring transactions like web-based auctions and reverse markets. In the latter form potential buyers indicate needs and preferred prices to sellers. These examples show that e-business combines previously unconnected parties by utilizing new transaction technologies and by creating new markets.

Amit and Zott (2001) give an innovative contribution to the understanding of e-business success. However, they do not specify the antecedents for these success factors. We argue that SMEs face a competence challenge in defining and utilizing external sources of competence. Complementarities, one of the success factors, also imply complementing competencies between e-business partners. In combination with the resource situation in SMEs, this underline the need for competencies in sourcing and alignment.

3. RESEARCH HYPOTHESES

The literature review has shown a definite need for more studies of factors that influence the success of e-business efforts in SMEs. As we have seen, the literature on competence is so far more influenced by studies of IT-investments related to traditional business than by e-business that highlights new and technology based interactions between companies. As a result, several models do not include competence on whether activities should be sourced within the company or within the e-business network. Such competence in sourcing and also in alignment can be viewed as secondary types of competence, where sourcing relates to obtaining and alignment relates to combining and using primary types of competence. We will here address the meta-competencies of sourcing and alignment, which describe activities related to the company's ability to involve and take advantage of competencies inside or outside of the company in order to develop and implement strategies related to e-business. Sourcing refers to whether competencies will be built up internally or accessed from external sources, either through insourcing of competencies or outsourcing of activities where competencies are needed. When competencies are not sufficiently represented internally, the company must decide what strategy it will utilize to close this gap. Identified competencies could be built up internally through education programs or through recruitment of new personnel. Alternatively, the activity where competencies are needed could also be partly or totally outsourced to partners in the business network. The company could also implement some combination of these strategies. Sourcing of competence describes the process of gaining access to competencies inside or outside of the company. Having access to sources of competence per se both inside or outside of the company is not enough for the successful use of competence. Work processes whether they are located inside or outside of the company, can be intertwined and dependent on each other. In the effort of designing new processes, or improving existing processes, the company must make competencies work together in order to realize the potential of information technology.

The concept of alignment describes both the process of how different competencies represented in different sources are brought together and put into action, and the outcome of this process as a resulting level of alignment. Previous work on the alignment between business and IT shows that departments, business and IT experts must be brought together and involved in the effort of combining IT possibilities with business needs and opportunities. Relationships must be built between IT professionals and business unit managers. Research on both alignment and sourcing has documented that both the choices of alignment strategy and sourcing strategy are related to company performance.

Based on the aforementioned discussion, an important question is the significance of sourcing for the success of e-business. The theoretical review showed indications that larger companies will tend to outsource activities more than small companies, while there are also arguments that we may see the opposite effect. We proposed the role of sourcing for the success of e-business efforts. In our study, we will address sourcing as a competence both in terms of the sourcing of activities as well as the sourcing of e-business competencies from partners. We propose the following hypothesis with respect to sourcing:

H1: Competencies in sourcing will be positively associated with e-business performance in SMEs.

Secondly, as we have seen above, the recent research literature has shown that business performance will increase with competence in alignment. This conclusion was based on empirical tests of IT-investments in large companies. We will test whether this conclusion is also valid in SMEs. We therefore advance the following hypothesis:

H2: Competencies in alignment will be positively associated with e-business performance in SMEs.

In addition, our literature review identified two other significant factors relating to the success of e-business in SMEs: size and governance. We view these two factors as so important that they deserve explicit attention in our survey. Governance in this context can be defined as the degree of autonomy over e-business investment decisions.

H3: Company size will be positively associated with e-business performance in SMEs.

H4: External governance of e-business efforts will be negatively associated with e-business success in SMEs.

If e-business decisions in SMEs are controlled by an external business partner it is likely that the capability of aligning IT with business will decrease both as a result of reduced need to focus on this issue, but possibly also as a result of force more than rational communication. We will therefor test the following hypothesis:

H5: External governance of e-business efforts will be negatively associated with alignment.

We define the dependent variable e-business success in line with Amit and Zott (2001) who describe the potential of value creation in e-business in four interrelated dimensions, which are efficiency, complementarities, lock-in, and novelty. Efficiency relates to possible reduction in transaction costs, whereas complementarities describe the value potential from combining products and services, technologies and activities. The third dimension, lock-in, describes the potential value in creating switching costs from arrangements that motivate customers and business partners to repeat and improve transactions and relationships. The last dimension, novelty, describes value as a result of innovations in the way business is conducted (e.g. web-based auctions, etc.).

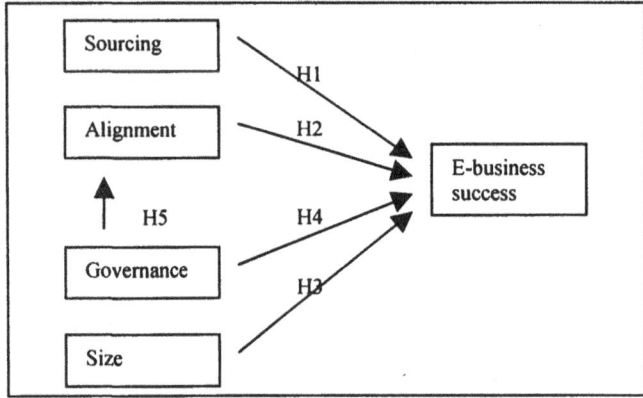

Figure 1. The research model

In addition to these variables we included a control variable measuring the perceived ability of the company to implement changes in its organization. It is clear that a sound e-business ambition is of no value if the company is not capable of implementing this ambition and the related organizational changes. Other control variables that we included are the type or types of e-business solutions the have been implemented, the knowledge of such solutions and their providers, along with the status regarding the process of implementing such e-business solutions. The research model is depicted in figure 1.

4. RESEARCH METHODS

4.1. Setting and operationalization of variables

To test the hypothesis we conducted a cross sectional study of SMEs from two industries in Norway. The industries were tourism and transportation. Based on a random sample of 150 SMEs executives were phoned, and if they could confirm that their company used web pages, e-mail or e-commerce systems for business purposes, they were invited to take part in the survey. 70 executives accepted and were subsequently interviewed.

The operationalizations of the variables discussed above were based on instruments and operationalizations previously documented in the literature. Sourcing and alignment were operationalized based on previous work by Sambamurthy and Zmud (1994), Bharadwaj, Sambamurthy and Zmud (2000), Basselier, Reich and Benbasat (2001), Feeny and Willcocks (1998), van der Heijden (2000), and Lee and Trauth (1995). 'E-business success' was operationalized according to the dimensions described by Amit and Zott (2001). This led to one item measuring 'perceived general e-business success' (q25), and four scales with items measuring the specific dimensions of e-business success. SME size was operationalized as the number of employees, and governance was operationalized as the degree to which a dominating customer or supplier dictated the e-business efforts in the respective SME. The type of e-business solutions utilized were included as a control variable and operationalized according to descriptions in Laudon and Laudon (2001).

The indicators were measured on a seven point Likert-type scale between 'totally disagree' and 'totally agree' with a 'not applicable' response option for most indicators.

We then conducted open-ended interviews with eight SME managers and two related consultants as an additional reality check of our model and as a test of the relevance, wording and response format of the indicators. The outcome of this process led to the resulting questionnaire, which is included in appendix 1.

5. DATA ANALYSIS

5.1. Measurements and measurement quality

All of our constructs have been measured with reflective items. To assess the validity of our constructs measured with two or more items, we examined the inter-item correlations. This revealed that one of the two items measuring sourcing (item 10) showed a higher correlation with items measuring alignment than with the other item measuring sourcing. This is indicative of low discriminant validity, and the item was therefore deleted from the subsequent analysis.

The final measures of variables with multi-item scales, i.e. alignment and the four dimensions of e-business success, were constructed as the average item score. Sourcing was measured with one item only, along with size, external governance, and perceived general e-business success.

The reliability of the constructs was assessed using Cronbach's alpha. Hair et al. (1998) suggest that an alpha of 0,60 is acceptable for exploratory research and 0,80 for confirmatory research. Reliability analysis for the multi-item scales showed the following coefficient alphas: Alignment 0,83 and e-business effectiveness 0,73, e-business complementarities 0,89, e-business lock-in 0,46, and e-business novelty 0,69.

5.3. Test of hypotheses

We generated partial correlations through stepwise forward linear regression analysis to test our hypothesis. Hypotheses H1 to H4 were first tested against 'perceived general e-business success' with control for the other variables and control variables respectively. Then we tested H1 to H4 against the specific dimensions of e-business success, i.e. a) efficiency, b) complementarities, c) lock-in and d) novelty. As the final step we tested hypothesis H5 with control for sourcing, alignment, size and the specific control variables. For all these analyses we investigated the regression assumptions in terms of normally distributed dependent variables, normally distributed and independent residuals. The results of these analyses are reported in appendix 3, and show that these assumptions are reasonable. Descriptive statistics are reported in appendix 2.

Table 1 shows the result of the regression analyses. As is evident from table 1, the hypotheses H1, H1b,c, d, H2a, and H5 were all supported. Hypothesis H1 states that sourcing and general e-business success will be positively related, which was supported (.30, $p \leq 0,05$). Hypothesis H1b states that sourcing will also be positively related to e-business success in terms of complementarities, which was strongly supported (.44, $p \leq 0,01$). H1c states a positive relationship between sourcing and lock-in, which was also supported (.34, $p \leq 0,05$), and finally H1d states that sourcing will be positively related to e-business success in terms of novelty, which was strongly supported (.45, $p \leq 0,01$). Hypothesis H2a states that alignment will be positively associated with e-business success in terms of efficiency, and this was supported (.38, $p \leq 0,05$). Finally, the last hypothesis H5, states that external governance will have a negative association with alignment, which was strongly supported (-.32, $p \leq 0,01$).

Table 1. Results of hypotheses test

Hypotheses	Independent Variable	Dependent Variable	Partial correlation
H1	Sourcing	General e-business success	.30*
H1a	Sourcing	Efficiency	n.s.
H1b	Sourcing	Complementarities	.44**
H1c	Sourcing	Lock-in	.34*
H1d	Sourcing	Novelty	.45**
H2	Alignment	general e-business success	n.s.
H2a	Alignment	Efficiency	.38*
H2b	Alignment	Complementarities	n.s.
H2c	Alignment	Lock-in	n.s.
H2d	Alignment	Novelty	n.s.
H3	SME Size	general e-business success	n.s.
H3a	SME Size	Efficiency	n.s.
H3b	SME Size	Complementarities	n.s.
H3c	SME Size	Lock-in	n.s.
H3d	SME Size	Novelty	n.s.
H4	External Governance	general e-business success	n.s.
H4a	External Gov.	Efficiency	n.s.
H4b	External Gov.	Complementarities	n.s.
H4c	External Gov.	Lock-in	n.s.
H4d	External Gov.	Novelty	n.s.
H5	External Governance	Alignment	-.32**

* Correlation is significant at the 0.05 level (1-tailed)
** Correlation is significant at the 0.01 level (1-tailed)

6. DISCUSSION

Most studies of IT competence within the IS field are generalizing their findings to companies in general based on the often implicit assumption that data from bigger companies are also valid for small companies. Our results both support and question this assumption. Based on literature developed from studies of bigger companies we found that sourcing and alignment were important for the success of e-business efforts in SMEs.

Our results also identified differences in influence between bigger and smaller companies, which question the assumption above and point to the need for SME specific variables. As companies grow bigger, their influence on business partners also increase. Stated differently, small companies will often experience a big supplier or customer who heavily influence or totally control the e-business decisions. Our results found support for this argument in that increased external governance was negatively associated with alignment, which again is positively associated with e-business success. An SME that experience external governance of e-business decisions will be less able to coordinate and align the choice of e-business efforts to the general business strategy and related projects. As the results indicate, this could as a secondary effect reduce the e-business success.

There is a clear need for more studies both on this issue in particular but also on the difference between SMEs and bigger companies in general. The lack of theoretical and empirical work on this topic and the results from our exploratory study, suggest that future research should devote much more resources on SMEs in order to make existing models more robust or in order to develop if necessary, specific theories for SMEs.

The descriptions of e-business as an opportunity to conduct business in new and improved ways may represent a threat to small and medium sized companies. If the success of e-business in SMEs depends on the ability to utilize and align the technology with business needs and characteristics, the influence of an external business partner could reduce this ability and hence the success of e-business. Descriptions of e-business benefits in the literature often assume that trust or dialog is present when small companies coordinate their e-business efforts with bigger partners. This may not hold true in general, but in particular not when small companies are involved. The importance of external governance in this study points to the importance of other variables like power, trust and effective communication that may represent one of many characteristics that discriminate relationships that include a small and a big company from relationships with equal partners.

There are several practical implications from this study. Firstly, programs that aim to stimulate e-business development and success in SMEs should notice the importance of competence in sourcing and alignment. E-business means cooperation, and hence it will be important to be able to identify and utilize sources of relative advantage inside or outside of the company. This ability should therefor be stimulated in the SME segment.

Secondly, our results question the practice of several e-business stimulation programs both in Norway and internationally who have used the power of bigger business partners in order to increase the adoption of e-business in SMEs. This practice could of course in the short run increase the adoption of e-business solutions in SMEs, but in the longer run the effects could be negative as a result of insufficient alignment with other strategies and initiatives in the SMEs. As such, this strategy could also be biased in the way that it creates more positive effects for the influential partner than for the SME. In the longer run for the e-business relationship, the effects could be negative for both parties as a result of reduced alignment for one of the partners.

To conclude, both research and practice should be careful in generalizing from big to small companies. Both theories and stimulation programs should be based on a better understanding of SME characteristics in general.

ACKNOWLEDGEMENTS

This paper benefited from work conducted under the ANTRA project that is partly financed by the EU Commission and the Leonardo da Vinci program.

REFERENCES

Amit, R. and Schoemaker, P., 1993, Strategic assets and organizational rent. *Strategic Management Journal*, 14(1): 33-46.
Amit, R. and Zott, C., 2001, Value creation in e-business. *Strategic Management Journal*, 22, 493-520.

Ballantine, J., Levy, M. and Powell, P., 1998, Evaluating information systems in small and medium enterprises. *European Journal of Information Systems*, 7: 241-251.

Basselier, G., Reich, B.H., and Benbasat, I, 2001, Information technology competence of business managers: a definition and research model, *Journal of Management Information Systems*, Spring 2001, 17(4): 159-182.

Bharadwaj, A.S., Sambamurthy, V. and Zmud, R.W., 2000, IT capabilities: theoretical perspectives and empirical operalization. *Proceedings of the International Conference on Information Systems*, 2000: 378-385.

Bode, S and Burn, J., 2001, Consultancy engagement and e-business development – a case analysis of Australian SMEs. *Proceedings of the 9th European Conference on Information Systems*, 2001: 568-578.

Brynjolfsson, E. and Hitt, L.M., 1998, Beyond the productivity paradox: computers are the catalyst for bigger changes. *Communications of the ACM*. 41(8): 49-55.

Dans, E., 2001, IT investment in small and medium enterprises: paradoxically productive? *The Electronic Journal of Information Systems Evaluation*, 4(1)

Feeny, D.E. and Willcocks, L.P., 1998, Core IS capabilities for exploiting information technology. *Sloan Management Review*, Spring 1998: 9-21.

Forrester Research, 1999, Referred to in *Connectis* 1, Europe's E-business Magazine.

E/1224, 1997: *Training processes in SMEs: practices, problems and Requirements.*

Hair, J.F., Anderson, R.E., Tatham, R.L. and Black, W.C., 1992, *Multivariate data analysis with readings*. 3rd edition. MacMillan, New York.

Henderson, J.C. and Venkatraman, N., 1993, Strategic alignment: leveraging information technology for transforming organizations". *IBM Systems Journal*, Vol. 32, No. 1: pp. 4-16.

Kraiger, K., Ford, K., Salas, E., 1993, Application of cognitive, skill based and affective theories of learning outcomes to new methods of training evaluation. *Journal of Applied Psychology* 78(2): 311-328 .

Laudon, K.C. and Laudon, J.P., 2002, *Management information systems, Managing the digital firm*. Seventh edition. Prentice Hall.

Lee, D.M.S. and Trauth E.M., 1995, Critical skills and knowledge requirements of IS professionals: a joint academic/industry investigation. *MIS Quarterly*, 19(3): 313-340.

Luftman, J., 2000, Assessing Business-IT alignment maturity. *Communications of the Association of Information Systems*, 4(14), December 2000.

Luftman, J.N., Papp, R. & T. Brier, 1999, Enablers and inhibitors of business IT alignment. *Communications of the Association for Information Systems*. 1(11).

Marcolin, B.L., Compeau, D.R., Munro, M.C. and Huff, S.L., 2000, Assessing user competence: conceptualization and measurement. *Information Systems Research*. 11(1): 37-60

Owens, I. and Beynon-Davies, P., 2001, A survey of electronic commerce utilization in small and medium sized enterprises in South Wales: [research in progress]. *Proceedings of the 9th European Conference on Information Systems*, 2001, 461-467.

Penrose, E.T., 1959: *The theory of the growth of the firm*, Oxford, Basil Blackwell.

Papke-Shields, K.E. and Malhotra, M.K., 2001, Assessing the impact of the manufacturing executive's role on business performance through strategic alignment. *Journal of Operations Management*, 19(1).

Rodwell, J. and Shadur, M., 1997, What's size got to do with it? Implications for contemporary management practices in IT companies, *International Small Business Journal*. 15(2): 51-63.

Sambamurthy, V. And Zmud, R.W., 1994, IT management competency assessment: a tool for creating business value through IT. *Financial Executives Research Foundation*, Morristown, New Jersey.

Skinner, W., 1969: Manufacturing – missing link in corporate strategy, *Harvard Business Review*. 47(3): 136-145.

Teece, D.J., Pisano, G. and Shuen, A., 1997, Dynamic capabilities and strategic management. *Strategic Management Journal*. 18(7): 509-533.

Van der Heijden, H., 2000, Measuring IT core capabilities for electronic commerce: results from a confirmatory analysis". *Proceedings of the International Conference on Information Systems*, 2000: 152-163.

Watson, R. and Brancheau, J., 1991, Key Issues In Information Systems Management: An International Perspective, *Information & Management*. 20: 213-23

APPENDIX 1. THE SURVEY INSTRUMENT

Background information

1. Approximately, how many employees are there in your company? ____ employees

2. In what type of industry is your company? ☐Tourism, ☐Transport

3. Does your company use web pages, e-mail or e-commerce systems for business purposes?

    ☐Yes ☐No

4. What types of e-business systems does your company use? (multiple responses possible)
    ☐ web pages whith information to individual customers about products or services
    ☐ web pages where individual customers can make orders
    ☐ web pages where companies can find information about products or services
    ☐ systems for electronic sales of products and services to other companies
    ☐ EDI solutions on the Internet
    ☐ systems that enables your suppliers to see information about your demand or production
    ☐ systems that integrates supply chains
    ☐ other

5. To what extent are IT activities in your company outsourced to external providers?

a very low extent	1	2	3	4	5	6	7	a very high extent

6. To what extent is your company informed about commercially available e-business systems?

a very low extent	1	2	3	4	5	6	7	a very high extent

7. To what extent is your company informed about providers of e-business related training?

a very low extent	1	2	3	4	5	6	7	a very high extent

8. To what extent has your company implemented its e-business intentions?

a very low extent	1	2	3	4	5	6	7	a very high extent

Sourcing competencies

9. Our company has a high level of knowledge on outsourcing of activities to other companies

    totally disagree 1 2 3 4 5 6 7 *totally agree* N/A ☐

10. Our company has a high level of knowledge on how to use competencies in our business partners

    1 2 3 4 5 6 7 ☐

Alignment competencies

11. In my company business and IT managers very much agree on how IT contributes to business value

    totally disagree 1 2 3 4 5 6 7 *totally agree* N/A ☐

12. In my company there is effective exchange of

ideas between business people and IT people   1 2 3 4 5 6 7   □

13. In general, my company is good at using the
    competencies it already has   1 2 3 4 5 6 7   □

14. In general, my company is good at using
    competencies represented in our
    business partners   1 2 3 4 5 6 7   □

E-business success

Efficiency

| | *totally* *disagree* | | | | | *totally* *agree* | | *N/A* |

15. Our e-business efforts have reduced costs by
    electronic order taking over the Internet   1 2 3 4 5 6 7   □

16. Our e-business efforts have made us able
    to deliver faster   1 2 3 4 5 6 7   □

17. Our e-business efforts have reduced costs
    in communication with suppliers
    and customers   1 2 3 4 5 6 7   □

Complementarities

18. As a result of our e-business efforts our
    products or services complement products
    or services from other suppliers   1 2 3 4 5 6 7   □

19. Our business efforts make it possible for
    other suppliers to complement our products
    or services   1 2 3 4 5 6 7   □

20. Our e-business efforts have made our supply
    chain strongly integrated to our partners'
    supply chains   1 2 3 4 5 6 7   □

Lock-in

21. Our e-business efforts make it more expensive
    for our customers or suppliers to replace us   1 2 3 4 5 6 7   □

22. Our e-business efforts have made our products
    and services more tailored to our
    customers' needs   1 2 3 4 5 6 7   □

Novelty

23. Our e-business efforts have made our company
    a pioneer in utilizing e-commerce solutions   1 2 3 4 5 6 7   □

24. Our e-business efforts have made us cooperating
    with our customers or suppliers in new and
    innovative ways   1 2 3 4 5 6 7   □

General

25. In general, my company has experienced very
    positive effects from its e-business efforts   1 2 3 4 5 6 7   □

Other/Control

Leader vs. Follower

26. There is a dominating customer or supplier |
    who dictates our e-business efforts   1 2 3 4 5 6 7   □

27. Our company is good at implementing
    changes in its organization   1 2 3 4 5 6 7   □

28. Overall, my company has a high level of
    competence for utilizing e-business technology   1 2 3 4 5 6 7   □

APPENDIX 2: DESCRIPTIVE STATISTICS

Variables	N	Mean	Std. Dev.	Vari-ance
1 Size	81	26,59	30,98	959,70
2 Type of Industry (1: Tourism, 2: Transport)	81	1,49	,50	,25
3 Does your company use web pages, email or ...	81	1,00	,00	,00
4_1 web pages with information to individual customers about products or services?	81	,89	,32	,10
4_2 web pages where individual customers can make orders	81	,52	,50	,25
4_3 web pages where companies can find information about products or services	81	,86	,35	,12
4_4 systems for electronic sales of products or services to other companies	81	,14	,35	,12
4_5 EDI solutions on the internet	81	,30	,46	,21
4_6 sytems that enable your suppliers to see information about your demand or production	81	,20	,40	,16
4_7 systems that integrate supply chains	81	,16	,37	,14
4_8 other	81	,23	,43	,18
5 To what extent are IT activities in your c...	81	2,94	1,87	3,48
6 To what extent is your company informed about commercially available...	81	3,51	1,80	3,20
7 To what extent is your company informed about providers of...	81	2,90	1,67	2,79
8 To what extent has your company implemente...	81	3,17	1,59	2,52
9 Our company has a high level of knowledge on outsourcing of outsourcing of activities to other companies	72	3,49	1,82	3,32
10 Our company has a high level of knowledge on how to use competencies in our business partners	77	4,21	1,45	2,09
11 In my company, business and IT managers very much...	69	4,72	1,51	2,29
12 In my company, there is effective exchange...	59	4,54	1,92	3,70
13 In general, my company is effective at using the comp. it already has	79	4,95	1,40	1,95
14 In general, my company is effective at using comp. in our bus. partners	76	4,21	1,34	1,80
15 Our e-business efforts have reduced costs ...	58	3,33	1,75	3,07
16 Our e-business efforts have made us able to deliver faster	51	3,24	1,80	3,22
17 Our e-business efforts have reduced costs in comm. with our bus. part.	74	4,08	1,74	3,03
18 As a result of our e-business efforts our ...	58	3,28	1,77	3,12
19 Our business efforts make it possible for ...	64	3,48	1,59	2,54
20 Our e-business efforts have made our suppliers...	69	3,39	1,68	2,83
21 Our e-business efforts make it more expensive...	67	2,28	1,48	2,18
22 Our e-business efforts have made our products...	66	3,36	1,95	3,80
23 Our e-business efforts have made our company...	71	2,31	1,67	2,79
24 Our e-business efforts have made us cooperate...	69	3,71	1,99	3,94
25 In general, my company has experienced very positive effects from...	73	4,22	1,71	2,92
26 There is a dominating customer or supplier who dictates our e-business	71	1,93	1,32	1,75
27 Our company is effective at implementing changes in its org....	73	4,12	1,54	2,36
28 Overall, my company has a high level of c...	79	3,14	1,61	2,58
ALIGNMNT Index q11-14	57	4,66	1,18	1,40
EFFICIENCY Index q15-17	42	3,41	1,37	1,89
COMPLEMENTARITIES Index q18-20	56	3,35	1,52	2,31
LOCK-IN Index q21-22	61	2,80	1,37	1,89
NOVELTY Index q23-24	66	3,04	1,63	2,66

APPENDIX 3: TEST OF REGRESSION ASSUMPTIONS

Hypothesis 1: Sourcing and general e-business success

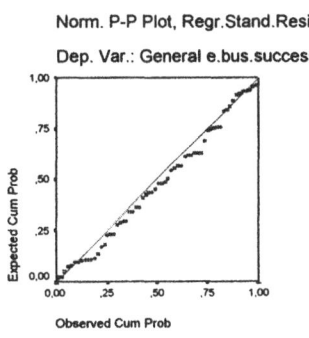
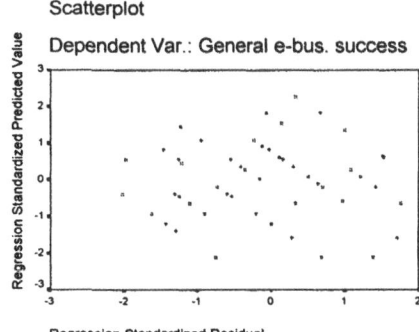

Hypothesis 1b: Sourcing and Complementarities

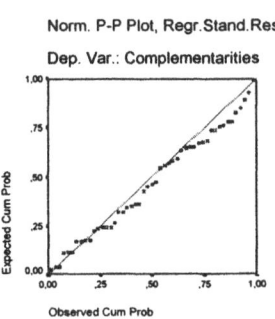
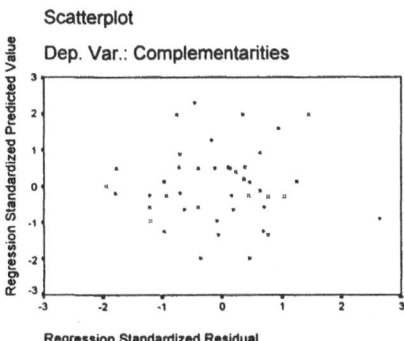

Hypothesis 1c: Sourcing and Lock-in

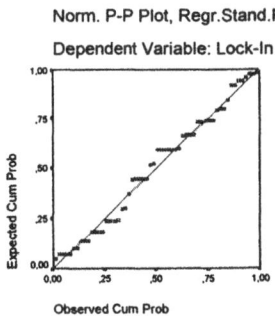
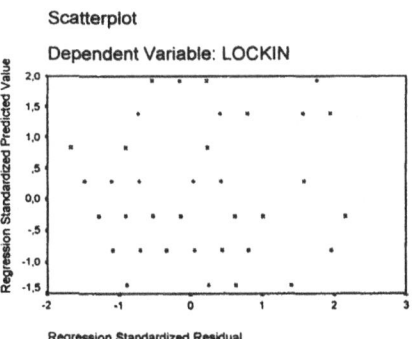

Hypothesis 1d: Sourcing and Novelty

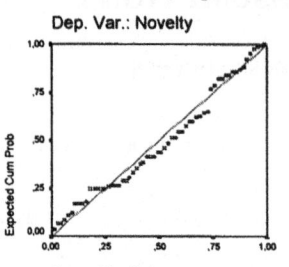

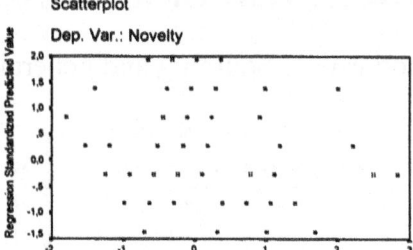

Hypothesis 2a: Alignment and Efficiency

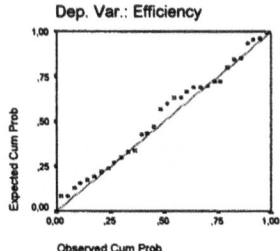

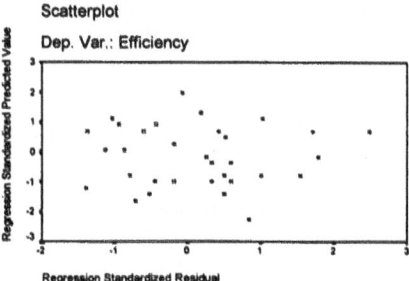

Hypothesis 5: External Governance and Alignment

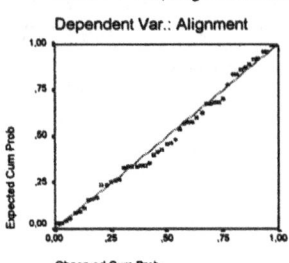

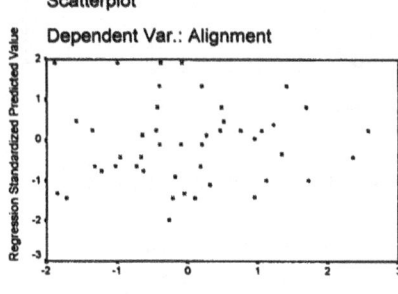

THE GAP BETWEEN RHETORIC AND REALITY:

Public reluctance to adopt e-commerce in Australia

Kitty Vigo*

1. INTRODUCTION

In Australia Electronic Service Deliver (ESD) and e-commerce have become increasingly important means for governments and citizens to engage in business transactions. The pressure for take-up of ESD and e-commerce and the transformation of Australia into a digital economy has come from both government and business (Chifley Research Centre, 2001). Nevertheless, in spite of these government and business inducements, Australian small to medium businesses (SMEs) are still slow to adopt e-commerce options. According to a report published in 2000 on a study commissioned by the NOIE into the take-up of e-commerce by SMEs, *Taking the Plunge. Sink or Swim?*, SMEs slow to adopt e-commerce "do so at their own peril" (NOIE, 2000).

The slow take-up of e-commerce options by SMEs occurs in spite of the fact that Australia is a global leader in internet use. In a paper titled "*The Intangible Economy and Australia*" Daley (2001) notes that Australia has more PCs per worker "than almost any country", has a better developed "web infrastructure" than any country except the United States and Iceland, as well as one of the highest global levels of computer and internet usage among private citizens. Daley warns that while these figures seem impressive, Australia is declining in terms of world exports and in living standards relative to the rest of the developed world. He suggests that this trend can only be reversed if Australia makes greater attempts to become "a new economy" which places a greater emphasis on electronic global business than on "local" manufacture.(Daley 2001)

Australian Local Governments have made significant investments in creating online business transactions opportunities. One example is The City of Manningham which has made a long-term commitment to the introduction of Electronic Service Delivery (ESD) and e-commerce inside, and outside its municipal boundaries. (Vigo et.al 2000) It has introduced the capacity for its citizens to conduct on-line transactions with the City of Manningham and other government instrumentalities through Maxi Multimedia and has introduced a number of initiatives to support its shopping centres to explore and adopt e-

---

* Media and Multimedia, Swinburne University of Technology, Lilydale, kvigo@swin.edu.au

Information Systems Development: Advances in Methodologies, Components, and Management
Edited by Kirikova *et al.*, Kluwer Academic/Plenum Publishers, 2002

commerce potentials. The Manningham On-Line Project was one such initiative. This project was designed to encourage City of Manningham traders and shoppers to explore e-commerce potentials. It was also anticipated that the project would act as a means by which Manningham citizens could be further introduced to on-line services and be encouraged to become effective on-line citizens.

The Manningham On-Line Project was a collaborative project funded by StreetLIFE, a State Government funding initiative to encourage employment opportunities in state shopping centres, the City of Manningham, CitySearch (www.CitySearch.com.au), a commercial online directory business, and Hewlett Packard. The project aimed to address the problem of slow up-take of e-commerce by small businesses within the City of Manningham. Its purpose was to facilitate the introduction of e-commerce in the City of Manningham, create new business opportunities for traders, offer Manningham citizens flexible on-line shopping, and to create new jobs associated with on-line business.

The project involved a partnership between Manningham Council and four community level shopping centres (Tunstall Square, Jackson Court, Macedon Square, and Templestowe Village) seeking to position themselves in the age of e-commerce. Other project participants included Swinburne University of Technology which would conduct ongoing research into the progress and outcomes of the project, and CitySearch.com.au, an on-line leisure and lifestyle guide and one of Australia's largest website developers. Funding for the research component of the project came from Hewlett Packard which has offices located in the City of Manningham.

The project was funded at $80,000 for two years, with half the money coming from StreetLIFE which sought to use the project as a pilot for adoption of e-commerce by other centres throughout Victoria, and half from the City of Manningham. Hewlett-Packard contributed $15,000 to fund a series of surveys during the life of the project and the production of research report. A key outcome from the project sought by StreetLIFE was the development of business/marketing plans by each of the four shopping centres incorporating e-commerce and the creation of jobs resulting from increased e-commerce business.

This paper briefly describes the outcomes of the Manningham Online Project and attempts to identify why trader and consumer take-up of the e-commerce options provided by this project was minimal.

2. MANNINGHAM ONLINE – ENCOURAGING E-COMMERCE TAKE-UP BY SMALL RETAILERS AND TRADERS

The Manningham Online Project specifically targeted traders and customers of four major City of Manningham shopping centers (Tunstall Square, Jackson Court, Macedon Square and Templestowe Village) with a view to helping them make the transition to electronic commerce.

The four participating shopping centres are open-air strip shopping centres and include a total of 233 retailers and other businesses. Tunstall Square Shopping Centre has 73 businesses and services and facilities offered include four major banks, (four ATM locations), Australia Post Office, and extensive parking. Tunstall Square has a particular local reputation for its fresh food retailers including large grocery chain store Coles New World which offers 24 hour, seven-day trading, gourmet delicatessens, bakers, fresh meat, fish and seafood, chicken, fruit and vegetables.

Jackson Court has over 90 businesses including a large variety of shops and services including major banking facilities, Safeway Supermarket, restaurants, a range of professional business services, travel agent and sports/hobbies stores.

Macedon Square has 19 businesses. Shops and services include Safeway Supermarket - open for 18 hours a day, seven days a week, specialty cake shops, Sushi Bar, legal services and specialist community health services.

Templestowe Village Shopping Centre has over 50 businesses, including 20 food retailers. Templestowe Village has seven restaurants, a supermarket, a butcher, a specialist fruit shop, a hot bread shop and a number of take-away outlets.

The Manningham Online Project involved the development of a strategy which would:

- introduce from each of the four shopping centres to the projects aims and to the opportunities offered by e-commerce;
- provide a basic webpage which included a business description and communications facilities for each of the traders from the four shopping centres;
- provide traders with the opportunity to develop more detailed webpages which included products for sale on-line;
- provide customers with opportunities to buy products on-line from local traders;
- create an environment which would introduce traders and customers to the advantages of conducting on-line transactions.

The strategy involved establishing the Tunstall Square shopping centre as the pilot site, with the other three shopping centres being able to take advantage of the Tunstall Square experience. By June 2001 each of the four shopping centres had a web presence accessible through City Search. Tunstall Square had 10 businesses with fully-developed web presences; Jackson Court, eight; Templestowe Village, seven; and, Macedon Square, 13.

The project also included a research component which sought to investigate key factors influencing take-up of e-commerce by traders and customers and to measure progress in terms of experience and acceptance of the model. The research focused around four surveys of traders and shoppers which were conducted at approximately six-monthly intervals. Questionnaires were developed for traders and customers with additional questions being added with each questionnaire. These questionnaires were distributed to traders and shoppers in December 1999, February 2001 and August 2001. In August 2000 a series of in-depth interviews were conducted with the Tunstall Square Centre Coordinator and seven traders who had developed extended websites that offered customers with opportunities to place on-line orders.

3. MANNINGHAM ONLINE: TRADER AND CUSTOMER ECOMMECE TAKE-UP

The research undertaken by this project indicates that while there was a steady increase in the take-up of e-commerce by traders and customers in the four shopping centres involved in the project, it was slow and not as dramatic as hoped for the by the project participants. For example, by the end of the project only 38 out of the 233 traders operating in the four shopping centres had a fully-developed e-commerce website, and there was only a 10 percent increase over the two years in the numbers of traders with internet connection between the time of the first survey in December 1999 and the final survey conducted in August 2001. Further, while there was a 20% increase in customer access to the internet between December 1999 and August 2001 (from 53% to 73%), and a 11% increase in the number of shoppers who had purchased goods online (from 1% - 11%), only five of the traders reported that they had made online sales to local customers.

While there was relatively little increase in the uptake of e-commerce opportunities, there was a significant shift in attitude towards being prepared to invest in e-commerce hardware and software. One explanation for this change in attitude may be that during the project period the Federal Government introduced a Goods and Services Tax (GST) with a small cash incentive to any small business wishing to invest in GST-related hardware or software (computers, financial management software, etc.). Many traders felt that investment in GST-related hardware and software might also be applied to e-commerce uses. Nevertheless, while more traders said they were prepared to invest in hardware and software there was an odd discrepancy between December 1999 and August 2001 in the amount of money respondents said they would be prepared to invest, with more traders in December 1999 stating they would be prepared to invest up to $2000 than in August 2001. This discrepancy may be explained by the fact that by the end of the survey period more traders had a better understanding of the real cost of purchasing relevant hardware and software requirements.

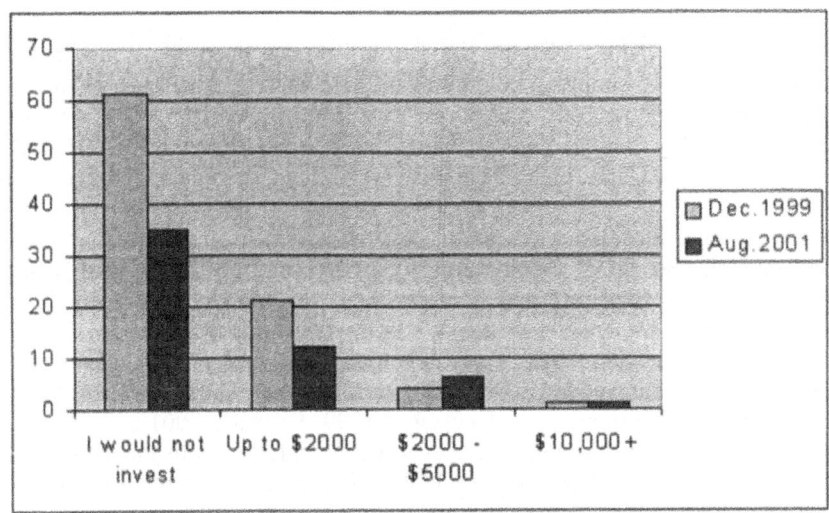

Graph 1. December 1999, August 2001 Trader willingness to invest in e-commerce.

This shift in attitude to investing in e-commerce correlates with the increase in the number of traders who said they believed that their business would benefit from offering their customers online shopping opportunities. In December 1999 41% of traders said they could see little or no benefit to their business from e-commerce, by August 2001 this figure had dropped to 21%.

The discrepancy between the shift in traders' acceptance of the principle of e-commerce and their reluctance to actually engage in it might be explained by the consistently high importance that traders placed on personal relationships with their customers. In December 1999 97% of traders claimed it was "*important*" or "*very important*" to develop a personal relationship with customers. Only one of the traders claimed that it was "*not important*". By August 2001 the figure had increased slightly, with 99% claiming their personal relationship with customers was "*important*" or "*very important*". Comments on why personal relationships with customers were important included:

> "In spite of what the Government is saying now, business is tough for us small players. This shopping centre [Tunstall Square] works hard to provide a high level of service to our customers. We try to create an atmosphere where they feel comfortable and feel that we are our friends. The relationships I form with our regular customers is about our best asset." (Butcher, Tunstall Square. December 1999)

and

> "Since the introduction of the GST our sales have gone down a bit. The backbone of my business is my regular customers. I don't think I can give them my best just through the internet." (Jeweller, Tunstall Square. August 2001)

City of Manningham customers proved to be more enthusiastic adopters of the internet and more willing to embrace the principle of e-commerce. In the time between the first and last survey periods there was a 10 % increase in the numbers of people who said they had access to a computer - from 75% to 85% of the 200 shoppers surveyed. Further, the majority of customers had access to high-end computers: There was also a 12% increase between the survey periods in the number of customers who said they accessed the internet every day:

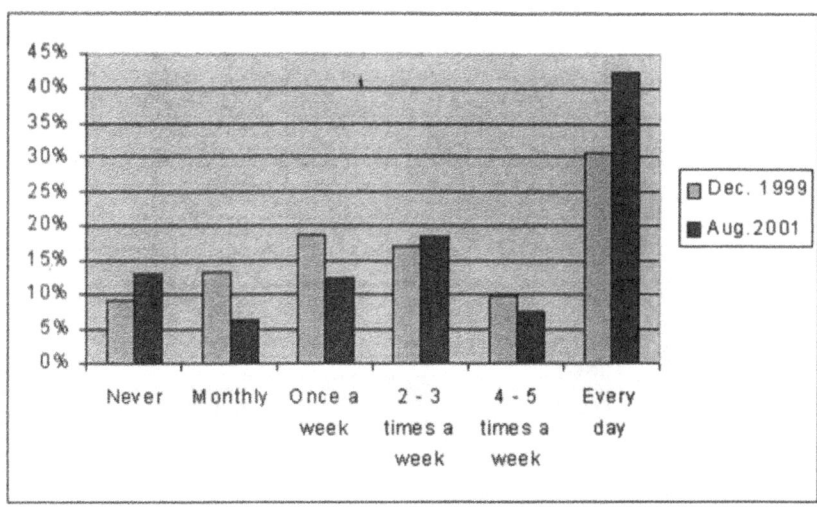

Graph 2. December 1999, August 2001 Customer internet access frequency.

There was also a 10% increase between December 1999 and August 2001 in the number of shoppers who said they used the internet for shopping.

Customers also proved to have an improved understanding of the term "e-commerce". In December 1999 the majority of customers had "*no idea*" or "*not much*" of an understanding of the term "electronic commerce", for most the term meant "*nothing*". Some respondents thought the term to mean "*internet banking*" or "*doing business using computers*", "*electronic trading*". Others answered that it meant, "*shopping via the internet*", "*paying bills*" or "internet banking". A few shoppers thought the term referred to "*ATM's*". By August 2001, however, the majority of shoppers understood the term to mean either "*shopping*", "*trading*", "*bill-paying*" or "*banking over the internet*".

In spite of this increased understanding of e-commerce, the majority of shoppers in both surveys said they were not willing to purchase products over the internet, citing concern about security, as well as a desire to do personal shopping and developing a personal relationship with their retailers. However, the number of shoppers who said they might purchase products via the internet increased between the two survey periods. In December 1999 only 30% said they would purchase gifts over the internet compared with 45% in August 2001 and 7% said they would purchase electrical goods in 1999 compared with 17% in 2001. The most significant increase occurred in the number of people who said they would buy food or liquor online – increasing from 15% in December 1999 to 54% in August 2001. Interestingly, most customers said that if they did purchased products over the internet it would only be to support their local retailers.

One of the key outcomes of the project was to increase awareness of City of Manningham citizens of electronic payment of Local Government and public utilities bills through the internet Maxi. Public awareness of these bill-paying possibilities increased significantly between the two survey periods, with many customers saying they learned of this opportunity from printed publicity material distributed in relation to the Manningham Online Project:

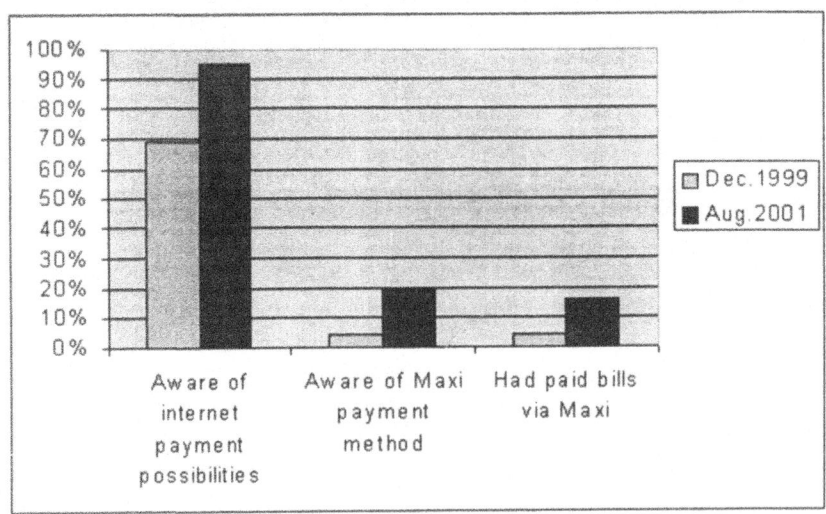

Graph 3. December 1999, August 2001 Customer awareness of internet bill payment options.

4. DISCUSSION

Although a number of factors have been identified for causing the relatively slow take-up of e-commerce by City of Manningham traders and customers, they should be understood in the context of Australian national take-up of e-commerce.

According to the Australian Bureau of Statistics, Australia is only second to the United States in internet take-up (2001). In spite of this, on-line purchasing goods by the general public is still limited by a continued concern by members of the general public about issues such as security and unfamiliarity with purchasing products from unknown traders. Harry Wendt, head of eStrategy at Wespac Banking Corporation told a recent internet conference, "Australians are normally quick to adopt new technology but they are reasonably cautious before they place their trust in new purchasing formats." (Wendt, 2000) Wendt suggested that e-commerce had been more quickly accepted by customers in the US because mail order was popular there and people were used to dealing with companies they has never heard of. This preference for a personal relationship between traders and customers appears to be strongly supported by the research conducted on the Manningham On-Line Project. In each of the surveys conducted with traders, a heavy emphasis was given to the importance of developing personal relationships with customers. Customers surveyed indicated that they preferred to do shopping in person and while by the end of the survey period 63.5% of customers indicated that they would be willing to buy products on-line, only 12% indicated that they ever done so. This was in spite of the fact that 73% of the surveyed customers said that they had internet access.

The research completed for this project, and research conducted elsewhere, would appear to indicate that considerable barriers continue to exist to the take-up of e-commerce by traders and customers. For example, research conducted by the Yellow Pages Business Index into the take-up of e-commerce by Australian SMEs, found that the most significant barrier was the lack of personal contact, followed by security concerns, and the belief that customers were not yet ready to accept e-commerce. (Yellow Pages Business Index, 2001).

The research conducted by the Yellow Pages Business Index also indicated that while most businesses surveyed by them indicated that they would move towards becoming on-line business, they were not doing so at that time. Ten percent of businesses surveyed by the Yellow Pages Business Index said they had no intention of taking-up e-commerce. These findings are reflected in the research conducted by the Manningham On-Line Project which found that 21% of trader said they did not believe that the internet and e-commerce had any relevance to their business. Further, 35 of the 55 traders surveyed in August 2001, said they would not invest in computer technology for their businesses. The Yellow Pages Business Index research findings are also supported by the research conducted for this project which indicated that while nearly 400 businesses associated with the four shopping centres involved in the project were presented with the opportunity to develop an extended web presence, only 38 chose to do so through the facilities offered by CitySearch.

While the traders involved in the project were slow to take advantage of access to e-commerce offered by this project, research indicates that City of Manningham citizens continue to have amoungst the highest level of access to computers and the internet in Australia. During the survey period, computer access increased by 10 per cent from 75 per cent December 1999, to 85 per cent in August 2001; internet access increased from 53 per cent in 1999 to 73 per cent in 2001; and, the use of the internet for shopping increased from 1 per cent in 1999 to 12 per cent in 2001. These statistics appear to contradict research findings by the Yellow Pages Business Index report that indicated that traders believed that customers were not yet ready for e-commerce. In the case of he Manningham On-Line project, this belief by traders that customers prefer personal shopping to on-line shopping was supported by the fact that in spite of inducements such as discounts, only a few customers ordered on-line. However, the research conducted for this project did not indicate that customers were not ready for e-commerce, rather that they preferred to shop in person.

As noted above, shoppers surveyed by this project have high level of access by to the internet as well as a high level of understanding (69 per cent) of the opportunity for conducting local government transactions such as paying bills. Given this, one might expect a commensurately high level of use of facilities such as Maxi. However, only four per cent of shopper surveyed said that they used Maxi facilities. Concern about security was the main reason given for the failure to use facilities such as Maxi. This finding reflects closely other research, including the findings of the Yellow Pages Business Index, that Australians need to be convinced that the risk of losing money through fraudulently e-commerce or hacking practices is low.

Another barrier to customer take-up of e-commerce facilities offered by the Manningham On-Line Project is the lack of awareness by both customers and traders of the Manningham On-Line Project. In the case of customers this situation was compounded by relative difficulty of finding the four shopping centre sites. There is no direct or clear link provided through the City of Manningham web site, nor through the Maxi Kiosks, nor through the CitySearch Melbourne home page. Customers must therefore rely on information provided directly by the centres or their retailers. In the case of traders, lack of awareness of the project was compounded by distractions caused by the introduction of GST.

5. CONCLUSION

E-commerce has been cited as being particularly important for the Australian retail industry. The Australian Centre for Retail Studies has stated:

> "Electronic commerce is widely regarded as one of the most important forces of change shaping the retail industry around the world. Retailing is likely to be impacted by e-commerce more than most other sectors of the economy because of the position that retailers occupy as the interface between product suppliers and end customers. As well as within their own operations, e-commerce has the potential to impact on retailers' relationships with their suppliers and customers." (Australian Centre for Retail Studies, 1999)

The experience of the project described in this paper suggests that there is a gap between the rhetoric relating to the take-up of e-commerce and the reality of the business place and a significant difference between organisations such as the Centre for Retial Studies and retailers about the preferred style of relationship between retailers and customers. While governments in Australia continually promote and support the take-up of e-commerce, SMEs are slow to accept it even though they support it in principle. Further, the retailers in this study clearly indicated that they believed that personal relationships between retailer and customer could not be achieved via e-commerce.

The gap between rhetoric and reality also applies to customers who may have high levels of internet access, support e-commerce in principle, but have a high level of reluctance to buy products or pay bills online.

This gap between rhetoric and reality exists in spite of significant government funding for internet education of both the public and business and in spite of repeated warnings that Australia must transform itself into a "new" online economy if it is to retain its place in OECD rankings.

The outcomes of the Manningham Online suggests that significant research needs to be done to investigate and develop more effective strategies for encouraging e-commerce take-up.

6. REFERENCES

Alston, R., 1999, *The Role of Communications & E-Commerce in Building the Nation*, speech to CEDA - Minter Ellison Nation Builders Awards, Melbourne, 1999. Available on: http://www.dcita.gov.au/graphics_welcome.html

Australian Bureau of Statistics, *ABS Business And Government Information Technology Use Surveys*, Paper presented to an Ad Hoc Technical Meeting Of Asia/Pacific Statisticians On IT&T Statistics". 29 - 31 May 2001. Brisbane, Australia.

Australian Bureau of Statistics. Use of the Internet by Householders, Australia. 1999 http://www.statistics.gov.au/

Australian Centre for Retail Studies, *E-commerce in retailing: a survey of the Australian retail industry*, Short Report, 1999.

City of Manningham. 2000. Available: http://www.manningham.vic.gov.au.

Coursivanos, J., *IT Investment Strategy for Development: an 'Instrumental Analysis' Based on the Tasmanian and News Brunswick Information Strategies*, Prometheus, Vol. 18. No.1, 2000, pp75-91.

Chifley Research Centre, *An Agenda for the Knowledge Nation: Report of the Knowledge Nation Taskforce*, 2001.

Daley, John., *The Intangible Economy and Australia*, Australian Journal of Management, pp3-15, August 2001.

NOIE, Taking the Plunge: Sink or Swim?, Special Report, 2000.

NOIE, Australia's E-Commerce Report Card, 1999. Available on: ftp://ftp.dcita.gov.au/pub/ecommerce/report.doc

Rohm, A.J., & Milne, G.R., *Consumer Privacy Concerns about direct-marketer's use of personal medical in formation*, In J.F. Hair, Jr. (Ed), Proceedings of the 1999 Health Care Research Conference. pp 27 – 37. Breckenridge, CO, 1999

Vigo, K., Dawe, R., Hutcheson, M. and Redwood, M., *Manningham On-Line – Using Global Technologies For Building Local Electronic Commerce Business*, paper presented to 2000 ISD Conference, Norway.

LEARNING AND ANALYSIS IN AN E-COMMERCE PROJECT

Sten Carlsson[1]

1. AN INTRODUCTION OF THE AIM OF THE PROJECT AND RESEARCH QUESTIONS

An electronic commerce project might have many different aims. The main aim of this particular project – called market place Värmland - was to strengthen small and medium enterprises (SME) power of competition and create a positive growth for the enterprises in the whole province Värmland. The target group was enterprises which have up to 50 employees. The idea was to develop the persons' competence in using e-commerce in their business. Therefore companies were invited to take part in a course about e-commerce. The members of the project group had assumed that persons from about 500 companies would follow the courses during the three years when the project was going on. The focus of the course was directed towards doing analysis of the usability of electronic commerce. By doing the analysis it was assumed that the participants would be able to apply this competence in the companies of their own. Further more the plan was to create a network between the companies and an IT-platform for information and exchange of knowledge. The project was also designed to support other projects which have been started to market the enterprises products and to support these enterprises possibilities to buy and sell to each other. The municipal councils in the province make great purchases. The trend also is, that these councils want to use electronic commerce when buying something. Until now many small companies have not been able to sign contracts in order to deal with the councils. Many of them do not know what to do or they do not have any equipment for it. Many of them also feel that there are many difficult hinder to climb over.

The project has been established firmly in a broad sense both locally and in the region. The partnership in the project is based on collaboration between the Chamber of Commerce (the head of the project), companies, Karlstad University, industrialist organizations, The County Administrative Board and different municipal councils.

---

[1] Sten Carlsson, Karlstad University, Karlstad, Sweden

Information Systems Development: Advances in Methodologies, Components, and Management
Edited by Kirikova *et al.*, Kluwer Academic/Plenum Publishers, 2002

401

Hopefully, which also was suggested, the collaboration in this network will go on to support the improvement of the companies business even after the project is finished.

1.1 My research questions for moment

What will happen to the companies' economical situation after having developped e-commerce in their companies? Will these investments strengthen the small and medium enterprises (SME) power of competition and create a positive growth for these companies? In the long run we are going to follow what will happen concerning these questions. In this report I am going to present some experiences of my study of the courses in the project concerning learning and self-analysis. My thesis in this study is that the willingness to use information technology is very much depending on the end-user's possibilities to understand the usability and the utility of computers. It is not just enough to understand to use a computer. The utility and the advantages of doing that related to some business must also be quite clear for you (Carlsson, 2000). Otherwise you will reject using that technology. The problem is that the end-users have big difficulties to realise both the usability and the utility (ibid). This is an issue of learning and understanding and the utility of computers has to be understood in the user's context (ibid). My first research question therefore is: *What can be done to help persons from small and medium companies to understand if they need to invest in e-commerce?*

My second research question is: *How do we perform systems analysis in small and medium companies?* When you talk to a company leader of a small and medium company about his working situation, he/she will tell you this. During daytime the work is about what the company produce for the customers and when the business is closed that work concerns paper work. These company leaders mostly work lot more than eight hours per day. If you ask them to go to a course it has to be very short and very inexpensive. They can think about it if the timetable of the course concerns a couple of times, four ours per session in the afternoon or in the evening. In that case they also must feel that the course will give them something back to their economical situation. They can afford about 160 £ for a course with a length of 12 hours and not much more.

According to what a company leader might be able to offer in time and money for something besides the business with the customers my question is: what time and money can he/she offer for the analysis of the company in order implement a new information system? If you have this in mind and than think of what time it takes to perform a study using the methods introduced in the market such as Coad & Youdon (1991), Downs et al. (1988) and so on, the small and medium company studied must be out of the market when the study is ready. Anyway, the company leader would not let you start if you tell him/her what time it takes and how much you as a consultant want to have for the work. Otherwise you have to have very strong arguments for a good pay off for the company by means of your study. These types of methods might perhaps be useful in bigger companies, which have special departments for information systems analysis. But I am not sure that they fit so very good for small and medium companies.

So the question is: How do we make information systems analysis in small and medium companies, which are more convenient concerning time and money spent in practise in these companies?

1.2 My path to present the contributions

In this report I am going to present the results of my study of the planning process of the project and the first three courses. At the beginning we planned that the learning processes would be performed according to something what we called workshop analysis (see chapter 3). Even if we thought that the idea was brilliant, we were not able go on with this type of analysis. The main problem was to get the companies together at the same time. After presenting our view on electronic commerce in chapter 2 I will present the idea of workshop analysis in chapter 3. This chapter also contains a description of the analysis method used and my experiences of this method according to my research questions. In chapter 4 the presentation will be about how we continued with the project according to a more education like model. Chapter 5 contains the conclusions.

2. DESCRIPTION OF THE VIEW ON ELECTRONIC COMMERCE

The aim of electronic commerce is to use Internet and Information Technology (IT) to enhance a company's possibilities to make better business. There are many definitions of electronic commerce. So, instead of formulating such a definition I will describe how the project group of Market Place Wermland is using the concept electronic commerce.

According to this view business by means of Internet and IT will improve the benefit for the business such as:

- To sell more to new markets, which you could not do so effectively before, such as you penetrate a new market by using Internet as a selling channel
- To be able to offer the customers better services, which create deeper relations to the customers. They are offered to do business with you throughout the twenty-four hours and have special prices and special possibilities to enter in your business systems
- Increased effectiveness by means of more rational managing of purchase, invoices, store routines and other business transactions

Lately there have been writings about electronic commerce in a negative sense. This has probably influenced many company leaders to avoid to take any steps to study the subject by themselves. It is therefore very important that their knowledge about electronic commerce will be better, so the company leaders and managers are able to estimate the benefits for the company to use electronic commerce. If so, they perhaps will not decide what to do from rumours but from a very well defined basis about using electronic commerce. The companies, which have had electronic commerce trouble, are those, which just used Internet as a selling channel to consumers. To make investment to start an Internet shop can be very expensive. It costs a lot to market a new trademark in order to change the customers buying behaviour. It is new companies, which have got these problems. Companies, which have been in business for years selling by means of mail orders, have not had any problem by just changing to sell via Internet. To take a step to buy via Internet is not so far for these companies' customers if they are used to buy by mail order.

Electronic commerce is something more than just a website for ordering things on the net. The development moves towards a deeper integration of the companies business systems, planning systems and so on. The goals might often be to integrate the customers and suppliers in the systems as much as possible. Small companies might have technical problems with this integration of business systems. They have not the capacity themselves to manage all technical problems, which follows in the tracks of implementing new technology. They very often have to trust consultants to do the job, which can be very expensive for a small company. Especially if no one in the company has knowledge enough about what the company really needs concerning electronic commerce. It is however important to strengthen that electronic commerce is not just a technical problem. Therefore it is not just technical knowledge the company leaders need. What is more important is knowledge about which benefits there are for the company to start a process in the perspective of electronic commerce, what strategy is needed and which business models will be the best to use.

The IT-maturity very often differs from company to company. The strategy in this project therefore is to meet the companies on their maturity level, so they accept to start to think of the possibilities to use electronic commerce and not rejecting the idea without trying to understand these possibilities. Their motivation is of great importance. And this motivation must arise from inside of their heads. They must have a conviction of that electronic commerce might be something for them and have curiosity enough to stand out fulfilling the process to pass though the narrow gate to their own enlightenment.

The focus of the project Market Place Wermland is on all the categories used concerning electronic commerce:

- B2C, Business to Consumer
- B2B, Business to Business
- B2A, Business to Administration
- C2B, Consumer to Business

At the beginning of the project we wanted to give priority to the category Business to Business. By means of the discussions with different persons in the companies before we started the project, we understood that they had the main interest to work with this issue of electronic commerce in this perspective. They wanted to start slowly in order to have control over the situation and did not want to throw themselves out in blindness towards something they did not so very much about. Great security was also a good wisdom they wanted to cultivate. The best way to obtain this for them was the alternative business to business. They also thought that this variant of using electronic commerce might give the best economical benefits in a short time.

3. WORKSHOP AS AN ANALYSIS METHOD

During the planning process of Market Place Värmland, I took part in a workshop together with small and medium companies in order to analyse their need to make investments in e-commerce. A consultant company was responsible for the workshop. The aim of the workshop as such was to educate the participants so that they would to be able to understand if there was any idea for them to start to introduce e-commerce in their business. But a consultant company does not work free. So, we decided that the first meeting in the project should be free. After that the companies had to pay a small sum for the analysis. By using the idea of a workshop together with about five companies the consultant company counted on that this type of workshop should be a good business both for them an the companies taking part in the workshops. The free part of the workshop I call arrangements before the analysis and second part for the analysis model. By the model I present how the analysis was performed during the workshop

3.1 Arrangements before the analysis

Some representatives for five companies were invited to participate in a learning and analysis process about the profitability of using e-commerce in their companies. The companies were chosen so that there were not any competitions between them on the market. I myself took part in the process as a listener and an interviewer in order to do research about what was going to happen in the process.

At the first meeting the representatives were informed about the project and how the analysis would be performed, why and the goals with it. They were also informed about the idea of e-commerce and some possibilities to hire software from Telia. The goals of the analysis during the workshop were presented by the following contents.

What was going to happen during the workshop was that they by themselves would make analysis of the companies' business models. From this starting point they were going to discuss about the possibilities for the companies to increase the sales, enhance the services and obtain a more effective company. The next step was to make calculations about costs and the investments they have to do in order to start their e-business. What income there will be must also be calculated. Other calculations should give them the answers about the savings they can do, what type of effectiveness they can obtain in the working routines. After these calculations they have to decide if they want to make the investments. How much, that depends on their willingness to do investments. Is it a low or a high risk of making these investments? When they have made the decisions then they had to make up the strategy to obtain the goals.

At the end of the meeting the model of analysis was presented. Concerning the first part of the model - making a map of the business of the company - the representatives of the companies have to think over as their home works when they got home. They had to think over the objectives of the analysis of this part of the model related to their own company. This has to be done so they are prepared to do a workshop together with each other the next time at the workshop.

3.2 The analysis model

The model of analysis performed consisted of five parts, such as:

- Making a map of the business of the company – the map will include the management inside the company and the relations to its customers and suppliers. One question is which has to be answered: What are the changes of the business of the company if you are going to implement e-commerce?
- Doing analysis of the consequences of selling/market, suppliers and the management of the company
- Making up the goals/aim of going in for Internet/e-commerce
- Doing calculations of costs, income and the willingness of investment
- Making conclusions of the outcome of the analysis

3.2.1 Making a map of the business of the company

Analyse step 1.1 concerns the administration of the company. You start this first step of analysis by describing the management of the company, what you in fact are doing there to perform your business. When you are doing something you must have some competence for it. Therefore you have to ask yourself some questions: What is your core competence? What can you do which your competitors cannot do? There you have your strength. How can you use this strength to enhance your business? Write down your answers of these questions and why you think so.

But you have also weaknesses. You perhaps know something about your competitors and their strength in the market. What is their strength? Can you do something to be able to get a stronger position against your competitors? The weaknesses perhaps cannot be explained due to your competitors. May be, you have not taken advantage of your position in the market in a way that you should have. Perhaps there is much more you can do. What can you do about this weakness? You should write down these answers reflecting on what you can do about your weaknesses and why your solutions can strengthen your competition on the market.

Making a map of the business also means that you have to analyse your core processes such as how these processes are organised, how your organisation handle these processes and use your internal systems to support the working situation. Such systems are, for instance, business systems, product planning systems and intranets. Why should you do this analysis? By doing it you should realise if your core processes are up to date. According to a process oriented view the processes should be effective related to the customer (Christiansson, 2001). That is very important. It is also important that your management processes are effective as such. But the aim of this effectiveness should be much more related to the customers than other internal demands of effectiveness. Other questions you also have to answer and document are if your organisation or the systems are suitable for handling the customers' demands. In case not what are the problems and what can you do and why?

When analysing your business in the perspective of e-commerce the logistics is of great importance to study. What you have to understand and document is how the transport system looks like for your company?

If you later on are going to start a change process to implement e-commerce you should know how prepared your employees are for this step. Are there any projects going on or already planned in your company? If the answer is yes, you have to think of if there is time enough to start another one. Perhaps you have to give priorities to the project, which seems to be the best one for your business. What answers you have to search for are: What project am I going to start and what are the consequences of doing that in relation to other projects? Are you going to start a project you also have to reflect about what experiences there are from earlier projects in your company? It is important to know if your employees have bad experiences from earlier projects. Such feelings will have a strong influence on them in the next project (Carlsson, 2000). You will otherwise have trouble in the new project if you do not do anything about the situation of these experiences (Carlsson, 2000). Their willingness to accept changes will be influenced negatively. The maturity to handle changes also influences the willingness and the possibilities to succeed. If the employees have a low maturity concerning changes in their working situation you perhaps must arrange some sort of education before the employees take part in the project work. Before you start the project you should make notes about how much time resources you need for the project and how much you can afford. Furthermore you also have to notify the employees' attitudes according the e-commerce and why they think so. Your documentation should also include some words about the employees' maturity in relation to what the project demands from them. Knowing this you can prepare yourself for a successful change process.

Analyse step 1.2 concerns the customers. Which are your company's customers? Where are they located? Are they located in the neighbourhood? Perhaps they are wide spread out all over the country and perhaps also located abroad. Where they are located can be of importance if there will be any advantages for you to communicate with you customers electronically. There is another important question according to the analysis of the issue of e-commerce, which you have to think about. What are the sales amount/result in percent concerning each customer? The more sales amount and the more regularly the customers are buying from you the better the chances are for you to succeed in your investments of e-commerce. What type of relation do you have to your customers? Is it good and informal? If you would be able to succeed according to your plans in e-commerce your customers have to help you. You must see your project like a joint project together with them. If they do not want to buy from you electronically, you perhaps cannot force them to do that. Some big companies can. You instead have to come to an agreement, which is good for both parties. If there will be any idea for you to start this discussion, you have to think over how you communicate with them today. Do you use phone, fax or Internet? If you are sending a lot of invoices by mail there is one reason for you to think of if you can do that electronically instead.

Which are the most profitable customers and why? And furthermore, which are the least profitable customers? If they are not so profitable you perhaps have to think over if e-commerce will make the situation any better. If not, you have to reject the whole idea concerning these customers. How many of them are profitable for you? The situation is the same. If e-commerce will not do any good for the relation to these profitable customers you have to drop the idea. If you have come to the conclusion that some of your customers will be more profitable than before you have to investigate these customers' maturity and willingness to support a change towards new business models and e-commerce.

Analyse step 1.3 concerns the suppliers. Which are your suppliers? Where are they located? Are they located in the neighbourhood, wide spread out all over the country or abroad? One important part of the studies is how much every supplier is selling to you in percentage of you total amount of purchase. Do you think that e-commerce together with them will give you any better situation concerning your purchases? What are your relations to the suppliers? The relations can be very stable or unstable. If they are stable it is much easier to get support from them to start e-commerce and to negotiate about a solution. If you are going to make any investments in e-commerce related to your suppliers, you have to analyse how you communicate with them at the moment. What do you use? Are you using such as fax, Internet, telephone or letters. If you will find your communication very complicated and time-consuming there might be advantages for you to do your business electronically.

The analysis of your suppliers also concerns who is the most profitable and who is not. Your choice of doing your purchase electronically depends on the profit by doing so. If the profit will not be any better, you should not do any investments in e-commerce equipment. You also have to explain why you should do that or why not. The result of the decision to use e-commerce can be that you have to finish buying from some of your suppliers. Why you should do that must also be explained. Therefore you have to know who is the most valuable and who is not. If someone is not so valuable you must explain to yourself why it is like that and if you can do anything about it. Before you start making investments in any e-commerce together with your suppliers you have to realise their maturity according to your situation. You cannot start any use of e-commerce without talking and negotiating with your suppliers about your situation and how you can co-operate concerning your routines and management. If they are mature enough in this business you have the possibilities to start developing solutions together with you.

3.2.2 Doing analysis of the consequences of market

Analyse step 2.1 concerns optimising selling in the market. One important issue of the analysis of your customer is if you are able to extent your sales for your existing customers. Can you do that by your investment in e-commerce? It is also important that you can secure your most important customers' loyalty. They must realise that there are some advantages for them making business with you and your future investments in e-commerce. The next issue in the analysis of the customers concerns if you are prepared to get new customers. What are you going to do in order to get new customers? Concerning e-commerce you must answer if you sell will directly to some of them. By selling directly they can order direct from you by Internet. In order to make your business efficient you have to arrange your logistics in an efficient way. At the end of the analyses of your customers you have to relate your company's goals to market and customers.

Analyse step 2.2 concerns optimising your purchase from the suppliers. During the analysis of your relation to the suppliers you have to realise what purchase is most important for you. This purchase should be analysed related to your most important customers and the result. Then you have to secure that your most important suppliers are loyal to you. What you also have to think about is if you are able to get new and cheaper supplier by means of Internet as a new business channel. To be able to be more effective you have to analyse and decide if you can cut some intermediary links between your company and the manufacturer. In order to make business with your suppliers you perhaps have to deal with expensive transports. Your logistics is therefore an important issue to analyse. By planning this logistics a little bit better, you might be able to reduce the cost for your transports. Now you can summarise the company's goals to suppliers and your purchase.

Analyse step 2.3 the consequences for your own company. After the analysis of selling and purchase and what you perhaps have to change concerning these matters, there are some issues to think of. What will happen to your kernel business? When you have thought of the possibilities concerning e-commerce your kernel business might be changed. What is new and what is abandoned in your business. New products perhaps demand new routines. The organisation of your company might therefore be influenced and so even the head processes. You also perhaps need new equipment like computers and software. These issues should also be analysed. When there are new routines and new equipment in your company your employees are not able to handle the situation. Your employees therefore might need new competence.

3.2.3 Making up the goals/aim for Internet/e-commerce

When making up your goals for making investments in e-commerce there are three questions to be answered. Why should the company make investments in e-commerce? What is the aim of this e-commerce and what are you going to obtain? In order to explain why you are doing the investments you have to be aware of, if there is good pay off concerning this investment and you can explain why it is like that. Concerning the goals and what you are going to obtain you have to try to quantify the goals such as:

- Making the purchase more effective by reducing the amount of suppliers, storage charges and by reducing the costs for transports by means of better co-ordination between better product planning, storage and purchase
- Reduce the selling time and cost for employees for routine selling and use more time for selling to new customers
- Making 10 % more visits to the customers per week by liberating more indoors sellers
- Reduce the costs for management by 15 %

3.2.4 Doing calculations of costs, income and the willingness of investment

When you are doing your calculation you have to start your conclusions from the analysis and the goals you have made. By this calculation you can be aware of the costs and what you are willing to invest. There are five issues, which you have to calculate:

- The cost for making the changes in your organisation
- The costs for the investments in new equipment (computers and software)
- What are your estimated income from your business
- What are your estimated savings in your company by making your organisation, business and purchasing more effective
- How do you find your willingness to make the calculated investments

3.2.5 Making conclusions of the outcome of the analysis

Making up your conclusion and summarising the situation you have to present and explain your new business model. What is new and what have you abandoned? The next step is to describe the goals you have made up related to the market and to the maturity of your business relations. It is of great importance that you know about both your suppliers' and your customers' maturity concerning e-commerce. Otherwise you cannot start making your investments. You must know if they support your start of this process and if they are willing to co-operate with you. Now you have to make up your economical calculation and describe under what circumstances you can afford to follow your strategy of investment. At the end the need of competence must be described. This description concerns both management and the employee.

3.3 Some impressions of the analysis

I participated in three analysis sessions together with two companies. When I have read the analysis model above I must say that this model contains much more than what stood out in the analysis conversation in the workshop. The characteristics of the contents of the conversation were that we just touched the surface of what the participants from the companies really should have known about their business. Let us say, that the talks stayed on a conceptual level. When one of the participants was asked about his business he mentioned three things such as stones, ground planning and concrete. There were not any deeper discussions about this business. No figures were mentioned about the sales. The same thing happened concerning the questions about the customers. He told that his customers are in the area of the building trade and tile-layers. Some customers were also private persons. No figures were asked for or mentioned by him. After that we got a very slim description of the business routines of the company and a conceptual description of the company's business systems. The systems were a salary system, an accounting system and an order/invoice system. He also used Internet in his business.

He also had his own site on the web. He also mentioned that he hired transports except for a special transport of stones. This type of transport he did by his own car. This person was also asked about his vision of the future and what he thought might happen with the organisation if he implemented electronic commerce in his company. As the selling organisation was one of the company's weaknesses he wanted to develop that organisation. If there was any job time saved because of a rationalisation, he said that he could not employ more people. Instead he was planning to use his competence about working with stones and rocks in order to start consulting in this area. He also thought that his suppliers were mature enough to start to sell to him by means of electronic commerce. This he thought might help him. It happened that his customers were asking for prizes and the possibilities to buy a special product when the suppliers' offices were closed. So if he had had the possibilities to check his suppliers store after closing time electronically this had helped him a lot.

My conclusion of this description of the participants' situation is that the person is doing it on a strategic level. The presentation from him was not any deeper. The other presentations were at the same level. An interesting discussion came up when a transport company tried to describe its product. What is a transport? It was not so easy to realise that. The company had a whole book of different types of transports. If a customer want to order a transport by internet he/she has to mark on the screen what type of lorry or truck he needs for the transport, which is not easy to understand. The conclusion was that company would try to develop a site for the biggest customers transports and build it according to what transports the company normally ordered.

4. THE COURSES DURING THE SECOND PART OF THE PROJECT

4.1 A short presentation of the course contents

After we have realised that we could not go on with the project according to the workshop model we decided to use a more traditional educational like model. The companies were invited to take part in a more traditional electronic commerce course. Before they sent the applications to the course they first took part in an open evening seminar about what was going to happen. This evening seminar was free. But if they accepted to take part in the course they have pay about 75 euros per person for 4 times 4 hours and there will be maximum 12 participants per course. If it is just 12 participators on a course it will be much easier to have discussions and not just lectures.

The course is divided in themes related to electronic commerce such as:

- How to analyse the companies situation (4 hours)
- Practical use of Internet and technical information about computers (4 hours)
- Electronic marketing, electronic commerce and bank business (4 hours)
- Legal questions, book keeping and audit issues 4 hours)

The teachers on the different parts of the course are all experts who work daily with the different themes in their companies. An interesting aspect is that the experts do not get any money for their teaching. They have assumed that they will get customers later on by marketing themselves by means of the course.

4.2 My experiences and comments about the courses

Concerning the participants judgments of the contents they are very satisfied both with the contents and the teachers. According to my view the contents give a broad description of the area of problems which must be important for a company thinking of investments of e-commerce. There were some certain remarks on the technical lessons. The contents were too technical. I understand the participants. What the teachers missed was the pedagogical rule number one to choose the strategy according to how the learner is thinking (Marton & Both,1997). To fulfil this strategy the teacher has to try to put the contents in relation to the learners' normal daily situation. In this technical case the teacher did not do that. They presented atomic facts. As the learner could not put these facts in a wholeness he/she is used to in her/his working life the facts were not understandable. It is also very difficult to get everyone to understand everything by means of a lecture, as the dialogue is the base for mutual understanding (Carlsson, 2000). You cannot have a dialogue with everyone in a lecture. What you can obtain by means of a lecture is something which is called an inner conception of relevance (Hodgson, 1995). In order to manage to get the learners to feel an inner conception of relevance they have to be able to understand the contents presented according to the working situation. If they do that they will be interested to learn more about the contents (ibid). As the other lecturers gave practical examples which were quite easy to understand, their lectures were much more accepted.

Some remarks from the lecture about E-commerce solutions were that it was boring to hear about so many concepts. It had been much better to get a wordlist containing these concepts. The Internet examples should be chosen from the learners' home sites if it is possible. What the teacher also had forgotten to present was an analysis about how such a site should be designed.

5. SOME CLOSING REMARKS CONCERNING MY RESEARCH QUESTIONS

Concerning the question about what we can do to help a person to understand if they need to invest in e-commerce, the course above is not enough. One of the participants, who was pleased with the course said: It has opened my eyes that e-commerce is something we have to think over for our business. The other written judgements did not contain any more evidence for that the other participants understood the situation any deeper. This is exactly what Hodgson (1995) says about lectures. If you get an inner conception of relevance you will be interested in learning more about it, but you have just got a surface understanding of the issue. To get a deeper understanding you have to do something else.

Therefore we have built up a studio where we can let the participants of the courses use e-commerce software like a prototype. Doing this they can follow the whole business case from when someone order something until the article is delivered. The meaning with the studio also is that a company can invite customers, purchasers, suppliers, managers to discussions about e-commerce cases. I will now go on studying the problem of understanding and learning in the studio and by interviews.

I have not come to any final conclusion about the analysis model problem either. The model used was understandable for the users. But it had probably taken too much time if the analysis had been performed more deeply according to the model. In this case the steps were performed in a more strategic level. In order o make strategic analysis the model might fit quite well. Because it does not take so long time to make the analyses at this level. The participants also found that this analysis contributed to a brief understanding of the company's situation related to electronic commerce.

My conclusion is: If we are going to make deeper analysis by means of the model the participants have to do analysis as homework. Otherwise it will take so long time to complete the analysis in a deeper sense. Another conclusion is that workshop analysis with a mixture of companies might be a good idea. But they are very difficult to carry out. The companies are afraid of showing up all secrets for strangers. But I think the model can be used inside of a company.

REFERENCES

Carlsson, S., 2000, *Learning in systems development and forms of co-operation – from communication to mutual understanding by learning and teaching.*, Karlstad University Studies (in Swedish).

Christiansson, M-T, 2001, Process orientation in inter-organisational co-operation – by which strategy, in: *On Methods for Systems Development in Professional Organisations,*. A.G. Nilsson and J.S. Pettersson , eds., Stiudentlitteratur, Lund, pp 67-87.

Coad, P. and Yourdon, E., 1991, *Object-Oriented Analysis*, Prentice-Hall, Inc, London.

Downs, E., Clare, P. and Coe, I., 1988, *Structured Systems Analysis and Design Method: Application and Context*, Prentice Hall, London.

Hodgson, V.,1995, To learn from lectures, in: *How we learn*, F. Marton,, D. Hounsell, and N.Entwistle, eds, Rabén Prisma (in Swedish), pp 126-142.

Marton, F. and Booth, S.,1997, *Learning and Awereness*, Lawrence Erlbaum Associates, New Jersey.

CONVERGENCE APPROACH : INTEGRATE ACTIVE PACKET WITH MOBILE COMPONENTS IN ADVANCED INTELLIGENT NETWORK

Soo-Hyun Park

1. INTRODUCTION

Traditional data networks transport their packet bits end-to-end from node to node passively. However, increasingly widespread use of the internet has placed new demands on the networking infrastructure. Novel applications continue to emerge rapidly and often benefit from new network services that better accommodate their modes of use.

Active networks are novel approach to network architecture in which the switches of the network perform customized computations on the messages following through them. This approach is motivated by both lead user applications, which perform user-driven computation at nodes within the network today, and the emergence of mobile code technologies that make dynamic network service innovation attainable.[1] An active network node is capable of dynamically loading and executing programs, written in a variety of languages. These programs are carried in the payload of an active network frame. The program is executed by a receiving node in the environment specified by the Active Network Encapsulation Protocol(ANEP). Various options can be specified in the ANEP header, such as authentication, confidentiality, or integrity.[2-3]

The concept of active networking emerged from the discussions within the broad DARPA research community in 1994 ~ 1995 on the future directions of networking system. When we think of recent network system, we can find several problems more easily. The first is the difficulty of integrating new technologies and standards into the shared network infrastructure. The second is poor performance caused by redundant operations at several protocol layers. The last is difficulty accommodating new services in the existing architectural model. Several strategies have been emerged to address these issues. The idea of messages carrying procedures and data is a natural step beyond traditional circuit and packet switching, and can be used to rapidly adapt the network to

Information Systems Development: Advances in Methodologies, Components, and Management
Edited by Kirikova *et al.*, Kluwer Academic/Plenum Publishers, 2002

change requirements. Coupled with a well-understood execution environment within network nodes, this program-based approach provides a foundation for expressing networking systems as the composition of many smaller components with specific properties. The first property is that services can be distributed and configured to meet the needs of applications. The second is that statements can be made about overall network behavior in terms of the properties of individual components.[1]

This paper provides how to approach to active network concept, Active Network Encapsulation Protocol (ANEP),[2-3] overview of active network architecture, such as SwitchWare, and current research areas of active network. The I-Farmer model[4] can provide the new services creation concept on the communication network by using Service Independent Building Blocks (SIBs)[4] that is already defined, and if it is impossible to create the new communication service for the previously defined SIB, it includes the function to store the newly made SIB into the database of Service Management Part(SMP) / Service Creation Environment Point (SCEP). [5-6] Not only for the new communication service, but in order to create the application program that composes the network components such as Network Element Network Management System (NE NMS) Agent,[7-9] the concept of Applicable SIB (ASIB) is needed. By applying this concept, the applications such as NMS agent designed by the I-Farmer model by using ASIBs [4] that is stored in SMP ASR can be composed. This paper also INTAS algorithm which is for the transformation from Interface Specification of Entity Node and loading block of Agent System which is designed by the I-Farmer Model [4] to Applicable SIB of AIN. The result output of this algorithm, ASIB, will be carried by the Active Packet with the type of payload to the target subsystem.

2. ACTIVE NETWORK CONCEPT

2.1 Approaches to Active Networks

This section provides an overview of active network that has the meaning of highly programmable networks that perform computations on the user data that is passing through them. The approaching to active network can be roughly divided into two approaches, discrete and capsule-integrated, depending on whether programs and data are carried discretely, such as, within separate messages, or in an integrated fashion.[1, 10-12]

In an active network, Network Elements (NEs) of the network, i.e., routers, switches and bridges, perform customized computations on the message flowing through them. These networks are active in the sense that nodes can perform computations on packet contents, and modify these packets. Also this processing can be customized on a per user or per application basis. On the contrary, the role of computation within traditional packet networks, such as, the internet, is extremely limited. Although routers may modify a packet's header, they pass the user data non-transparently without examination or modification. Furthermore, the header computation and associated NE actions are specified independently of the user process or application that generates the packets.[1, 10-12]

2.1.1 Discrete Approach

This is called programmable switches approach alternatively. The processing of messages may be architecturally separated from the business of injecting programs into the node, with a separate mechanism for each function. Users would send their packets through such a "programmable" node much the way they do today. When a packet arrives, its header is examined and a program is dispatched to operate on its contents. The program actively processes the packet, possibly changing its contents. A degree of customized computation is possible because the header of the message identifies which program should be run - so it is possible to arrange for different programs to be executed for different users or applications. The separation of program execution and loading might be valuable when it is desirable for program loading to be carefully controlled or when the individual programs are relatively large. This approach is used, for example, in the Intelligent Network being standardized by ITU. In the Internet, program loading could be restricted to a router's operator who is furnished with a "back door" through which they can dynamically load code. This back door would at minimum authenticate the operator and might also perform extensive checks on the code that is being loaded. Note that allowing operators to dynamically load code into their routers would be useful for router extensibility purposes, even if the programs do not perform application- or user-specific computations. [1, 10-12]

2.1.2 Capsules Approach

This approach is called alternatively an integrated approach. A more extreme view of active networks is one in which every message is a program. Every message, or capsule, that passes between nodes contains a program fragment (of at least one instruction) that may include embedded data. When a capsule arrives at an active node, its contents are

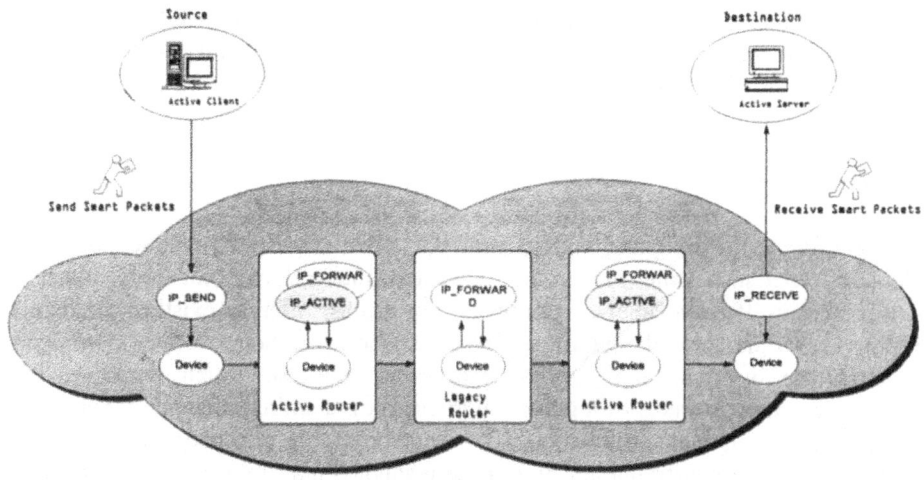

Network

Figure 1. Active Router in Active Network [1]

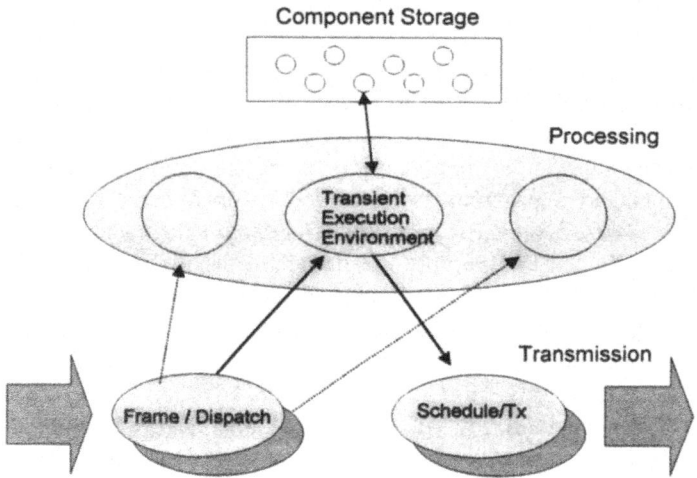

Figure 2. Active Node Organization [12]

evaluated, in much the same way that a PostScript printer interprets the contents of each file that is sent to it. Figure 1 provides a conceptual view of how an active node might be organized. Bits arriving on incoming links are processed by a mechanism that identifies capsule boundaries, possibly using the framing mechanisms provided by traditional link layer protocols. The capsule's contents are dispatched to a transient execution environment where they can safely be evaluated. We hypothesize that programs are composed of "primitive" instructions, that perform basic computations on the capsule contents, and can also invoke external "methods", which may provide access to resources external to the transient environment. The execution of a capsule results in the scheduling of zero or more capsules for transmission on the outgoing links and may change the non-transient state of the node. The transient environment is destroyed when capsule evaluation terminates. [1, 10-12]

2.2 What is Active Node ?

A key difficulty in designing a programmable network is to allow nodes to execute user-defined programs while preventing unwanted interactions. Not only must the network protect itself from runaway protocols, but it must offer co-existing protocols a consistent view of the network and allocate resources between them. Active nodes export an API for use by application-defined processing routines, which combine these primitives using the constructs of a general-purpose programming language rather than a more restricted model, such as layering. They also supply the resources shared between protocols and enforce constraints on how these resources may be used as protocols are executed. We describe our node design along these two lines[12] Figure 2 shows active node organization.

Active network paradigm is motivated by user "pull", as well as technology "push".

The "pull" comes from the ad hoc collection of firewalls, Web proxies, multicast routers, mobile proxies, video gateways, etc. that perform user-driven computation at nodes "within" the network. These nodes are flourishing, suggesting user and management

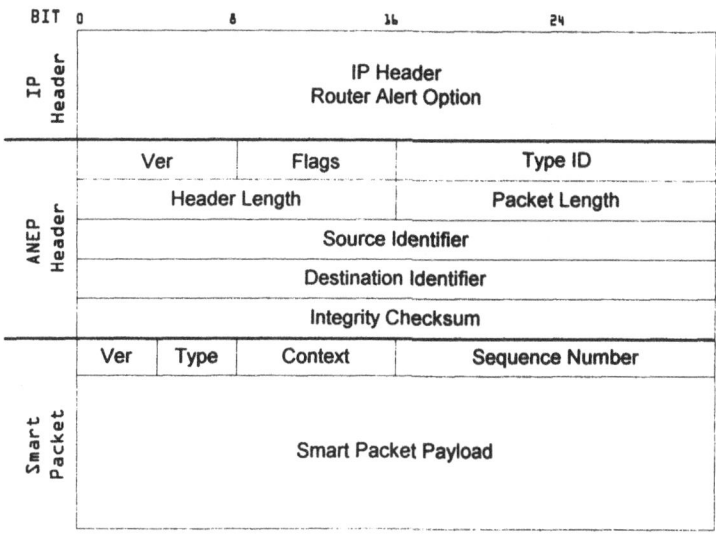

Figure 3. Smart Packet with IP and ANEP Encapsulation. [3,13]

demand for their services. One goal of our work is to replace the present collection of ad hoc approaches with a generic capability that allows users to program their networks. The technology "push" is the emergence of "active technologies", supporting the encapsulation, transfer, interposition, and safe and efficient execution of program fragments. Today, active technologies are applied above the end-to-end network layer; for example, to allow clients and servers to exchange program fragments. Our innovation is to leverage and extend these technologies for use within the network - in ways that will fundamentally change today's model of what is "in" the network. [1, 11-13]

2.3 Active Network Encapsulation Protocol (ANEP)

Smart Packets is a DARPA-funded Active Networks project focusing on applying active networks technology to network management and monitoring without placing undue burden on the nodes in the network.

A smart packet consists of a Smart Packets header followed by payload. The smart packet is encapsulated within an ANEP packet which, in turn, is carried within IPv4, IPv6, or UDP (within an Active Networks test bed).

Smart packets contain either a program, resulting data, or messages wrapped within a common Smart Packets header and encapsulated within ANEP. Figure 3 shows the format of a smart packet. The Smart Packets header has four fields: version number, type, context and sequence number. The version number is used to identify language upgrades and packet format changes. The type field identifies the message as one of four types: a Program Packet, a Data Packet, an Error Packet, or a Message Packet.

The Program Packet carries the code to be executed at the appropriate hosts. The Data Packet carries the results of the execution back to the originating network management program. The Message Packet carries informational messages rather than executable code. Error Packets return error conditions if the transport of a Program Packet or execution of its Spanner code encounters exceptions. The context field holds a value that identifies the originator of the smart packet. The context value is generated by the ANEP Daemon for each client, and is unique for that client within that host. The value is placed into outgoing Program Packets. As Program Packets traverse the network and generate one or more responses (Data, Error, and Message Packets), the context value is used to identify the client to which the responses must be delivered. The sequence number field holds a value that is used to differentiate between messages from the same context. This value allows a client to match response packets with injected programs. Like the context field value, response packets echo the sequence number value of the Program Packet.[13]

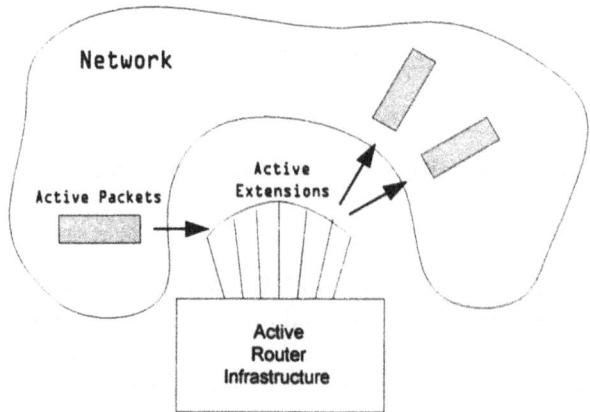

Figure 4. The SwitchWare Architecture.[14-15]

2.4 Active Network Architecture

In order to explain the architecture of active network, in this paper, we show the representative system of Univ. of Pennsylvannia, SwitchWare. The SwitchWare[14-15] uses a layered architecture to provide a range of different flexibility, safety and security, performance, and usability tradeoffs. These layers allow us to employ a variety of different approaches to meeting the challenge of providing security in a programmable network, while still gaining the flexibility of programmability and leaving the network usable. These approaches include providing some functionality that is inherently safe and secure, in large part because it is not powerful enough to cause harm, using cryptography-based security to establish and maintain trust relationships, and using verification technologies such as type-checking and program verification to prove that other functionality is safe and secure. As shown in figure 4, this system employs three main

Table 1. Concept-Mapping between the I-Farmer Model and the ISM

The I-Farmer Model	Interface Specification Model
Entity node	Interface
Aspect entity node	Aspect Interface
Uniformity entity node	Interface which has the same name in the interface catalog
Uniformity aspect entity node	Aspect Interface which has the same name in the aspect interface catalog
IM-Component	Interface which supports static Interface Invocation(SII), and attribute Type_Of_BasicComponent has "ILB" value
OM-Component	Interface which supports Dynamic Interface Invocation(DII), and attribute Type_Of_BasicComponent has "OLB" value
Decomposition	Decomposition
Specialization	Specialization
Multiplicity Abstraction	Refinement & Representative interface

layers: active packets, active extensions, and a secure active router infrastructure.

Active packets replace the traditional network packet with a mobile program, consisting of both code and data. The code part of an active packet provides the control function of a traditional packet header, but does so much more flexibly, since it can interact with the environment of the router in a more complex and customizable way than the simple table lookup provided by headers. Similarly, the data in the active packet program replaces the payload of a traditional packet, but provides a customizable structure that can be used by the program. Because even the most performance critical aspect of the network, basic data transport, is done by executing programs, it is crucial that executing these programs be as lightweight as possible.

Node-resident extensions form the middle layer of our architecture. They can be dynamically-loaded active extensions, or they can be part of the base functionality of the router. They are not mobile: to communicate with other routers they use active packets. Because they are only invoked when needed, there is no inherent need for extensions to be lightweight.[14-15]

3. TRANSFORMING THE INTELLIGENT FARMER MODEL TO THE INTERFACE SPECIFICATION MODEL

In order to apply the mobile features of the active packet to Advanced Intelligent Network, this paper shows the transformation methodology from the concepts defined in the I-Farmer model [4] to the construction elements of the to Interface Specification Meta Model(ISM)[16-17] which is proposed in the Component Based Development (CBD),[16] and also explains ITI (I-Farmer model To Interface specification model) algorithm.[16-17] Furthermore, we can see the real example of transformation of Telecommunication

Management Network (TMN) agent design that is executed by from the I-Farmer model to Interface Specification Meta Model. [16-17]

For the purpose of implementing practically the system that is designed by the Farmer model, there is need to have the Interface Specification Model that explains specification about the functions of entities of the I-Farmer model, such as, entity node, aspect node and ILB/OLB. [16-17]

This paragraph suggests the transformation methodology from the concepts of the I-Farmer model to the mapping notions of the ISM and proposes ATI algorithm[17] that executes the transformation. Also in reality, TMN agents system which is designed by the I-Farmer model is transformed to the ISM system design.

Table 1 shows conceptual mapping relationships between two models. In order to support aspect nodes of the I-Farmer model, the ISM has aspect interface. Especially, aspect interface catalog keeps specific information of aspect interfaces. Uniformity entity node of the I-Farmer model is transformed into the interface having the same identity in the interface catalog. In the case of uniformity aspect entity node of the I-Farmer model, it is mapped to the aspect interface having the same identity in the aspect interface catalog.

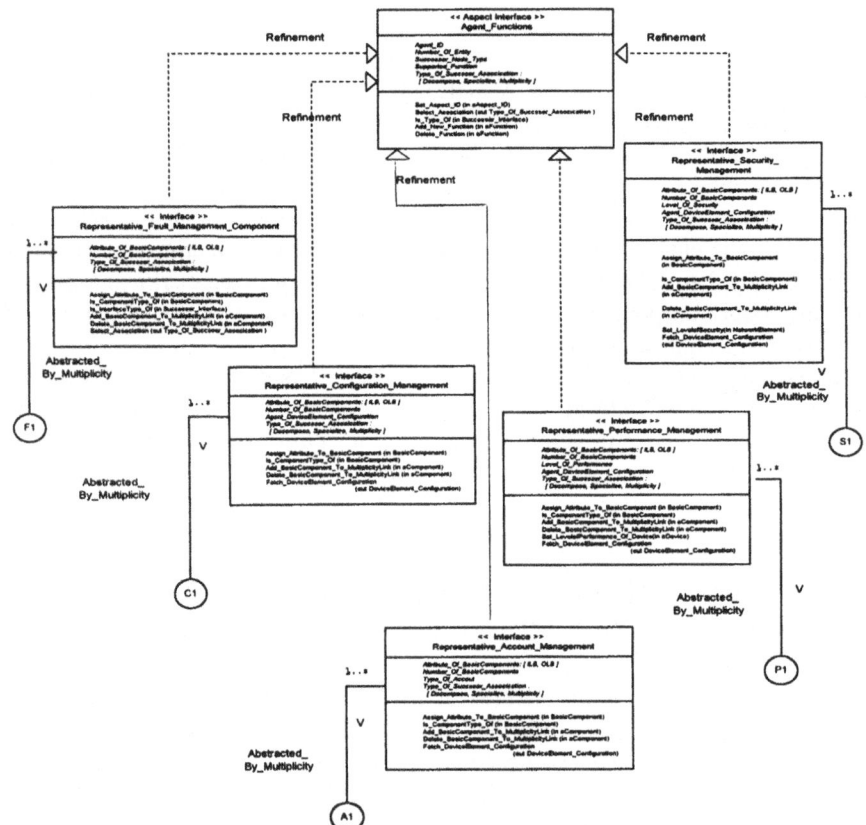

Figure 5. Telecommunication Management Network Agent Design by the ISM Model

Representative entity nodes generated by the multiplicity abstraction of the I-Farmer model are transformed to representative interfaces of the ISM. ILBs in the I-Farmer model are mapped to interfaces of the ISM which support static interface invocation (SII) and these interfaces set their attribute value, Type_Of_BasicComponent, into "ILB". In the case of OLBs of the I-Farmer model, they are transformed into interfaces of the ISM which support dynamic interface invocation (DII) and these interfaces set their attribute value, Type_Of_BasicComponent, into "OLB".

Figure 5 shows this kind of example of aspect interface. We can find that through the ITI algorithm[17] TMN agents system designed by the I-Farmer model which is shown is transformed to interface specifications designed by the ISM which is explained in figure 5 partially.

4. CONVERSION FROM ISM META SPECIFICATION TO APPLICABLE SIB BY USING INTAS ALGORITHM

INTAS (INterface To Applicable SIB) algorithm is an algorithm that transforms entity node and loading components among the agent system designed by the I-Farmer model into ASIB in Global Functional Plane (GFP) of AIN and ASIB that is the creation component for the system application program.

[Algorithm] INTAS (N : Node) // Interface to ASIB
// This algorithm is for the transformation from Interface Specification of Entity Node and loading block of Agent System which is designed by the I-Farmer Model to Applicable SIB of AIN. The result output of this algorithm, ASIB, will be carried by the Active Packet with the type of payload to the target subsystem.
 The basic concept of INTAS algorithm is originated from the NTS algorithm.
// The Abstraction Concepts of the I-Farmer Model Diagram are assigned to Interface Specification Model(ISM) Conversion mechanism.

1.   WHILE End of FMD DO
// FMD : I-Farmer Model Diagram
1.1   Traverse the Farmer Model Diagram from node E to leaf node by the Breadth First Search(BFS)
1.2   Read Current Node
1.3   CASE type_of(current_node) OF
// If the current node of the Farmer Model Diagram is entity node type
1.3.1 Entity Node Type :
// In case former relationship of current Entity Node is multiplicity relationship, skip.
IF Former relationship of current Entity Node = multiplicity
THEN
        Exception (Multiplicity)
 END IF
// In case the next relationship of current Entity Node is multiplicity relationship, generate representative Entity Type Node.

IF the next relationship of current Entity Node ≠ multiplicity
THEN
// Generate SIB
      Get name of entity node e (e.Eid)
      Get attribute set of entity node e (e.A)
Set name of entity node e (e.Eid) to name of SIB (b.Sid)
      According to attribute set of entity node e,
Define service feature set (b.Sf)
Assign Service set to SIB
      Construct SSP(Service Logic Program)
      Construct GSL(Global Service Logic)
END IF
1.3.2 Aspect Entity Node Type :
      Exception(Aspect Entity Node Type)
1.3.3 Uniformity Entity Node Type :
Find any interface having same name of current Entity Node in the interface catalog
IF check_if_current_node = exist THEN
      Entity_Node ← Current Uniformity Entity Node Type
      Call NTS(Entity_Node)
ELSE
      Exception(No_Exist)
ENDIF
// In case the current node of the Farmer Model Diagram is Uniformity Aspect Entity
Node Type
1.3.4 Uniformity Aspect Entity Node Type :
      Exception(Uniformity Aspect Entity Node Type)
// In case the current node of the Farmer Model Diagram is ILB Multiplicity Component
Type Node
1.3.5 ILB Multiplicity Component Type Node:
OLB Multiplicity Component Type Node:
// Generate ASIB(Applicable SIB) for ILB/OLB
Get name of entity node e (e.Eid)
Get attribute set of entity node e (e.A)
Set name of entity node e (e.Eid) to name of ASIB (b.Sid)
According to attribute set of entity node e,
      Define service feature set (b.Sf)
      Assign Service set to ASIB
IF ASIB can be applicable to BCP(Basic Call Process)
THEN
      Construct SSP(Service Logic Program)
      Construct GSL(Global Service Logic)
ENDIF
1.4   END CASE
1.5   Call Register_SIB(b) // b : instance of SIB or ASIB
2.   END WHILE

It is very trivial to prove the correctness of the INTAS algorithm. We can find that through the NTS algorithm, ILB/OLB generated by the I-Farmer model which is shown in figure 2 is transformed to Applicable SIB in the form of platform independent component.

5. CONCLUSION

Active networks are an approach to providing a programmable network infrastructure based on such a programmable interoperability layer. The ability to download new services into the infrastructure will lead to a user-driven innovation process, in which the availability of new services will be dependent on their acceptance in the marketplace, and not be delayed by vendor consensus and standardization activities.

In this paper, we provide how to approach to active network concept, Active Network Encapsulation Protocol (ANEP), overview of active network architecture, such as SwitchWare, and current research areas of active network.

As transformation from the concepts defined in the I-Farmer model to the interfaces of the ISM which is proposed in the Component Based Development, we can have versatile implementation methodologies of the I-Farmer model design through CORBA and JAVA packages. Furthermore, we can maximize reusability of the loading entity type nodes through the mapping from the loading blocks to interfaces.

IMT-2000 network based on AIN CS-3 can easily apply the services of the wired intelligent network that have been developed in the past or that will be developed in the future by combining and integrating with the components of intelligent network. It is difficult to develop the standard for Q3 interface implementation of the agent in TMN system that may occur in the development or the maintenance for the different platforms of IMT-2000 and other networks that are integrated with IMT-2000. In order to solve this kind of problems, this paper suggest the Intelligent Farmer model based on AIN ASIB. IMT-2000 network based on AIN CS-3 can easily apply the services of the wired intelligent network that have been developed in the past or that will be developed in the future by combining and integrating with the components of intelligent network. The I-Farmer model signifies the front step for developing the network management system and application system based on TINA in the future. By using INTAS algorithm, we can easily transform Interface Specification of Entity Node and loading block of Agent System which is designed by the I-Farmer Model to Applicable SIB of AIN. The result output of this algorithm, ASIB, will be carried by the Active Packet with the type of payload to the target subsystem.

REFERENCES

1. David L. Tennenhouse, Jonathan M. Smith, W. David Sincoskie, David J. Wetherall, and Gary J.Minden, A Survey of Active Network Research, IEEE Communications Magazine, 25(1):80-86, January (1997)
2. ANEP-RFC, Active Networks Group Request for Comments, (1997)
3. D.Scott Alexander, Bob Braden, Carl A. Gunter, Alden W. Jackson, et al, Active Network Encapsulation Protocol(ANEP), http://www.cis.upenn.edu/switchware/ANEP/docs/ANEP.txt
4. Soo-Hyun Park, "Intelligent Farmer Model for Network Management Agent Design of IMT-2000",

Perspectives on Information Systems Development: Theory, Methods and Practice, Kluwer Academic / Plenum Publishers, edited By G. Harindranath, Duska Rosenberg, John A.A. Sillince, Wita Wojtkowski, W. Gregory Wojtkowski, Stanislaw Wrycza and Joze Zupancic, (unpublished)

5. ITU-T Recommendations - AIN Q.1210 ~ Q.1229

6. K.B. Choi, et al, "AIN Technology", (Hong-Reung Press, Korea, 1997)

7. ITU-T Recommendation M.3010, Principles for a TMN (1992)

8. ITU-T Recommendation M.3020, TMN Interface Specification Methodology (1992)

9. ITU-T Recommendation M.3100, Generic Network Information Model (1992)

10. David L. Tennenhouse and David J. Wetherall, Towards and Active Network Architecture, ACM Computer Communications Review, 26(2) (1996)

11. David L. Tennenhouse, S.J. Garland, L. Shirira et al, From Internet to ActiveNET, IEEE Network Special Issue on Active and Controllable Networks, vol. 12 no. 3 (1996)

12. David Wetheral, Ulana Legedza, and John Guttag, Introducing New Internet Services: Why and How, IEEE NETWORK Magazine Special Issue on Active and Programmable Networks (1998)

13 Beverly Schwartz, Wenyi Zhou, Alden W.Jackson, et al, Smart Packets for Active Networks, Draft, BBN Technology (1997)

14. D.Scott Alexander, Bob Braden, Carl A. Gunter, Alden W. Jackson, et al, The SwitchWare Active Network Architecture, IEEE Network Special Issue on Active and Controllable Networks (1998)

15. D.Scott Alexander, Michael W.Hicks, et al, The SwitchWare Active Network Implementation, ML Workshop, Baltimore (1998)

16. Soo-Hyun Park, Sung-Gi Min, Tai-Suk Kim, FTI Algorithm for Component Interface Meta Modeling of the TMN Agents, Contemporary Trends in System Development, Kluwer Academic / Plenum Publishers, pp.129 - 145, Edited by W.Gregory Wojtkowski, Wita Wojtkowski (2001)

17 Soo-Hyun Park, Applicable SIBs of ASR for Network Management Agents Design, In Proceedings of The 2001 International Technical Conference on Circuit/Systems, Computers and Communications (ITC-CSCC'2001), pp. 494 - 497, Tokushima, Japan (2001)

INTELLIGENT TRANSPORT SYSTEMS AND SERVICES (ITS)

New challenges for system developers and researchers

Owen Eriksson

1. INTRODUCTION

Today there is a rapid development of information technology which can be used to support the mobility of people, vehicles and goods (Ertico, 2001; Francica, 2001). The systems which are built with this new technology combine:

- mobile units for communication, e.g. units which are built into vehicles, cellular phones and Personal Digital Assistants (PDAs);
- wireless telecommunication, e.g. 3G-mobile telecommunication and radio communication;
- positioning, e.g. Global Positioning Systems (GPS) and cellular phone triangulation;
- Geographical Information Systems GIS-technology.

These systems can be used in different mobile use situations and for mobility management, i.e. for transport- and travel- management. These systems and services are of great interest because travel and transport activities are becoming increasingly important for industry and society.

1.1. Intelligent Transport Systems and Services

Intelligent Transport Systems and Services (ITS) is the concept used to describe how the new mobile information technology can be used in the transport sector (Ertico, 2001). The idea with ITS is that the services should bring extra knowledge to travellers and operators in order to improve transport and travel activities. In cars, ITS is used to help drivers navigate, avoid traffic hold-ups and collisions. On trains and buses ITS is used for managing and optimising fleet operations and to offer passengers automatic ticketing and

Information Systems Development: Advances in Methodologies, Components, and Management
Edited by Kirikova *et al.*, Kluwer Academic/Plenum Publishers, 2002

425

real-time traffic information. At the roadside ITS is used for co-ordinating traffic signals, detecting and managing incidents and to display information for drivers, passengers and pedestrians. Today a number of ITS-services are available and some of these applications are described below.

1.1.1. Car navigation services

Car navigation systems is a type of ITS-application which can be bought and used in vehicles. The systems give drivers advice on how to find their destination.

1.1.2. On trip information services

On trip information is another service which is available and used in the ITS-sector. The aim of this type of service is to provide drivers with dynamic information about congestion, accidents and road conditions in order to improve decisionmaking during the trip (Rodseth et. al., 2001).

1.1.3. Parking information services

The rise in individual transport and the use of cars has also become a big problem, especially in the large cities in Europe (Hinz, 2001). As a consequence ITS-services have been developed to inform drivers about the parking situation in their destination area.

1.1.4. Public transport management services

Today a lot of money is spent on developing ITS-services to support public transport passengers with better information (Van Ross and Hearn, 2001). The aim of these services is to provide passengers with better information about:
- public transport routes and time tables;
- positions of buses and trains and their predicted arrival times based on prevailing traffic conditions;
- advice and guidance to help the passengers choose the most convenient public transport facilities.

1.1.5 Pedestrian support services

Pedestrian information systems support the movement, safety and convenience of pedestrians (Anttila et. al., 2001). These systems provide facility and town information, e.g. weather information, shopping information, tourism information and navigation services.

1.1.6. Tracking services

To locate people, vehicles and goods is another application in which the new technology makes possible. This is e.g. important for people who want to co-ordinate and manage mobility.

1.1.7. Security and emergency services

Security and emergency services are important in the ITS-sector. In the USA the Federal Communication Commission has required that wireless carriers have to provide the location of emergency callers to 911 operators. In Europe the European Commission is promoting location determination of all wireless emergency calls to the 112 number (Jaaskelainen, 2001). In Japan Mayday Services (Masatomo et. al., 2001) have been developed, which provide assistance in the case of traffic accidents, sudden illness and vehicle breakdowns.

1.2 ITS provides researchers and systems developers with new challenges

An important aspect of the development of ITS is that a number of general IT-based services are developed which support mobile activities, and this development creates new and interesting opportunities for using information technology, e.g. the possibility to provide targeted information to mobile actors based on time and location. However a problem is that applications and services are developed from a technical perspective, and is driven by the vendors of technical equipment. Although the technology is important the big challenge will be to make ITS useful and understandable for people. This implies that ITS provide researchers and systems developers with a business, organisational and user perspective with new and interesting challenges, and that there is a need for knowledge development and research in a number of areas, some of these areas will be discussed in the paper, and they are are presented in the list below.

- theories and methods for the analysis of mobile activities (section 2);
- transport network modeling and analysis (section 3);
- location based information (section 4);
- information systems infrastructures (section 5);
- development performed in an organisational network context; (section 6);
- mobile usability (section 7).

2. THEORIES AND METHODS FOR ANALYSING MOBILE ACTIVITIES

So far has the development of ITS been driven by technology and has not primarily been based on the needs and requirements of users and customers. This implies that there is a challenge to learn more about mobile activities and the need for IT-support for mobile avtivities. In order to identify user and customer needs it is important to analyse the behavior of mobile actors, and activities where planning and co-ordination of mobility is essential. To be able do that there is a need for theories and system development methods which can be helpful to analyse the notion of mobility. This is important because the concept of mobility is crucial in order to develop ITS, and traditional systems development and theories are not focused upon this concept.

When we talk about the notion of mobility in the ITS context it is the mobility of people and objects in physical space that is of interest. An interesting theory that could be used to analyse physical mobility is time-geography (Hägerstrand, 1991). The focus of time-geography is to perform contextual analysis of human action based on the space-time dimension. In order to do that geographical and spatial analysis is important. However, the advocates of time geography claim that in order to analyse physical mobility the spatial dimension must be analysed together with the time dimension. Hägerstrand (1991) the inventor of time geography claims that *"We need to rise up from the flat map with its static patterns and think in terms of a world on the move"*.

An important idea in time geography is to analyse how actions and choices of the actor are constrained by physical restrictions. Hägerstrand claims *"Even if many constraints are formulated as general and abstract rules of behaviour we can give them a physical shape in terms of location in space, areal extension, and duration in time"*. According to Hägerstrand (1991, p. 146) there are three large aggregations of constraints: capability, coupling and authority constraints. Capability constraints are those that limit the activities and choices because of biological constitution, and/or the tools that are available, e.g. the tools used for transportation and communication. Coupling constraints has to do with the need for, and problems of, coordinating human activities. In order to perform activities, actors, tools and materials have to be brought together, and the time-space dimension affects the coordination of these activities. The authority constraints imply a space-time perspective on authority, power and control. This aspect has to do with who is in control over a certain space, e.g. a building or vehicle, and who can access it. The authority constraints can also concern the time dimension, e.g. opening and closing hours for service facilities has a major impact on the activities and choices of actors. Time-geography is of interest for systems analyses because it can be used to help systems developers to understand and analyse the concept of physical mobility. The fact that time geography describes events and actions in a situational context with a focus on the time-space dimension can be helpful in a number of ways. Time-geography can e.g. be used to analyse the constraints imposed by the time-space dimension on human action and choices, and how actors can overcome these restrictions. This can be helpful in order to identify the demand for, and the usability of ITS-services, because the use of ITS can eliminate time and space constraints but also impose new ones.

3. TRANSPORT NETWORK MODELING AND ANALYSES

The concept of the transport network is important in order to analyse mobility, because mobile activities are performed along different paths, roads and routes. A transport network is a set of interconnected links through which people travel and goods are transported. The transport network can be described on a conceptual level as a network model constituted by a number of links and nodes. The nodes represent start- and end nodes and junctions in the transport network. This implies that the basic links show how navigation can be performed through the network (Rigaux et. al., 2002) see figure 1 below.

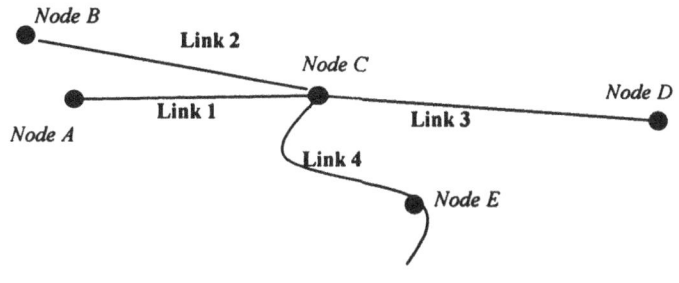

Node ID	Link ID	Start Node	End Node	Road ID	Road ID	Road Name
A	LINK 1	A	C	R1	R1	Main road
B	LINK 2	B	C	R2	R2	City road
C	LINK 3	C	D	R1		
D	LINK 4	C	E	R2		
E						

Figure 1. Transport network model

Transport network models are used for describing roads, railways routes and paths by combining the links in different sequences, e.g. figure 1 shows that Road R1 is constituted by Link 1 and 3, and Road R2 by Link 2 and 4. The challenge for system developers is to learn more about transport network modelling and analyses (Axhausen, 2000), because the core concept of many ITS-applications is the transport network. This implies that the basic information structure of many ITS-applications will be the transport network model (see figur 1), and most of the information that is gathered and communicated has to be related to the network.

It is important to be able to use and implement (see figure 1 above) the conceptual network model in a consistent and understandable way in systems and databases. It is also essential to be able to analyse and gather information about the activities performed on the network. This means that in order to develop ITS-services system developers must be able to analyse and model:

- the structure of the transport network (see figure 1 above);
- the activities performed on the network, i.e. the traffic, transport and travel activities that takes place on the network;
- the impedance which is the cost associated with traversing the network which will include traffic volume, traffic control systems, wheather conditions, traffic accidents and parking conditions.

4. LOCATION BASED INFORMATION

To be able to develop ITS applications *location* is a key concept because virtually all information used is related to locations. Information about locations and the spatial dimension is complex, and this type of information have traditionally been handled in GIS-systems (Bernardsen, 1999; Worboys, 1995). However in many ITS-applications we have to combine:
- information about locations;
- transport network and traffic information;
- business information.

This implies that ITS-applications cannot be developed with the help of traditional GIS-systems. Instead we have to develop information systems where the information presented in the list above is integrated. For example a fleet management system used by a truck driver could contain the information which is described in figure 2 below.

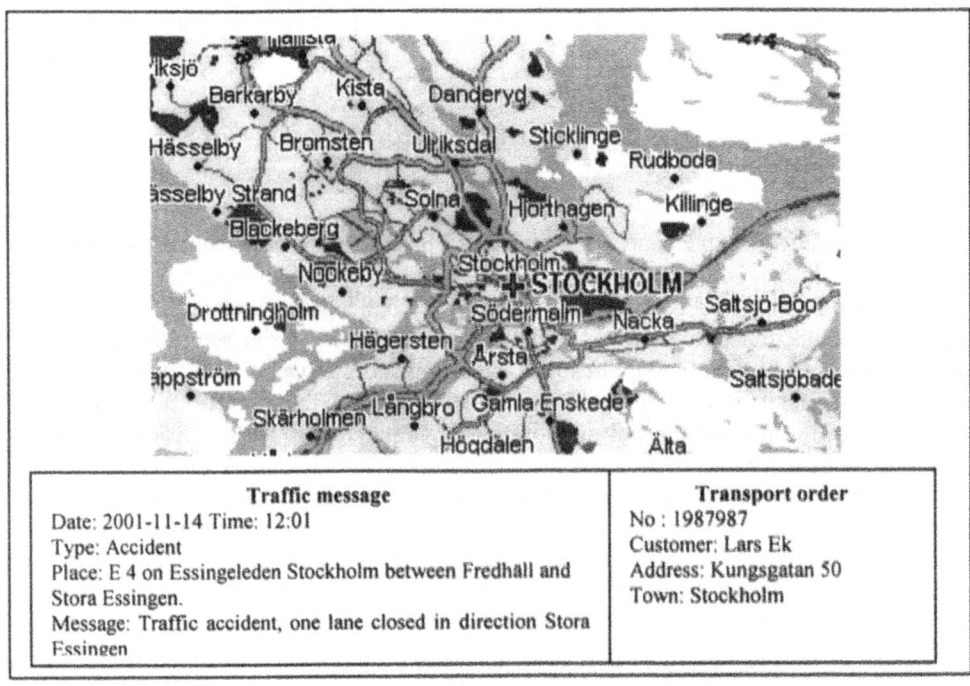

Traffic message	Transport order
Date: 2001-11-14 Time: 12:01	No : 1987987
Type: Accident	Customer: Lars Ek
Place: E 4 on Essingeleden Stockholm between Fredhäll and	Address: Kungsgatan 50
Stora Essingen.	Town: Stockholm
Message: Traffic accident, one lane closed in direction Stora	
Essingen	

Figure 2. ITS-applications combine business information, transport networks and traffic, based on location.

In this case we have a transport order which tells the truck driver the *location where to deliver the goods, the location is also represented with a cross on the map.* We have transport network information which is *location based* because the road is located in the landscape. Furthermore there is a traffic message which tells the driver that there has been an *accident on a certain location on the road.*

Figure 2 illustrates that information about locations and positions is very important in order to develop and use ITS-applications. However in research projects we have performed (see e.g. Eriksson and Axelsson, 2000; 2001) we experienced that people who traditionally work with information systems development have little experience to work with location based information. The challange in this area is to learn more about how to model information about locations and to be able to combine information about locations, transport and traffic information, and business information based on the spatial dimension.

Another challenge that has to be met is the problem that people tend to see location as something that only can be represented by maps and coordinates (Couclelis, 1992). This problem is important to recognise because we do not normally communicate locations in terms of coordinates (Couclelis 1992), instead we use geographical identifiers (ISO/DIS 19112), e.g. names of places and addresses. In figure 2 above we can see that address information, e.g. "Kungsgatan 50" and names of places, e.g. "Essingeleden", "Fredhäll" and "Stora Essingen" are used. If we look at these concepts we can see that they have a meaning that cannot be represented only by coordinates on the map (Fitzpatrick et. al., 1996). "Essingeleden" is a road in Stockholm and "Fredhäll" and "Stora Essingen" are names which are related to certain places on that road. Coordinates are very important if we want to represent the geometry of a location on a map, or if we have technical equipment which we can use to establish our exact position, but the concept of location is a complicated phenomenon. I maintain that if we are going to succeed in developing ITS-applications there is a need for research and analysis of concepts that we use to describe locations.

5. INFORMATION SYSTEMS INFRASTRUCTURES (ISI)

A special feature of the ITS-applications are that they are related to important transport network infrastructures, e.g. roads and rail-roads (see above), they are also dependent on location based information. This implies that in order to develop ITS-applications, systems developers must have access to databases which contain information about transport networks, traffic information, address information, maps and so on. The reason for that is the systems developers and service providers who want to develop ITS-services cannot gather this information by themselves (Dueker and Tu Ton, 2000). For example, if a service provider wants to develop a navigation system or a fleet management system he needs a road map database which includes maps and road network information, and it is not feasible to gather and structure all of this information on his own. He has to have access to a basic information systems infrastructure which can provide this kind of information.

The National Road Database (NVDB) which has been developed by the Swedish National Road Administration (SNRA) in co-operation with the National Land Survey Administration (NLS), the municipalities (there are 278 municipalities in Sweden) and the forest industry, is an example of this (Lundgren, 2000). The NVDB contains information about the swedish road network. This implies that the NVDB together with a number of other databases and IS, e.g. databases which contain information about postal and street addresses, constitutes an important part of the ISI in the ITS-sector in Sweden.

The things listed below are important features of the ISI illustrated in figure 3:
- The information about the road network.

- The functionality used to communicate the information between the data suppliers in the data acquisition process.
- The functionality used to communicate the NVDB-information to the service providers, i.e. the companies who develop ITS-services, e.g. navigation and fleet management applications.
- The standards which are used to facilitate the communication of information between different actors, business units and information systems.
- How responsibilities should be divided between different actors and business units, e.g. between data suppliers, the NVDB-organization and service providers.

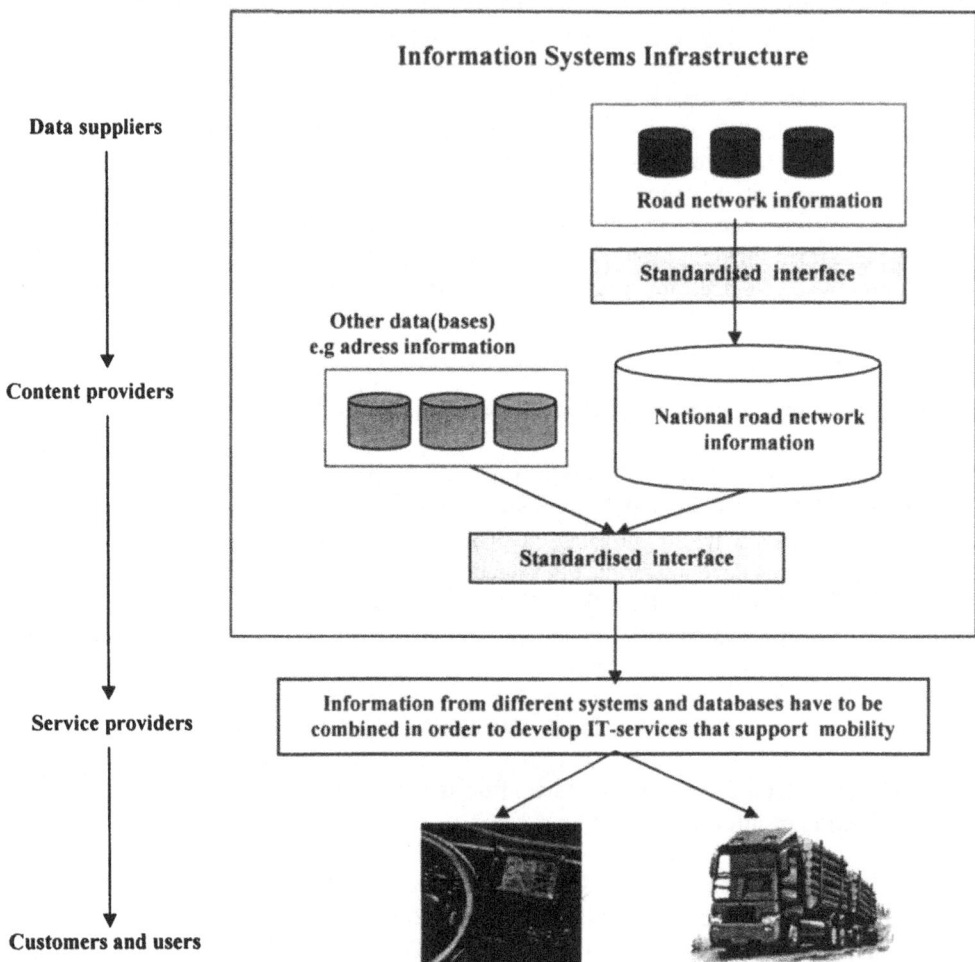

Figure 3. The NVDB is a part of an ISI that is a prerequisite for a number of ITS-services

In a research project we have evaluated and analysed the NVDB (Eriksson and Axelsson, 2001). The analyses show that there are already a number of databases, both at the local and national level, that contain information about the road network.

The municipalities of Sweden store information about the roads and streets in the cities.

The SNRA keep the information about the roads that connect the cities.

The forest industry keeps information about the forest roads.

This is a problem because the service providers got to have access to information about all types of roads. To accomplish this the road network information has to be standardised and co-ordinated. This means that the NVDB-organisation which has the main responsibility to establish the NVDB has to cooperate with a number of data suppliers, content and service providers.

The challange in this area is to learn more about how to develop and maintain these types of information systems infrastructures because this is different compared to conventional systems development:

1. It is important to work strategically and with a long-term perspective, because it takes time to build an ISI (Hanseth, 2000).
2. The ISI should have an enabling function (ibid.), e.g. the ISI decribed in figure 3 should be designed to support a number of ITS-applications, it should not be tailored to one type of application.
3. The development, use and implementation of standards is essential in order to develop an ISI (ibid.). In two research projects we have investigated the use of national and international standards in the ITS-sector (Eriksson and Axelsson, 2000; Forsman 2001), and we found that there is a lack of knowledge how to use and implement the standards.

6. DEVELOPMENT PERFORMED IN AN ORGANISATIONAL NETWORK CONTEXT

A typical feature of the development processes performed in the ITS-sector is their inter-organisational character. For example, the organisation of the data acquisition process in the NVDB-project is a complicated inter-organisational co-operative process (Eriksson and Axelsson, 2001). Furthermore a service provider who wants to develop an ITS-service is dependent on a number of other actors because there is no single actor who controls the infrastructure that is needed to develop and deliver the service. This means that several private companies and authorities have to co-operate in a network organisational context. Co-operation in such a context is built on communication and relationships between different actors. Furthermore, the actors involved in the cooperation usually have other commitments with other actors. This implies that the development process has to be understood as acts of co-operation in a complex organisational network. Such development projects are often coordinated (the SNRA in the case of the NVDB), but not controlled, by a single actor as the driving force, and the expected effects from these types of developing processes are more problematic to anticipate in comparison to systems development projects that is controlled by one single organisation.

The challenge in this area is to learn more about how these processes emerge, are performed and how they should be managed. To be able to understand the complexity of these co-operative processes it can be fruitful to analyse them from a network organisational perspective which is discussed in network relationship theory (Håkansson

& Snehota, 1995). In this theory it is emphasized that the business processes of a single company are performed in a complex network of interorganisational relationships between several interacting companies and organisations. The network theorists also claim that actors can influence the development of the network but that the network is too complex to be controlled by a single actor. This implies that the development processes that take place in a network organisational context cannot be seen as a structured design process that is controlled by a single company or actor. It can be better described as a process of *evolution* and *dynamics*.

If we consider the *evolution aspect,* it is important to realise that evolution takes time and that the effects from a specific development activity (e.g. a design activity) is not possible to fully anticipate. This means that the network perspective emphasise the long term perspective of business development. The *dynamic aspect* of the development process concerns the web of actors and their interests, which can influence the development process. This implies that within the ITS-sector it is important that we analyse and understand the development of applications and information infrastructures in an evolutionary long-term perspective.

7. MOBILE USABILITY

Questions dealing with the users ability, motives and willingness to use ITS is of major interest. An important issue when we talk about user's motives and willingness to use ITS is the balance between empowerment and control. One reason why the balance between empowerment and control will be focused in ITS is the possibilities that the positioning technology creates. One important aspect of ITS is to deliver Location Based Services (LBS). LBS are services for mobile users that take the current position of the user into account (Francica 2001). This implies that users can obtain driving directions and see local traffic conditions based on their actual location. In order to do that it must be possible to track and control the movements of people, but this is sensitive information which could be misused. It can e.g. be used for controling people's mobility against their will. This implies that information about locations of people and objects must be handled carefully. The challenge will be to allow flexible access to location information without violating privacy.

Another interesting feature of many ITS-applications are that they are used in mobile use situations. The mobile use situation will change the prerequisites and impose new constraints for developing usable systems. The reason for that is that the mobile use situations and the tools which are used are quite different compared to stationary use situations (Kristoffersen and Ljungberg, 2000).

In a mobile use situation you would e.g. expect that there may be a lot of disturbance from the environment compared to a stationary use situation. Using a system when driving a vehicle implies disturbance from the traffic, and the problem of using the system and controlling the vehicle at the same time. This implies that new questions will araise when usability is discussed in this context, e.g. traffic safety issues.

Another constraint is that mobile clients have low limited processing and storage capacity, small input and output devices compared to stationary computers. This implies that the mobile clients will impose restrictions, which are quite different compared to the use of stationary computers, and this will affect the possibilities to build usable systems.

In the mobile use situation there will also be a need for interfaces which are adjusted to mobile activities. For example digital maps are very important interfaces in these types of applications, but also voice interfaces, because looking at a computer screen can be difficult and dangerous when users are on the move, e.g. when they are driving a vehicle.

8. CONCLUSION

ITS is an emerging and interesting area for business and systems developers, because ITS creates new and interesting opportunities for using mobile information technology. In previous sections a number of areas have been discussed which we have to learn more about in order to be able to develop ITS. The reason for this is that ITS have special features which will make the business and systems development challenging in a number of ways, which also means that the development of these services will be different compared to traditional systems development. This implies that there is a need of knowledge development in these areas, but most of all there is a need for people who are able to integrate knowledge from these areas. There are specialists in GIS, transport modeling and information systems development but few people who are able to integrate knowledge from different areas. This is a major problem because knowledge and systems integration will be a key factor in order to develop ITS.

REFERENCES

Anttila, V., Raino, A., Penttinen, M. (2001). User Need Research in Personal Navigation Programme NAVI 2000 – 2002. In *Proceedings of the Eight World Congress on Intelligent Transport Systems*, 30 Sep – 4 Oct, Sydney, Australia.

Axhausen K. W (2000) Definition of Movement and Activity for Transport Modeling, In (Henscher D. A., Button K. J.) Handbook of Transport Modelling, Pergamon; Amsterdampp. 271-283

Bernhardsen T. (1999) *Geographic information systems an introduction*, second edition, Wiley, New York

Couclelis C. People Manipulate Objects (Cultivate Fields) : Beyond the Raster-Vector Debate in GIS, *Theories and Methods of Spatio-Temporal Reasoning in Geographic Space : proceedings / International Conference GIS - From Space to Territory: Theories and Methods of Spatio-Temporal Reasoning*, Pisa, Italy, September 21 - 23, 1992

Dueker K. J., Tu ton (2000) Geographical Information Systems for Transport, In (Henscher D. A., Button K. J.) Handbook of Transport Modelling, Pergamon; Amsterdampp. 253-268

Eriksson O. Axelsson K. (2000) ITS Systems Architectures - From Vision to Reality, In *Proceedings of The Seventh World Congress on Intelligent Transport Systems*, Turin, Italy, November 6-9, 2000.

Eriksson O., Axelsson K. (2001). *The Swedish National Road Database – Analysing governing principles for an Information Systems Architectures, Proceedings of the Eight World Congress on Intelligent Transport Systems*, 30 Sep – 4 Oct, Sydney, Australia

Ertico (2001). *What is ITS*. (WWW document) http://www.ertico.com/what_its/what_is.htm.

Fitzpatrick, G., Kaplan, S., Mansfield, T. (1996). Physical Spaces, Virtual Places and Social Worlds: A study of work in the virtual. *Proceedings of CSCW'96*. ACM Press

Forsman A. (2001). *Standardisering som stöd för informationssamverkan i samband med trafikinformation*, In Swedish, Högskolan Dalarna, Borlänge.

Francica J. (2001). *Location-Based Services Where Wireless Meets GIS*. (WWW document), URL http://www.geoplace.com/bg/2000/1000/1000spf.asp

Håkansson H., Snehota I. (eds, 1995). *Developing Relationships in Business Networks*. Routledge, London

Hägerstrand T. (1991) *Om tidens vidd och tingens ordning* , Carlestam G., Sollbe B. (red.), Statens råd för byggnadsforskning , Solna, (in Swedish)

Hanseth O. (2000) Infrastructures- From systems to infrastructures, In Braa K., Sörensen C., Dahlbom B. (eds.) *Planet Internet*, Studentlitteratur, Lund, pp. 179-216

Hinz R. (2001) Parking Space at the Push of a button, *Proceedings of the Eight World Congress on Intelligent Transport Systems*, 30 Sep – 4 Oct, Sydney, Australia

ISO/DIS 19112. *Geographic information- Spatial referencing by geographic identifiers*. (WWW document), URL http://www.statkart.no/isotc211/scope.htm#19112

Jaaskelainen, J. (2001). Towards the Mobile Information Society – First Results of E-Europe2002 Action Plan Paving the Way to Q-Mobility. *Proceedings of the Eight World Congress on Intelligent Transport Systems*, 30 Sep – 4 Oct, Sydney, Australia.

Kriegl, J. (2000). *Location in Celluar Networks – Automated Vehicle Monitoring based on GSM Positioning*. Diploma Thesis, University of Technology Graz, Austria

Kristoffersen S. (2000) Mobility, In Braa K., Sörensen C., Dahlbom B. (eds.) *Planet Internet*, Studentlitteratur, Lund, pp. 137-156

Lundgren, M-L. (2000). The Swedish National Road Database – Collaboration Enhances Quality, *Proceedings of the Seventh World Congress on Intelligent Transport Systems*, 6-9 November, Turin, Italy

Masatomo, S., Kurokawa, H., Yoshida, S., Yamaguchi, K. (2001). Development of a Navigation System Adapted to "Helpnet" Emergency Call Services. *Proceedings of the Eight World Congress on Intelligent Transport Systems*, 30 Sep – 4 Oct, Sydney, Australia

Rigaux P., Sholl, M., Voisard, A. (2002) Spatial Databases – with applications to GIS, Morgan Kaufman Publishers, San Francisco, USA

Rodseth, J., Jenssen, G. D., Moen, T., Asmundvaag, B. (2001). ICT in Road Traffic, *Proceedings of the Eight World Congress on Intelligent Transport Systems*, 30 Sep – 4 Oct, Sydney, Australia.

Van Ross, P., Hearn, B. (2001). Smart Bus Passenger Information Systems, *Proceedings of the Eight World Congress on Intelligent Transport Systems*, 30 Sep – 4 Oct, Sydney, Australia

Worboys M.F. (1995) *GIS a computing perspective*, Taylor & Francis, London

A COMPONENT-BASED FRAMEWORK FOR INTEGRATED MESSAGING SERVICES

George N. Kogiomtzis[*] and Drakoulis Martakos[† ‡]

1. INTRODUCTION

In the last decade companies focus on building enterprise-wide messaging and communication infrastructures in order to meet the ever-increasing requirements for inter- and intra-enterprise communications. In this perspective, enterprise services are crafted to facilitate and improve the communications within the enterprise, with the goal of reducing the overall cost of communication for the enterprise as a single entity. Towards this direction, they continue to evolve towards a core set of standards and capabilities that will offer organizations benefits such as universal interoperability, network convergence, end-to-end media fidelity and high reliability.

Communications and messaging in information systems development, an ever growing and rapidly changing milieu itself, have nowadays become the fastest growing kind of middleware and are constantly being extended to support new technologies and infrastructures as they emerge. The exchange of application data that support day-to-day transactions and activities lies beyond a simple file transfer or transmission of an email message. As a result, the existence of a sophisticated communications infrastructure is often highly desirable by both small companies and large enterprises with numerous business units dispersed all over the world.

In this paper, we propose a well-defined communication management system architecture that supports application integration across diverse business environments and applications. It has been developed as part of a wider Enterprise Business Operating System (EBOS) architecture, with the aim of integrating business-related artefacts (modelled as messages) with any information modelled in EBOS, such as purchase orders, customer records and invoices. It should be noted however that it represents a generic, open middleware architecture that can be applied in a variety of organizational

[*] E-mail: g.kogiomtzis@interworks.gr
[†] E-mail: martakos@di.uoa.gr
[‡] Department of Informatics and Telecommunications, National and Kapodistrian University of Athens, Athens, Greece, 15784, Tel: +30-10-9715103, Fax: +30-10-6456596.

Information Systems Development: Advances in Methodologies, Components, and Management
Edited by Kirikova *et al.*, Kluwer Academic/Plenum Publishers, 2002

437

models, similar to EBOS, regardless of the purpose, content and structure of the information to be transmitted. Any further reference of the EBOS architecture specifics is outside the scope of this paper. However, a detailed treatment of these issues may be found in Karakaxas et al.[1].

The rest of the paper is structured as follows. Section 2 introduces the reader to the challenge of integrated communications and messaging in information systems development and outlines the requirements of the proposed solution. Section 3 provides a detailed overview of the proposed architecture and presents the conceptual framework of the system. Section 4 evaluates the architecture and discusses the main trends and issues behind its implementation. The paper concludes in Section 5, where we summarize the presented work and identify areas of further research.

2. THE CHALLENGE OF MESSAGE-ORIENTED MIDDLEWARE IN THE DEVELOPMENT OF INFORMATION SYSTEMS

The importance of communications, messaging and queuing systems at the enterprise level is difficult to overstate[2]. Messaging is undeniably becoming an essential part of enterprise-wide IT infrastructures enabling disparate software applications to communicate by sending discrete pieces of information back and forth–a process that differs from other systems designed to share data, such as client/server models and central databases[3]. This section aims at reflecting on these requirements in an attempt to propose a feasible and highly effective solution to inter-company communications and messaging.

The ultimate goal in integrated communication services is to achieve reliability, reduced cost of ownership, manageability, scalability and seamless connectivity. Towards this end, integrated communications have been largely associated with Enterprise Application Integration (EAI) leading to the convergence of three important trends in enterprise distributed computing and information systems development[4]:

- The growing need among organizations to deploy sophisticated electronic messaging infrastructures based on client/server technology.
- The increasing importance of the Internet as a tool for electronic commerce and inter-enterprise collaboration.
- A growing awareness of the competitive advantages gained by implementing collaborative applications that meet strategic business objectives (i.e., groupware).

Components of enterprise systems include workflow, business process applications, document databases, and support for RDBMS integration[5]. This convergence drives a growing requirement for a synthesis of these technologies in everyday business solutions. For example, today's Customer Service solutions not only require a database and forms support for storing and tracking customer requests, but also call for integration with the messaging system for workflow routing and problem escalation. It is therefore evident from the above that architectures able to provide seamless integration between applications in enterprise distributed environments are gaining wide respect. Furthermore, the ability to provide a high-level of connectivity in heterogeneous environments is a key feature for competitive advantage.

Specific requirements that Internet connections place on messaging middleware systems and gateways vary according to the type of messaging methodology used. In general terms, messaging can be divided into the following two types[3]:

- *Store-and-Forward Messaging:* With store-and-forward messaging, either side can initiate data transfer and the message may pass through multiple servers along the way (e.g. SMTP servers). The system must be able to incorporate a message storage and message management mechanism either through a queuing management system or through built-in functionality. Typical examples of store-and-forward messaging include SMTP/POP3 mail, MAPI and cc:Mail.

- *Point-to-Point Messaging:* For point-to-point messaging the system must be able to initiate, and respond to externally initiated sessions. The downside of point-to-point messaging is that when one side wants to "talk", the other side has to be ready to "listen"[6]. Examples of point-to-point messaging include sessions of the TCP and FTP protocols.

Until recently, most conventional distributed communication technologies were tightly coupled, requiring sending and receiving applications to be online at the same time. The existing loosely coupled operational environment of many distributed applications however, renders such technologies inappropriate for broad-based adoption because applications in different physical locations are not always "online" at the same time and networks are not always available and reliable. This fact makes the requirements for enterprise communications more complicated and demanding. Message queue-based communication services are nowadays widely accepted as the solution to address this complexity because they provide reliable, asynchronous and loosely coupled communications. They enable applications to send requests to other applications that are not expected to be running or be reachable at the same time and provide guaranteed message delivery as soon as network connections become available and receiving applications begin processing[6].

One of the challenges of communication middleware is the ability to handle both connection-oriented and connectionless communication. In connection-oriented communication, the application must direct a request at a given instance of the receiver and wait for a response, while in connectionless communication applications can forward messages to other applications regardless of whether they are currently running. Undeniably a robust messaging system must support both the above types of messaging and must be equipped with a complete kit that allows the system to look like a peer, server or client to a host of applications and messaging systems. A well-defined communications and messaging infrastructure that can act as a central point of all messages being passed between applications and provide high level connectivity to accommodate changing business requirements can deliver a feasible solution. In this context, a highly flexible architecture that will improve performance, increase scalability and availability while leveraging established technologies and infrastructures is likely to be an effective approach to business integration and application-to-application communication.

3.   CONCEPTUAL OVERVIEW OF THE ARCHITECTURE

The proposed Communication Management System (hereafter abbreviated as CMS) is an intelligent communications and messaging architecture based on interrelated business objects and components, designed to support business-to-business communication. Its main objective is to introduce a framework that allows the implementation of reusable components that can be automatically plugged-in to the system and used as cartridges on demand, according to the communication requirements of the application that uses them. One such class of components is designed to wrap the functionality of standard communication, networking and messaging protocols into a middleware component with a defined set of properties and methods conforming to a standard interface–components that manage FAX, SMTP, TCP, FTP and HTTP sessions for example[7, 8]. This implementation only needs to package program functionality in a way such that the capabilities of the component are discoverable by the container during assembly or at run time[9]. As a result, the proposed architecture seeks to exploit the quality attributes of component-based development such as:

- *Reusability,* by reusing the legacy code through wrapping.
- *Scalability,* by adding components and functionality on demand to accommodate the changing communication and messaging requirements.
- *Flexibility,* by dynamically adding components at runtime.
- *Adaptability,* by implementing software that is readily extendible to meet new requirements.
- *Maintainability,* by localizing software problems at component level to ensure robust design.

The CMS philosophy is to construct a scalable and flexible system architecture that enables incorporation of new services at minimum effort. New services are implemented independently from the core architecture and are incorporated dynamically to the system. This approach offers the following distinct advantages:

- *Complete control on service customization:* Each installation is customized with services that are relevant to the requirements of each application independently. Duplication and use of unnecessary cartridges is abolished.
- *Increased service reliability and minimum disruption to business operations:* Each component is designed, constructed and tested away from the core business operations of the organization. Once the cartridge development process is completed, the cartridges can easily be deployed in a *plug-and-play* fashion.
- *Increased system flexibility and easy adaptation to custom requirements:* Every application may impose different requirements even if they require the same communication protocol. The CMS service-based architecture allows the implementation of multiple cartridges even if they are based on the same communication protocol.

3.1. Conceptual Framework of CMS

Over the last few years, it has become apparent that creating an *n-tier* environment, where developers can be shielded from platform and data storage specifics, in order to concentrate on implementing the necessary business logic, is an effective solution[10]. As such, the CMS is modeled as a *four-layer* architecture, depicted in Figure 1, separating the low-level messaging service components and communication protocols from the business logic and data layers. These layers and their corresponding subsystems are explained in subsequent sections.

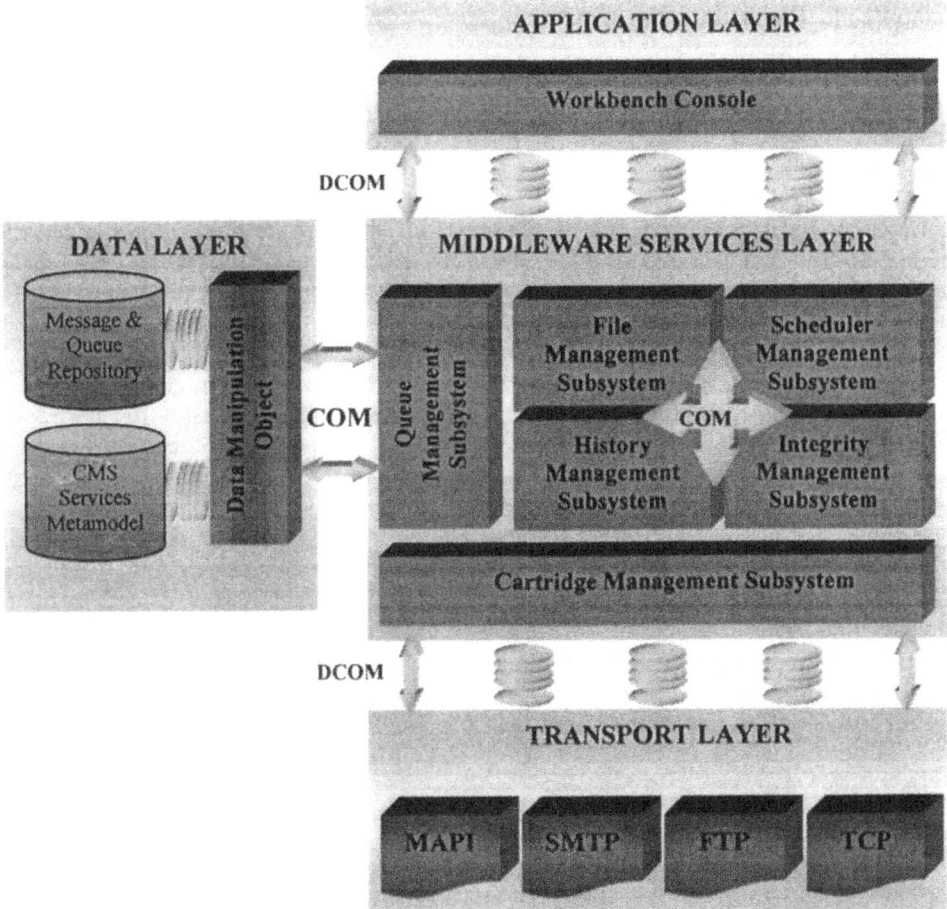

Figure 1. Conceptual overview of the Communication Management System (CMS) architecture. System modules are implemented as separate components, fully cooperating together in the common container architecture.

3.2.1a. The Workbench Console. This subsystem is a Graphical User Interface (GUI) responsible for manipulating the objects in the system and allowing the user to organize the different system views, set configuration information and user preferences and manage the system's organizational structure. As far as CMS is concerned, this console constitutes the user's single point of administration where queues can be added and removed and communication services can be plugged-in and assigned to queues.

3.2.2. The Middleware Services Layer

This layer implements the main engine of the communication system and the Cartridge Support Services and is responsible for co-coordinating all processes and modules under CMS. The subsystems incorporated in this layer are discussed below:

3.2.2a. The Core Engine Controller. This component is vital to the architectural consistency of CMS in that it ensures interoperability between the different subsystems and acts as a Component Manager, similar to a *CORBA Object Request Broker* to provide a single point of authentication and engine session management[11, 12, 13]. It also exposes the *ICore* public interface, shown in Figure 2, which allows the user to access and initialize the CMS-defined processes. Upon user interaction, the Core Engine Controller uses a *Queue Parser* mechanism to access the QMS subsystem through the Message and Queue Interface (MQI) that it exposes, retrieve all messages pending for transmission and create the *System Outbox* at run time. The system outbox is a CMS-defined object type that packages all outgoing messages according to destination queues and services and includes service, message and queue-related information for the outgoing messages. Upon creation of the system outbox, the Core Engine Controller passes the message-related files and information to the File Management subsystem to form the message package container file (attachment) and perform compression and encryption functions on it. It then initializes the Cartridge Manager, providing all transmission-related information and initiates the transmission. A similar to the above operation is followed for incoming communications, where the Core Engine Controller informs the Cartridge Manager to start incoming message detection, by initializing all supported incoming services, and upon receipt of a new message package, the attachment file is decompressed and decrypted and the message header is parsed for transport and sender service identification as well as data integrity operations. Upon confirmation that the package is received in the correct order, the Core Engine Controller appends the message under the corresponding queue in the message and queue repository.

InitialiseCMS()

StartDeamon(DType as DeamonTypes, Optional ByVal SrvName as String)

StartIcomingServices(OpType as OperationTypes)

StopAutoIncomingServices()

CreateInbox() as MessageInbox

CreateOutbox() as MessageOutbox

InitMsgReceipt(InboxIndex as Long)

InitMsgTransfer(ByVal OutboxIndex as Long, ByVal HistoryIndex as Long,
                ByVal IntegrityCounter as Long)

Figure 2. The ICore Public Interface.

3.2.2b. The Queue Management Subsystem (QMS). Today's MQM products are being used along with message and protocol translators to bridge the gap between dissimilar application architectures[2, 14] and Message-Oriented Middleware (MOM) has proved to be a time-honored mainframe approach in this context[11]. Realizing the growing importance of message queuing middleware, the CMS philosophy relies on a ubiquitous message queue-based approach. In the proposed architecture, QMS and CMS reside on the server and are closely interrelated. However, some of the main mechanisms of QMS have been incorporated in CMS in an effort to decentralize some core functionality in CMS service modules, such as the data integrity mechanism. The QMS still maintains the message and queue management mechanisms but it is not responsible for message routing or protocol convergence. As such, it exposes an API that allows application modules to control the flow of messages between queues and constitutes a common access system in which the user or higher-level systems (such as workflow) can deposit messages and from which CMS can collect messages for transmission to the destination queues. The QMS conceptual framework is depicted in Figure 3. In QMS we can identify the following entities:

- *Queue*. In an organizational structure, every physical business unit that supports communication is modeled as a queue. Every queue is associated with one or more messages for transmission to that queue and is related to a list of supported communication services. QMS exposes a common API allowing the configuration of queues upon definition of the system's organizational structure. At every installation the current business unit maintains information on all other queues in the organizational structure for outgoing communication. QMS operates a First-In-First-Out (FIFO) model, where messages are passed on to receiving queues based on the time that they were received by the queue manager.

- *Message*. Messages are defined as the communication entities containing user data. The user data may be simple messages containing an addressing header, a subject and a body text, or files containing application-specific data. The current implementation of CMS makes provision for both the above cases, although the later is more common in enterprise business applications. Depending on the type of the supported communication services, files may be sent as attachments (Internet mail and MAPI for example) or as binary information (TCP, FTP). CMS is not concerned with the content and structure of information but rather concentrates on achieving guaranteed message delivery, ensuring data integrity and applying cost-effective routing.

- *Message and Queue interface (MQI)*. This is a common, application programming interface (API), allowing applications to perform operations to messages and queues. The core engine of QMS relies on a queue manager that provides the service for the MQI calls issued by the applications. All queues representing the communicating physical business units are created in the Workbench Console GUI upon definition of the overall organizational structure. If a queue joins or leaves the group, all queue managers must be automatically informed as to the resources that it is bringing or taking away from the group.

- *Message and Queue Repository*. The message and queue repository is a database holding all relevant information of queues and messages and is integrated with the rest of the CMS data in a common Relational Database Management System (RDBMS). Access to the repository is achieved through an abstract database implementation template contained in QMS, which encapsulates a set of database operations in a database management component and exposes a standard API for database navigation and editing.

3.2.2c. The File Management Subsystem. This subsystem is primarily responsible for converting the transmission data to a CMS-specific standard format while protecting the data's integrity by ensuring that the original data is not changed in transit. To maintain the CMS architectural consistency at the highest level, a common standard format has been derived to which all transmission data must conform. The proposed standard specifies that all message data be wrapped up into a message package that contains the original message data and a message header (in XML file format) with all relevant communication information. By definition, a message package may contain one or more messages addressed to the same queue using the same communication service but will only contain a single common message header. Both the message package and the message header are constructed and assembled at run time whenever a message is picked up for transmission and are temporarily persisted in a physical storage, which is destroyed following transmission. In this subsystem, CMS also seeks to make provisions for data security and cost-effective routing. The former is handled by an integrated mechanism which applies a DES encryption algorithm to all outgoing message packages to ensure that they cannot be viewed or changed in transit, while the later by a data compression mechanism which compresses the message packages to minimize transmission costs and network traffic. At the receiving end, similar encryption and decryption mechanisms are applied to all detected incoming message packages to restore the data to its original format.

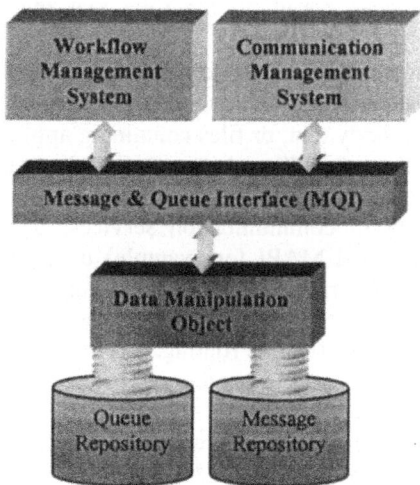

Figure 3. Conceptual overview of the Queue Management Subsystem (QMS) architecture. A Message and Queue Interface (MQI) is used to access data in the separate repositories maintained for messages and queues in the system. This is achieved through a gateway -the Data Manipulation Object- which contains data access functions specific to various legacy database systems.

3.2.2d. The Data Integrity Management Subsystem. The proposed CMS infrastructure provides assured message delivery, even after application or network failure by incorporating a Data Integrity Management Subsystem. This subsystem ensures that messages arrive at the destination intact and in the correct order. In doing so, the system incorporates a message tracking mechanism with which all CMS installations across the enterprise know at any time what message number to expect from any given queue. For as long as the correct message number is being received at the destination, no further action is taken. When however the wrong message number is received, the system implements an algorithm, which checks the message counters and issues a warning to the source application in case one or more messages have gone missing.

3.2.2e. The Scheduler Management Subsystem. Owing to the fact that CMS implements diverse services for synchronous and asynchronous communication, applications may stand idle for relatively long periods of time[15]. To avoid the unnecessary system resource overhead that it entails, CMS provides a scheduling mechanism that will force the application to start when certain specified conditions are met. This mechanism exposes an API allowing the user to specify service and queue based scheduling conditions.

3.2.2f. The History Management Subsystem. CMS also implements a History Management Subsystem responsible for maintaining a history of messages in the queue. Every message is allocated a history number prior to transmission so that new messages can be kept in history for as long as needed and existing messages can be deleted gradually, starting from the oldest message in the queue.

3.2.2g. The Cartridge Management Subsystem. In the proposed *plug-and-play* approach to communication components, a challenging design issue is to enable the set-up of components that implement diverse requirements and use different settings without ever modifying the source code that manages the cartridges and configures their settings. Some mechanism is therefore required, intelligent enough to feed each cartridge with its associated settings data, while remaining completely ignorant of the purpose, type and meaning of each setting. It is also vital that this mechanism is completely unaware of issues that can only be interpreted by each cartridge independently. The CMS Cartridge Manager is tuned to perform this task by providing a single point for the administration, authentication and management of Cartridges and a place for components to register and provide their services. This subsystem consults the service metamodel to obtain a reference to the required service and collects all data required for the initialization of the cartridge. It then forwards the metadata to the cartridge, which will evaluate the data and configure itself accordingly for operation. This subsystem is finally responsible for automatically adding and removing cartridges through a mechanism, which automatically detects a cartridge upon installation and persists all the required cartridge information on the service metamodel for future use.

3.2.3. The Communication Services (Transport) Layer

As indicated previously, one of the most important aims of the CMS philosophy has been to construct a flexible system architecture that is capable of introducing communication services dynamically on demand without the need for source code modifications in the core architecture. This layer maintains a collection of software components that encapsulate the low-level messaging services and communication protocols, while conforming to a common set of properties, methods and interfaces. We define these independent communication modules as *"cartridges"*. The CMS implementation distinguishes between the following two types of cartridges, depending on their operation:

- *Outgoing Transport Cartridge:* A component that supports communication through a single protocol and is dedicated to the transmission of outgoing messages from one queue to another.
- *Incoming Transport Cartridge:* A component that supports communication through a single protocol and is dedicated to receiving incoming messages.

CMS makes provisions for synchronous communications and point-to-point messaging by separating cartridges into manual and automatic. Manually operated components are invoked on demand, whereas automatic cartridges are implemented as service daemons, which maintain connection-oriented sessions and "*listen*" on a permanent connection to the Internet. This distinction is based on the assumption that each installation may use a different communication protocol for outgoing transmissions and a different one for incoming communications and aims to provide system flexibility and increase the component's granularity to the highest possible level.

In order for a cartridge to be CMS compliant, it needs to conform to the common interface specification standard. One such specification exists for each cartridge type (i.e. incoming, outgoing, manual and automatic). This common API exposed by the cartridges remains consistent even if they implement different communication protocols or messaging services. Conformance to such standards protects the integrity of the application and maintains a high level of architectural consistency.

Because different communication cartridges require different implementation methods to cope with the demands and characteristics of individual protocols, CMS allows for variations of different cartridges to be implemented by internal private methods, as long as they conform to the common public interface, namely the *IPlugin* interface, depicted in Figure 4. It is also reasonable to expect that each component needs to be supplied with several settings and properties prior to execution. For this reason, the cartridge settings are configured in the Workbench Console and persisted as XML metadata in the corresponding storage location in the RDBMS upon installation of a new cartridge.

CartridgeIsValid() As Boolean

LoadViewer()

ProduceReport()

SetMsgSettings(ByVal MsgHeader As String, ByVal FileName As String)

SetSrvSettings(ByVal SrvHeader As String)

Send()

Receive()

Figure 4. The IPlugin Public Interface.

3.2.4. *The Relational Database Management System (Data) Layer*

This is where CMS persists all information required for its normal day-to-day operation. At this layer, CMS maintains a common Message and Queue Repository, which can be accessed by QMS through a *Data Manipulation Object* containing an abstract database implementation template for database connection, navigation and other operations in the RDBMS. Finally, the CMS maintains repositories for each of the core engine services such as the Data Integrity Management, the History Management and the Scheduler Management Services as well as a metamodel of all communication services.

4. EVALUATION

The CMS architecture has been implemented in a four-tier model in an attempt to take advantage of the load-distributing capacities of n-tier architectures. This allows for the design and implementation of the data store and data manipulation functions away from the core architecture, while centralizing the business rules in a middleware data-centric tier. Furthermore, communication and messaging service-related implementations are designed, implemented and tested away from the core architecture in a transport tier[16, 17], limiting in this way disruptions to business operations. The idea has been to expand the COM model and integrate it with messaging technologies in a *Message-Oriented Middleware (MOM)* application[18, 10]. Bridging a distributed, object-oriented programming model with an intelligent transport mechanism, can deliver enhanced functionality such as load balancing, message management and quality of service[7]. Therefore, by combining the best of both technologies, a more reliable and scalable system has been created that moves both COM and the message transport capabilities into the enterprise level of applications.

The extensible foundation upon which this architecture is built allows for the coexistence of heterogeneous messaging components whereby multiple transport modules can reside on the same installation and provide seamless and transparent communication services, while offering full interoperability and cooperation with the core engine of the architecture. The architecture implements an intelligent Cartridge Manager mechanism, similar to an Object Request Broker (ORB)[13], for initializing, tracking and managing transport service cartridges and ensuring quick, reliable and smooth interaction between them and the rest of the system. As such, the current architecture implementation includes a rich set of external communication connectivity capabilities by providing migration with heterogeneous messaging systems such as SMTP/MIME, Lotus cc:Mail, MHS, X.400 and VIM as well as communication protocols such as TCP/IP, FTP and UDP[4, 19, 20, 21].

What differentiates the CMS approach is that the architecture is pioneering an extensible foundation to provide seamless connectivity in multi-messaging environments. It provides a *plug-and-play* mechanism in which all pluggable transport modules are treated as *black box* components conforming to a defined universal standard interface and are responsible for performing transport service-related operations, authentication and error handling mechanisms, yet fully cooperating with the rest of the CMS components. Furthermore, the approach ensures a high level of reliability, by restricting limitations to only those imposed by communication protocols and messaging systems themselves.

5. CONCLUSIONS AND FUTURE WORK

For years companies have been interested in integrating enterprise applications to offer solutions that range from improved operating systems to relational databases, bridges and APIs that made it possible to exchange data between applications. In this paper, we presented an enterprise framework for inter-enterprise co-operation in an attempt to demonstrate the implementation of an intelligent, object-oriented, component-based architecture from the perspective of information systems development. The idea behind the proposed system architecture emerged in an attempt to provide a different approach to application integration by providing a Message-Oriented Middleware Platform for collaborative computing environments. The architecture is based on an n-tier component-based, object-oriented model, which provides various disparate transport services and infrastructure support for automatic service discovery and configuration, security policies and integrated message and queue management.

The CMS architecture has been implemented as an integral part of the EBOS Platform[1] to support communication between groupware applications in collaborative computing environments. As such, it is implemented as a middleware system to provide communication facilities between diverse application modules. In this real application environment, a commercial implementation of the proposed system architecture is currently being tested and evaluated for use in the Shipping Industry to support satellite communication infrastructures and legacy protocols such as those provided by *Rydex*, *Marinet* and *Inmarsat*.

Towards this direction, further work will focus on tuning and evaluating the system under real-life conditions in the shipping industry, a rather complex and demanding area by itself in the field of enterprise communication and application integration.

REFERENCES

1. A. Karakaxas, B. Karakostas and V. Zografos, A Business Object Oriented Layered Enterprise Architecture, *Proceedings of DoME Workshop DEXA 2000*, Greenwich, England, 2000.
2. M. Thomson, Technology Audit, MQSeries Product Family, Butler Direct Limited, September 1999.
3. D. Johnsson, The Changing Requirements of Inter-Company Messaging: Electronic Commerce Trends, GE Information Services Inc., 1998.
4. B. Burch, Notes Release 5 Technical Overview, Iris Associates, 1999.
5. Novell Corporation, GroupWise Architecture Overview, Novell Corporation, 1997.
6. M. F. Parker, G. R. Brunson, W. J. Barnett and G. M. Vaudreuil, Unified Messaging: A Value-Creating Engine for Next-Generation Network Services, *Bell Labs Technical Journal*, April-June 1999.
7. R. Lhotka, *Visual Basic 6 Business Objects* (Wrox Press Ltd., 1998).
8. J. Moniz, *Enterprise Application Architectures* (Wrox Press Ltd, 1999).
9. T. Spitzer, Component Architectures, *DBMS*, September 1997.
10. A. Rofail and T. Martin, *Building n-Tier Applications with COM and Visual Basic 6.0* (Wiley Computer Publishing, 1999).
11. Expersoft Corporation, Integrating CORBA with Message-oriented Middleware (MOM), Expersoft Corporation, 1997.
12. R. Orfali and D. Harkey, *Client/Server Programming with JAVA and CORBA*, (Wiley Publishing, 1997).
13. S. Vinoski, *Distributed Object Computing With CORBA* (Hewlett-Packard Company, 1993).
14. Microsoft Corporation, Microsoft Exchange Server: Microsoft's Vision of Unified Messaging, Microsoft Corporation, 1999.
15. F. Domagoj, The EMA System: A CTI-Based E-Mail Alerting Service, *IEEE Communications Management*, 38(2), February 2000.
16. M. Roy and A. Ewald, Inside DCOM, Miller Freeman Inc., April 1997.
17. A. Caric and K. Toivo, New Generation Network Architectures and Software Design, *IEEE Communications Management*, 38(2), February 2000.
18. A. Pharoah, Creating Commercial Components (COM) Technical White Paper, ComponentSource, February 2000.
19. B. Burch, Domino Release 5 Technical Overview, Iris Associates, 1999.
20. S. Radicati, *Electronic Mail: An Introduction to the X.400 Message Handling Standards* (McGraw-Hill Inc, 1992).
21. P. Vervest, *Electronic Mail and Message Handling* (Frances Pinter Publishers, London, 1985).

IMPROVED EFFICIENCY IN ELECTRONIC CASH SCHEME

Amornrat Pornprasit, and Panpiti Piamsa-nga[*]

1. INTRODUCTION

By the increasing number of participants in the worldwide computer networks, like the Internet, the importance of electronic communication and electronic commerce grows rapidly. In most existing electronic payment systems over the Internet, security aspect is an interesting issue, especially on privacy of customer. Every time customer made a purchasing and paid by his credit card all purchase transactions will be stored in a database of credit-card company. If the information is collected into a list, a dossier could be compiled of your consumer habit and financial status. This information is valuable of interest parties. Your consumer habits would be valuable information to a mail-order company as specific products could be targeted at you. Companies would be able to assess your credit worthiness from your financial history.

Eventually, there is a new alternative payment that has been developed for implementing the process that provides a way of ensuring anonymity in electronic financial transactions which is called "Electronic cash". Electronic cash, sometimes called Digital cash, refers to electronic records or messages, which serve as money and are hard to be counterfeited. Digital cash just like paper cash, in that they do not reveal any information about the person spending the money. Obviously, the uniqueness of electronic cash is protection of the customer's privacy and providing customer's anonymity. Several electronic cash systems have been proposed[1-7]. There are two types of electronic cash systems; on-line untraceable electronic cash systems and off-line untraceable electronic cash systems.

Some on-line untraceable electronic cash systems have been proposed; however, the on-line cash systems are not practical from viewpoints of turn-around-time, communication cost, and database-maintenance cost. Therefore, the off-line cash systems are preferable from practical viewpoint. Stefan Brands presented a very efficient off-line electronic cash scheme[2] working with a smart card, which has a tamper-resistance unit, storing electronic cash to prevent double spending. Using cryptographic techniques, in particular blind signature schemes, it allows the customer to remain anonymous and untraceable, without affecting the other security requirements of the system.

However, in general, the size, the thickness, and bend requirements for the smart card are designed to protect the card from being spoiled physically. Unfortunately, this

[*] Amornrat Pornprasit, Department of Computer Science, Thammasat University, Thailand. Panpiti Piamsa-nga, Department of Computer Engineering, Kasetsart University, Thailand.

Information Systems Development: Advances in Methodologies, Components, and Management
Edited by Kirikova *et al.*, Kluwer Academic/Plenum Publishers, 2002

also limits the memory and processing resources that may be placed on the card. Moreover, the current cryptographic standard[8] recommends size of the private key and public key should be at least 160 and 1024 bits, respectively. The existing scheme required more storage on smart card and had computation costs to conform this security standard; as the result, this will affect system efficiency.

As mentioned above, this paper proposes an efficient electronic cash scheme for protecting the customer's privacy. As for the majority of such systems, it is based on the concept of blind signature which is "blind signature based on the modification of DSA" proposed by Cemenisch-Piveteau-Stadler[9]. This protocol describes an off-line system, which presents similar features as Brands' scheme[2]. The major advantage of our system is that the size of the electronic coin can be reduced around 30% (see Section 6 for a detailed analysis). Furthermore, this system is more secured with current cryptography standard. Hence the off-line electronic cash proposed in this paper is reasonably securing payment system over open networks such as the Internet.

The remainder of this paper is organized as follows. Section 2 reviews related works on electronic cash schemes from the academic literature, and blind signature schemes. Section 3 presents new off-line electronic cash schemes based on modification of DSA. Section 4 presents the prevention of double-spending with observer. Section 5 discusses the security of the scheme. Section 6 discusses the efficiency of the scheme. Finally, Section 7 is conclusions.

2. RELATED WORKS

This section presents a summary of the main off-line electronic cash schemes from the academic literature. There are three: those of Chaum-Fiat-Naor[4], Brands[2], and Ferguson[7]. The first electronic cash scheme is proposed by Chaum-Fiat-Naor[4], and is the simplest conceptually. The prevention of double spending is accomplished by the cut-and-choose method, for this reason, this scheme is relatively inefficient. But the innovation of Chaum-Fiat-Naor inspired many electronic cash schemes that were proposed after that. However, the Brands' scheme[2] is considered by many to be the best one because it avoids the awkward cut-and-choose technique. Brands' scheme is Schnorr[10] based, i.e. it is based on zero knowledge proof of possession of a secret key without revealing it. Ferguson's scheme is RSA[11] based like Chaum-Fiat-Naor, but it uses the two-points-on-a-line principle like Brands. Although Ferguson's techniques avoids the cut and choose technique, it is considered the most complicated of the three. The proposed scheme in this paper is conceptually related to the coin paradigm, as introduced by Brands.

An electronic cash system consists of three parties (a bank B, a user U, and a shop S) and four main procedures (account establishment, withdrawal, payment, and deposit). The user U first performs an account establishment protocol to open an account with the bank B. To obtain a coin, U performs a withdrawal protocol with B. During a purchase U spends a coin by participating in a payment protocol with the shop S. To deposit a coin, S performs a deposit protocol with the bank B. The system is off-line because during payment the shop S does not need to communicate with the bank B and anonymous because the bank B, in collaboration with the shop S, cannot trace the coin to the user. In case a coin is doubly spent, the user's identity is revealed. An observer[10] is a tamper-resistant module that is incorporated in the user's computer. The observer is incorporated in the electronic cash protocols to prevent double-spending by providing prior restraint against the double-spending fraud. The combination of an electronic cash system with observers provides the best of both worlds: the security is only dependent on cryptographic assumptions and double-spending is prevented by a tamper-resistant device.

The important part of electronic cash schemes is a blind signature protocol to provide anonymity and user's privacy. Blind signature protocol allows a verifier to obtain a valid signature for a message from a signer without seeing the message or its signature. Signer is unable to link the message-signature pair to a particular instance of the signing protocol, which has led to the pair. Chaum has introduced the concept of a blind signature scheme in 1982[12]. This paper brought a blind signature scheme, which is derived from a variation of the DSA[13] (Digital Signature Algorithm), called "Blind signature on a modification of DSA" proposed by Cemenisch-Piveteau-Stadler[9] produced 320 bits of blind signature that is less than Schnorr signature in Brands' scheme[2].

3. THE IMPROVEMENT OF ELECTRONIC CASH SCHEME

This Section describes the improvement of Brands' scheme with blind signature on modification of DSA[9].

3.1. The bank B's set up protocol (performed once)

A 1024-bit prime p and 160-bit q are chosen such that $q|(p-1)$. Afterwards, generators g, g_1, g_2 of G_q are defined, where G_q is the subgroub of prime order q of the multiplicative group Z_p. A private key $x \in_R Z_q$ is created. B computes its public key $y = g^x$ hash functions H_1, H_0, from a family of collision intractable hash functions such as SHA-1[13] are also defined.

$$H_1: G_q \times G_q \times G_q \times G_q \times G_q \rightarrow Z_q \tag{1}$$

$$H_0: G_q \times G_q \times \text{Shop-Id} \times \text{Date/Time} \rightarrow Z_q \tag{2}$$

B publishes p, q, g, g_1, g_2, H_1, H_0.

3.2. The account establishment protocol

When U opens an account at B, B requests U to identify himself by actual identifier. U randomly generates a number $u_1 \in_R Z_q$, and computes $I = g_1{}^u$. Then U transmits I to B, and keeps u_1 secret. B transmits $z = (Ig_2)^x$ to U. Alternatively, B publishes $g_1{}^x$ and $g_2{}^x$ as part of his public key. Therefore, U calculates z for himself. B stores identifying information of U in the account database together with I.

3.3. The withdrawal protocol

1. a) B randomly generates a number $k \in_R Z_q$, and computes $\tilde{a} = g^k$.

    b) B checks whether $\gcd(\tilde{a}, q) \overset{?}{=} 1$. If this is not the case, B goes back to step a). Otherwise, B computes $\tilde{b} = (Ig_2)^k$ and sends $\tilde{a}$ and $\tilde{b}$ to U.

2. U checks that $\gcd(\tilde{a}, q) \overset{?}{=} 1$. U then randomly generates three numbers $v \in_R Z_q{}^*$, $x_1, x_2 \in_R Z_q$ and computes $A = (Ig_2)^v$, $B = g_1{}^{x_1} g_2{}^{x_2}$, and $z = \tilde{z}^v$. U also randomly generates two numbers $\alpha, \beta \in_R Z_q$ and computes $a = \tilde{a}^\alpha g^\beta \pmod q$ and $b = \tilde{b}^{v\alpha} A^\beta \pmod q$. U then computes the challenge $m = H_1(A, B, a, b, z)$ and sends the blinded challenge $\tilde{m} = \alpha m \tilde{a} a^{-1} \pmod q$ to B.

3. B computes the response $\tilde{s} = k\tilde{m} + \tilde{a}x \pmod q$ and sends it to U, and debits the account of U.

4. U accepts signature of B if and only if equations (3) and (4) are satisfied. If this verification holds, U computes $s = \tilde{s}a\tilde{a}^{-1} + \beta m \pmod{q}$.

$$\tilde{a} \overset{?}{=} (g^{\tilde{s}} y^{-\tilde{a}})^{\tilde{m}^{-1}} \pmod{q} \tag{3}$$

$$\tilde{b} \overset{?}{=} ((Ig)^{\tilde{s}} \tilde{z}^{-\tilde{a}})^{\tilde{m}^{-1}} \pmod{q} \tag{4}$$

User   Bank

$$k \in_R Z_q$$
$$\tilde{a} \leftarrow g^k$$
$$\gcd(\tilde{a}, q) \overset{?}{=} 1$$
$$\tilde{b} \leftarrow (Ig_2)^k$$

$$\xleftarrow{\quad \tilde{a}, \tilde{b} \quad}$$

$$v \in_R Z_q^*$$
$$A \leftarrow (Ig_2)^v$$
$$z \leftarrow \tilde{z}^v$$
$$x_1, x_2, \alpha, \beta \in_R Z_q$$
$$B \leftarrow g_1^{x_1} g_2^{x_2}$$
$$a \leftarrow \tilde{a}^\alpha g^\beta$$
$$b \leftarrow \tilde{b}^{v\alpha} A^\beta$$
$$m \leftarrow H_1(A, B, a, b, z)$$
$$\tilde{m} \leftarrow \alpha m \tilde{a} a^{-1} \pmod{q}$$

$$\xrightarrow{\quad \tilde{m} \quad}$$

$$\tilde{s} \leftarrow k\tilde{m} + \tilde{a}x \pmod{q}$$

$$\xleftarrow{\quad \tilde{s} \quad}$$

$$\tilde{a} \overset{?}{=} (g^{\tilde{s}} y^{-\tilde{a}})^{\tilde{m}^{-1}} \pmod{q}$$
$$\tilde{b} \overset{?}{=} ((Ig)^{\tilde{s}} \tilde{z}^{-\tilde{a}})^{\tilde{m}^{-1}} \pmod{q}$$
$$s \leftarrow \tilde{s}a\tilde{a}^{-1} + \beta m \pmod{q}$$

Figure 1. The withdrawal protocol

3.4. The payment protocol

When U wants to spend his coin at S, the following protocol is performed:

1. U sends (A, B, a, b, z, s) to S.
2. S computes challenge $m_p = H_0(A, B, I_s, Date/Time)$, where *Date/Time* is the date and time of the transaction. S sends m_p to U.
3. U computes the responses s_1 and s_2, and forwards them to S.

S accepts the coin if and only if equations (5), (6), and (8) are satisfied.

$$a \overset{?}{=} (g^s y^{-a})^{m^{-1}} \pmod{q} \tag{5}$$

$$b \overset{?}{=} ((Ig_2)^s z^{-a})^{m^{-1}} \pmod{q} \tag{6}$$

$$r = A \pmod{q} \tag{7}$$

$$r \overset{?}{=} (g_1{}^{s_1}g_2{}^{s_2}B^{-r})^{m_p{}^{-1}} \pmod{q} \tag{8}$$

User   Shop

$$A, B, (a, b, z, s)$$
$$\longrightarrow$$

$$m_p \leftarrow H_0(A, B, I_s, Date/Time)$$

$$m_p$$
$$\longleftarrow$$

$r \leftarrow A \pmod{q}$
$s_1 \leftarrow vum_p + rx_1 \pmod{q}$
$s_2 \leftarrow vm_p + rx_2 \pmod{q}$

$$s_1, s_2$$
$$\longrightarrow$$

Verify Sign(A, B)
$$r \leftarrow A \pmod{q}$$
$$r \overset{?}{=} (g_1{}^{s_1}g_2{}^{s_2}B^{-r})^{m_p{}^{-1}} \pmod{q}$$

Figure 2. The payment protocol

3.5. The deposit protocol

The shop deposits the coin by providing the bank with a transcript of the deposit. If the user U associated with I spends the coin twice, the bank obtains two transcripts (m_p, s_1, s_2) and (m_p', s_1', s_2') and B can compute user's identity by equation (9).

$$(s_1-s_1')/(s_2-s_2') = u \pmod{q} \tag{9}$$

At payment protocol, U supplies a point on a "line" to S. Two such points determine U's identity. Hence, double spenders are identified.

4. PREVENTION OF DOUBLE SPENDING WITH OBSERVER

In previous Section, off-line double spending is detected when the coin is deposited and compared to a database of spent coins. The identity of the double-spender can be revealed when the abuse is detected. Detection after the fact may be enough to discourage double spending in most cases, but it will not solve the problem. If someone were able to obtain an account under a false identity, or were willing to disappear after re-spending a large sum of money, they could successfully cheat the system. Therefore, the remaining problem is that double spending cannot be prevented in off-line electronic cash schemes.

In order to prevent double spending in off-line payment, the scheme needs to rely on physical security. The extension of this scheme requires observer, which embeds "tamper-resistance" unit, to ensure that customer does not spend electronic cash more than once.

4.1. The setup of the system

This is the same as in the basic cash system.

4.2. The account establishment protocol

After U transmits g_1'' to B, B then provides U with and observer O, with embedded in its Read-Only Memory (ROM) a randomly chosen number $o_1 \in_R Z_q^*$ which is unknown to U. B computes $I_o = g_1^{o_1}$, $I = I_u I_o$, and $\tilde{z} = (Ig_2)^x$, then transmits I_o and $\tilde{z}$ to U.

4.3. The withdrawal protocol

1. O randomly generates a number $o_2 \in_R Z_q$, and computes $B_o = g_1^{o_2}$. O sends B_o to U.

2. a) B randomly generates a number $k \in_R Z_q$, and computes $\tilde{a} = g^k$.

   b) B checks whether $\gcd(\tilde{a}, q) \overset{?}{=} 1$. If this is not the case, B goes back to step a). Otherwise, B computes $\tilde{b} = (Ig_2)^k$ and send $\tilde{a}$ and $\tilde{b}$ to U.

Observer	User	Bank
		$k \in_R Z_q$
		$\tilde{a} \leftarrow g^k$
		$\gcd(\tilde{a}, q) \overset{?}{=} 1$
		$\tilde{b} \leftarrow (Ig_2)^k$

$o_2 \in_R Z_q$ $\tilde{a}, \tilde{b}$

$B_o \leftarrow g_1^{o_2}$ $\longleftarrow$

          B_o

       $\longrightarrow$

$v \in_R Z_q$

$A \leftarrow (Ig_2)^v$

$z \leftarrow \tilde{z}^v$

$x_1, x_2, \alpha, \beta, \varepsilon \in_R Z_q$

$B \leftarrow g_1^{x_1} g_2^{x_2} I_o^{v\varepsilon} B_o$

$a \leftarrow \tilde{a}^\alpha g^\beta$

$b \leftarrow \tilde{b}^{v\alpha} A^\beta$

$m \leftarrow H_1(A, B, a, b, z)$

$\tilde{m} \leftarrow \alpha m \tilde{a} a^{-1} \pmod q$

                    $\tilde{m}$

                $\longrightarrow$

                         $\tilde{s} \leftarrow k\tilde{m} + \tilde{a}x \pmod q$

                   $\tilde{s}$

                $\longleftarrow$

$\tilde{a} \overset{?}{=} (g^{\tilde{s}} y^{-\tilde{a}})^{\tilde{m}^{-1}} \pmod q$

$\tilde{b} \overset{?}{=} ((Ig)^{\tilde{s}} z^{-\tilde{a}})^{\tilde{m}^{-1}} \pmod q$

$s \leftarrow \tilde{s} a \tilde{a}^{-1} + \beta m \pmod q$

Figure 3. The withdrawal protocol

3. U checks that $\gcd(\tilde{a}, q) \stackrel{?}{=} 1$. U then randomly generates four numbers $v \in_R Z_q^*$, $x_1, x_2, \varepsilon \in_R Z_q$ and computes $A = (Ig_2)^v$, $B = g_1^{x_1} g_2^{x_2} I_o^{v\varepsilon} B_o$ and $z = \tilde{z}^v$. U also randomly generates two numbers $\alpha, \beta \in_R Z_q$ and computes $a = \tilde{a}^{\alpha} g^{\beta}$ (mod q) and $b = \tilde{b}^{v\alpha} A^{\beta}$ (mod q). U then computes the challenge $m = H_1(A, B, a, b, z)$ and sends the blinded challenge $\tilde{m} = \alpha m \tilde{a} a^{-1}$ (mod q) to B.
4. B computes the response $\tilde{s} = k\tilde{m} + \tilde{a}x$ (mod q) and sends it to U, and debits the account of U.
5. U accepts signature of B if and only if equations (3) and (4) are satisfied. If this verification holds, U computes $s = \tilde{s} a \tilde{a}^{-1} + \beta m$ (mod q).

Observer	User	Shop
	$\xrightarrow{A,\ B,\ (a,\ b,\ z,\ s)}$	
	$\xleftarrow{m_p}$	$m_p \leftarrow H_0(A, B, I_s, Date/Time)$
	$r \leftarrow A$ (mod q) $\tilde{m}_p \leftarrow v(m_p + r\varepsilon)$ (mod q)	
$\xleftarrow{r,\ \tilde{m}_p}$		
o_2 still in memory? $s_o \leftarrow \tilde{m}_p o_1 + r o_2$ (mod q) $\xrightarrow{s_o}$		
	$B_o \stackrel{?}{=} (g^{s_o} I_o^{-\tilde{m}_p})^{r^{-1}}$ $s_1 \leftarrow s_o + vum_p + rx_1$ (mod q) $s_2 \leftarrow vm_p + rx_2$ (mod q) $\xrightarrow{s_1,\ s_2}$	
		Verify Sign(A, B) $r \leftarrow A$ (mod q) $r \stackrel{?}{=} (g_1^{s_1} g_2^{s_2} B^{-r})^{m_p^{-1}}$ (mod q)

Figure 4. The payment protocol

4.4. The payment protocol

1. U sends (A, B, a, b, z, s) to S.
2. S computes the challenge $m_p = H_0(A, B, I_s, Date/Time)$, where *Date/Time* is the date and time of the transaction. S sends m_p to U.
3. U computes $r = A$ (mod q) and $\tilde{m}_p = v(m_p + r\varepsilon)$ (mod q), and sends them to O.
4. If o_2 is still in memory, then O computes the response $s_o = \tilde{m}_p o_1 + r o_2$ (mod q) and sends it to U. Then O erases o_2 from its memory.
5. U verifies that $B_o \stackrel{?}{=} (g^{s_o} I_o^{-\tilde{m}_p})^{r^{-1}}$. If this verification holds, U computes $s_1 = s_o + vum_p + rx_1$ (mod q) and $s_2 = vm_p + rx_2$ (mod q). U then sends (s_1, s_2) to S.

S accepts the coin if and only if equation (5), (6), and (8) are satisfied.

4.5. The deposit protocol

This is similar to the basic system. In case the coin was double spent, B can reveal the identity of double spender by equation (10).

$$g_1^{(s_1-s_1')/(s_2-s_2')} = I. \tag{10}$$

5. SECURITY ANALYSIS

This section illustrates the security of the electronic cash scheme proposed in previous section where Brands' scheme is improved by blind signature based on modification of DSA. Security of the proposed scheme based on Brands[2] should satisfy three conditions as listed below:

1. Unforgeablility: Only banks are able to issue valid digital coins.
2. Untraceability: The relationship between a digital coin and a user is untraceable for the bank.
3. Unreusability: User cannot spend the coin more than once.

5.1. The account establishment protocol

The security in this part is only physical security for activities between a card reader and a smart card.

5.2. The withdrawal protocol

The exact form of the signature on the coin is sign$(A, B) = (a, b, z, s)$,which differs from the signature that bank provides (a, b, z, s), thus in bank's view, the signature and blind signature pair $((a, b, z, s), (a', b', z', s'))$ are independent.

Even though B does not have any information of coin's owner, the protocol guarantees that no one can create any counterfeit coin. If the forger wants to create a valid looking coin (A, B, a, b, z, s), he must sign bank's signature on $m = H_1(A, B, a, b, z)$. Even though the forger can create a temporary variable k by himself, he does not have enough information, which is x, to compute signature $\tilde{s} = k\tilde{m} + \tilde{a}x \pmod{q}$. The forger must try to extract x from y which is in the form of exponential over modulo, he must calculate a discrete logarithm over y which is classified as an uncomputable problem[14-18].

Moreover, It is a difficult task to modify some parameters that will allow the undetected double-spending. The signature on the coin is $\tilde{s} = k\tilde{m} + \tilde{a}x \pmod{q}$, it is impossible to modify the value of m while the signature is still correct. This assumption relies on the collision-resistance of the hash function $H(.)$ as defined by Brands[2].

Hence, this blind signature ensures that user's privacy is protected from bank tracing and bank is protected from fraudulent users.

5.3. The payment protocol

The electronic coins are stored on smart cards, if O cannot be compromised physically, then its response $s_o = \tilde{m}_p o_1 + r o_2 \pmod{q}$ in the payment protocol cannot be used by user to spend the coin once more. Because O will delete o_2 after it has been sent $\tilde{s}_o$ to U at first time the coin is spent. Thus, user cannot pay the same coin more than once.

5.4. The deposit protocol

When the shop deposits a coin to the bank in the case that the shop doubly spends a coin, bank will know whether a user or a shop doubly spends the coin. If *Date/Time* and response (s_1, s_2) of two transactions are the same, then this shop deposits the coin more than once. Otherwise, if the *Date/Time* and response of two transactions are different, then user doubly spends the coin and bank can reveal user's identity by equation (10).

6. EFFICIENCY ANALYSIS

We now compare the scheme proposed in this paper and the scheme previously proposed by Brands[2]. First, the data storage, in the scheme of the latter type a coin is a tuple of (A, B, a, b, z, r). In order to achieve a reasonable security level the lengths of the element in Z_q and Z_p must be at least 160 and 1024 bits, respectively, the coin then requires 5280 bits.

In contrast, the coin in the proposed scheme is a tuple of (A, B, a, b, z, s). With a similarity of lengths in Z_q and Z_p, the coin's length can be quite shorter than Brands scheme with 3552 bits. Hence the proposed scheme can be reduced about 32.73% of coin's length in bits, 28.42% of data stored on smart card, and 30% of data stored on Merchant database. For the communication costs, this scheme can reduce the cost of data transmission about 32.73% which is coin sent by customer to merchant in payment protocol, and 30% of the coin and its response sent by merchant to bank in deposit protocol.

Second, the computational effort of generating the random numbers and computing images of $H(.)$ will be neglected. The computation of the modular exponentiation is the most time consuming operation in comparison to other operations within Z_p. Assume that a modular multiplication is performed in $O((\log p)^2)$ steps, then a bit complexity of modular exponentiation require $O((\log q)(\log p)^2)$ bit operations[17]. Therefore, only the computation of modular exponentiation is considered. The Table 1 shows the estimate CPU-time performed on a Personal Computer with 80586 microprocessor running at 150 MHz; a Java-implementation for both schemes performs 40 experiments.

Table 1. Execution times of Modular Exponentiations

Brands		This scheme	
Modular Exponentiation	Time (s)	Modular Exponentiation	Time (s)
$a = g^w$ $b = (Ig_2)^w$	0.38	$\tilde{a} = g^k$ $\tilde{b} = (Ig_2)^k$	0.39
$A = (Ig_2)^s$ $z' = z^v$ $a' = a^u g^v$ $b' = b^{su} A^v$ $B = g_1^{x_1} g_2^{x_2}$	1.54	$A = (Ig_2)^v$ $z = \tilde{z}^v$ $a = \tilde{a}^\alpha g^\beta$ $b = \tilde{b}^{v\alpha} A^\beta$ $B = g_1^{x_1} g_2^{x_2} I_o^{v\varepsilon} B_o$	1.54
$g^r \overset{?}{=} h^{c'} a'$ $A^r \overset{?}{=} Z^{c'} b'$	0.83	$a \overset{?}{=} (g^s y^{-a})^{m^{-1}} \pmod q$ $b \overset{?}{=} ((Ig)^s z^{-a})^{m^{-1}} \pmod q$	0.88
$g_1^{r_1} g_2^{r_2} \overset{?}{=} A^d B$	0.66	$r \overset{?}{=} (g_1^{s_1} g_2^{s_2} B^{-r})^{m_p^{-1}} \pmod q$	0.66

The results in Table 1 show, that the execution times of two schemes are not different. Furthermore, for both schemes, the execution time of other modular operations are neglected from the experiments, even though the time of modular multiplication on Z_p is close to zero.

Thus, regarding the data storage and the computational costs it can be concluded that this scheme has improved efficiency of Brands' scheme.

7. CONCLUSIONS

The method described in this paper has improved efficiency on electronic cash scheme due to the size of coin is reduced about 32.73% of coin's length in bits, 28.42% of data stored on the smart card, and 30% of data stored on Merchant database. Furthermore, this scheme can reduce the size of data transmission about 32.73% in payment protocol, and 30% in deposit protocol. The proposed scheme also provides more security and privacy with private key size 160 bits and public key size 1024 bits. To achieve higher security against fraudulent parties, the protocol can be extended to incorporate the certificate authority.

8. REFERENCES

1. S. Brands, An Efficient Off-line Electronic Cash System Based on The Representation Problem, Technical Report CS-R9323, CWI, 1993, p. 77.
2. S. Brands, Untraceable Off-line Cash in Wallets with Observers, Advances in Cryptology -Proceeding of CRYPTO'93, 1993, pp. 302-318.
3. D. Chaum, Blind Signatures for Untraceable Payments, Advances in Cryptology - Proceeding of CRYPTO'82, 1982, pp. 199-203.
4. D. Chaum, A. Fiat, and M. Naor, Untraceable electronic cash, Advances in Cryptology-Proceeding of CRYPTO '88, 1988, pp. 319-327.
5. D. Chaum, and T. P. Pedersen, Transferred Cash Grows in Size, Advances in Cryptology - Proceeding of EUROCRYPT '92, 1993, pp. 390-470.
6. D. Chaum, T. P. and Pedersen, Wallet Databases with Observers, Advances in Cryptology -Proceeding of CRYPTO'92, 1993, pp. 89-105.
7. N. Ferguson, Single Term Off-line Coins, Advances in Cryptology - EUROCRYPT'93, 1993, pp. 318-328.
8. IEEE P1363, Standard Specifications for Public Key, 1999, p. 238.
9. J. Camenisch, J. Piveteau, and M. Stadler, Blind Signature Based on the Discrete Logarithm Problem, Advances in Cryptology-EURO'94, 1994, pp. 428-432.
10. C. P. Schnorr, Efficient Signature Generation by Smart Cards, Journal of Cryptology, 1991, pp. 161-174.
11. R. L. Rivest, A. Shamir, and L. M. Adelman, A Method for Obtaining Digital Signatures and Public-Key Cryptosystems, Communication of the ACM, 1978, pp. 120-126.
12. D. Chaum, Security without identification: Transaction systems to make big brother obsolete, Communications of the ACM, 1985, pp. 1030-1044.
13. NIST., Digital Signature Standard (DSS) (Federal Information Processing Standards Publications, 1994).
14. B. Schneier, *Applied Cryptography Protocol, Algorithms, and Source Code in C. 2 ed.* (John Wiley & Sons, Inc., New York, 1996), p. 759.
15. W. Stalling, *Cryptography and Network Security: Principles and Practice. 2 ed.* (Prentice Hall International, Inc., New Jersey, 1995), p. 379.
16. D. R. Stinson, *Cryptography Theory and Practice* (CRC Press, Florida, 1995), p. 434.
17. A. J. Menezes, P. C. van Oorchhot, and S. A. Vanstone, Handbook of Applied Cryptography (CRC Press, Florida, 1996), p. 780.
18. W. Diffie, and M. E. Hellman, Multiuser cryptographic techniques, Proceedings AFIPS 1976 National Computer Conference, 1976, pp. 109-112.

AUTHOR INDEX

The manufacturer's authorised representative in the EU is Springer
Nature Customer Service Centre GmbH, Europaplatz 3, 69115 Heidelberg,
Germany. If you have any concerns regarding our products, please
contact ProductSafety@springernature.com

Printed and bound by CPI Group (UK) Ltd, Croydon, CR0 4YY
23/04/2026
02095628-0016